Lecture Notes in Production Engineering

Lecture Notes in Production Engineering (LNPE) is book series that reports the latest research and developments in Production Engineering, comprising:

- Biomanufacturing
- Control and Management of Processes
- Cutting and Forming
- Design
- Life Cycle Engineering
- Machines and Systems
- Optimization
- Precision Engineering and Metrology
- Surfaces

LNPE publishes authored conference proceedings, contributed volumes and authored monographs that present cutting-edge research information as well as new perspectives on classical fields, while maintaining Springer's high standards of excellence. Also considered for publication are lecture notes and other related material of exceptionally high quality and interest. The subject matter should be original and timely, reporting the latest research and developments in all areas of production engineering. The target audience of LNPE consists of advanced level students, researchers, as well as industry professionals working at the forefront of their fields. Much like Springer's other Lecture Notes series, LNPE will be distributed through Springer's print and electronic publishing channels.

Indexed in Scopus* *Indexed in EI Compendex To submit a proposal or request further information please contact Anthony Doyle, Executive Editor, Springer (anthony.doyle@springer.com).

Mihai Dragomir · Daniela Popescu ·
Chin-Yin Huang · Shun-Fung Chiu ·
Luis Quezada
Editors

Building Resilience into Production: Contemporary Challenges for the Future

Proceedings of the 27th International Conference on Production Research

Springer

Editors
Mihai Dragomir
Department of Design Engineering and Robotics
Technical University of Cluj-Napoca
Cluj-Napoca, Romania

Daniela Popescu
Department of Design Engineering and Robotics
Technical University of Cluj-Napoca
Cluj-Napoca, Romania

Chin-Yin Huang
Industrial Engineering
Tunghai University
Taichung, Taiwan

Shun-Fung Chiu
College of Engineering
De La Salle University
Manila, Philippines

Luis Quezada
Department of Industrial Engineering
University of Santiago of Chile
Santiago, RM - Santiago, Chile

ISSN 2194-0525 ISSN 2194-0533 (electronic)
Lecture Notes in Production Engineering
ISBN 978-3-031-92081-3 ISBN 978-3-031-92082-0 (eBook)
https://doi.org/10.1007/978-3-031-92082-0

This Springer imprint is published by the registered company Springer Nature Switzerland AG
The registered company address is: Gewerbestrasse 11, 6330 Cham, Switzerland

Foreword

The 27th edition of the International Conference on Production Research (ICPR) was held between July 23rd, 2023, and July 28th, 2023, in Cluj-Napoca, Romania, organized by the International Foundation for Production Researched (IFPR) and hosted by the Technical University of Cluj-Napoca (TUCN). The scientific event included captivating keynotes, paper presentations of authors from over 30 countries, a Doctoral Training Workshop, an Early Career Researchers Workshop, the IFPR Women Forum and an Industry Leaders Panel. The overall theme of the conference was "Building resilience into production", and the contributions presented covered a wide range of topics from smart manufacturing to sustainable production.

The current book, included in the Springer series "Lecture Notes in Production Engineering", presents a selection of papers from the ICPR conference, that we hope can be of interest to researchers and practitioners in this field working towards a more resilient and competitive manufacturing sector across the globe. Although regional conditions might differ, the challenges of creating more robust companies and processes, and more innovative and competitive products, while safeguarding the environment and society for the future, bring about exciting and demanding opportunities. Researchers in the field of production engineering and management are called upon to deliver improvements that can contribute to the long-term development of the production sector that has important strategic, economic and social implications.

We express our gratitude as editors, along with the organizers of the conference, to all authors, participants and supporters of the 27th ICPR conference, and we hope that the current volume will represent an important contribution to the knowledge base that IFPR is disseminating internationally for fostering production resilience.

The Editors

Organization

The 27th **edition of the International Conference on Production Research** was organized in 2023 in Cluj-Napoca, Romania, by the International Foundation for Production Research and was hosted by the Technical University of Cluj-Napoca, under the theme "Building resilience into production".

Editors

Mihai Dragomir, Daniela Popescu, Chin-Yin Huang, Shun-Fung Chiu, Luis Quezada

International Scientific Committee

Barad Miryam, Israel
Bauer Wilhelm H., Germany
Berman Sigal, Israel
Bidanda Bopaya, USA
Bullinger Hans, Germany
Ceroni Jose, Chile
Chan Hing Kai, China
Chiu (Anthony) Shun Fung, Philippines
Dagli Cihan H., USA
Dekkers Rob, UK
Dolgui Alexandre, France
Fertsch Marek, Poland
Golany Boaz, Israel
Gouvea da Costa Sergio, Brazil
Grubbstrom Robert W., Sweden
Hallikas Jukka, Finland
Huang Chin-Yin, Taiwan
Ichimura Takaya, Japan
Irohara Takashi, Japan
Ishii Kazuyoshi, Japan
Kang Suk-Ho, Korea
Kaspi Moshe, Israel
Kengpol Athakorn, Thailand
Lee Rongbin W.B., HKG, China
Li Ming, China
MacCarthy Bart, UK
Matsui Masayuki, Japan

Mejia Gonzalo, Colombia
Minner Stefan, Germany
Moon Ilkyeong, Korea
Nakashima Kenichi, Japan
Nof Shimon, USA
Nose Toyokazu, Japan
O'Brien Christopher, UK
Park Jinwoo J., Korea
Pasquino Raimondo, Italy
Pinheiro de Lima Edson, Brazil
Popescu Daniela, Romania
Quezada Luis, Chile
Riedel Oliver, Germany
Shayan Ebrahim S., Australia
Sihn Wilfried, Austria
Sohal Amrik, Australia
Sormaz Dusan, USA
Spath Dieter, Germany
Stachowiak Agnieszka, Poland
Tabucanon Mario T., Thailand
Takahashi Katsuhiko, Japan
Tang Ou, Sweden
Velasco Nubia, Colombia
Villa Agostino, Italy
Virolainen Veli Matti, Finland

Organizing Committee

The team of the Technical University of Cluj-Napoca coordinated by Daniela Popescu & Mihai Dragomir: Călin Neamţu, Diana Dragomir, Florin Popişter, Diana Blagu, Denisa Szabo, Emilia Câmpean, Ştefan Bodi, Grigore Pop, Răzvan Curta, Zsolt Buna, Raul Rozsos, Radu Comes, Paul Ciudin, Horea Goia, Flavia Jascău, Mihaela Vaida, Nicoleta Gavrilă, Maria Nădăşan, Ştefania Chiribău and TUCN students & staff.

Contents

EU Carbon Border Adjustment Mechanism Under Economies of Scale

Tanzin Ahmed(✉) and Esko Niemi

Department of Mechanical Engineering, Aalto University, Puumiehenkuja 3, 02150 Espoo, Finland
tanzin.ahmed@aalto.fi

Abstract. As part of the European green deal European Union (EU) has proposed a Carbon Border Adjustment Mechanism (CBAM). Its purpose is to compensate for emission taxes within the EU to avoid carbon leakage to non-taxing countries. In CBAM, border tariffs for products imported from non-taxing countries to EU countries are set equal to emission taxes. This appears fair, but the economies of scale effects are not considered. This paper examines the effects of realistic cost functions and taxes on the resulting production allocation. We experiment with a single product, four-country, equilibrium model where three countries are potential producers, and all four countries are markets. In our model, demand and supply (producer cost) function parameter values and transportation costs are equal for all countries, and total producer profit is maximized. Our producer long-run cost function is a power curve entailing economies of scale. The results show that the amount of economies of scale strongly affects the profit maximizing production allocation. It appears that under economies of scale the EU border tariff levels are far too low to create a fair marketplace. Low border tariffs under economies of scale lead to centralized production by a non-taxed producer probably using polluting technology, whereas sufficiently large border tariffs decentralize production to different countries, and the amount of polluting production is reduced significantly. On the other hand, economies of scale benefit and utility to society is reduced.

Keywords: Carbon border adjustment mechanism · border tax · producer tax · consumer tax · economies of scale · production location

1 Introduction

European Union (EU) is planning to implement Carbon Border Adjustment Mechanism (CBAM) from October 2023, with a three-year transition period, with full implementation beginning on January 1, 2026, and gradually progressing until 2035 (European Comission 2021, Benson et al. 2023). It is a system to charge tax on carbon intensive products. The goal of this system is to reduce carbon emissions. CBAM will ensure that the carbon price of imports is equivalent to the carbon price of domestic production, and that the EU's climate objectives are not undermined. Primarily CBAM will apply to the most carbon intensive products such as iron and steel, aluminium, cement, fertilizers, and electricity. EU published a general plan about identifying different products with their

M. Dragomir et al. (Eds.): ICPR 2023, LNPE, pp. 1–16, 2025.
https://doi.org/10.1007/978-3-031-92082-0_1

unique codes (European Comission 2024). Importers must collect CBAM data for their products and understand the overall policy. They should reveal their product emissions to the EU within 3 months intervals. EU will set a price tag for each product's carbon emission every week, which will be called the CBAM certificate. Importers have to buy this certificate for their products. There will be an accredited verifier who will verify the authenticity of the importers and adjust the price of the CBAM certificate according to that. By a specific time of the year (31st May) verified data should be submitted to the authority of CBAM maintenance. Now, assessing the above things some problems are raised. The objective of the paper is to answer the following questions:

- How the CBAM should be crafted so that there will be no carbon leakage, i.e., transfer of production to countries with no carbon tax?
- What could be the scenario of production allocation, profit, and welfare under a specified CBAM?

2 Literature Review

Balistreri et al. (2019) worked with two good, two-country hypothetical models to measure the effect of border carbon adjustment. They numerically simulate the models to analyze the equilibrium theory. Both the theoretical model and simulation say that if the policy's objective is to protect the environment, carbon border adjustments should, in practice, be set much lower than the domestic carbon price. Because border tax is set higher than the domestic carbon price or equal to the domestic carbon price, it will increase the price of the carbon intensive goods, which will indirectly encourage the unregulated regions to consume those products. Gros and Egenhofer (2011) also analyzed the border carbon adjustment effect with a general equilibrium model. Their analysis showed that if the EU imports goods from a country where there is no climate policy and carbon pricing, in that case, CBAM will increase global welfare because it shifts the price of carbon to regions of the world where governments have not yet imposed domestic measures, at least in part through trade flows. In addition, Jakob et al. (2013) also worked with a 2×2 general equilibrium model where they showed that CBAM can cause carbon leakage if the CBAM is based on production rather than consumption, because in that case, energy-intensive industries might migrate to countries where there is no regulation of emissions. Markusen (1975) also worked with a 2×2 general equilibrium model and compared three different types of taxes (producer tax, consumer tax, border tax). He concluded that an optimal solution would be a consumption-based border tax rather than a production-based. It will in the end increase the overall welfare and reduce emissions. Fang et al. (2020) made a stylized global supply chain model where with the basic model they found no guarantee of imposing the carbon tax will lower global emissions. With their extended model, they found that carbon tariffs would increase carbon emissions instead of reducing because of the carbon leakage. They suggested sharing technology and financial support instead of tariffs, which will give both the exporter and importer a win-win situation.

The results we obtain from our microeconomic optimization models differ from the above, because we consider a more complex market scenario. In our models, there are more than two factories and multiple markets. In addition, the above and other research generally assumes a fixed unit cost of production. In reality, production cost depends on the production volume. Particularly we examine the effect of economies of scale of producers in the short-run and in the long-run on the CBAM system. In a recent thorough review on CBAM related papers (Zhong and Pei 2023), this aspect is not considered at all. Generally, in this paper the focus is on the effects of CBAM on producers and countries, not so much on emissions.

3 Models

The study is done with three different optimization models. These models are different from each other mainly because of the different time horizons. In each model, there are similar factories and markets situated in different countries which are named Home (H), Foreign 1 (F1), Foreign 2 (F2), and Foreign I (FI), see Fig. 1. Several quantities of interest are analyzed from these models using numerical experimentation, such as amount of production, profit, welfare, revenue, and taxes.

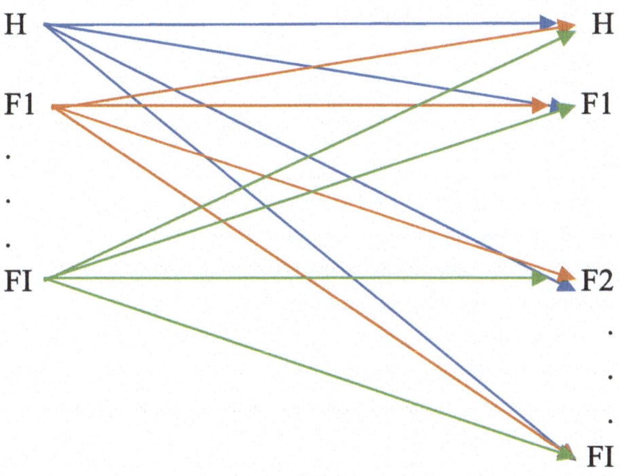

Fig. 1. Product flow

The optimization objective is total profit maximization of the suppliers. The demand in the models works linearly so that when the price increases, demand decreases. There is a maximum price at which demand is zero. This means that consumers are not willing to buy at that price. In our models, we have one type of average product, and any substitution is assumed to be reflected in the demand function.

Our first model is called the *basic model* where there is no fixed cost or capacity limit. Only costs involved are variable production costs, transportation costs, and taxes. The total production varies depending on the demand, taxes, and production cost. In our study, total profits and taxes together are a proxy for welfare. Profit, welfare, and revenue will depend on the amount of production, price, and total cost.

Our second model is called the *short-run* model where there are fixed costs and capacity restrictions in addition to the variable production costs, taxes, and transportation costs.

Our third model is called the *long-run* model. Here, both the fixed cost and variable cost depend on the amount of production and economies of scale. The main difference between short-run and long-run models is the fixed cost. In the short-run there is at least one fixed cost but in the long-run fixed cost itself is variable depending on the technology used and economies of scale (Niemi 1998). There are no capacity restrictions in the long-run model.

There are some common assumptions for all models:

- Border adjustment and transportation costs are only applicable when the product is moving to a market in a different country from its factory country.
- Demand dependence on product price is similar in all models.
- Parameters for the demand function are the same for all models.
- Summations are done for all factories and markets (i and j).

One must note that the numerical results obtained solving the different models cannot be directly compared with each other, because the cost functions are different.

The variables and parameters in the models with default values used and justified in the experimentation in Sect. 4 are listed in Table 1.

Table 1. List of notations and default values

Symbol	Description	Default value
z_{ij}	Products manufactured in country i and transported to market in country j	Variable
y_i	Binary variable indicating that production at factory i is active	Variable
d_j	Demand parameter indicating maximum price in the consumer countries j	1000
k_j	Price parameter for per unit consumer price in market j	0.1
a_i	Production cost per unit at factory i	500
F_i	Fixed cost for each factory i for the short-run model	100 000
M	Big M in constraints for activating fixed cost for each factory i in the short-run	100 000
L_i	Production capacity limit	∞
G_i	Fixed cost for each factory i for the long-run model	10 000
e_i	Level of economic scale for factory i in the long-run	–0.6
n_{ij}	Transportation cost per unit from factory i to market j	5
b_{ij}	Border adjustment per unit from factory i to market j	0 / 20
l_i	Producer tax per unit at factory i	0 / 20
m_j	Consumer tax per unit in market j	0 / 20
p_i	Profits earned by each factory i	
Q_j	Average consumer price for market j	
c_i	Total production cost for each factory i	
P_{tot}	Total profits earned by all the factories	
R_{tot}	Total revenues earned by all factories	
$Ctot$	Total production costs for all factories	
W_{tot}	Total welfare of the system	
T_{tot}	Total taxes collected by all countries	

The main objective of all our models is to maximize the overall profit P_{tot} where the number of products to be determined are $z_{ij} \geq 0$ produced at factories i for markets j:

$$\text{Maximize } P_{tot} = R_{tot} - C_{tot} \tag{1}$$

Here, R_{tot} is total revenue which can be calculated with the below equation:

$$R_{tot} = \Sigma_j \left(Q_j \Sigma_i z_{ij} \right) \tag{2}$$

Average consumer price per product for market j (Q_j) can be calculated with the below equation, which represents the price-demand relation for all models:

$$Q_j = d_j - k_j \Sigma_i z_{ij} \tag{3}$$

Basic Model

Total production cost *Ctot* for the basic model is calculated by summing up all cost items per product and multiplying by *zij*

$$C_{tot} = \Sigma_i c_i = \Sigma_i \Sigma_j z_{ij} \left(a_i + b_{ij} + l_j + m_j + n_{ij} \right) \tag{4}$$

By solving this, profit for each factory can be obtained:

$$p_i = \Sigma_j z_{ij} \left(Q_j - c_i \right) \tag{5}$$

Total welfare:

$$W_{tot} = P_{tot} + T_{tot} \tag{6}$$

Total taxes:

$$T_{tot} = z_{ij} \Sigma_i \Sigma_j \left(b_{ij} + l_j + m_j \right) \tag{7}$$

Our basic model is a convex quadratic program (QP) and therefore easily solvable with common algorithms and software.

In those cases where one factory - one market area pairs exist, taking the derivative of profit (P) with respect to production amount allows direct calculation of the optimal allocation, see profit curve in Fig. 2. This is the allocation where marginal revenue is equal to marginal cost.

Short-Run Model

All formulas in the short-run model are similar to the basic model apart from the production cost function, which contains a fixed cost in addition to other costs. So, the cost function (C_{tot}) is:

$$C_{tot} = \Sigma_i c_i = z_{ij} \Sigma_i \Sigma_j \left(a_i + b_{ij} + l_i + m_j + n_{ij} \right) + \Sigma_i F_i y_i \tag{8}$$

Here, y_i is a binary variable that takes the value 1 if there is production at factory i. This is realized by an auxiliary constraint:

$$\Sigma_i z_{ij} \leq M y_i \tag{9}$$

where M is a large number.

The production amounts ($\Sigma_i\, z_{ij}$) are constrained to the factory capacities (L_i); which in practice largely depend on fixed cost (F_i).

$$\Sigma_i z_{ij} \leq L_i \tag{10}$$

By solving this problem, production allocation, profit, welfare, revenue, and taxes for each factory and market can be obtained like for the basic model. The model is quadratic and binary-constrained, but solvable in reasonable time using an appropriate enumeration algorithm.

For more trivial cases, like the 2 × 2 factory-market setting, transition points between one and two factories operating can be analytically solved. Here the interesting threshold values are the sums of transportation costs and border taxes.

Long-Run Model

The objective and constraints of this model are again similar to the first and second models. However, the cost function is different because of the economies of scale (ES). The total cost function (C_{tot}) is:

$$C_{tot} = \Sigma_i c_i. = \Sigma_i \Sigma_j \left(G_i z_{ij}^{e+1} + z_{ij}(a_i + b_{ij} + l_i + m_j + n_{ij}) \right) \tag{11}$$

and the average cost function (C_{ave}) is:

$$C_{ave} = \Sigma_i c_i / \Sigma_i \Sigma_j z_{ij}. = \Sigma_i \Sigma_j \left(G_i z_{ij}^{e} + a_i + b_{ij} + l_i + m_j + n_{ij} \right) \tag{12}$$

Here, e is the level of the economic scale. In the long-run, investment cost is not fixed because of the economic scale. As the economic scale changes the technology used changes and the value of fixed cost is also changed. So, there is no fixed cost in the long-run model in that sense (Niemi 1998). Figure 2 illustrates the effect of e on the shape of the cost curve. There is no big M constraint or capacity limit either for the long-run model. Profits, taxes, welfare, revenues for each factory, total welfare of the system, total taxes, and so on can be obtained by solving this problem. The model is nonlinear smooth within any active factory setting, but further analysis of derivatives of the functions reveals that it is non-convex. Heuristic or multi-start gradient algorithms can be applied for solving. However, the problems in this paper are very small.

Both the short-run (SR) and long-run (LR) cost curves have similar features (Fig. 2). The SR average cost curve is a simple hyperbola, but the LR curves are power curves, the slope of which depends on the amount of economies of scale (e).

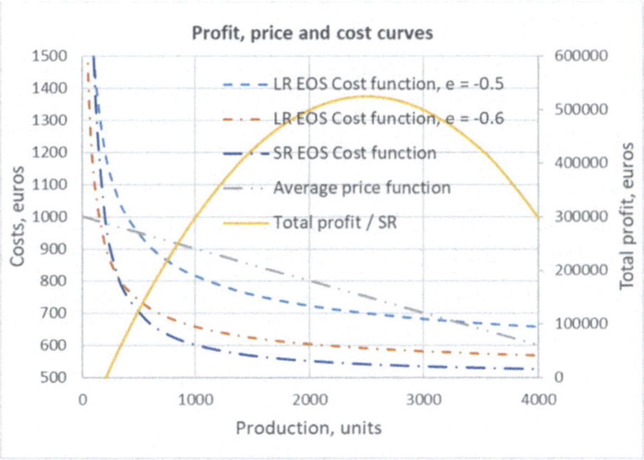

Fig. 2. Examples of different cost, profit, and price curves in relation to production amount

4 Experiments and Results

This chapter introduces the experimental producer-market system and default values of parameters used in different models, whether they are kept constant or adjusted. We justify the values and provide different ranges. For seeing the effects of the taxes with a range of values for different parameters we solve the three models several times. We use the shareware solvers found at Opensolver (2024). All the countries, markets, factories, transportations etc. are assumed to be alike, unless otherwise stated.

A constant value for direct production cost is assumed, which is 500 euros and the consumer price is set at zero demand to 1000 euros. So initially, it is assumed that the production cost in the basic model is half of the maximum price. The price parameter (k_j) is fixed as well, and it is set to 0.1 euros. A constant transportation cost is also assumed, which is 5 euros/unit. Initially, it is assumed that there is no producer tax, border tax or consumer tax. Then in the next phases, producer tax is set to 20, consumer tax to 20, and border tax to 20. These values are justified by per ton of steel production CO_2 emission 2 tons and CO_2 emission tax 100 euros/ton for a hypothetical steel construction weighing 100 kg. The production cost at 5 euros/kg should be realistic for a simple steel construction. But the values for the taxes can be adjusted, from 0 upwards. The fixed cost (F_i) for the short-run model is set to 100 000 euros which is equal for all factories. Production capacity limit (L_i) in the short-run model is equal for all factories and it is varied depending on the experimental setting. For the long-run model, there is also the investment cost parameter (G_i). Initially, it is set to 10 000 euros. The level of economic scale e is set to –0.60 and variable cost a_i is set to 500 euros per unit for the LR model, too. Experimentation with different values of the input parameters is done. Our aim is to gain general insight of the system behaviour; particularly the effects of the taxes under different circumstances to observe:

- Profit maximizing production location
- Profit, welfare, and revenue of the factories, market locations, and the whole system.

Parameters and their default values are listed in Table 1 in Sect. 3.

4.1 Experiments Without Tax

In the experimentation part, at first experimentation with the no-tax system is done.

Basic Model

From the allocation results in Table 2, we can see that under the basic model, production to markets H, F1, and F2 is domestic and equal when there is no tax. Production to market F3 is slightly lower and the sales price higher due to the transportation cost to that market. The production to market F3 could take place at any factory. All the production amounts and sales are only constrained by the profit maximization objective.

Table 2. Profit-maximizing production allocation from factories to markets for basic model and short-run model

Markets	Factories		
	H	F1	F2
H	2500	0	0
F1	0	2500	0
F2	0	0	2500
F3	825	825	825
Total	3325	3325	3325

Short-Run Model

The production allocation scenario is similar to that in Table 2 for the basic model if a production capacity limit 3325 is set. If we could technically increase the capacity limit from the 3325, then the production allocation could be uneven for market F3, but the total profit maximizing production volume or profit would not increase until capacity limit of about 4040. Exceeding this would concentrate the profit-maximizing sales amounts for each market to be produced first on two and after about 8054 finally on one factory only.

If we would decrease the capacity limit from the 3325, the consumer price would go up, but the decreased production would decrease the overall profit. Similarly, adjusting capacity limit within ranges 4040 – 4975 and 8054 – 9925 directly affects production amount and profit. Limits 4975 and 9925 are the profit maximizing production volumes for two and one factories correspondingly. In practice the fixed cost would increase with such large production increases and the direct cost would be lower. This is taken into account in the long-run model.

Long-Run Model

We can see from Table 3 below that we only have production in one factory in the long-run using our default values. All productions are moving from the H factory to H, F1, F2, and F3 markets.

Economies of scale favor centralization of production. Similarly to the short-run case, artificially limiting capacities would decentralize the profit-maximizing production allocation.

Table 3. Allocation of production to markets in the long-run when no taxes are implemented

Markets	Factories		
	H	F1	F2
H	2418	0	0
F1	2393	0	0
F2	2393	0	0
F3	2393	0	0
Total	9598	0	0

4.2 Effect of Producer Tax

When a carbon emission penalty, a producer tax, is imposed on a factory, it will reduce the quantity that will be sold in the markets because per unit production cost is increased. Therefore, consumer price is increased, and demand is decreased. Producer tax goes to the country where the factory is located. A more expensive, less polluting technology would have the same effect.

Experimentation with producer tax case is done initially with a producer tax of 20 euros per product at all three factories.

Basic Model
From Table 4 we can see that due to the producer tax, the profit-maximizing production amounts are lower, but the allocation is similar to the system with no taxes.

Table 4. Allocation of production from each factory to markets for basic model and short-run model with producer tax

Markets	Factories		
	H	F1	F2
H	2400	0	0
F1	0	2400	0
F2	0	0	2400
F3	792	792	792
Total	3192	3192	3192

Short-Run Model
For short-run model, production allocation scenario is similar to the no-tax case, but all the allocation transition production volumes are lower. The limited-capacity profit-maximizing production volumes (see Table 4) are the same as those for the (unlimited capacity) basic case, as the fixed cost does no affect the marginal cost of production in the short run. Total profit is lower due to the fixed cost.

Long-Run Model
In the long-run, when there is producer tax, the production volume is lower than in the no-tax case (Table 3). But the characteristics of the result are similar to that system. Total profit is lower, but total welfare is almost unchanged from the no-tax case.

4.3 Effect of Consumer Tax

Consumer tax is charged while purchasing a good or a service. It goes to the market where the product is purchased by customers whereas producer tax is collected in the country where the factory is located. Therefore, if it is implemented in all countries, the cost effects of consumer tax in our models are the same as that of the producer tax, and consequently the results concerning production allocation are likewise the same. From the political point of view the situation is clearly different, due to the tax being collected in different countries.

4.4 Effect of Cross-Country Border Tax

Border tax means that when product is moving to a different country rather than the producer country there is a tax collected at the border. Experimentation is done with a cross-country border tax of 20 euros per product in line with the EU CBAM. Here we assume that the same tax is collected on all borders.

Basic Model
In the basic model there is no fixed production cost and no incentive to centralize production. Therefore, in the profit-maximizing allocation the market areas H, F1, and F2 have factories producing to the home market and border tax is avoided. On the other hand, for market F3 the border tax is realized, and the tax increases the optimal price, decreases total profits and slightly reduces total welfare. The resulting allocation is similar to the no-tax case in Table 2, except that for market F3 the allocation is the same as for producer tax case in Table 4.

Short-Run Model
The production allocation for the short-run model is similar to the no-tax case for countries H, F1, and F2, and similar to the producer-tax case for market F3. The transition production volumes toward more centralized production allocation are somewhat higher than for the no-tax and producer-tax cases. This is because the border tax penalizes for the increase of cross-border transportation from the fewer factories to the markets abroad. Increasing the border tax to 29 euros would make the decentralized allocation most profitable without any assumptions about capacity limits. In the decentralized three-factory allocation the total profit is 6.0% lower and total welfare 3.9% lower than in the no-tax case.

Long-Run Model
In the long-run, production allocation with 20 euros cross-border tax is similar to the scenarios where there is no tax or there is producer tax. In the profit-maximizing allocation the home market H having a factory produces to the home market with no tax. On the other hand, for markets F1, F2 and F3 the cross-border tax is realized, and the tax

increases the optimal price and decreases total profits, but the total welfare is almost the same as in all the other previous cases.

If the border tax is increased significantly, at a value of 80 euros, the production allocation becomes decentralized with factories in countries H, F1, and F2 (Table 5). Economies of scale diminish, profits reduce by 24.2% and total welfare by 16.7% compared to the no-tax case. Market F3 is served by one factory (here H) only, because economies of scale favor centralization of production.

Table 5. Allocation of production from each factory to markets in the long-run with high cross-border tax (80 euros)

Markets	Factories		
	H	F1	F2
H	2368	0	0
F1		2308	0
F2		0	2308
F3	1943	0	0
Total	4251	2308	2308

4.5 Effect of Combination of Producer or Consumer Tax and Border Tax

Experimentation with both producer tax and border tax in all countries is not very interesting. The behavior of production allocation with all models is similar to the no-tax, producer-tax, and border-tax models. The production amounts and profits reduce while the optimal prices increase. The amount of possible harmful emissions reduces, and total welfare generally stays almost unchanged.

If we implement only a producer tax, higher than the transportation cost, at one factory, the competitiveness of that factory is destroyed even at the home market. In this case, consumer tax is better, as it increases the cost of products only at the home market, and it affects all producers equally.

However, examining the effects of one or some countries introducing producer or consumer tax to reduce harmful emissions and border tax at the same time to avoid carbon leakage is the main focus of this study. In the following, we study these scenarios. We assume that the border tax is implemented by all countries at all borders.

Basic Model

If a producer tax is implemented at one factory together with an equal cross-border tax, we get decentralized production allocation in all factories.

Table 6 shows the production allocation with 20 euros producer tax in home factory only and cross-country border tax of 20 euros. The production at the home factory is lower than that at the other factories, and market F3 is not served by the home factory.

Table 6. Production allocation from each factory to markets for the basic model with producer tax 20 in home factory (H) and 20 euros tax for all cross-border transportation

Markets	Factories		
	H	F1	F2
H	2400	0	0
F1	0	2500	0
F2	0	0	2500
F3	0	1187	1188
Total	2400	3687	3688

This is because producer tax applies only to the home factory production. With consumer tax implemented, factory H would be competitive in market F3.

Short-Run Model
For the short-run model, implementation of producer tax 20 euros in country H only and cross-border tax 20 euros centralizes production to one foreign factory only or with capacity constraints to two foreign factories, see Table 7.

Table 7. Allocation of production with producer tax 20 euros at home factory, cross-border tax 20 euros and limited capacity

Markets	Factories		
	H	F1	F2
H		2375	0
F1	0	2500	0
F2	0	0	2500
F3			2375
Total		4875	4875

Increasing cross-border tax to 70 or assuming lower capacities decentralizes production to all factories, see Table 8. The increased fixed cost of three factories requires high cross-border taxes for protection of low volume domestic production. The total profits in this case are 9.2% lower than in the case of Table 7 and total welfare is 4.3% lower.

Implementing consumer tax in market H affects imports like domestic production and does not burden exports. Therefore, a cross-border tax of 40 euros decentralizes production to all factories and the total profits are 3.3% higher than in the Table 8 scenario, while the total welfare is about equal (Table 9).

Implementing producer or consumer taxes in two factories has the same effect concerning production allocation as one: The cross-border tax would need to be risen to 70 and 40 euros in order to make fully decentralized production more profitable than

Table 8. Decentralized production allocation with producer tax 20 at home factory and cross-border tax 70

Markets	Factories		
	H	F1	F2
H	2400	0	0
F1	0	2500	0
F2	0	0	2500
F3	0	1063	1063
Total	2400	3562	3562

Table 9. Decentralized production allocation with consumer tax 20 euros at home factory and cross-border tax 40 euros

Markets	Factories		
	H	F1	F2
H	2400	0	0
F1	0	2500	0
F2	0	0	2500
F3	825	725	725
Total	3225	3225	3225

production in only two countries. Again, this would reduce both total profitability and welfare significantly.

Long-Run Model

Producer tax in one factory only allocates all production to one of the non-taxing countries with the same amounts as in the no tax case. When a cross-border tax of 20 euros is implemented, the resulting profit-maximizing production allocation is still the same as in the cross-border tax only case, but in this case the only factory is in a non-taxing country. If the factory would for some reason exist only in the taxing country, the profit disadvantage of the taxing country factory would be about 9% (no border tax) or 9.4% (20 euros border tax), whereas the total welfare would be about equal. Increasing cross-border tax up to 80 euros causes decentralization of production to both non-taxing countries, see Table 10. Factory in the other one (F2) serves that market only, and all the other markets are served by the other factory (F1). This distribution is due to economies of scale, that favor concentration of production.

Further increase of cross-border tax to 95 euros decentralizes profit-maximizing production to all countries, see Table 11.

In terms of profit, welfare, and revenue it is the best point for the country with producer tax that can be obtained with the profit-maximizing objective. If the producer

Table 10. Profit-maximizing production allocation to markets with producer tax 20 euros in one (H) factory and cross-border tax 80 euros

Markets	Factories		
	H	F1	F2
H	0	1970	0
F1	0	2395	0
F2	0	0	2308
F3	0	1970	0
Total	0	2355	6730

Table 11. Decentralized profit-maximizing production allocation to markets with producer tax 20 in one (H) factory and cross-border tax 95 euros

Markets	Factories		
	H	F1	F2
H	2203	0	0
F1	0	2367	0
F2	0	0	2308
F3	0	1867	0
Total	2202	4726	2308

tax at home country is taken to be proxy for non-polluting technology, the environment benefits, too. However, at this point the cross-border tax eliminates the economies of scale effects of centralized production and the total profits and welfare are 22.5% and 16.9% lower than in the foreign-centralized-production case with 20 euros border tax. Compared to the decentralized cross-border tax only case (Table 5), the total profit is 4.7% lower and total welfare 0.4% lower.

Implementation of consumer tax in one country only has similar effects concerning production allocation as producer tax. The transition from centralized to fully decentralized production allocation takes place at 80 euros cross-border tax. Total profits are reduced by 19.1% and total welfare by 16.5% from the 20 euros cross-border case. Total profit is 2.9% lower than in the decentralized cross-border tax only case, while total welfare is about the same.

If producer tax or consumer tax is implemented in two countries, the production allocation transition takes place directly from one factory to three factories at 100 euros and 80 euros cross-border tax respectively. Profits and welfare are only slightly lower than in corresponding one-country-tax production allocation configuration.

For the long-run model, if the fixed cost, economies of scale factor, or variable cost is increased individually and other parameters remain the same, increasing the border tax is also needed to get decentralized allocation. This is because economies of scale are effectively increased either by changing the unit cost curve steeper or by moving the curve so that the profit maximizing operating quantities take place at steeper parts of the curve. So, in the long-run model, all the cost curve parameters affect the economies of scale significantly. For all above cases, economies scale, $e = -0.6$ is used. If the value of the economic factor (e) is increased, production cost (per piece) decreases first less with volume and then more (see Fig. 2). On the other hand, with other parameter values being equal, the $e = -0.5$ curve in Fig. 2 is above the other. Economic factor –0.5 in the above example would require border tariff of 185 euros for distribution of production to the two foreign (polluting) factories and 200 euros for full decentralization. The total profit in this case is 11.1% lower than in the corresponding decentralized border-tax-only case, where the border tax is 175 euros. The total welfare is 3.4% lower.

5 Conclusions

To fulfil EU´s goal to reduce carbon emissions from carbon-intensive products, CBAM is a way to tackle the situation. The idea is to implement cross-border tax to compensate for cost of domestic carbon tax or cost of technical reduction of emissions while avoiding transfer of production to countries with no carbon tax. Our microeconomic study demonstrates the effects of implementing CBAM in different tax scenarios. Experiments were performed using realistic values for different taxes in the models. Correspondingly, depending on the model, different technologies were represented by the parameter values of the cost functions. Based on our numerical results we make the following conclusions:

- If the per unit production cost is constant, i.e., there is no fixed cost involved, the CBAM together with producer or consumer tax works as intended. It prevents carbon leakage without any significant disadvantages.
- Here and generally, the consumer tax is somewhat better than the producer tax, as it affects imports like domestic production and does not penalize exports. Producer tax benefits the producing factory country, whereas consumer tax is collected at the consuming market.
- Implementing carbon taxes to manufacturing industries, where investments, i.e. fixed costs, are involved, reduces the competitiveness of domestic production. Border taxes protect local producers against imports, but in our example scenarios the EU CBAM tax level is too low. With consumer tax the disadvantage is relatively modest.
- In the long run used production technology adapts and emission free-riders may get a significant advantage in the market place. Our results show that the amount of economies of scale strongly affects the profit maximizing production allocation and carbon taxes make matters worse. It appears that under economies of scale effects, the proposed EU border tariff levels are far too low to create a fair marketplace. Our long-run model indicates that under economies of scale centralized allocation in a non-taxing country is not avoided with low border tariffs, whereas with sufficiently high (5 to 10 times the EU CBAM tariff level) border tariffs decentralized allocation is obtained to different countries and the amount of polluting production is reduced significantly. A major side effect of obtaining the fair marketplace is that economies of scale are reduced, and the profits of the producers and taxes collected are reduced significantly.

References

Balistreri, E.J., Kaffine, D.T., Yonezawa, H.: Optimal environmental border adjustments under the general agreement on tariffs and trade. Env. Res. Econ. **74**(3), 1037–1075 (2019). https://doi.org/10.1007/s10640-019-00359-2

Benson, E., Majkut, J., Reinsch, W.A., Steinberg, F.: Analyzing the European Union's Carbon Border Adjustment Center for Strategic and International Studies (CSIS) (2023). https://www.csis.org/analysis/analyzing-european-unions-carbon-border-adjustment-mechanism. Accessed 22 Dec 2024

European Commission Proposal for a regulation of the European parliament and of the council establishing a carbon border adjustment mechanism (2021). https://eur-lex.europa.eu/legal-content/en/TXT/?uri=CELEX%3A52021PC0564. Accessed 23 Dec 2024

European Commission Carbon Border Adjustment Mechanism (2024). https://taxation-customs.ec.europa.eu/carbon-border-adjustment-mechanism. Accessed 23 Dec 2024

Fang, Y., Yu, Y., Shi, Y., Liu, J.: The effect of carbon tariffs on global emission control: a global supply chain model. Transp Res E **133**, 101818 (2020). https://doi.org/10.1016/j.tre.2019.11.012

Gros, D., Egenhofer, C.: The case for taxing carbon at the border. Clim Policy **11**(5), 1262–1268 (2011). https://doi.org/10.1080/14693062.2011.592669

Jakob, M., Marschinski, R., Hübler, M.: Between a rock and a hard place: a trade- theory analysis of leakage under production-and consumption-based policies. Environ. Res. Econ. **56**(1), 47–72 (2013). https://doi.org/10.1007/s10640-013-9638-y

Markusen, J.R.: International externalities and optimal tax structures. J. Intl. Econ. **5**(1), 15–29 (1975). https://doi.org/10.1016/0022-1996(75)90025-2

Niemi, E.: Economies of scale in make-to-order production with applications. Dissertation, Helsinki University of Technology (1998)

Opensolver (2024) Opensolver.org

Zhong, J., Pei, J.: Carbon border adjustment mechanism: a systematic literature review of the latest developments. Clim. Policy **24**(2), 228–242 (2023). https://doi.org/10.1080/14693062.2023.2190074

Skill-and-Knowledge Sharing by Augmented Reality: HUB-CI Model

Praditya Ajidarma[1,2](✉) and Shimon Y. Nof[1]

[1] PRISM Center and Edwardson School of Industrial Engineering, Purdue University, West Lafayette, USA
ajidarmap@gmail.com
[2] Faculty of Industrial Technology, Bandung Institute of Technology, Bandung, Indonesia

Abstract. Advancements and the adoption of productive human-robot collaborations have im pacted every aspect of modern manufacturing. Although automation has been increasingly utilized on the shop floor for decades, a significant portion of manufacturing operations remains manual primarily due to the dominating ability of human operators to perform these operations well, compared to robots. These manual or semi-automated operations involve timely complex manipulations and reasoning, and depend highly on human labor skills, intelligence, and expertise. To prepare the workforce of the future, manufacturers are increasingly turning to augmented reality (AR) to prepare, train, guide, and refine the skills of their less-experienced workers by allowing skilled workers to collaboratively guide and mentor less-experienced workers, collaboratively and on the fly. In this paper, we propose a new skill- and- knowledge- -sharing model in manufacturing systems with a hub for collaborative intelligence (HUB-CI) that supports an effective learning protocol. The learning curves of operators are modeled as a function of time and the complexity of skills and knowledge, and analyzed for the learning quality. A case- study of assembly kitting workflow is used to conduct the experimental simulations and analyses. Several insights are drawn from the preliminary simulation analyses to inform the design of future AR technologies for skill and knowledge sharing among manufacturing workers, working collaboratively with robots on complex manipulation and reasoning tasks.

Keywords: Augmented reality (AR) · HUB-CI · learning protocols · skill and knowledge sharing · workforce preparation

1 Introduction

With highly variable job and task requirements in modern work, as well as highly variable levels of preparedness of workers and robots under such conditions, a major concern of researchers of future work and factories has been: How to dynamically share skills and knowledge as soon as needed, with workers and robots. Collaborative Control Theory (CCT) can optimize such collaboration and integration.

A widely acceptable taxonomy for objective assessment in the cognitive domain is termed Bloom's Taxonomy which categorized skill into an ordinal scale of cognitive

M. Dragomir et al. (Eds.): ICPR 2023, LNPE, pp. 17–25, 2025.
https://doi.org/10.1007/978-3-031-92082-0_2

ability (Bloom, 1956). This taxonomy has been extended further to a taxonomy of skill and knowledge relationships for cyber-physical manufacturing systems (Krachtt, 2019). In the collaborative work and factories of the future (Moghaddam & Nof, 2018), multiple systems, including humans as participants and as clients, are designed to cooperate, collaborate, and integrate skills and knowledge toward accomplishing given objectives. Extensive technological applications have been employed, as reviewed in the subsequent sections.

To enable better human augmentation, the development of an effective skill and knowledge-sharing model is crucial. As the market requirement for skills changes dynamically, shortening the learning curve and enabling faster skill sharing on the fly are crucial. The key challenge is, with the limited availability of resources and training time, to identify and execute the optimal protocol for skill and knowledge sharing in the cyber-collaborative physical systems (CCPS).

Despite these recent advances in AR applications, limited research has focused on enabling skill and knowledge-sharing instances on the fly. This study aims to fill this gap by proposing a preliminary model and simulations that utilize HUB-CI to enable skill and knowledge sharing within an augmented reality system.

2 Background

2.1 Augmented Reality-Based Skill and Knowledge Sharing

Augmented reality (AR) has enabled a multitude of applications in manufacturing and industry in general. Several survey articles have summarized the state-of-the-art of AR applications by their problem domain, intelligence source, utilized devices, and their key benefits in industries. A survey of AR applications that enabled skill and knowledge sharing is presented in Table 1.

2.2 HUB-CI

Hubs for collaborative intelligence (HUB-CI) focused on improving human collaboration through e-collaboration tools and services in CCPS. It has been proven to enhance the timely synthesis, integration, and delivery of knowledge and discoveries (Seok & Nof, 2011) and enable human-robot collaborative control for physical automation and assembly in a variety of manufacturing applications (Zhong & Nof, 2013). A HUB-CI for Collaborative Telerobotics (CTR) is developed where operators with differing levels of expertise work alongside robots to execute a set of aggregated protocols with a human-robot feedback loop (Zhong et al., 2014).

HUB-CI was also developed for Agricultural Robotics Systems (ARS) to optimize human-robot collaboration and managed to improve system efficiency by between 210% to 255% (Nair et al., 2019). Furthermore, HUB-CI has been designed to enable collaboration requirement planning (CRP) for a HUB-CI within factories of the future (Dusadeerungsikul et al., 2019), which showed two protocols reduced the makespan by 45% compared to the baseline scenario. Furthermore, HUB-CI also aims to enable collaboration within a company, by optimizing information flow between workers on the shop floor (Moghaddam & Nof, 2022), and between companies, by participating in knowledge sharing, which enables better demand and capacity sharing in a collaborative supply network (Ajidarma et al., 2022).

2.3 Learning Curves and AR Elements

The learning curve is described by performance time and is formalized as a power function of the number of trials since the start of learning (Newell & Rosenbloom, 1993). A multitude of approaches have been utilized to shorten the learning curve, including the implementation of AR and cyber-collaborative manufacturing systems. One study concluded that learning outcomes and learning experiences are improved by AR in a gamified primary and secondary education setting (Pellas et al., 2019). Another study suggested that it also enabled faster psychomotor skills training, increased visualization capacity, and higher conceptual understanding of causality in medical education and improved operators' learning for motherboard assembly operations (Kamphuis et al., 2014).

These instances of learning are enabled by AR elements, which are generated by the AR platform and channeled through operators' primary modalities: visual, auditory, and tactile. Multiple studies have examined these cross-modalities. For instance, visual and auditory cues increased accuracy, and both have a more significant effect compared to tactile cues (Föcker et al., 2022). Tactile and auditory cues improved performance in terms of faster response times for both visual and visual-auditory scenarios. However, certain levels of audio or tactile in an element may become ineffective due to noise and disturbance in the shop floor environment (Hopkins et al., 2017).

Table 1. State of the Art of Augmented Reality Systems for Skill and Knowledge Sharing

Article	Research Domain	Application of Skill and Knowledge Sharing
(Kim et al., 2020)	Knowledge Benchmark	Enable knowledge sharing to view, annotate, classify, and analyze mechanical components
(Wang et al., 2021) (Nassar et al., 2021)	Human-to-human tutoring	Platform to transfer workers' experiences and learning behaviors, as well as workload sharing, and skills in general, for procedure-specified tasks
(A. M. Villanueva et al., 2021) (A. Villanueva et al., 2021)	Electronics and circuitry	AR to enable teleconsulting for circuitry assembly, micro-skills assessment, and curriculum implementation
(Kim et al., 2021) (Ipsita et al., 2021)	Shop floor and object simulators	AR platform to simulate operator tracking, object synthesis, part retrieval, and manipulation within shop floor operations
(Nikolaidis & Shah, 2013) (Huang et al., 2021)	Collaborative human-robot	Human-robot workflow protocols for skill and knowledge sharing for flexible jobs, as well as for procedure-specified executions

Table 2. Properties of AR Elements

i	AR element (AR_i)	Visual Cost	Auditory Cost	Tactile Cost	Time Reduction
1	Particular troubleshooting steps video guidelines	High	High	Low	High
2	Operation process video guidelines	High	High	Low	Moderate
3	Haptic feedback on wrists when a step requirement is violated	Low	Low	Moderate	Low
…	…	…	…	…	…
n	Auditory guide on the next step of a semi-automated operation	Low	High	Low	Moderate

3 Methodology

The skill and knowledge sharing with HUB-CI are modeled for a semi-autonomous kitting operation. Kitting operation is a time-consuming task, which is mostly done manually by human operators, mostly due to the complex recognition, and grasping of a large variety of objects, as well as the precise positioning of parts into kitting boxes that is necessary for kitting operations. To streamline the kitting operations, HUB-CI is equipped to feed-forward information to operators via the elements within the AR platform (Fig. 1).

Each element is given via AR platform impact operators differently, depending on the stimulus type, size/duration, and the relevancy of information it carries. Based on the previous literature review, several AR elements are defined within the context of a semi-autonomous kitting operation, as shown in Table 2. Each element has its corresponding costs and benefits with regard to processing time reduction, and therefore, a chosen set of stimuli can be optimized, given the availability of deterministic cost and benefit data. The skill sharing with augmentation is formulated as a 0/1 Multidimensional Knapsack Problem (0/1 MKP).

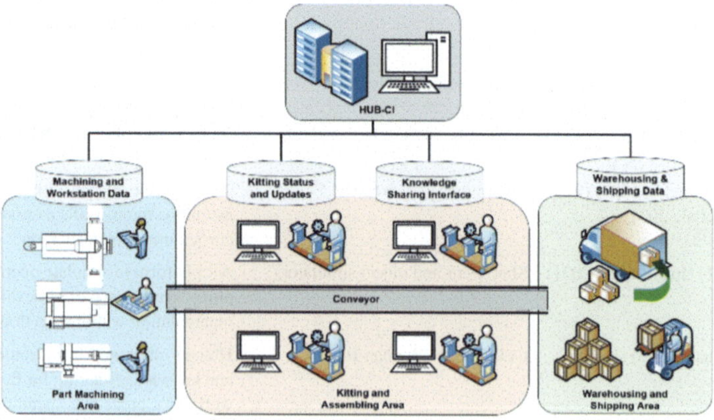

Fig. 1. HUB-CI for AR Systems with Semi-Autonomous Kitting Operations

Each AR element AR_i has a corresponding ${w_i}^A$ for its auditory cost, ${w_i}^V$ for its visual cost, and ${w_i}^T$ for its tactile cost. To model the cognitive load of each operator, the maximum capacity of auditory, visual, and tactile input for each operator is defined as C^A, C^V, and C^T, respectively. The reward given by each element AR_i is the reduction in processing time in the semi-autonomous kitting operations or r_i. The decision variable x_i is a binary variable indicating if stimulus i is chosen ($x_i = 1$) or not ($x_i = 1$). Thus, the 0/1 MKP for deterministic skill and knowledge sharing is defined as follows:

$$\max \sum_{i=1}^{n} p_i x_i \tag{1}$$

$$\sum_{i=1}^{n} w_i^A x_i \leqslant C^A \tag{2}$$

$$\sum_{i=1}^{n} w_i^V x_i \leqslant C^V \tag{3}$$

$$\sum_{i=1}^{n} w_i^T x_i \leqslant C^T \tag{4}$$

$$x_i \in \{0, 1\}; \forall i = 1, 2, ..., n \tag{5}$$

This instance of skill and knowledge sharing enabled by HUB-CI reduces the processing time over time. The instance is further demonstrated in an operator assignment problem for a semi-autonomous kitting operation.

4 Numerical Examples and Analysis

The skill and knowledge-sharing model with HUB-CI is simulated in a variety of scenarios. The results are analyzed in terms of the reduction of processing times and the effectiveness of operator assignment for semi-autonomous kitting operations.

4.1 Reduction of Processing Times

The skill and knowledge sharing model with HUB-CI is simulated based on the parameters as shown in Table 3. The reduction in processing times is simulated over the different components of input data, varying numbers of AR elements, and differing lengths of sharing iterations. The input data is generated based on previous literature, as specified in Sect. 2.3.

The simulation results on the learning curve and the reduction of processing times are presented in Figs. 2 and 3. Based on Fig. 2, the reduction in processing times due to skill and knowledge sharing is more significant compared to the baseline scenario when the number of AR elements is more considerable, i.e., 100 in Fig. 2 (b), compared to 10 in Fig. 2 (a). Furthermore , from Fig. 3, it can be concluded that over a larger span of

Table 3. Simulation Parameters

Parameters	Levels	Values
AR elements input data	3	80% audio-based; 80% visual-based; 80% haptic-based
AR elements (*n*)	2	10, 100
Sharing iterations (*k*)	2	100, 500

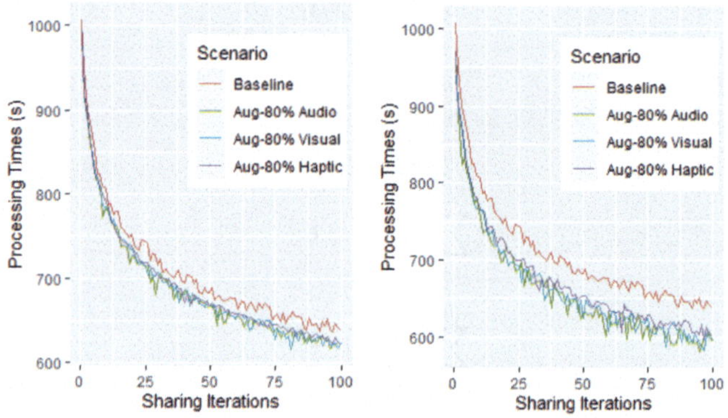

(a) (b)

Fig. 2. SaKS Learning Curves for (a) n = 10, (b) n = 100, over 100 Iterations

sharing iterations, the reduction appears to approach their corresponding asymptotes. It can also be inferred generally, from Figs. 2 and 3, that skill and knowledge sharing with the majority of audio-based and visual-based elements have a more significant reduction compared to skill and knowledge sharing with the majority of haptic-based elements.

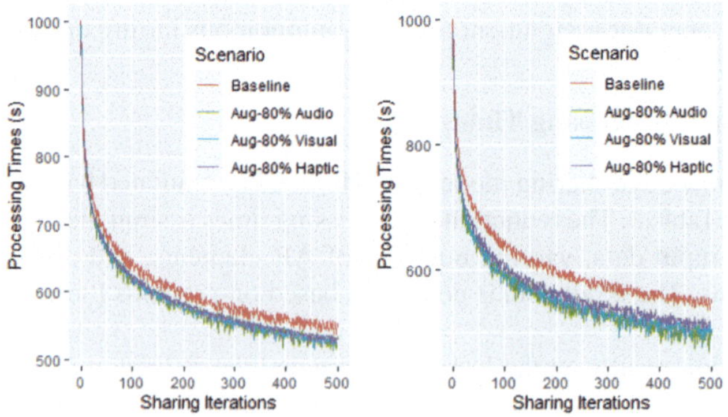

(a) (b)

Fig. 3. SaKS Learning Curves for (a) n = 10, (b) n = 100, over 500 Iterations

4.2 Operator Assignment for Kitting Operations

The processing times after skill and knowledge augmentation are simulated for operator assignment in kitting operations. In each scenario, ten operators are assigned ten kitting tasks; each has a processing time generated from a Poisson distribution. A similarly constructed constraint matrix is imposed for each scenario. The makespan reduction results are shown in Table 4.

Table 4. Comparison of Kitting Operation Makespan

Scenario	Makespan of Kitting Operations (s)			
	$k = 100$		$k = 500$	
	$n = 10$	$n = 100$	$n = 10$	$n = 100$
Baseline	6082		5091	
HUB-CI 1	5905	5620	5033	4750
HUB-CI 2	5903	5619	5014	4729
HUB-CI 3	5913	5687	5006	4827

Based on the simulation results, it can be generally concluded that a longer duration of skill and knowledge-sharing instances improves makespan, even for baseline scenarios. Furthermore, a higher number of AR elements included in each iteration ($n = 100$) also leads to a more significant reduction in makespan. The makespan reduction varies from 1.14% ($p_{value} < 0.01$) in the HUB-CI 1 scenario to 7.61% ($p_{value} < 0.01$) in the HUB-CI 2 scenario. The makespan reductions are statistically significant.

5 Conclusion

This study proposed a new skill-and-knowledge-sharing model in manufacturing systems with a hub for collaborative intelligence (HUB-CI). The reduction in processing times due to the shortening of learning curves has been demonstrated. A preliminary case study of assembly kitting workflow is used to demonstrate the efficacy of the proposed model. Future research direction will also address the potential of error and conflict due to the untimely delivery of AR elements, as well as the trade-off between collaboration costs and makespan reduction associated with skill and knowledge sharing.

Acknowledgments. This article was supported by the PRISM Center at Purdue University and by NSF Grant #1839971: Pre-Skilling Workers, Understanding Labor Force Implications and Designing Future Factory Human-Robot Workflows Using Physical Simulation Platform. We also thank colleagues who gave valuable comments to improve this study.

References

Bloom, B.S., Krathwohl, D.R.: Taxonomy of educational objectives; the classification of educational goals by a committee of college and university examiners. Handbook I: Cognitive Domain. New York (1956)

Krachtt, N.: The workforce implications of Industry 4.0: manufacturing workforce strategies to enable enterprise transformation, University of Wisconsin, Platteville (2018)

Moghaddam, M., Nof, S.Y.: Collaborative service-component integration in cloud manufacturing. Int. J. Prod. Res. **56**(1–2), 677–691 (2018)

Seok, H., Nof, S.Y.: The HUB-CI initiative for cultural, education and training, and healthcare networks. In: Proceedings of the International Conference on Production Research (ICPR). Stuttgart (2011)

Zhong, H., Nof, S.Y.: Collaborative design for assembly: the HUB-CI Model. In: 22nd International Conference on Production Research (2013)

Zhong, H., Wachs, J.P., Nof, S.Y.: HUB-CI Model for collaborative telerobotics in manufacturing. IFAC Proc. Volumes **46**(7), 63–68 (2013)

Nair, A.S., Bechar, A., Tao, Y., Nof, S.Y.: The HUB-CI model for telerobotics in greenhouse monitoring. Procedia Manuf. **39**, 414–421 (2019)

Dusadeerungsikul, P.O., et al.: Collaboration requirement planning protocol for HUB-CI in factories of the future. Procedia Manuf. **39**, 218–225 (2019)

Moghaddam, M., Nof, S.Y.: Information Flow Optimization in Augmented Reality Systems for Production & Manufacturing. Proc. of ICPR_AR Conf., Curitiba (2022)

Ajidarma, P., Nof, S.Y., Pradana, R.A., Nugroho, W.A., Halim, A.H.: Preemptive demand and capacity sharing learning protocols using long short-term memory (LSTM) neural network autoencoders, Proc. of 10th IFAC MIM, Nantes (2022)

Newell, A., Rosenbloom, P.: Mechanisms of skill acquisition and the law of practice. In: Cognitive Skills and Their Acquisition, vol. 1 (1993)

Pellas, N., Fotaris, P., Kazanidis, I., Wells, D.: Augmenting the learning experience in primary and secondary school education: a systematic review of recent trends in augmented reality game-based learning. Virtual Reality **23**(4), 329–346 (2019)

Kamphuis, C., Barsom, E., Schijven, M., Christoph, N.: Augmented reality in medical education? Perspect. Med. Educ. **3**(4), 300–311 (2014)

Föcker, J., Atkins, P., Vantzos, F.-C., Wilhelm, M., Schenk, T., Meyerhoff, H.S.: Exploring the effectiveness of auditory, visual, and audio-visual sensory cues in a multiple object tracking environment. Atten. Percept. Psychophys. **84**(5), 1611–1624 (2022)

Hopkins, K., Kass, S.J., Blalock, L.D., Brill, J.C.: Effectiveness of auditory and tactile crossmodal cues in a dual-task visual and auditory scenario. Ergonomics **60**(5), 692–700 (2017)

Kim, S., Chi, H., Hu, X., Huang, Q., Ramani, K.: A large-scale annotated mechanical components benchmark for classification and retrieval tasks with deep neural networks, pp. 175–191 (2020)

Wang, K.J., Rizqi, D.A., Nguyen, H.P.: Skill transfer support model based on deep learning. J. Intell. Manuf. **32**(4), 1129–1146 (2021)

Nassar, A.K., Al-Manaseer, F., Knowlton, L.M., Tuma, F.: Virtual reality (VR) as a simulation modality for technical skills acquisition. Ann. Med. Surg. **71**, (2021)

Villanueva, A., et al.: Towards modeling of human skilling for electrical circuitry using augmented reality applications. Int. J. Educ. Technol. High. Educ. **18**(1), 39 (2021)

Villanueva, A.M., Liu, Z., Zhu, Z., Du, X., Huang, J., Peppler, K.A., Ramani, K.: RobotAR: An augmented reality compatible teleconsulting robotics toolkit for augmented makerspace experiences. In: Proceedings of the 2021 CHI Conference on Human Factors in Computing Systems, pp. 1–13 (2021)

Kim, S., Chi, H., Ramani, K.: Object synthesis by learning part geometry with surface and volumetric representations. Computer-Aided Des. **130**, 102932 (2021)

Ipsita, A., Li, H., Duan, R., Cao, Y., Chidambaram, S., Liu, M., Ramani, K.: VRFromX: from scanned reality to interactive virtual experience with Human-in-the-Loop. In: 2021 CHI Conference on Human Factors in Computing Systems, pp. 1–7 (2021)

Nikolaidis, S., Shah, J.A.: Human-robot cross-training: computational formulation, modeling and evaluation of a human team training strategy. In: 2013 8th ACM/IEEE International Conference on Human-Robot Interaction (HRI), pp. 33–40 (2013)

Huang, Y., Xiao, W., Wang, C., Liu, H., Huang, R., Sun, Z.: Towards fully autonomous ultrasound scanning robot with imitation learning based on clinical protocols. IEEE Robot. Autom. Lett. **6**(2), 3671–3678 (2021)

Comparative Study of Seru Production Systems and Traditional Assembly Lines

Ciro Alberto Amaya(✉) and Melissa Pajaro

Department of Industrial Engineering, University of Los Andes, Bogotá, Colombia
ca.amaya@uniandes.edu.co

Abstract. The implementation of the Seru production system has meant a significant improvement for several Japanese manufacturing companies. Currently, different types of Seru are identified: Divisional, Rotary and Yatai. In this study, different production metrics were compared between the assembly line and the different types of Seru. For this purpose, a worker allocation model was developed for the divisional Seru and the assembly line. Subsequently, experiments were conducted in a laboratory setup, where different production systems were implemented. Finally, simulations of these experiments were developed over a longer time horizon. From the results, a positive impact of the implementation of the different types of Seru on cycle time, throughput and in-process inventory was observed.

Keywords: Seru Production System · Assembly Line · Worker Allocation Model · Learning Curve

1 Introduction

Seru means cell in Japanese and was so named to differentiate it from the traditional CM (cellular manufacturing) cell production system (Yin et al. 2017). Seru is an assembly unit that includes simple equipment and one or several workers who perform most production processes (Yu and Tang 2019). Workers then can be multi-skilled (can complete several tasks) or fully skilled (can complete all tasks). In fact, Seru production system (SPS) is a worker-centered manufacturing system with a high degree of Jiritsu, which means autonomy, self-management, learning, and self-evolution (Yin, Stecke, and Li 2018). Various case reports of SPS implementation have concluded that SPS can adapt to high variety and low and variable volume demands, and also can be flexibly reorganized following dynamic demands, and can enhance workers' sense of responsibility. However, SPS cannot improve productivity in all cases. For example, in mass productions, the efficiency of SPS is worse than that of the assembly line (Yin et al. 2017, 2018).

So, the question that has been asked by several researchers is when it is appropriate to use a Seru production system. Johnson studied the conversion from assembly line to SPS in the production of sheet metal products, and concluded that when the lead time is short and the products are customized by customers the Seru system should be employed (Johnson 2005). Kaku et al. concluded that companies should implement SPS

M. Dragomir et al. (Eds.): ICPR 2023, LNPE, pp. 26–32, 2025.
https://doi.org/10.1007/978-3-031-92082-0_3

in a high variety and low volume environment after studying a multi-objective makespan and labor hour minimization model (Kaku et al. 2009). Also, Yu et al. deduced that a pure SPS should be applied in high variety and low volume cases from an analysis of a factorial design of computational experiments (Yu et al. 2012). On the other hand, Yin et al. stressed that Seru production should be proposed in response to an environment with a dynamic market and high costs (Yin et al. 2017).

Considering that there are different types of SPS, this paper compares different types of Seru and assembly line considering different production metrics. In this paper a set of simulations are carried out using data collected at the laboratory level. These simulations will allow comparing the performance of the different types of Seru and the assembly line.

2 Methodology

2.1 Construction of Models for Divisional and Assembly Line

In the case of divisional Seru and assembly line, a worker allocation model to find the most appropriate distribution of workers within the production line (Wu et al. 2018) was constructed. In the divisional Seru, each worker is assigned certain tasks, and this set of tasks constitutes the workstation. The model assumptions taken into account were:

- Model assumptions
- All workers are capable of performing all tasks.
- The skill level of the workers for the tasks is given by a random variable (processing time for each task).
- All available workers should be utilized.
- Each station includes different tasks that must comply with the sequence required to assemble the final product.
- Each station can only have a single worker due to space limitations, so no tasks are shared between workers.
- Travel time and assembly time are not taken into account.
- In the manufacturing line only the tasks for the assembly of a specific product are performed.
- Since the processing time in each task for each worker is a stochastic random variable, it was decided to develop this model in a deterministic way by means of several scenarios, which are equally weighted in the objective function.

2.2 Laboratory Experiments

In order to fit a simulation model, a series of real experiments were carried out in the industrial engineering laboratory at the Universidad de los Andes where a manufacturing system with conveyor belts was simulated, the final product was a Lego cart shown in Fig. 1. The assembly process of this item starts with individual parts and was divided into 7 tasks, which were placed along the manufacturing line.

Fig. 1. Final product of the experiments

The sequence of the tasks is shown in Fig. 2, where the necessary precedencies for each of the tasks performed in the production line are identified.

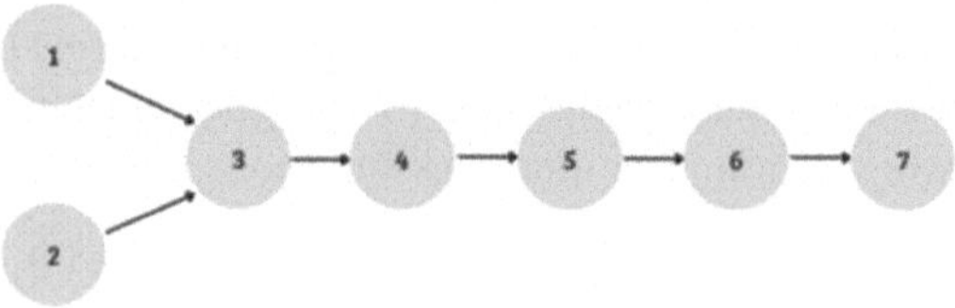

Fig. 2. Sequence of tasks

The experiments were carried out with two workers. First, the skill level of each worker was determined by measuring the time spent on each task. At this point, the effect of the learning/forgetting curve on the skill level of each worker became evident. Once the processing time began to stabilize, which was observed from 10 fully assembled products, the processing time of each task was measured with 30 replicates to adjust an appropriate distribution of time in minutes for each worker. Also, the travel time between each task, which influences the production cycle time, was measured in order to fit an appropriate statistical distribution and use it in the simulation.

Finally, cycle time measurements of the production line were carried out in different configurations, taking into account the types of production systems to be compared. For each of the experiments, two replicates were carried out in order to confirm the data obtained. For each of the experiments, two replicates were carried out in order to confirm the data obtained.

2.3 Computational Simulations

Once the processing and travel times were adjusted to the distributions, and with the configurations obtained from the results of the divisional Seru model, simulation models were built for each of the production types. These models included both the processing time in each task and the travel time between tasks. The Simmer library in R was used for the implementation and analysis of the simulations, which aim to evaluate the differences for the different types of configurations studied. It should be noted that once the models were built in the software, they were verified and validated with the experiments developed in the laboratory.

The configurations that were implemented in the simulations are as follows:

- Assembly line
- Seru divisional PUSH system
- Seru divisional CONWIP
- Rotating Seru
- Seru Yatai

Each of these simulations were developed for 4 h of continuous work with the two workers, and additionally 100 replicates of each case were implemented. The generation of arrivals to the system was not set as a random variable, instead a part arrives to the system each time the worker in charge of task 1 is available in that task. This was implemented in this way in order to perform a more appropriate validation of the simulation model. In the case of the pull system (CONWIP divisional Seru), the activation of an arrival to the system was configured each time a part leaves the last task, i.e. task 7.

3 Computational Simulations Results

This paper presents the results obtained from the computational simulation, other results were obtained and presented in an article that is subject to publication.

3.1 Cycle Time

A box plot associating the average cycle time for each of the simulated configurations is shown in Fig. 3. It is evident that the assembly line and the divisional Seru in push system present a higher variability than the other Seru types. Additionally, the cycle time of these two configurations tends to have a higher value than the others. This makes sense since it is related to the bottleneck generated in both configurations. In fact, the cycle time distribution of the divisional Seru divisional in push system tends to have higher values than the assembly line, since a larger bottleneck is formed as observed in the simulation previously. As for the Seru divisional in pull system is the one that has a tendency to a lower cycle time since it is the one that optimizes the processing time by means of the allocation model and does not present bottlenecks.

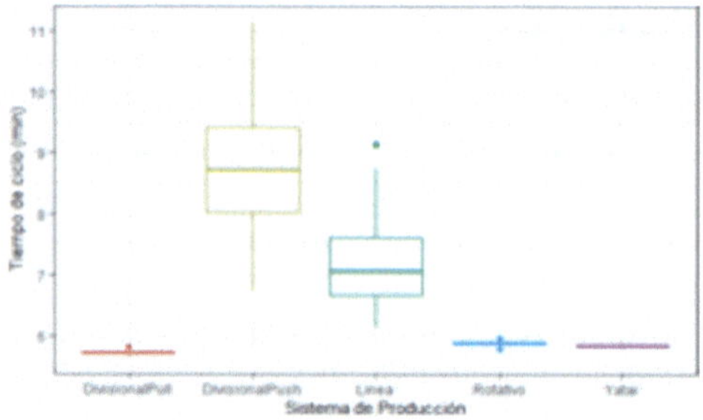

Fig. 3. Cycle time comparison

3.2 Throughput

Figure 4 compares the production rate (throughput) in units per hour for the different cases analyzed. Evidently, the divisional Seru in pull system presents a very low rate since the objective of this system is to reduce production costs and inventories by limiting the same production rate in response to the customer's needs. The other configurations present a very similar throughput value, but the Seru Yatai stands out, which tends to have the highest rate. It should be noted that the difference in magnitude of the throughput rate is not very high, but these results are the product of laboratory experiments. In an industrial environment where tasks take a greater amount of time, the difference may be more noticeable.

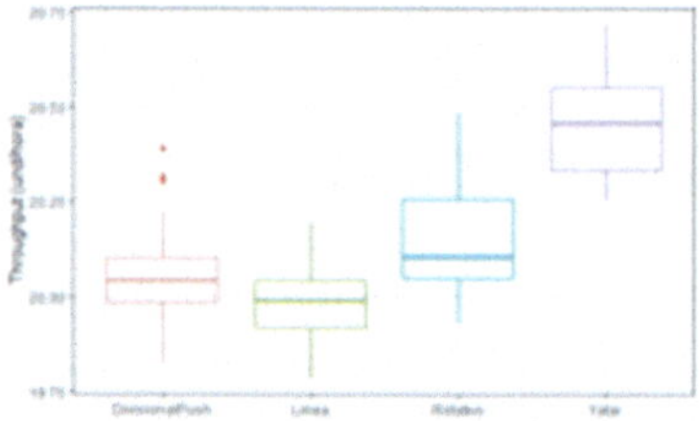

Fig. 4. Throughput comparison

3.3 WIP

Finally, the in-process inventory was calculated by means of Little's Law relating cycle time and throughput. Due to the high variability and high cycle time values in the divisional Seru with push system and in the assembly line, there is a high and variable in-process inventory in these two configurations, which is again related to the bottleneck. For its part, the Yatai Seru maintains an in-process inventory equal to 2, since there are two Yatais in the system; and the rotary Seru an in-process inventory ~2 since the two workers handle two parts simultaneously. Clearly the WIP of the divisional Seru in pull system is equal to 1 since the whole line takes care of only one piece until it is finished. See Fig. 5.

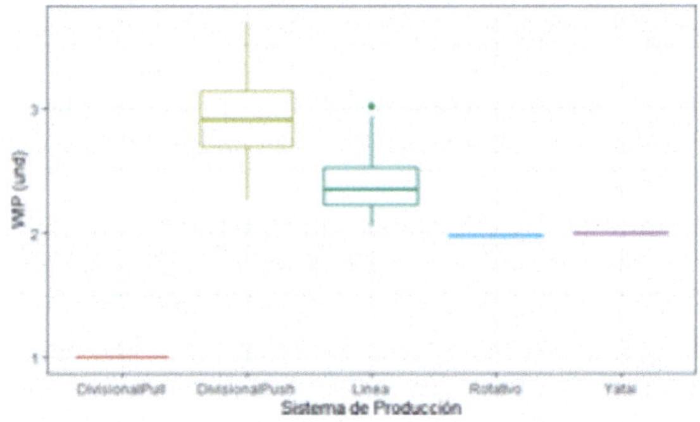

Fig. 5. WIP comparison

4 Analysis of Results

When analyzing together the metrics evaluated for the different configurations studied, it is evident that each configuration has its advantages and disadvantages. The assembly line and the divisional Seru in push system are constrained by bottlenecks, which leads to an increase in lead times and in-process inventory. It is important to note that even if the assembly line is switched to a push divisional system, in-process inventory costs may increase. However, a non-paired test of the mean difference between the throughput of the divisional Seru in push system and the assembly line shows that with a confidence level of 95%, the production rate in the Seru is higher.

The problem of both the assembly line and the divisional Seru is also related to their low responsiveness. When workers are assigned to specific tasks and cycle time is subject to bottlenecks, less efficient workers can block the passage of more efficient workers, resulting in significant losses in uptime (Gai et al. 2020). Additionally, these two systems are inflexible and in an environment with high dynamism are not appropriate. However, it has been found that the assembly line is suitable when there is no cross-training, i.e. when the skill level among workers differs significantly (Aboelfotoh, Süer, and Abdullah 2018).

5 Conclusions and Future Work

The Seru production system is proposed as a production mode to overcome the low flexibility of the assembly line (Yu and Tang 2019). In Japanese companies in the electronics industry such as Canon and Sony it has had good results such as increased productivity, reduced assembly line length, reduced space requirements, reduced costs, improved product quality, among others (Zwierzyński and Ahmad 2018). In this study, it was intended to compare the performance of different types of Seru with computational simulations built from laboratory level experiments. From the experiments it was found that it is very important to take into account the learning and forgetting curve, since it is a source of variability in the system. It is also important to consider the travel time between tasks or workstations, since the flow time depends on this variable.

A worker assignment model was built seeking to minimize the processing time of the workers and restricting the problem to a maximum balancing ratio. It was possible to implement this model having random variables as processing time parameters in the tasks, and a sensitivity analysis was performed depending on the number of scenarios evaluated and the balancing ratio.

Based on the solution of the allocation model and the data collected in the experiments, each of the configurations was modeled in a simulation software. Considering the results of these simulations, it can be concluded that the production throughput improves with each step of the conversion between assembly line and Seru. In other words, the throughput of the Yatai Seru is higher than that of the rotary Seru, and that of the rotary Seru is higher than that of the divisional Seru. The cycle time also decreases for rotary Seru and Yatai Seru compared to the assembly line, as bottlenecks are decreased/eliminated. Finally, the in-process inventory of the rotary Seru and Yatai are lower and have lower variability than those of the assembly line.

References

Aboelfotoh, A., Süer, G.A., Abdullah, M.: Selection of assembly systems; assembly lines vs. seru systems. Procedia Comput. Sci. **140**, 351–358 (2018). https://doi.org/10.1016/j.procs.2018.10.304

Gai, Y., Yin, Y., Tang, J., Liu, S.: Minimizing makespan of a production batch within concurrent systems: seru production perspective. J. Manage. Sci. Eng. (xxxx) (2020). https://doi.org/10.1016/j.jmse.2020.10.002

Johnson, D.J.: Converting assembly lines to assembly cells at sheet metal products: insights on performance improvements. Int. J. Prod. Res. **43**(7), 1483–1509 (2005). https://doi.org/10.1080/00207540412331299594

Kaku, I., Gong, J., Tang, J., Yin, Y.: Modeling and numerical analysis of line-cell conversion problems. Int. J. Prod. Res. **47**(8), 2055–2078 (2009). https://doi.org/10.1080/00207540802275889

Wu, L., Chan, F.T.S., Niu, B., Li, L.: Cross-trained worker assignment and comparative analysis on throughput of divisional and rotating seru. Ind. Manag. Data Syst. **118**(5), 1114–1136 (2018). https://doi.org/10.1108/IMDS-07-2017-0303

Yin, Y., Stecke, K.E., Li, D.: The evolution of production systems from Industry 2.0 through Industry 4.0. Int. J. Prod. Res. **56**(1–2), 848–861 (2018). https://doi.org/10.1080/00207543.2017.1403664

Yin, Y., Stecke, K.E., Swink, M., Kaku, I.: Lessons from seru production on manufacturing competitively in a high cost environment. J. Oper. Manage. **49**, 67–76 (2017). https://doi.org/10.1016/j.jom.2017.01.003

Yu, Y., Gong, J., Tang, J., Yin, Y., Kaku, I.: How to carry out assembly line-cell conversion? a discussion based on factor analysis of system performance improvements. Int. J. Prod. Res. **50**(18), 5259–5280 (2012). https://doi.org/10.1080/00207543.2012.693642

Yu, Y., Tang, J.: Review of seru production. Front. Eng. Manage. **6**(2), 183–192 (2019)

Zwierzyński, P., Ahmad, H.: Seru production as an alternative to a traditional assembly line **10**(3) (2018). https://doi.org/10.2478/emj-2018-0017

Ontology Model for Mapping Terms and Relations in Plastic Manufacturing – A Case Study

Riad Al Hasan Abir, Mandvi Malik, Dušan Šormaz(✉), Peter Adjei, Felix Asare, David Koonce, and Saruda Seeharit

Ohio University, Athens, OH 45701, USA
sormaz@ohio.edu

Abstract. In modern manufacturing, ontology provides a standardized and structured way of organizing and sharing information. It enables better communication and collaboration between different entities involved in the manufacturing process. Acknowledging these benefits, we establish an ontology for manufacturing plastic components using the Industrial Ontology Foundry (IOF) framework. The IOF framework, derived from the Basic Formal Ontology (BFO), is an essential component of smart manufacturing. It facilitates semantic integration, enabling data exchange and reuse between different systems and organizations in the existing IOF framework. Our research employs a case study approach to develop the ontology for plastic manufacturing, which involves two main operations: sheet extrusion, and thermal forming operation, each with specific sequences of processes that must be followed to produce the desired final products. Our goal is to integrate the necessary process steps into the existing IOF framework in plastic manufacturing. To achieve this, we utilized Protégé software version 5.5 and used ontographs to visually depict the relationships between different categories and subclasses. By doing so, we aim to showcase how ontology reasoning can be applied effectively in the manufacturing sector. This will demonstrate the practical use of logical inference and computational techniques, ultimately improving the overall manufacturing process.

Keywords: Ontology · Plastic Manufacturing · BFO · IOF

1 Introduction

Automation has become an indispensable aspect of various industries, such as agriculture, manufacturing, healthcare, and transportation, due to its ability to enhance efficiency, increase productivity, and reduce cost [1–3]. However, it can also present new challenges, notably in the effective communication between machines. In the manufacturing sector, researchers have identified several challenges, such as inconsistent terminology, information complexity, a lack of widely accessible knowledge repositories, and insufficient information retrieval tool. Sharing distributed knowledge across

M. Dragomir et al. (Eds.): ICPR 2023, LNPE, pp. 33–41, 2025.
https://doi.org/10.1007/978-3-031-92082-0_4

cultures and levels of expertise is also a significant challenge for organizations. Fortunately, ontology, like the Industrial Ontology Foundry (IOF), can play a pivotal role in addressing these issues. Ontology is a formal representation of knowledge that defines the relationships between different entities in a given domain [4]. IOF-based ontology provides a standardized vocabulary and framework for describing components, processes, and materials in industrial settings, enabling machines to understand and communicate with each other through a common language [5]. This enhances the interoperability and collaboration between different systems and platforms, ultimately leading to improved efficiency, reduced costs, and increased quality. To demonstrate the use of ontology reasoning in manufacturing, we have established an ontology for manufacturing plastic products using the IOF framework. Our research centers on showcasing the application of ontology reasoning in the manufacturing sector, specifically in the production of plastic products. We have developed an ontology for manufacturing plastic containers and lids using the IOF framework. Through a case study methodology, we have examined the processes involved in manufacturing these plastic products, namely sheet extrusion and thermal forming. Sheet extrusion encompasses a series of process sequences, including melting, sheet extrusion, cooling, and rolling. In contrast, thermal forming involves processes such as feeding, heating, thermal forming, cooling, and trimming. To incorporate these operations and processes into the IOF framework, we utilized Protege software version 5.5. Additionally, we employed Ontographs as visual representations to elucidate the relationships between different classes and subclasses representing the processes and operations involved in plastic manufacturing, providing a clear understanding of the plastic manufacturing processes. Our primary aim is to enhance communication and collaboration among the various entities engaged in plastic product manufacturing. By leveraging ontology, our research endeavors to enhance the efficiency and effectiveness of the plastic manufacturing process, ultimately leading to improved outcomes.

2 Literature Review

The goal of ontology in manufacturing is to provide a standard platform to store information so that there are fewer ambiguities in communication regarding product manufacturing and there is less room for error. This section discusses the previous research done in creating ontology in manufacturing. Sarkar and Sormaz created ontology for model-based product design, the goal of this paper was to increase interoperability in product design information and facilitate integration with domain-specific design features from a manufacturing perspective [6]. Fumagalli et al. created ontology for the execution system of manufacturing equipment. The authors said that created ontology would reduce the assumptions that can happen when various people collaborate on a project [7]. The authors said that ontology will help in increasing flexibility, and adaptability and will make response easy to new changes in the market demands. Macris used ontology on the ERP system. The objective of their paper was to reduce semantics inconsistencies, provide explicit definitions of constructs used in ERP system, increase knowledge reusability to improve effectiveness of ERP training and reduce the time and resources required for training the system [8]. Hu et al. created ontology for the development of an aircraft industrial system. The aircraft industry faces the challenge of digital discontinuity and the paper addresses it. The authors said that proposed ontology based system reduces development time, cost and enhances design quality [9]. Adjei et. al created ontology for computer aided process planning. The authors showed that ontology can significantly improve the manufacturing planning and execution, providing a deeper understanding of the process and optimizing operations in the context of Industry 4.0 [10].

The research in ontology in the plastic industry has been gaining attention by researcher recently to improve semantic interoperability. Leite et al. created ontology for the development of plastic injected products such as mold design, mould manufacturing, etc. The goal of the paper is to increase the efficiency and effectiveness by decreasing the mistakes exchange in information error [11]. They Proposed an interoperable product design and manufacturing framework to address manufacturing industry challenges in global competitiveness, lead-time, pricing, quality, and customization. The authors created ontology using Protégé to address the interoperability issue. The authors used plastic rotation products to demonstrate the effectiveness of the proposed ontology framework [12]. This research introduces manufacturing operations and link the manufacturing operation information to the processes of plastic manufacturing and machines used in the process. To our knowledge this is the first research in this area that creates the relationships between these two areas in the plastic manufacturing.

3 Methodology

This research introduces the relationship between manufacturing operations, plastic manufacturing processes, and material artifacts that are used in it. To build the relationships in the protégé the first step taken was to obtain the definitions of the major processes used in plastic manufacturing and the second step was to add them into the protégé.

3.1 Major Definitions in Plastic Manufacturing Operations

The two major operations that are part of plastic manufacturing are sheet extrusion and thermal forming operation. Their definitions are described below:

- Sheet Extrusion: Sheet extrusion is a manufacturing operation where molten plastic is forced through a flat die (shown in Fig. 1) to form continuous plastic sheets.

- Thermal Forming: Thermoforming is an operation in which a thermoplastic sheet is heated to a temperature above its softening point and then formed into a desired shape (shown in Fig. 2) by applying pressure.

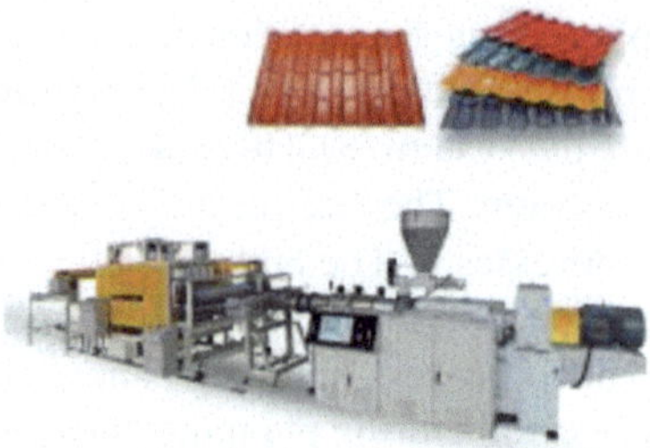

Fig. 1. Sheet extrusion operation [13]

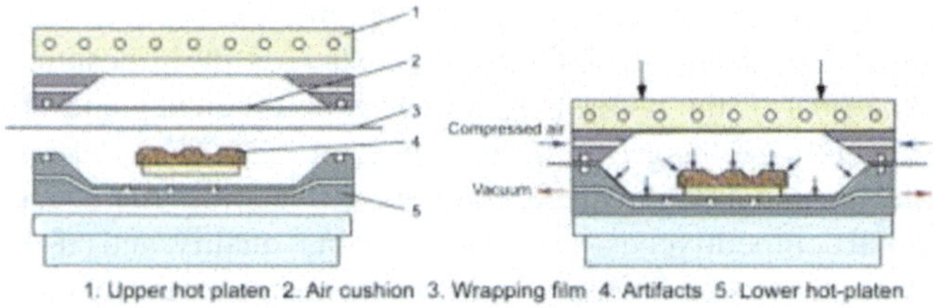

Fig. 2. Thermal forming operation [14]

3.2 Adding Classes in Protégé

The second step is to create the classes in the protégé. A new file was created in the protégé, this new file imports MfgPlanning and IOF core which is an ongoing project by the IOF community. The sublasses were created in the existing inherited classes of MfgPlanning and core. In following classes subclasses were created:

- **Manufacturing Operation:** Manufacturing operation is a subclass of manufacturing enterprise process. In this class subclass of operations that happen in plastic manufacturing were created. The two subclasses created are sheet extrusion operation and thermal forming operation shown in Fig. 3.
- **Manufacturing Process:** Manufacturing process is a subclass of planned process. The three major subclasses were created in it: cooling process, heating process and plastic manufacturing. Since cooling and heating are involved in other processes too apart from plastic manufacturing, so they were created as separate subclasses. The class hierarchy tree is shown in Fig. 4. In the plastic manufacturing subclasses were created for the process that are part of it. Following subclasses were created:
  - o **Melting Process:** This class represents melting process where the plastic granules are heated, melted, and being transported towards the sheet extrusion process.
  - o **Sheet Extrusion Process**: This class represents the sheet extrusion process, which involves the passage of molten plastic through a die. The die makes the molten plastic into a flat sheet with the desired width and thickness. Following this step, the extruded sheet undergoes a cooling process to bring down its temperature.

- o **Rolling Process**: This class represents the rolling process that occurs after the sheet extrusion and cooling process. During this process, the sheets are carefully wound into rolls.
- o **Feeding Process:** This class signifies the primary feeding phase of the thermal forming operation. In this phase, a roll of sheet material is unwound and introduced into the heating system to undergo the heating process. Subsequently, the heated sheet progresses to the thermal forming process.
- o **Thermal Forming Process:** This class represent the thermal forming process where the actual shaping of the plastic sheet occurs. Once the desired shape is formed, the cooling process helps cool down the plastic product to solidify it.
- o **Trimming Process:** This class represent the trimming process where after cooling, excess material is trimmed from the formed product.

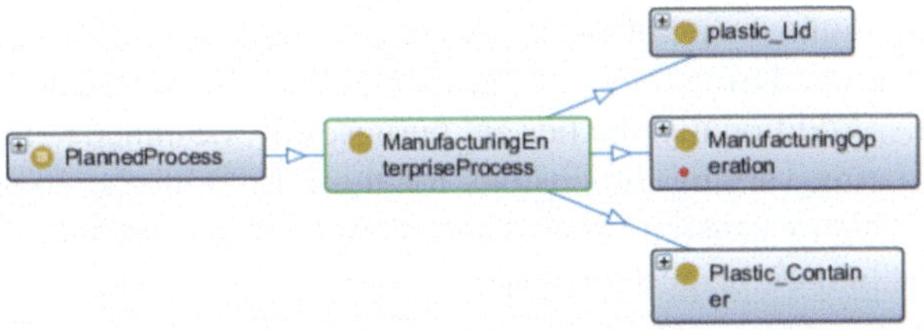

Fig. 3. Plastic manufacturing operation hiereachy tree from Protégé software

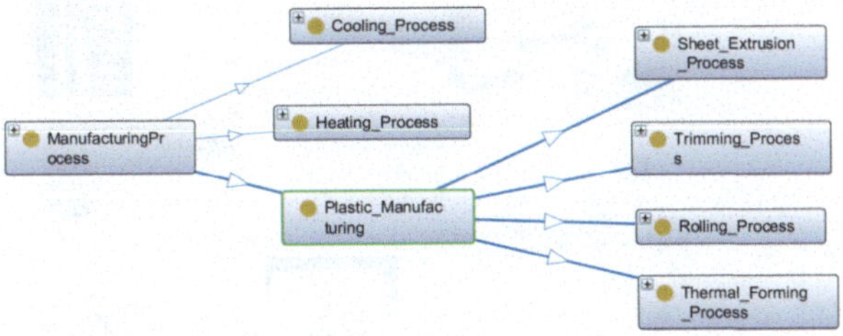

Fig. 4. Plastic manufacturing process hiereachy tree from Protégé software

3.3 Material Artifacts

Any physical object is a part of a material artifact. Machines, plastic containers, and plastic lids are subclasses of material artifacts since they are physical objects. Two machines subclasses were added under the machine: sheet extrusion machine and thermal forming machine. Plastic containers and their lids are the outputs of all plastic manufacturing processes. The hierarchy tree of material artifacts is shown through Ontograph in Fig. 5.

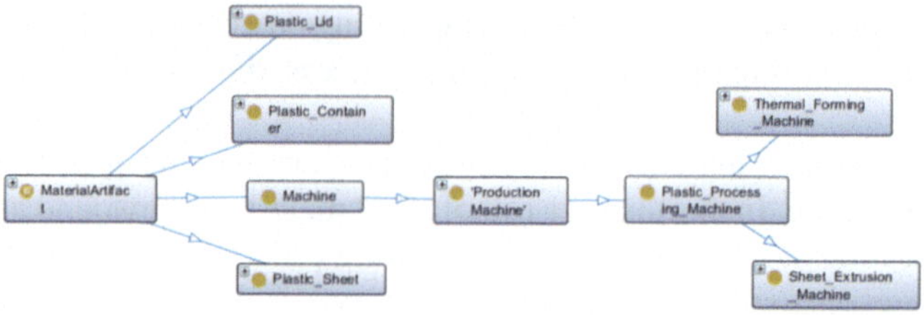

Fig. 5. Material artifact hiereachy tree for plastic manufacturing from Protégé software

4 Case Study

4.1 Container Making Process

This section presents an outline of the case study employed in this research. The case study entails the manufacturing of two plastic containers and their corresponding lids. Despite being constructed from the same plastic material, the containers and lids possess unique dimensions. Additionally, the lids are purposefully designed to fit each container precisely, necessitating varying dimensions for the lids. Both the containers and lids undergo a shared sheet extrusion process, but they undergo separate thermal forming operations specific to their respective shapes.

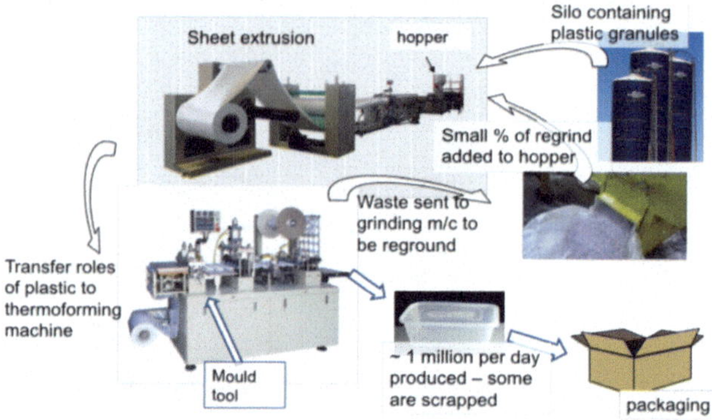

Fig. 6. Plastic Manufacturing Process and Operations for making plastic containers and lids.

In the production process of the containers and lids, the initial step involves the plastic granules, which serve as the raw material. These granules undergo a sheet extrusion operation, which comprises several consecutive stages: melting, sheet extrusion, cooling, and rolling. Throughout these processes, the raw material undergoes changes in its form, resulting in the production of plastic rolls. These rolls serve as the input material for the subsequent thermal forming operation. The thermal forming operation encompasses five sequential stages: feeding, heating, thermal forming, cooling, and trimming. During this operation, the plastic rolls are utilized. At the end of the thermal forming process, the output obtained is either a plastic container or a plastic lid. The overall process flow for manufacturing the plastic containers and lids is depicted in Fig. 6.

4.2 Ontology of Plastic Manufacturing Process

The instances and the relationships between the instances were created based on the case study. The relationships were created to manufacture two containers and two lids. Figure 7 shows the class diagram for the instances and the hierarchy of their relations. In the Fig. 7 the circles represent the instances. For example, Manufacturing Enterprise Process is a class and it has two instances plastic container and plastic lid, then manufacturing operation which is subclass of manufacturing enterprise process has three instances sheet extrusion, thermal forming operation for lid and, thermal forming operation for container. There is one sheet extrusion machine and one thermal forming machine which are instances of machine class. According to Fig. 7 Manufacturing process class has ten instances. In our research we considered same thickness of objects so, we created common instances for lids and containers that take part in the sheet extrusion process, the instances created are: melting and convening process, sheet extrusion process and cooling process, these instances take part with sheet extrusion operation as shown in the Fig. 7 and they are common for all containers and lids, there were two instances created for each feeding, heating and thermal forming and trimming. The arrows in the figure represent the relationship between two instances for example the third rectangle box arrows are in relationship with the middle rectangle instances and the middle are in relationship with the top rectangle box.

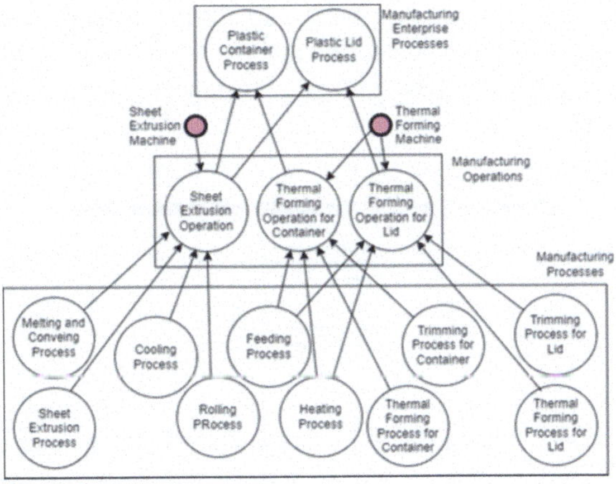

Fig. 7. Instance Diagram for plastic manufacturing based on the case study

Figure 8 shows the relationship of an instance TFM1 that belong to thermal forming machine class in Protégé. This relationship is represented in the middle rectangle in Fig. 7. The Fig. 8 shows that TFM1 is resource in TFOL2, TFOL1, TFLOC1 and TFLOC2 which are instances of the manufacturing operations classes.

Figure 9 shows the relationship for an instance that belongs to the lower rectangle in Fig. 7. The Figure shows MP1 which is an instance of melting process class is an occurrent part of SEO1 instance which is an instance of sheet extrusion operation that belongs to the middle rectangle.

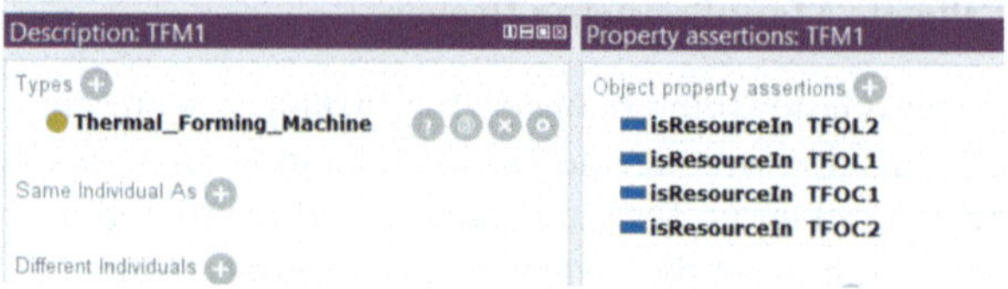

Fig. 8. Relationship view of an instance in Protégé

Fig. 9. Relationship view of melting process instance

4.3 Ontology Verification

To verify the relationship between the objects are correct, reasoner was run. As you can see in the Fig. 10, the reasoner detected the inverse relationship of MP1. It shows now all the relationships that it is involved with other instances. This means that the ontology relationships are correct.

Fig. 10. Reasoner detecting inverse relationship of SEO1 in Protégé software

5 Conclusion

This research shows the ontology for plastic manufacturing case. The research connects the manufacturing enterprise process, manufacturing operation, manufacturing process and machines. The research shows that ontology was successful in creating the hierarchy of its classes and was able to detect the inverse relationships. The future work in this research needs that ontology created for much bigger example, where more different type of objects are created. Also this research does not focus on shipping and packing. The future work also needs to be done on that.

References

1. Chang, T.C., Wysk, R.A., Wang, H.P.: Computer-aided Manufacturing. Prentice Hall (1998)
2. Asfahl, C.R.: Robots And Manufacturing Automation, 2nd ed. Wiley India Pvt. Limited (2010)
3. Roy, A., Abir, R.A.H., Hashemy, A.M.: Design and performance test of an automatic IRRIGATION system. In: International Conference on Mechanical, Industrial and Materials Engineering, pp. 28–30 (2017)
4. Guarino, N., Oberle, D., Staab, S.: What is an ontology?. In: Staab, S., Studer, R. (eds) Handbook on Ontologies, pp. 1–17. Springer Berlin Heidelberg (2009). https://doi.org/10.1007/978-3-540-92673-3_0
5. Kulvatunyou, B., Wallace, E., Kiritsis, D., Smith, B., Will, C.: The industrial ontologies foundry proof-of-concept project. In: Moon, I., Lee, G.M., Park, J., Kiritsis, D., Von Cieminski, G. (eds) Advances in Production Management Systems. Smart Manufacturing for Industry 4.0, in IFIP Advances in Information and Communication Technology, vol. 536, pp. 402–409. Springer, Cham (2018). https://doi.org/10.1007/978-3-319-99707-0_50
6. Sarkar, A., Šormaz, D.: On semantic interoperability of model-based definition of product design. Procedia Manuf. **38**, 513–523 (2019). https://doi.org/10.1016/j.promfg.2020.01.065
7. Fumagalli, L., Pala, S., Garetti, M., Negri, E.: Ontology-based modeling of manufacturing and logistics systems for a new MES architecture. In: Grabot, B., Vallespir, B., Gomes, S., Bouras, A., Kiritsis, D. (eds) Advances in Production Management Systems. Innovative and Knowledge-Based Production Management in a Global-Local World in IFIP Advances in Information and Communication Technology, pp. 192–200. Springer, Berlin, Heidelberg (2014). https://doi.org/10.1007/978-3-662-44739-0_24
8. Macris, A.M.: Enhancing enterprise resource planning users' understanding through ontology-based training. Comput. Hum. Behav. **27**(4), 1450–1459 (2011). https://doi.org/10.1016/j.chb.2010.09.013
9. Hu, X., et al.: Ontology-based system to support industrial system design for aircraft assembly. IFAC-PapersOnLine **55**(2), 175–180 (2022). https://doi.org/10.1016/j.ifacol.2022.04.189
10. Adjei, P., et al.: Application of Ontology Reasoning in Machining Process Planning – Case Study, 32th Flexible Automation and Intelligent Manufacturing Conference, FAIM2023, June 18–22. Porto, Portugal (2023)
11. CurideMouraLeite, A.F., Canciglieri, M.B., Szejka, A.L., CanciglieriJunior, O.: The reference view for semantic interoperability in Integrated product development process: the conceptual structure for injecting thin walled plastic products. J. Ind. Inform. Integr. **7**, 13–23 (2017). https://doi.org/10.1016/j.jii.2017.06.002
12. Szejka, A.L.: Contribution to interoperable products design and manufacturing information: application to plastic injection products manufacturing (2016)
13. Plastic Sheet Extrusion Line for Sale - Sevenstars https://www.sevenstarsgroup.cn/products-category/plastic-sheetboard-extrusion-line.html. Accessed 24 Jun 2023
14. Cao, P., Guo, X., Li, R.: 16 - Architectural (decorative) natural fiber composites for construction. In: Fan, M., Fu, F.(eds) Advanced High Strength Natural Fibre Composites in Construction, pp. 425–445. Woodhead Publishing (2017). https://doi.org/10.1016/B978-0-08-100411-1.00016-9

Unlocking Innovation: A Pathway to Success for Early-Stage Ventures

Jelena Borocki, Vladimir Djaković, Aleksandar Vekić(✉), and Olivera Cikota

Department of Industrial Engineering and Engineering Management, Faculty of Technical Sciences, University of Novi Sad, Trg Dositeja Obradovića 6, 21000 Novi Sad, Serbia
vekic@uns.ac.rs

Abstract. Innovation is crucial for the survival, resilience, and success of early-stage ventures in a complex market. It enables them to develop new products, services, and business models that meet customer needs and create value. Encouraging and supporting innovation by startup companies is essential to foster economic growth and progress in various industries. However, innovation is a complex and challenging process that requires significant resources, including financial, human, and technological. Through a comprehensive literature review and research involving startup founders, we identify key factors that facilitate effective innovation in early-stage ventures. These include creating a culture of innovation that encourages creativity and experimentation, leveraging technology to streamline innovation, collaborating with external partners such as universities, research institutions, and accelerators, and securing funding from various sources, including government grants, venture capitalists, and angel investors. Our findings contribute to understanding the challenges and opportunities in pursuing innovation for early-stage ventures, with practical implications for entrepreneurs, policymakers, and investors in this field.

Keywords: innovation · entrepreneurship · startup · success · resilience

1 Introduction

In today's rapidly evolving business environment, early-stage ventures face unprecedented challenges and opportunities. Amidst fierce competition and market uncertainties, these ventures must navigate a complex business landscape to ensure their survival and pave the way for long-term success. One of the most critical drivers of success for early-stage ventures is their ability to recognize and harness the power of innovation. Companies should view innovation as a most important element for developing groundbreaking products, services, and business models that address growing customer needs and create significant added value. Furthermore, innovation often serves as a catalyst for the emergence of new business ventures [1].

Through a comprehensive examination of existing literature and empirical research, it is possible to identify the key factors and strategies that can unlock the full potential of innovation in these ventures. By delving into the experiences and perspectives of

M. Dragomir et al. (Eds.): ICPR 2023, LNPE, pp. 42–49, 2025.
https://doi.org/10.1007/978-3-031-92082-0_5

startup founders, this research highlights the challenges and opportunities associated with fostering innovation in the early stages of a venture. Also, it explores various aspects that contribute to the development and innovation implementation, including the establishment of an innovation-driven culture as the foundation for this type of business activity.

Understanding the key factors that influence innovation and their interactions is crucial for startup founders and all those who aim to support and promote the success of early-stage ventures. It lays the groundwork for cultivating creativity and encouraging experimentation by the founders of these companies. Furthermore, by identifying the main factors that affect innovation, other ecosystem participants, such as governments and institutions can devise essential policies and frameworks that facilitate growth and economic development driven by innovation.

By providing a comprehensive analysis of the role of innovation in the success of early-stage ventures, this paper contributes to a broader understanding of entrepreneurial dynamics and gives an insight into fully exploiting the potential of innovation.

For startup companies operating in complex and dynamic conditions, fostering innovation becomes a crucial requirement for progress, adaptation, and the transformation of existing challenges into opportunities. Encouraging innovation can be viewed as a driving force for economic growth, industry transformation, and overall societal progress.

2 Theoretical Background

Entrepreneurship as a multidimensional concept has become the cornerstone of the development and sustainability of the economy as a whole, as well as a prerequisite for achieving a higher economic growth rate. Entrepreneurship can be observed as the process of discovering, evaluating, and exploiting opportunities for introducing new products, services, and processes [2]. In one of the more recent definitions of entrepreneurship, Coulter states that entrepreneurship is a dynamic process in which an individual or individuals utilize innovation through an organized process of seeking opportunities in which certain value can be created and delivered to the market [3]. For the development of entrepreneurship and sustainability of entrepreneurial ventures, it is necessary to continuously encourage innovation as a crucial characteristic in all business activities of these business subjects.

Innovation is a multidimensional and dynamic concept that has garnered significant attention across various disciplines, including business, economics, sociology, and technology. It refers to the creation, development, and implementation of new ideas, processes, products, or services that bring about value and positive change. Understanding innovation is significant for all parties involved in the innovation processes and activities and holds critical theoretical and practical significance. Innovative activities have always been one of the characteristics of human society, enabling its progress and further development.

Innovations have been mentioned in business context in the works of authors in the early 20th century [4], and certainly the most important perspective on innovations was given by Joseph Schumpeter. His approach to economic development in 1934 defines innovation as the creation of novel products or services, novel processes, novel raw materials, novel markets, and novel organizations [5]. In all these instances, to qualify as innovative and adhere to the concept of new combinations, it is essential to consider that no similar entity has been previously introduced, experienced, known, or existed. Schumpeter's theory posits that innovation is the primary driving force behind economic growth and development. Also, he coined the term "creative destruction" to describe how innovation replaces outdated technologies and structures, leading to long-term economic progress. Creative destruction involves the introduction of new and more efficient methods of production, as well as the creation of entirely new industries and markets. Therefore, creative destruction is a concept that highlights the constant transformation and evolution of economies through the replacement of old, less efficient structures with new and innovative ones, resulting in progress and increased productivity.

Most contemporary definitions of 'innovation,' viewed as the result of a process, are based on two fundamental characteristics: the degree of novelty or newness in a change and the degree of usefulness or success in the application of something new. The term 'new' can encompass various levels of novelty, such as being new to the world, new to a nation, new to a firm, etc. [6]. Today, innovation is not just a predisposition for creative engagement and experimentation; it also encompasses the introduction of new products or services and assumes technological leadership through research and development activities in novel processes [7].

According to their nature (degree of novelty), innovations can take the form of something completely new (radical) or improvements to existing ones (incremental), and a prerequisite for their emergence are changes [8]. Authors Hisrich and Kearney define innovation as "the process of creating and introducing something new or improved with the aim of generating appropriate value and benefit" [9].

In the field of entrepreneurship and innovation, authors Tidd and Bessant are highly esteemed. In their research, they emphasize innovation as a central question and state that it is "driven by the ability to see connections, identify opportunities, and exploit their advantages" [10]. Innovations enable business sustainability and lead to the successful fulfillment of a company's vision and mission, as well as increased market success.

Schumpeter was the first to propose a classification of innovations, outlining the following five types of innovation [5]:

1) Introduction of a new product or a new type of an existing product.
2) Application of new methods of production or product sales (not yet confirmed in the industry).
3) Opening of a new market (a market where a particular industry branch has not yet been represented).
4) Discovery of new sources for raw material or semi-finished goods procurement.
5) New industry structure, such as the creation or destruction of a monopolistic position.

Today, the most commonly used classification of innovations is established by the OECD (Organization for Economic Co-operation and Development) and presented in the publication called the Oslo Manual.

In Fig. 1, the OECD classification of innovations is depicted, which can be divided into four basic types [12]:

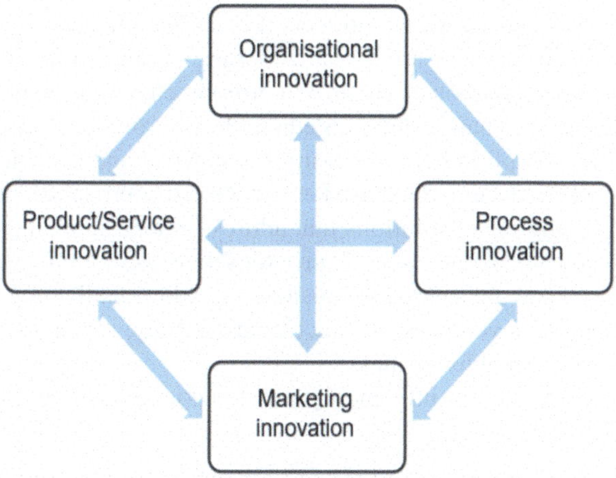

Fig. 1. The classification of innovations according to the Oslo Manual [11]

• *Product/Service Innovation* - involves a new or significantly improved product or service, encompassing substantial enhancements in technical specifications, components and materials, product software, ease of use, or other functional characteristics.

• *Process Innovation* - entails a new or significantly improved method of production or delivery, encompassing significant changes in techniques, equipment, and/or software.

• *Marketing Innovation* - encompasses new marketing methods that involve significant changes in product design or packaging, product placement, product promotion, or pricing.

• *Organizational Innovation* - involves the introduction of a new organizational method in business practices, workplace organization, or external relations.

According to this classification, product/service innovations and process innovations can be considered the most significant and are commonly implemented within organizations.

2.1 Innovation and Early-Stage Ventures

Early-stage ventures and innovation have a strong and symbiotic connection. These pioneering enterprises, characterized by their entrepreneurial nature and innovative ideas, play a crucial role in driving innovation and shaping the future of industries.

Innovation is at the core of such early-stage ventures or startups, which thrive on the spirit of creativity and forward thinking. So, startup companies are always associated with innovation [13]. One of the early definitions of startups was given by Eric Ries in his book "The Lean Startup" in 2011. According to his definition, startup companies are "organizations designed to create new products or services under conditions of extreme uncertainty" [14]. This definition emphasizes the uncertainty that typically accompanies the work of startups, particularly in relation to adopting new ideas, technologies, or market opportunities.

Startups often work on innovative projects that can be pioneers in their industry or introduce new approaches. These are forward-looking organizations fueled by the passion, creativity, and determination of visionary entrepreneurs eager to take risks and push boundaries. Operating in dynamic and unpredictable environments, they recognize that innovation is not just an option but a necessity for survival and growth. Continually on the lookout for unmet needs, untapped markets, and fresh approaches, these ventures are committed to developing novel solutions that address emerging challenges. Innovation fuels the success of early-stage ventures. It provides them a competitive edge, differentiation, and the ability to create a value for customers. Additionally, innovation enhances the ability of new business ventures to attract funding, forge strategic partnerships, and build a sustainable and scalable business.

The most important factors influencing the development of innovation in startup companies are [15]:

- Creating a culture of innovation - Startup companies should establish an organizational environment that nurtures creativity, encourages the generation of new ideas, and embraces a mindset of experimentation. This culture should promote the acceptance of risk-taking and learning from failure, enabling team members to think innovatively and explore unconventional solutions [16].
- Leveraging technology - Startups can optimize their innovation processes by effectively utilizing technology [17]. This involves adopting appropriate tools and platforms that facilitate efficient collaboration, prototyping, and testing novel concepts. Embracing digital advancements can expedite development and implementation of innovative products or services.
- Collaborating with external partners - Collaboration with external entities, such as universities, research institutions, and accelerators, can be highly beneficial for startups. These partnerships provide access to specialized knowledge, additional resources, and valuable networks that support the innovation process. Collaborative efforts may include joint research projects, knowledge sharing initiatives, or mentorship programs [18].
- Securing funding from various sources - Access to funding is crucial for startups to fuel their innovation endeavors. Startups can seek financial support from various sources, including government grants, venture capitalists, angel investors, crowdfunding platforms, initial public offerings, etc.[19]. Diversifying funding sources reduces reliance on a single channel and enhances the likelihood of acquiring the necessary resources for innovation.

All these factors significantly influence startup companies' emergence, development, and sustainability, particularly regarding their resilience during challenging periods and the high-risk nature inherent in their operations.

3 Methodology and Research Results

This research is part of a wide-reaching study on the innovative startup ecosystem of the Republic of Serbia. The research utilized an online questionnaire created through the Google Forms tool, with the founders of startup companies appearing as respondents. The data collection period for the research spanned from May to October 2021 and encompassed startup companies located throughout the entire Republic of Serbia, including all five statistical regions. The total number of respondents in this study is 106. The research sample includes startup companies that use innovation in their business.

Implementation of innovations associated with startups' market survival, risk resilience, and business success were subject to test using the X^2 test for one variable.

The X^2 test for one variable was chosen, as it allows the examination of differences in the frequency of individual answer categories within the context of a single question or variable.

Research Question (RQ): What types of innovations are applied in startup companies in the Republic of Serbia?

The results of the research are presented in the following Table 1. And Fig. 2.:

Table 1. Implementation of innovation in startup companies in Republic of Serbia [20]

Category	Observed N	Expected N	Residual
Product innovation	51	21,2	29,8
Service innovation	42	21,2	20,8
Marketing innovation	6	21,2	–15,2
Organizational innovation	3	21,2	–18,2
Process innovation	4	21,2	–17,2
	X^2 (4) = 102,7, p = 0,000		

Legend: N- Number of responses, Residual – the difference between the observed and expected number of responses

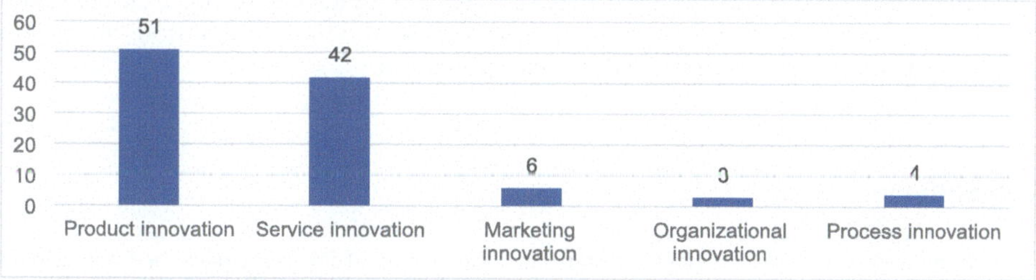

Fig. 2. Implementation of innovation in startap companies in Republic of Serbia [20]

This research result indicates that product and service innovations are the most developed innovation type within startup companies. The connection between product and service innovation and startup survival, resilience, and business success is very strong. Other types of innovations are less prevalent in startup companies, and although they have some impact on the survival, resilience, and business success of startups, it is negligible. The special importance of product and service innovations has been established in many previous studies, and according to one definition, it is considered that the mission of startups is fundamentally to develop product and service innovations, thereby bringing about revolutionary changes in the market [21], where they will play the role of leaders.

4 Conclusions

Innovations are now the driving force behind the growth and development of startup companies, with a significant impact on the ecosystem in which they operate and the economy as a whole. Innovations don't only contribute to the survival and success of startups, but also enable resilience in the face of risks and challenges, becoming a strategic focus for these companies. Implementing the innovations on which startup companies rely is of great importance not only for the startups themselves but also for other participants in the innovative startup ecosystem, such as government, educational institutions, support organizations, other companies, etc. In order to achieve optimal results and effects of innovation for all participants in the innovative startup ecosystem, it is necessary to establish a mechanism for supporting the strengthening of all factors that influence the positive development of innovation in startup companies. This enables accelerated economic development at all levels and creates a foundation for better networking and collaboration among ecosystem participants.

It can be concluded that in the future, the significance of innovation for startup companies will continue to grow, and it will become a defining measure of their success, enabling the sustainability of their business operations. Future research could explore specific industries or sectors within early-stage ventures and their unique challenges in generating and implementing innovation. Additionally, as the sample size is limited, future studies should expand the sample to include companies of different sizes and conduct comparative analysis based on pre-established innovation criteria.

Acknowledgments. This research (paper) has been supported by the Ministry of Science, Technological Development and Innovation through project no. 451-03-47/2023-01/200156, "Innovative scientific and artistic research from the FTS (activity) domain.

References

1. Abrahamsson, J., Vanyushyn, V., Boter, H.: Organisational innovativeness of international new ventures: evidence from Swedish firms. Eur. J. Int. Manag. **13**(4), 479–495 (2019)
2. Shane, S., Venkataraman, S.: The promise of entrepreneurship as a field of researc. Acad. Manag. Rev. **25**(1), 217–226 (2000)
3. Coulter, M.: Entreprenuership in Action, Upper Saddle RIver: Prentence-Hall (2001)
4. Veblen, T.: The Theory of the Leisure Class: An Economic Study of Institutions. Unwin, London (1899)
5. Schumpeter, J.: The Theory of Economic Development: An Inquiry into Profits, Capital, Credit, Interest, and the Business Cycle, Cambridge, Harvard University Press, vol. XLVI (1934)
6. Granstrand, O., Holgersson, M.: Innovation ecosystems: a conceptual review and a new definition. Technovation **90**, 102098 (2020)

7. Rauch, A., Wiklund, J., Lumpkin, G.T., Frese, M.: Entrepreneurial orientation and business performance: an assessment of past research and suggestions for the future. Entrep. Theory Pract. **33**, 761–787 (2009)
8. Norman, D., Verganti, R.: Incremental and radical innovation design research vs technology and meaning change. Des. Issues **30**(1), 78–96 (2014)
9. Hisrich, R.D., Kearney, C.: Managing Innovation and Entrepreneurship, SAGE Publications (2014)
10. Tidd, J., Bessant, J.: Managing innovation: Integrating technological, market and organizational change,Wiley (2016)
11. Bigliardi, B., Bottani, E., Montanari, R., Vignali, G.: Successful new product development in the food packaging industry: evidence from a case study. Int. J. Eng. Sci. Technol. **2**(9) (2010)
12. OECD/Eurostat, Oslo Manual: Guidelines for Collecting and Interpreting Innovation Data, 3rd edition, The Measurement of Scientific and Technological Activities, OECD Publishing, Paris (2005)
13. Socorro Márquez, F.O., Reyes Ortiz, G.E.: Theoretical analysis of the misconceptions and disenchantments of entrepreneurship and start-up. Int. J. Entrepreneurship **25**(7), 1–13 (2021)
14. Reis, E.: The lean startap. Crown Business, New York (2011)
15. Brettel, M., Cleven, N.J.: Innovation culture, collaboration with external partners and NPD performance. Creativity Innovation Manage. **20**(4), 253–272 (2011)
16. Johannessen, J.A., Olsen, B.: Projects as communicating systems: creating a culture of innovation and performance. Int. J. Inf. Manage. **31**(1), 30–37 (2011)
17. White, M.A., Bruton, G.D.: The management of technology and innovation: a strategic approach. Cengage Learning (2010)
18. Vekic, A., Borocki, J.: The role of institutions in supporting startup companies. In: International Scientific Conference on Industrial Systems, Novi Sad, pp. 486–492 (2017)
19. Hofstrand, D.: Types and sources of financing for start-up businesses (2022). https://www.extension.iastate.edu/agdm/wholefarm/pdf/c5-92.pdf
20. Vekic, A.: Phd Disertation, Development factors of innovative startup ecosystem (original. in Serbian), Fakultet tehničkih nauka Novi Sad (2023). https://nardus.mpn.gov.rs/bitstream/handle/123456789/21308/Disertacija_13404.pdf?sequence=1
21. Moroni, I., Arruda, A., Araujo, K.: The design and technological innovation: how to understand the growth of startups companies in competitive business environment. Procedia Manufact. **3**, 2199–2204 (2015)

Industrial Symbiosis, a Key Strategy for Building Resilient and Sustainable Supply Chains

M. Bouillon[1], L. M. Gelsomino[1,2(✉)], and C. de Goeij[1]

[1] Windesheim University of Applied Science, Zwolle, Netherlands
l.m.gelsomino@rug.nl

[2] University of Groningen, Groningen, Netherlands

Abstract. In a world where companies are under rising pressure to be more resilient while also improving sustainability, industrial symbiosis is gaining interest among academics and practitioners. This research investigates these concepts through a case study of an Amsterdam-based bakery network which turned a simple bread chain to a sustainability beacon, nurturing mutually beneficial, resilient, and sustainable operations, following without realising it industrial symbiosis principles. Our research aims to dissect the operations, strategies and inherent collaborations or conflicts involved in the management of such a network. Through interviews of network partners, we wish to uncover and construct empirically based archetypes connecting industrial symbiosis, resilience, and sustainability, providing an enhanced understanding of their collective impact on extended networks. This case study offers unique insights that may shows the way towards more resilient and sustainable supply chains.

Keywords: Industrial symbiosis · supply chain · sustainability · resilience

1 Introduction

Companies part of supply networks are facing increasing complexity in the management of their operations. On the one hand, they are under rising pressure to be more resilient in face of major crises. In the last three years, profound supply chain disruptions, from the Covid-19 pandemic to the Ukrainian conflict or the Evergreen shipping vessel blocking the Suez Canal have impacted operational performance and behaviour of supply networks, as well as influenced general perception and management policies related to supply chain risk [7]. Already five years ago [6] called supply chains to address increasing vulnerability and uncertainty; such a call seems more relevant today than ever.

At the same time, sustainability has become a critical component of business success for most firms. This is extended throughout the supply network as well, as most sustainability issues do spread beyond the boundaries of the single organisation [8, 9]. Most supply chains are under pressure from stakeholders to better monitor and improve what suppliers, or affiliated third parties in general, do. Recent years have bought up the realisation that sustainability goes beyond reputational considerations but has clear

M. Dragomir et al. (Eds.): ICPR 2023, LNPE, pp. 50–56, 2025.
https://doi.org/10.1007/978-3-031-92082-0_6

and direct implication for supply chain performance. For example, early in 2022 the Intergovernmental Pannel on Climate change (IPCC) report revealed that supply chains that rely on specialised commodities and key infrastructure can be disrupted by weather and climate extreme events [11]. Earlier this year, the same IPCC was adamant about the urgent need to reduce greenhouse gas emissions and achieve sustainability in all sectors, including supply chains.

With the numerous challenges faced by supply chain partners, collaboration has become increasingly essential. While not a new concept, industrial symbiosis, a cooperative approach where waste and byproducts of one company become resources for another [10], is gaining traction as an effective tool to revolutionize sustainability practices and enhance operational resilience for companies. This promising news comes with a caveat though: implementing industrial symbiosis remains challenging due to its complexity and requiring extensive coordination and knowledge sharing among partners. As such, it may pose difficulties for firms accustomed to working independently or in competition with each other despite having tremendous potential to reduce waste production while promoting net-zero emissions goals. For companies, comprehending the interdependence of industrial symbiosis, sustainability, and resilience in their supply chains is a vital aspect. Neglecting this crucial element may lead to substantial disruptions, damage to environment or local communities or missed opportunities. Despite widespread awareness regarding the significance of these aspects across industries, attaining equilibrium between sustainability and resilience can be challenging due to inherent conflicts and struggles. Achieving equilibrium between these competing demands proves challenging with set goals often remaining unfulfilled. In this paper, we wish to progress knowledge regarding the potential of implementing industrial symbiosis to enhance sustainability and resilience in supply chains. To do so, we are looking at the supply network of an Amsterdam-based network of bakeries which is implementing noteworthy projects around the concepts of sustainability, resilience, and industrial symbiosis, involving farmers and customers in Europe and Africa. We structure this article as an analysis of initiatives in place at the focal firm. Without pretences of strong generalizability, this allows us to provide a unique example that can highlight a successful strategy in finding a balance between resilience, sustainability, and industrial symbiosis. This would prove useful, as such trade-off might be faced with increasing regularity over the next years.

The remainder of this article is structured as follows: Sect. 2 presents a comprehensive literature review, establishing the groundwork of our study and summarizing previous research on industrial symbiosis, sustainability, and resilience in supply chains. Section 3 presents our case study, outlining the focal firm, its supply network, interest for the study and the methods used during our research. Section 4 discloses the results of our study, highlighting the behaviours and strategies aligning with industrial symbiosis principles and leading the focal firm and its network to success. Section 5 is dedicated to the discussion and conclusion, where we interpret our findings within the broader context of literature, highlighting theoretical and practical contributions, limitations, and future research consequent to our research. Lastly, Sect. 6 will list all the references contributing to our understanding and analysis of this intricated issue.

2 Literature Review

Industrial symbiosis is a concept that has gained importance in the context of supply chain resilience and sustainability. It involves reuse, recycling and processing of byproducts and intermediates within a system of organisations [1]. By promoting the reuse and recycling of materials, industrial symbiosis reduces waste generation and minimizes the environmental impact of production processes [1]. Additionally, industrial symbiosis enhances supply chain resilience by reducing dependence on external suppliers and increasing the availability of critical resources within the system [2]. The collaboration and coordination among organisations in industrial symbiosis foster community resilience and contribute to the economic prosperity of the region [1]. However, the implementation of industrial symbiosis in supply chains is a challenge with the need of close collaboration and integration across diverse organisations [1]. Further research is needed to explore the implementation, relationship, and impact of industrial symbiosis on supply chain resilience and sustainability [3].

Supply chains, on their side, are displaying a complex and multifaced relation between resilience and sustainability. On one hand, supply chain resilience focuses on the ability of supply chain to support and recover from disruptions such as natural disasters or pandemics [4]. On the other hand, sustainability in supply chain looks to support and develop the integration of environmental, social, and economic considerations to minimize negative impacts and promote long term ecosystem viability [3]. If both practices are compatible, their implementation in real life has proven to be a challenge in several industries. Firstly, there is confusion and lack of clarity regarding implementation and practices of sustainable and resilient supply chains [3]. Both concepts have different focuses, sustainability emphasising on efficiency of resource and resilience seeking effectiveness [3]. This creates a conflict in objective and approaches. Furthermore, the measurement and assessment of supply chain sustainability and resilience performance requires different the fitting conception of performance measurement systems [3]. Overall, the integration of supply chain sustainability and resilience requires addressing conceptual confusion, clarifying practices, and developing measurement frameworks to effectively balance the objectives of both areas [3]. In conclusion, the interplay of supply chain sustainability and resilience could be soothed by the development of industrial symbiosis, but application of these theories to real-world context is scarce, with empirical studies further analysing the resulting dynamics lacking [3]. The interconnection of flexible collaboration, mutual benefit, resilience, and sustainability in supply networks necessitates exploration through practical lenses [3]. With this in mind, the proposed research addresses these gaps by analysing the flexible collaboration in a supply network providing the ground for an understanding of how such relationships can contribute to resilience and sustainability in supply chains. This study will not only build on theoretical knowledge but will also offer valuable insights from a practical perspective.

3 The Case Study

The primary approach employed in this research is a single case study methodology. This qualitative approach is particularly fitted to our research as it enables us to conduct an in-depth analysis of a complex dynamic in the real-world. Not only does a case study brings practical proof and relevance to theoretical studies, but this research is also based on research gaps calling for more empirical research.

The case selected for this study is a bakery network engaged in making their operation more sustainable. While mentioning industrial symbiosis in their strategy or in their ambitions, many aspects from this promising method can be found in their operations, relationships with their suppliers and ambitions. Picking this network became obvious when looking into their mix of local operations, close collaboration with their network stakeholders, strong sustainability ambitions and prosperity in times of crisis, making it a fertile ground for exploring our research question.

Data collection involved conducting semi-structured interviews with key members of the bakery and its network partners. These interviews provided first-hand insights into the workings of the network, strategies for sustainability and their views of supply network management.

Analysis of the data involved a compiling of interview data and interpretation. Through this process, we were able to identify and understand key themes related to the impact of their strategies on supply chain resilience and sustainability.

This methodology provided the foundation for an in-depth exploration of the research question, contributing both to theory and practice in the field of supply chain management.

4 Results

The analysis of the interviews revealed a several factors contributing significantly to the resilience and sustainability of the bakery's network.

Symbiotic network and flexible collaboration

Although never mentioned in the answers and not known from the bakery's management, this network operates following principles of industrial symbiosis. Together with their partners, they can thrive during crisis thanks to the establishment of an operation model where every action will benefit two or more network partners. Though business strategies and modus operandi differ across partners, they all show collaborative efforts, creating a symbiosis relationship strengthening the overall network. The use of the bakeries' wastes for the making of sandwiches by a network partner is one example. The collective thinking towards ways of removing meat from their offering another. Finally, if all those businesses operate in a competitive environment, they all put their operation at the service of more than just transforming raw materials in finished products. One example of that is the bakery, creating a community of farmers, millers and bakers dedicated to the making of higher quality bread. This initiative, coming from the bakery firm, shows their intention to share their practices and making them available beyond their own network.

Quality-driven approach

Across the network, several partners emphasised the importance of product quality in their business strategies. For most, focusing on high-quality bread and bakery products was a way to differentiate themselves from other players in the market, enabling them to charge a premium price for their products. This commitment led them the network's stakeholders to investigate all their supplies, making sure to use traditional, simple, and trusted ingredients.

Local operations and direct sourcing
What could be seen as a consequence of the quality-driven approach, stakeholders all had an aim to source as much as their supply locally. For the focal firm, this meant sourcing above 90% of their raw materials within the Netherlands, with the rest being nuts and chocolates, coming mostly from suppliers of the global south. This local focus came in handy when navigating the consecutive crisis from 2020 to 2023, for most of their operations could go on as before, unaffected by the geopolitical turmoil that led many businesses with similar activities to bankrupt.

Emphasis on sustainable practices
The cause behind most of the aforementioned results, sustainability forms a key part of the network's stakeholders. The focal firm being ambitious and daring, they surrounded themselves through the years by partners with similar ambitions, attracted to those or resulting from them. Already mentioned above, the community of farmers, millers and bakers is a clear proof of an intention to for more sustainable practices in the farm to brood chain that resulted in a new initiative with the potential to make healthier, local grain accessible to more customers.

Challenges and overcoming strategies
Despite the successes, all stakeholders faced challenges in maintaining their sustainability commitments. For their strong drive and ambition, the focal firm encountered resistance both with their employees and customers. Through a focus on the team, pedagogy, and long-term benefits, they overcame those and came out stronger. Network partners, on their side, faced unexpected struggles when their success led them to running out of wasted bread to use for their sandwiches. This led them to weave connections with new bakeries, adding new partners in this ever-evolving network.

Conclusion

In conclusion, the focal firm, together with its network provides a unique example of how flexible collaboration, a quality-driven approach, local operations, a strong focus on sustainability, and effective overcoming strategies can contribute to resilience and sustainability. Although never named by any member of the network, these results underline the potential of industrial symbiosis as a valuable tool in achieving resilience and sustainability in supply networks, despite the complexities and coordination challenges involved.

5 Discussion and Conclusions

The findings from our study contribute to significantly to the understanding of the relation between industrial symbiosis, resilience, and sustainability in supply chains. Existing literature has theorised the importance of this factors in ensuring the operational efficiency ad success, particularly in times of crisis. Our case study, thanks to our focal company and its network of stakeholders offers a unique perspective on the successful implementation of these principles in a mutually beneficial, thriving business network.

The network symbiosis underlines the possibilities of harvesting diverse strategies and operational processes in a supply network while benefiting more than one firm. This success story provides empirical evidence in favour of network dynamics following industrial symbiosis' principles, still a relatively underexplored area in current literature [3].

Our findings have practical implications for businesses and supply chains. All stakeholders, while showing sometimes fundamentally different approach to becoming more sustainable, suggests that businesses can leverage their unique strength and work collaboratively, enhancing resilience and sustainability. In times were many wants to fill their plates with healthier and better ingredients [5], a focus on quality, sustainability together with a commitment on local sourcing can serve as key differentiation and strengthen the supply network.

Our study shows however limitations. A strongly agri-food network, our findings might not be generalised to all industry sectors. Furthermore, the geographic location is one of a wealthy western capital common in countries of the north but not a geographic, political and business situation to be generalised for the whole world. Therefore, future research could explore similar concepts in different industries, geographical locations, and business contexts. Longitudinal studies could also help track the evolution of such relationships and their long-term impacts on supply chain resilience and sustainability. Expanding the amount of tier level interviewed within the same network, finally, might provide insightful results as to the impact of such strategies and relationships across global supply networks.

In conclusion, our research sheds light on the potential of relationships following the lines of industrial symbiosis principles as tools for enhancing resilience and sustainability in supply chains. It illustrates how a flexible collaboration between supply network partners can successfully navigates crises, even in the face of reluctant collaborators, differing strategies, and operational processes. By focusing on quality and sustainability and promoting local operation –an offer perfectly fit to their market needs– businesses can enhance their resilience while contributing to a more sustainable future, taking employees, customers, and network partners in the process. We hope that our findings encourage further exploration into the complex interplay of industrial symbiosis, resilience, and sustainability in supply chain networks, and inspire more businesses to incorporate these principles into their operations.

References

1. Bansal, P., McKnight, B.: Looking forward, pushing back and peering sideways: analyzing the sustainability of industrial symbiosis. J. Supply Chain Manag. **45**(4), 26–37 (2009). https://doi.org/10.1111/j.1745-493x.2009.03174.x
2. Rahman, S., Kim, J., Laratte, B.: Disruption in circularity? impact analysis of covid-19 on ship recycling using weibull tonnage estimation and scenario analysis method. Resour. Conserv. Recycl. **164**, 105139 (2021). https://doi.org/10.1016/j.resconrec.2020.105139
3. Negri, M., Cagno, E., Colicchia, C., Sarkis, J.: Integrating sustainability and resilience in the supply chain: a systematic literature review and a research agenda. Bus. Strat. Env. **30**(7), 2858–2886 (2021). https://doi.org/10.1002/bse.2776
4. Mari, S., Lee, Y., Memon, M.: Sustainable and resilient supply chain network design under disruption risks. Sustainability **6**(10), 6666–6686 (2014). https://doi.org/10.3390/su6106666
5. Munt, A., Partridge, S., Allman-Farinelli, M.: The barriers and enablers of healthy eating among young adults: a missing piece of the obesity puzzle: a scoping review. Obes. Rev. **18**(1), 1–17 (2016). https://doi.org/10.1111/obr.12472

6. Ali, A., Mahfouz, A., Arisha, A.: Analysing supply chain resilience: integrating the constructs in a concept mapping framework via a systematic literature review. SCM **22**(1), 16–39 (2017). https://doi.org/10.1108/scm-06-2016-0197
7. Ju, C., Zhao, J., Li, K., Bao, F., Xu, C., Ran, J.: Supply chain recovery strategies for high demand products based on the impact of capacity and transportation disruptions. Systems **10**(4), 88 (2022). https://doi.org/10.3390/systems10040088
8. Hofmann, H., Busse, C., Bode, C., Henke, M.: Sustainability-related supply chain risks: conceptualization and management. Bus. Strat. Env. **23**(3), 160–172 (2013). https://doi.org/10.1002/bse.1778
9. Ahi, P.: Sustainability analysis and assessment in the supply chain (2021). https://doi.org/10.32920/ryerson.14662254
10. Neves, A., Azevedo, S., Pimentel, C., Matias, J.: The potential of industrial symbiosis: case analysis and main drivers and barriers to its implementation. Sustainability **11**(24), 7095 (2019). https://doi.org/10.3390/su11247095
11. Pörtner, H.O., et al.: Climate Change Impacts, Adaptation, and Vulnerability in Sixth Assessment Report of the Intergovernmental Panel on Climate Change, Cambridge, Cambridge Univ. Press, pp. 3–33 (2022). https://doi.org/10.1017/9781009325844.001

The Value of Context in the Construction of Storytelling for Tacit Knowledge Management in Complex Organizations

Felisa Córdova[1](✉), Cecilia Montt[2], and Margareth Gutiérrez[3]

[1] Facultad de Ingeniería, Universidad San Sebastián, Santiago, Chile
felisa.cordova@uss.cl
[2] Facultad de Ingeniería, Universidad de Santiago de Chile, Santiago, Chile
cecilia.montt@usach.cl
[3] Center for Research and Innovation, Viña Concha y Toro S.A, Chile, Santiago, Chile
margareth.gutierrez@conchaytoro.cl

Abstract. This study presents an analysis of the importance and value of the correct definition of the context in the construction of Storytelling, for its application in organizations, as well as a tool to manage knowledge within it. In addition, it describes the emergent property that occurs between the phenomenon or the facts that the subjects live together with the environment that surrounds it and the "butterfly effect" that produces a small change of the context in the expected results within the organizations, specifically in the tasks that are carried out by their members. Finally, we propose a way to treat the context of a Storytelling applying the methodology of soft systems, from the point of view of the actors, who are those who live the reality of the story and the readers, who face a situation like that told in the story, but in different contexts. The tour of its seven phases is made to face the problem of accurately identifying the real and perceived context in Storytelling and reaching a solution by the reader.

Keywords: Storytelling · Tacit Knowledge · Context · Soft Systems Methodology

1 Introduction

Knowledge management has had a strong boom in recent years producing important changes that have transformed the reality of organizations. With the help of various tools, the knowledge generated in companies by partners, suppliers, customers or among the employees themselves is shared. However, the processes that occur in reality differ greatly from the way they are described in procedure manuals or reports (Zipes 2013). The experience of the members of the organization, their ideas and decisions must remain within the organization and knowledge is not always recorded. The experience of the members of the organization, their ideas and decisions are presented as "the tacit knowledge" by (Nonaka 1995). This knowledge is difficult to make explicit, it is subjective, based on the experience of each person and is strongly linked to the action of

M. Dragomir et al. (Eds.): ICPR 2023, LNPE, pp. 57–63, 2025.
https://doi.org/10.1007/978-3-031-92082-0_7

individuals in a given context. Everything that each person knows, is difficult to express in words or data.

One of the ways to solve this problem is through stories, using narrative as a mechanism for make explicit and transmitting tacit knowledge (Valle 2003). Storytelling is particularly effective in helping to change business and providing practical knowledge describing informal and formal day-to-day tasks within the organization. However, this aid is not enough. There are elements that can remain in the air when it comes to the construction of Storytelling and this is the "context".

The key elements of the environment, which help to understand why things happened, are not easy to identify as they depend on the different perceptions of the actors. Learning is directly related to the context in which the organization occurs, as well as how it understands that context. Normally, the context is determined perceptually by employees who are immersed in a reality and becomes more approachable from a soft system methodology (MSB) which is based on systems ideas and a systems approach, but not in the conventional sense raised by (Checkland 1989).

The objective of this work is to analyse the true value of how the context or environment that surrounds individuals can affect the outcome of the tasks or projects they are carrying out and how they can understand and capture the environment in a Storytelling from the MSB (Segel 2010).

2 Storytelling

2.1 Concepts

Storytelling as a means of communication has been used by many cultures around the world since ancient times. This is where the tool of the "narrator" becomes important, since historically it has been used as a way of propagating or disseminating the tacit knowledge of each individual in different environments, which must be detailed within a story. This mechanism has led organizations to adopt it as a support for knowledge management. It turns out to be very attractive to the members of an organization, because the stories are very familiar to them having worked with them since they were in their schools. Stories help bring about change and foster collaboration in an organizational setting.

The stories help to humanize the environment, creating a favourable scenario to share knowledge and build new communities. This is due, mainly, to the fact that the stories awaken feelings in the readers, which produces a personal commitment in the individual that stimulates their learning, which makes it, in turn, a tool that goes beyond an instruction manual or procedures of many pages that may formally exist within organizations (Mahyar 2015). In this sense, the narrative can be used as a tool in which the activities carried out within an organization are told or narrated, built by an individual or a group of people who were actors of some activity, leaving their experiences reflected in the performance of tasks already completed. In this way, the stories would be a support for the externalization of the set of tacit knowledge that can be detached from each story, something essential for the success of "knowledge management" of the organization. For example, a business model can be identified through stories that explain how companies work. However, very little is said about its optimal construction to achieve the goal you want to achieve with readers (Kleiner 1997).

2.2 Building Storytelling

In this form of organizational learning, three parties can be observed: the one who lives the story (actor of reality), the mediatized and the reader of the story. The people who star in the narrative are immersed in a lived experience that leads them to a kind of cognitive blindness when they reach a final consensus of the story. This, mainly due to an individual perception of each actor, since each one considers relevant different aspects of the lived reality. The "mediator" may or may not exist as such in the construction of Storytelling, since it can be a function made by the actors. He has the role of understanding the context and the facts in a consensual manner among the different actors, so he must coordinate the actions to create the story and write the final text in such a way as to meet the objectives for which a Storytelling is used. On the other hand, the reader is in the situation of an observed experience from which he will detach certain tacit elements and turn it into explicit knowledge. For him, every aspect of the story is important to be able to draw his conclusions (Mahyar 2015).

When building a story, it is necessary to be very careful with the description of certain situations that the mediator does not know how they arose. So, he must pay attention to those variables of which he does not have total control and that affect the daily situations that involve the activities or projects of an organization.

Two elements are fundamental in the experiences or events of each day to capture what is lived in a story: the phenomena as such and the context in which they develop. Phenomena refer to the things that the observer perceives, the first thing he experiences with things or with the fact itself. The context has to do with the external environment surrounding the events. These two elements together make sense of it and bring out the essence of the experience that makes each situation unique. There is no explanation of how and why it happens the way it does, because it just happens. So, there is no knowledge without such context. Thus, Storytelling, through narration, allows us to describe the emergence of events from a context (Chao 2018).

2.3 The Value of the Context

Context is defined as the circumstances on which the meaning and value of a word, action, story or other element that may be affected depends. It is a determining factor to understand what a story means, which is why you must leave well defined what objective you want to achieve with the story as said above, in order to describe the context and transform the narrative in a visually shared story based on that objective (Segel 2010).

In the context of organizations, there are elements of the immediate environment that can affect and change some conditions, modifying the course of the organization to meet its objectives. This is the so-called "microenvironment" oriented to the study of suppliers and their offer with the best competitive advantage over the product or service to be developed. Also, the competition, of which the company wants to study who they are, what they sell, to whom they sell them, what advantages they have, what deficiencies they have and why they achieve success or not. It is of interest to know the behaviour of potential customers/users to know their needs and expectations, as well as their intermediaries who are the ones who sell directly to customers. The study of the "microenvironment" is essential since organizations can influence each external actor by defining strategies with each of them.

A preliminary formulation of a more particular analysis of the context of a subject or group of individuals who participated in some project and/or activity within the organization can be made. It is a kind of "mini-environment" that involves individuals when carrying out their work and that has to do with the situation of the company at a given time. Here it does not influence the immediate environment of the organization, but rather it is the same organization that affects the person, its internal factors, which for the individual are external, since it affects him in a particular way that influences his final performance such as the work environment, his available resources given by the organization and other aspects that vary each time and makes each given situation unique.

2.4 The Butterfly Effect of Context

The "butterfly effect" refers to how small variations in the initial conditions of a study significantly affect the expected results. The political, economic, social and technological context can be defined as a system with a high degree of uncertainty, in which current conditions cannot explain future conditions, as well as the "micro-environment" and the "mini-environment" in which the person or the work group is subjected within the organization. Small changes in any of the three defined environments can affect the performance and final result of the tasks performed by individuals within the organization. The broader environment called "macro-environment" is composed of factors that affect all organizations and where they cannot influence their factors, such as macroeconomic, technological, political, sociocultural, legal and environmental elements.

3 Applying Soft Systems Methodology to Context

The Soft Systems Methodology helps to face situations/problems in which clear actions that lead to a solution cannot be established, due to the fact that they are complex situations where there are different and even contradictory perceptions on the part of the actors involved (Checkland 1989). It is essential to be able to build and interpret the environment from the perspective of the person(s) who are the protagonists of the Storytelling, who will be called actors. They are not necessarily the ones who write the story, but they are the ones who live it and the perspective of the ones who read the stories. Next, the different stages of the soft systems methodology applied to actions in the solution of problems in their context are described considering the participation of actors and readers.

3.1 Unstructured Situation/Problem Perception

Actor(s): In this initial stage, we try to identify as many perceptions or experiences as possible experienced by each of the different people involved in a task or project in the organization. This is the starting point to prepare the writing of the Storytelling. With this you can have in detail a mental construction of what is happening in a particular reality.

Reader(s): The reader, on the other hand, must identify each element of the story that belongs to the environment or environment, without necessarily establishing interrelations between the external elements that participate in reality, so he must be able to separate the phenomena of each element that makes up the context.

3.2 Perception of the Situation in a Structured Way

Actor(s): In this initial stage, we try to identify as many perceptions or experiences as possible experienced by each of the different people involved in a task or project in the organization. This is the starting point to prepare the writing of the Storytelling. With this you can have in detail a mental construction of what is happening in a particular reality.

Reader(s): The reader, on the other hand, must identify each element of the story that belongs to the environment or environment, without necessarily establishing interrelations between the external elements that participate in reality, so he must be able to separate the phenomena of each element that makes up the context.

3.3 Development of Basic Definitions of Relevant Systems

Actor(s): Once there is agreement on the aspects of the context that were relevant, the actors or participants of reality must jointly determine which elements of the context affected the problems of the situation and how they acted on that in order to solve it at the time the experience was lived. It must be stipulated in the narrative, both the problem with the possible solutions that were thought and the solution finally chosen that they put into practice at that time that managed to change reality, emphasizing the context that affected it directly or indirectly.

Reader(s): The reader should compare the problem in which he is currently involved with the one narrated in the story, so he must identify similarities in the type of problem. In addition, you must compare the environments in which both situations occurred to know if the solution that appears in the story can be used at the moment in which the reader is faced with his reality to add it to other possible solutions that change the situation.

3.4 Development and Testing of Conceptual Models

Actor(s): They must build, from the previous definition, a model of systems of activities that carried out the transformation of the situation/problem, where both the phenomenon and the context described step by step converge.

Reader(s): The reader, on the other hand, must be able to identify the system of activities described by the actors in the story and, in addition, define a definitive system of activity to be carried out, to carry out the transformation of reality. This decision must be made by analysing the context in which their reality developed, with the reality perceived in the story.

3.5 Comparison of Conceptual Models with Reality

Actor(s): The actors must compare their model of activity system with the lived reality, mainly the context in which they developed, since it is the most difficult thing to have actors with different perceptions of what happened in their environment or simply what one person lived was not lived by another. This means that no agreement is reached on the aspects that involved reality.

Reader(s): Like the actors, readers should compare the system of activities they determined in the previous phase with their reality, emphasizing the context found at the time of the problem to determine what actions were propitious in the story due to the context and what actions cannot be carried out by the reader due to the environment he currently faces. With this, it will be possible to give coherence to reality and the conceptual model by making the corresponding modifications where reality differs from the conceptual.

3.6 Execution of Feasible and Desirable Changes

Actor(s): In this phase they must apply the measures adopted in the previous stage, so the modifications, product of the comparison, must be part of the story that is constructed. These measures of change are made according to the context in which you are immersed.

Reader(s): The reader should do the exercise for his previous stage and identify from the reading the key points that were written in the execution of the actors to improve the situation/problem.

3.7 Implementing Real-World Changes

Actor(s): Once the corresponding changes have been made by consensus among the actors, it can be reflected in the final version of the text immediately.

Reader(s): In the case of readers, they must carry out the changes in the real world, which although it would solve their initial problem, could lead to new problems that they must solve in an analogous way to the whole process described.

4 Conclusions

One of the challenges is capturing tacit knowledge from individual members of the organization who participated in a complex task or project. In fact, well-structured stories are a good way to tell what happened, in great detail, in a task or project, externalizing the tacit knowledge that the group or individual involved in that reality had. The value of the context within the stories was mentioned and how well it must be expressed within the story for the correct understanding by the reader or readers who are subjected to a task similar to that told in the story. The soft system methodology was used to address the problem of context obtaining satisfactory results for each of its seven phases, covering both the point of view of those involved of the reality that is told in the story and that of the reader who must be able to detach tacit elements of the text. The next steps are to be able to massif Storytelling in organizations by providing support for its construction within companies, applying the MSB to determine the phenomena objectively and help determine the context with a systems approach that contributes to this situation/problem, so we could consider the complexity of the real world that is constantly changing.

Acknowledgements. This work was supported by the Facultad de Ingeniería, Arquitectura y Diseño, Universidad San Sebastián, Chile and the Facultad de Ingeniería, Universidad de Santago de Chile.

References

Chao, T., et al.: Storytelling and visualisation: an extended survey. Information **9**, 65 (2018)

Checkland, P.B.: Soft systems methodology. J. Human Syst. Manage. 273–289 (1989)

Kleiner, A., Roth, G.: How to make experience your company's best teacher. Knowl. Manage. Harvard Bus. Rev. 75(5) (1997)

Mahyar, N., Kim, S.H., Kwon, B.C.: towards a taxonomy for evaluating user engagement in information visualization. In: Proceedings of the Workshop on Personal Visualization: Exploring Everyday Life vol. 3, Chicago, IL, USA (2015)

Nonaka, I., Takeuchi, H.: The Knowledge-Creating Company: How Japanese Companies Create the Dynamics of Innovation. Oxford University Press, Ox-ford (1995)

Segel, E.: Heer J (2010) narrative visualization: telling stories with data. IEEE Trans. Vis. Comput. Graph. **16**, 1139–1148 (2010)

Valle, C., Raybourn, E.M., Prinz, W., Borges, M.R.S.: Group storytelling to support tacit knowledge externalization. In: Proceedings of the 10th International Conference on Human – Computer interaction. Crete, Greece (2003)

Zipes, J.: Creative Storytelling: Building Community/Changing Lives. Routledge, Abingdon (2013)

Strategic Corporate Sustainability Management (ESG) Based on the Materiality Process: A Systematic Literature Review

I. Dallabona[1], I. Gil[1], P. Carpejani[1(✉)], E. Pinheiro de Lima[2], U. Tortato[1,3], S. E. Gouvea da Costa[2], and F. Deschamps[1,4]

[1] Industrial and Systems Engineering, Pontifical Catholic University of Parana, 1155 Imaculada Conceicao, Prado Velho, Curitiba, Parana, Brazil
p.carpejani@pucpr.br
[2] Industrial and Systems Engineering, Federal University of Technology, 3165 Sete de Setembro, Reboucas, Curitiba, Parana, Brazil
[3] Business School, Pontifical Catholic University of Parana, Imaculada Conceicao, Prado Velho, Curitiba, Parana, Brazil
[4] Mechanical Engineering, Federal University of Parana, 230 Coronel Francisco Dos Santos, Jardim das Americas, Curitiba, Parana, Brazil

Abstract. Corporate sustainability has been a popular topic amongst organizations since the 90s, but now it's even more in the spotlight due to the UN's sustainable development goals/2030 Agenda. To keep track of the sustainability "issues," a lot of different development measurement systems have been created, and companies use the Global Reporting Initiative to adopt those practices, report their results, coordinate ESG indicators, and others. This paper aims to understand the connection between material topics and companies' sustainability reports based on stakeholders' demands. The technique used is the systematic review and qualitative analysis (content analysis), with support from ATLAS.ti software. The results show how organizations use material themes to manage their stakeholders via analysis of sustainability reports. They also present a categorization of the central studies present in the literature.

Keywords: Sustainability report · Stakeholders · Materiality · Systematic literature review

1 Introduction

In 1990, John Elkington published his book 'Cannibals with Forks' and emphasized the need for companies to incorporate this ideology into their processes. He presented the Triple Bottom Line (Sustainability Triple Bottom Line) model to meet the demands of stakeholders who were increasingly interested in environmental issues at that time (Elkington 1998).

M. Dragomir et al. (Eds.): ICPR 2023, LNPE, pp. 64–77, 2025.
https://doi.org/10.1007/978-3-031-92082-0_8

The relevance of the sustainability topic becomes even more apparent when discussing the United Nations' Sustainable Development Goals, 17 goals aimed at achieving the 2030 Agenda, a plan of action to eradicate poverty and hunger, protect the planet, and ensure prosperity and peace for all human beings (UN 2015).

During the 1970s, multiple authors criticized existing performance measurement systems, claiming that they encouraged short-term thinking, minimization of variance instead of continuous improvement, lacked a strategy or external focus, and even 'destroyed' the competitiveness of the American manufacturing industry. In this context, these systems were replaced by new multidimensional frameworks that focused more on non-financial information, including external as well as internal success measures, providing insights into both future performance and past achievements (Norton and Kaplan 1997). However, these frameworks did not offer specific guidance for implementing a performance measurement system.

Therefore, Neely (2000) proposed dividing the development of a performance measurement system into three parts: creating performance measures (identifying objectives and creating the measures themselves), implementing the measures (preparing systems and processes for regular data collection), and using the performance measures (evaluating strategy implementation by challenging assumptions).

To adopt sustainability practices, companies can make use of the Global Reporting Initiative (GRI 2023), the standard for sustainability reporting, which coordinates ESG indicators (environmental, social, and corporate governance).

It is recommended to document the process of defining topics within the company, including the chosen approach, decisions, considerations, analyzed sources, and evidence gathered (GRI 2023).

2 Methodology

This research consists of a Systematic Literature Review (SLR), based on the steps presented by Xiao and Watson (2019), (1) formulating the research problem; (2) developing and validating the review protocol; (3) searching the literature; (4) screening for inclusion; (5) assessing quality; (6) extracting data; (7) analyzing and synthesizing data; and (8) exporting the findings. Steps 2 to 6 were done along in a research group, but the analysis and findings happened separately, according to each research topic.

The research problem is 'identify whether the GRI materiality process is strategic for the business.' the chosen databases, Scopus, Web of Science, and SciELO, a qualitative analysis in the form of content analysis, will be caried out using the ATLAS.ti software. Content analysis is an important investigative technique for analyzing documents such as reports. Through content analysis, it is possible to codify the variables; that is, through cycles, it is possible to base inferences for the construction of a robust conclusion. (Saldaña 2015). The use of software, on the other hand, provides digitization of content analysis, being able to operationalize more documents and use more complex tools (Saldaña 2015). The ATLAS.TI software was chosen for its processing power and prior knowledge of the authors.

The research in the databases started with the use of combinations of words, which could be present in both the article titles and their abstracts and keywords. To further refine the results, filters were applied to the search, including document type (article), languages (English and Portuguese), publication stage (final), research area (business, engineering, etc.), and keywords associated with the topics. After the first filters, duplicate articles were removed (leaving 466 documents). Then, another language filter was applied (removing 45 documents), followed by a filter to eliminate journals from unrelated research areas (removing 44 documents). A total of 377 abstracts were read, and based on these readings, articles that did not fit the proposed theme were eliminated, resulting in 169 articles. In an effort to restrict the research to only 'general' and replicable approaches, articles from specific countries and industries found not relevant were excluded, resulting in only 79 articles remaining after this phase. Congress articles were also eliminated, and a filter that limited the year of publication from 2018 to 2023 was applied, leaving 23 as the final number of articles (Appendix A).

3 Systematic Literature Review

This section is based on results from the systematic literature review conducted on the ATLAS.ti software, contributing so organizations can build strategic corporate sustainability management. For this paper, the ATLAS.ti cloud version was initially used. As the project progressed and the need for increasingly complex options, it was necessary to

Fig. 1. Project overview

download the software for its offline version. Then, due to teamwork, ATLAS.ti version 23 was used on the Windows 11 operating system and later on MacOS Monterey. The images of software in this chapter are taken from the MacOS version. Figure 1 presents the project's overview, showing the total of documents, codes (and quotations), memos and networks created to instrumentalize the content analysis and be the basis for the discussion presented in this chapter.

It is possible to observe in Fig. 1 that the strategy code was used 29 times; that is, 29 quotations were created that have this connection.

3.1 Fundamentals of Sustainability Management

Back in 1987, on the World Commission of Environment and Development, the world was introduced to the concept of 'Sustainable Development,' that is 'the development that meets the needs of the present without compromising the ability of future generations to meet their own needs'.

Ever since, sustainability has been a relevant topic amongst corporations, which play a significant role in achieving the goals related to it, managing business policies and practices that prioritize the well-being of shareholders, employees, and society while minimizing negative impacts on the environment. (Jadoon et al. 2020).

Sustainability can be divided into three main pillars: economic, social, and environmental. According to Ferreira et al. (2019) 'the social pillar involves issues related to the surrounding community, training and education, job creation for minorities and philanthropy, workplace safety and products. The economic pillar is related to income and, in the case of sustainability, includes environmental and social issues. The environmental pillar is related to natural resources such as wood, soil, water, flora, fauna, biodiversity, greenhouse gas emissions, use of renewable resources, and additional wealth that sustains an ecosystem.' (p. 4).

But those three pillars shouldn't be the only ones considered in sustainability management. The environmental, social and governance (ESG) approach includes corporate governance, dealing with the firm's management and control to provide transparency in reporting, risk management and stakeholder rights. This approach states that ESG performance reflects the market value of firms, being an intangible asset for investors (Jadoon et al. 2020).

When analysing the relevance of corporate sustainability, the quadruple approach of corporate sustainability is the most proper. (Sidhoum and Serra, 2018, cited by Jadoon et al. 2020).

Yamane e Kaneko's (2022) study suggests that 'investing in the SDGs is beneficial for businesses, as stakeholders support companies that both contribute to the SDGs and earn profits from those contributions.', but it also comes with challenges such as needing to improve a communication process regarding the effects of the organization's actions to interest groups (Gray, Owen, & Maunders, 1987, cited by Torelli et al. 2019).

3.2 Stakeholders and Their Demands

According to Paula and Gil-Lafuente (2018), companies pursuing sustainable development must consider stakeholder engagement and management in their strategies, since they are ‘essential to achieve corporate sustainability; however, that is not an easy task for companies due to constant changes and competitiveness in the markets.’ (p. 1995).

The stakeholder theory suggests that organizations should prioritize the interests of a wider range of stakeholders to achieve success, focusing on primary stakeholders, like employees, customers, shareholders, government, and the secondary ones, such as environmentalist groups (Donaldson and Preston, 1995; Freeman, 1984; Schaltegger, Hörisch, and Freeman, 2019, cited by Jadoon et al. 2020), accordingly to Choi and Wang 2009 (cited by Stocker et al. 2021), who identified that by meeting the interest of various stakeholders, companies in a poor financial situation can improve their results, maximizing performance.

Stakeholder engagement is highly associated to the level of implementation of the materiality principle in a positive way, shows the empirical analysis done by Torelli et al. (2019). That’s why understanding material topics is such an important point for building a relationship with stakeholders.

3.3 Materiality Principle

As mentioned by Paula and Gil-Lafuente (2018), “The GRI defines material topic as the ‘topic that reflects a reporting organization’s significant economic, environmental and social impacts; or that substantively influences the assessments and decisions of stakeholders. (p. 1991).

By implementing a materiality analysis, organizations can achieve their highest level of equivalence between stakeholders’ interests. Informed Torelli et al. (2019), ‘With a careful and well-planned materiality analysis, company management is able to detect which interests and issues are dear to company stakeholders, and which have a significant impact on the company in addition to those, stemming from the company itself, that have an external impact.’ (p.473) promoting productive collaboration with stakeholders, and enhancing operational performance and decision-making (Ngu and Amran 2018).

Determination of material topics can encompass ‘business strategy, business model, products and services, type of stakeholder, size of business, geographical presence and business risk’ (Ngu and Amran, 2018).

Research highlighted that in most cases, without proper stakeholder engagement, companies can’t reach higher levels of the materiality principle implementation and cannot enjoy well-targeted communication (Torelli et al. 2019).

Materiality holds significant importance when it comes to corporate sustainability management, but also in sustainability reporting. In accordance with Taubken and Feld (2018), ‘This is exactly where a materiality assessment that is oriented towards the impacts can play a key role. When used as a strategic tool, it offers companies the opportunity to enhance the quality of their reporting, sharpen their target-setting and substantially improve their sustainability performance.’ (p. 97).

3.4 Sustainability Reporting with GRI Guidelines

Sustainability reporting is oriented towards the future, to safeguard and ensure preservation of resources for generations to come (Ngu and Amran 2018). For that reporting process, is recommended that the organization involves stakeholders (Paula and Gil-Lafuente, 2018).

According to Fernandez-Feijoo et al. (2014), the GRI is a tool used in communicating the environmental, social and economic aspects of an organization (cited by Ferreira et al. 2019), and it represents nearly 90% of the world's sustainability reports publication, establishing criteria that guide how ESG information ought to be disclosed to multi-stakeholders interested (Stocker et al. 2021), that being mean for 'understanding and managing the effects of sustainability developments on the organizations activities and strategy' (GRI 2015a, cited by Beske et al. 2019).

Indicators suggested by the GRI are used to evaluate corporate environmental profiles, measure and reporting results of sustainability management, informing stakeholders about the efforts being made to contribute to sustainable development goals (Tsalis et al. 2020).

However, organizations should use the materiality principle to share information and involvement of the company in actions that are of interest to the involved parts (Stocker et al. 2021) since the application is required when reporting in accordance with the GRI standards (Taubken and Feld 2018).

3.5 Materiality and GRI Reporting

Materiality analysis is still a very subjective process, and as mentioned by Calabrese et al. (2019), 'few contributions have proposed a structured approach for materiality analysis, through which stakeholders can participate directly in defining the reporting content (Bel- lantuono et al. 2016; Hsu, Lee, & Chao 2013).' (p. 1020), leaving companies facing difficulties when defining material topics.

For starters, the organization must define its key stakeholders.

Beske et al. (2019) observed during an archival research approach and content analysis that 'Usually the order was first, to identify key stakeholder and second, to identify relevant topics. However, one company proposed a reversed way for stakeholder identification. In this case, a preselection of CSR related topics was used to evaluate whether these topics are relevant for different stakeholder groups, so the agenda of the company determines whether a particular stakeholder group was relevant or not.' (p. 176).

Calabrese et al. (2019), proposed a method called 'Zone matrices', introducing an adequacy matrix used in conjunction with the GRI Materiality matrix. In this case, the determination of the stakeholders was done 'based on their representativeness of stakeholder categories (e.g. clients, employees, etc.) and their capability to offer relevant information on sustainability topics (Manetti, 2011)' (p. 1022).

Adib et al. (2020) proposed that to define key stakeholders, a company should map them adopting a practical framework suggested by Krick et al. (2005), grouping between 'people you have legal, financial or operational responsibilities to; Stakeholders who are affected by your organization's operations and People who are likely to influence your organization's performance'. But since this map is not analytically comprehensive, a second process should be conducted, positioning stakeholders based on the risk they present on the three attributes of the stakeholder salience model, prioritizing the ones who appear on the stripped quadrant, as presented in Fig. 2.

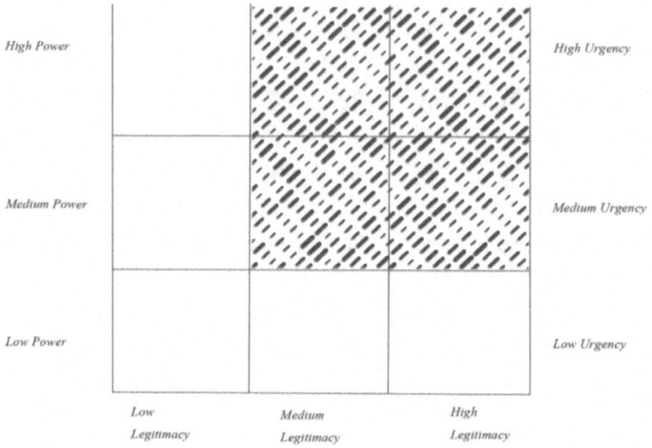

Fig. 2. Stakeholder prioritization matrix (Adib et al. 2020).

After identifying key stakeholders through an adequate method, companies must find the material topics.

GRI Materiality Matrix is created considering that a materiality topic is defined by 'its significance from a stake- holder perspective and evaluation of the impacts of the company's operations (GRI 2016b, cited by Taubken and Feld, 2018) and the intersection of those, as shown in Fig. 2.

From this matrix, material topics can be inferred and included on the sustainability report.

4 Discussion

As shown in Fig. 1, there were 29 occurrences of meanings (quotations) related to the strategy within the sample of this SLR. This result shows that the materiality process is increasingly being considered strategic for organizations. In addition, it is possible to relate the 'strategy' code with words present in all documents; this link can be seen in Fig. 3.

The larger the word available in Fig. 3, the greater its occurrence within the strategy code. With this it is possible to observe the importance of the stakeholders for the elaboration of the materiality process. The results show that the elaboration of a corporate sustainability strategy and/or ESG strategy must consider the stakeholders as part of the process, reinforcing the results of the literature present in this RSL study.

Observing the synchronism or conflict between the GRI Standards document and the RSL articles is still possible. The GRI has undergone changes over time, including the change from Guidelines to Standards. These changes brought greater strategic significance to materiality, including with sectorial norms; that is, the GRI itself is providing material topics for some sectors of activity, including gradually.

Fig. 3. Word cloud.

Thus, the Standard is in line with the results shown in Fig. 3 of the RSL, such as the item 'performance' and 'stakeholders'. For the company to deliver more and more results that generate impact, it must listen to its stakeholders to measure and report its performance. That is, in other words, the function of the materiality process.

5 Conclusion

In conclusion, this research has highlighted the importance of sustainability management and reporting in organizations. The concept of sustainable development, introduced in 1987, emphasizes meeting the needs of the present without compromising the ability of future generations to meet their own needs. Sustainability encompasses economic, social, and environmental aspects, and organizations must consider the triple bottom line to achieve long term success, as explained by Ferreira et al. 2019.

As mentioned by Paula and Gil-Lafuente (2018), stakeholder engagement plays a crucial role in corporate sustainability. Organizations should prioritize the interests of a wide range of stakeholders, including employees, customers, shareholders, and environmentalist groups, as seen in Stocker et al. 2021. By meeting the interests of various stakeholders, companies can improve their financial performance and maximize their overall performance.

The materiality principle is a key aspect of sustainability management and reporting. According to Torelli et al. 2019, through materiality analysis, organizations can strategically identify the topics that have a significant impact on their operations and stakeholders. Findings by Fernandez-Feijoo et al. (2014), Ngu and Amran (2018), Tsalis et al. (2020), and Taubken and Feld (2018) explain that the Global Reporting Initiative (GRI) provides guidelines for sustainability reporting, including the use of materiality to disclose relevant information to stakeholders. However, materiality analysis remains a subjective process, and more structured approaches are needed to involve stakeholders directly in defining reporting content.

A good result in sustainability reportings can help organizations stand out, as seen in the study conducted by Jadoon et al. (2020), 'These results indicated that higher sustainability performance would result in the higher market value of firms, and investors have a positive perception about sustainability performance.'

The next step should be the analysis of a real company's sustainability report, to understand in a practical way how the GRI tool is being used.

Future research can contribute to this field by further developing structured approaches for materiality analysis, exploring innovative methods for stakeholder identification, and refining the integration of sustainability principles into performance measurement systems. Additionally, studies can focus on the challenges and opportunities that organizations face in implementing sustainable practices and reporting frameworks, as well as the impact of sustainability efforts on financial performance and market value.

By addressing these areas, future research can provide valuable insights and practical recommendations for organizations striving to incorporate sustainability into their strategies and operations. This will contribute to the overall goal of achieving sustainable development and creating a better future for all. These review results will foster further research as initial contextual data, including other literature reviews. This paper also presents the evolution of the field according to previous studies, where the materiality process was underestimated (Carpejani, P. et. al. 2017). For future studies, it recommends expanding field and theoretical research exploring agile methodologies in the development of materiality (Carvalho, P. et. al. 2021).

Acknowledgments. To the Laboratory for teaching, practice, and analysis of qualitative research data with the support of ATLAS.ti maintained by PUCPR - CIDES-NAPQuali, for providing the license to conduct the research.

APPENDIX A: Systematic Literature Review Articles

TITLE	AUTHORS	JOURNAL	YEAR OF PUBLICATION	NUMBER OF CITATIONS
AccountAbility's AA1000AP standard: a framework for integrating sustainability into organisations	Farooq, M., et al	Sustainability Accounting, Management and Policy Journal	2021	-
Algorithm applied in corporate sustainability	Paula, L., Gil-Lafuente, A	Kybernetes	2018	4

(*continued*)

(*continued*)

TITLE	AUTHORS	JOURNAL	YEAR OF PUBLICATION	NUMBER OF CITATIONS
Analyst coverage and forecast accuracy when CSR reports improve stakeholder engagement: The Global Reporting Initiative-International Finance Corporation disclosure strategy	García-Sánchez, I., et al	Corporate Social Responsibility and Environmental Management	2019	-
Assessing the materiality of university G4-sustainability reports	Lubinger, M., et al	Journal of Public Budgeting, Accounting & Financial Management	2019	-
Creating sustainability reports that matter: an investigation of factors behind the narratives	Al-Shaer, H., et al	Journal of Applied Accounting Research	2021	-
Do institutional investors drive corporate transparency regarding business contribution to the sustainable development goals?	García-Sánchez, I., et al	Business strategy and the environment	2020	-
Doing Well by Doing Good, Uniquely: Materiality and the Market Value of Unique CSR Strategies	Nardi, L., et al	Strategy science	2022	-
Environmental, social, and governance (esg) practice and firm performance: an international evidence	Shaikh, I	Journal of Business Economics and Management	2021	-
Evaluating the corporate environmental profile by analyzing corporate social responsibility reports	Tsalis, T., et al	Economics Analysis and Policy	2020	2

(*continued*)

(*continued*)

TITLE	AUTHORS	JOURNAL	YEAR OF PUBLICATION	NUMBER OF CITATIONS
Impact measurement and the concept of materiality—new requirements and approaches for materiality assessments	Taubken, N., Feld, T	Springer	2018	4
ISO 14001:2015 and ISO 9001:2015: analyse the relationship between these management systems standards and corporate sustainability	Ferreira, C., et al	GESTÃO & PRODUÇÃO	2019	3
Management control system for corporate social responsibility implementation – a stakeholder perspective	Adib, M., et al	Corporate Governance	2021	1
Materiality analysis in sustainability and integrated reports	Beske, F., et al	Sustainability Accounting, Management and Policy Journal	2019	2
Materiality analysis in sustainability reporting: a tool for directing corporate sustainability towards emerging economic, environmental and social opportunities	Calabrese, A., et al	Technological and Economic Development of Economy	2019	2
Materiality disclosure in sustainability reporting: fostering stakeholder engagement	Ngu, S., Amran, A	Strategic direction	2018	4

(*continued*)

(*continued*)

TITLE	AUTHORS	JOURNAL	YEAR OF PUBLICATION	NUMBER OF CITATIONS
New challenges for corporate sustainability reporting: United Nations' 2030 Agenda for sustainable development and the sustainable development goals	Tsalis, T., et al	Corporate Social Responsibility and Environmental Management	2020	1
"Sell" recommendations by analysts in response to business communication strategies concerning the Sustainable Development Goals and the SDG compass	García-Sánchez, I., et al	Journal of Cleaner Production	2020	-
Stakeholders' perceptions of factors affecting the credibility of sustainability reports	Xiao, X., Shailer, G	The British Accounting Review	2021	-
The impact of sustainability reporting quality on the value relevance of corporate sustainability performance	Jadoon, I., et al	Sustainable Development	2020	5
The Influence of Board Structure and Ownership Concentration on GRI Report	Mascena, K., et al	Revista Brasileira de Gestão de Negócios	2020	-
The materiality assessment and stakeholder engagement: A content analysis of sustainability reports	Torelli, R., et al	Corporate Social Responsibility and Environmental Management	2020	5

(*continued*)

(*continued*)

TITLE	AUTHORS	JOURNAL	YEAR OF PUBLICATION	NUMBER OF CITATIONS
The role of multi-stakeholders in market orientation and sustainable performance	Stocker, F., et al	Marketing Intelligence & Planning	2021	4
The Sustainable Development Goals as new business norms: A survey experiment on stakeholder preferences	Yamane, T., Kaneko, S	Ecological Economics	2022	1

References

Elkington, J.: Cannibals with Forks., New Society Publishers, Canada (1998)

Kaplan, R., Norton, D.: A estratégia em Ação. Balanced scorecard**.,** 1st edition, Elsevier, Brasil (1997)

Neely, A., et al.: Designing, implementing and updating performance measurement systems. Int. J. Oper. Produ. Manage. **20**, 754–771 (2000)

Consolidated GRI Norms. Global Sustainability Standards Board (GSSB) (2021)

Xiao, Y., Watson, M.: Guidance on conducting a systematic literature review. J. Plann. Educ. Res. **39**, 93–112 (2019)

SALDAÑA, J. The Coding Manual for Qualitative Researchers. 2. ed. London: SAGE, (2015)

Jadoon, I., Ali, A., Ayub, U., Tahir, M., Mumtaz, R.: The impact of sustainability reporting quality on the value relevance of corporate sustainability performance. Sustain. Dev. **29**, 155–175 (2019)

Yamane, T., Kaneko, S.: The sustainable development goals as new business norms: a survey experiment on stakeholder preferences. Econol. Econ. **191** (2022)

Torelli, R., Balluchi, F., Furlotti, K.: The materiality assessment and stakeholder engagement: a content analysis of sustainability reports. Corpo. Soc. Responisbility Environ. Manage. **27**, 470–484 (2020)

Paula, L., Gil-Lafuente, A.: Algorithm applied in corporate sustainability. Kybernetes **50**, 1982–2000 (2021)

Ngu, S., Amran, A.: Materiality disclosure in sustainability reporting: fostering stakeholder engagement. Strateg. Dir. **34**, 1–4 (2018)

Taubken, N., Feld, T.: Impact measurement and the concept of materiality—new requirements and approaches for materiality assessments. Springer **26**, 87–100 (2018)

Ferreira, C., Poltronieiri, C., Gerolamo, M.: ISO 14001:2015 and ISO 9001:2015: analyse the relationship between these management systems standards and corporate sustainability. Gestão e Produção **26**, 1–14 (2019)

Stocker, F., Zanini, M., Irigaray, A.: The role of multi-stakeholders in market orientation and sustainable performance. Mark. Intell. Plan. **39**, 1091–1103 (2021)

Beske, F., Haunstein, E., Lorson, P.: Materiality analysis in sustainability and integrated reports. Sustain. Account. Manage. Policy J. **11**, 162–186 (2019)

Tsalis, T., Malamateniou, K., Koulouriotis, D.: New challenges for corporate sustainability reporting: United Nations' 2030 Agenda for sustainable development and the sustainable development goals. Corpo. Soc. Responisbility Environ. Manage. **27**, 1617–1629 (2020)

Tsalis, T., Nikolau, I., Konstantakopoulou, F., Zhang, Y., Evangelinos, K.: Evaluating the corporate environmental profile by analyzing corporate social responsibility reports. Econ. Anal. Policy **66**, 63–75 (2020)

Calabrese, A., Costa, R., Ghiron, N.L., Menichini, T.: Materiality analysis in sustainability reporting: a tool for directing corporate sustainability towards emerging economic. Environ. Soc. Opportunities Technol. Econ. Dev. Econ. **25**, 1016–1038 (2019)

Adib, M., Zhang, X., Zaid, M., Sahyouni, A.: Management control system for corporate social responsibility implementation – a stakeholder perspective. Corp. Gov. **21**, 378–396 (2021)

GRI Standards. Global Reporting Initiative (2023)

Carpejani, P., de Lima, E.P., da Costa, S.E.G., Machado, C.G., Tortato, U.: The contribution of the materiality process to a clean production. In: 24th International Conference on Production Research, ICPR 2017, 663–665. ISBN 978–160595507–0, Publisher DEStech Publications (2017)

Carvalho, P.B., Carpejani, P., Deschamps, F., de Lima, E.P., da Costa, S.E.G., de Freitas Rocha Loures, E.: Architecture-oriented agile approach for sustainability reporting. In: World Sustainability Series (2021). https://doi.org/10.1007/978-3-030-59975-1_3

Keynote: Why 'Business and Management Studies' Do not Get 'Design and Engineering'?

Rob Dekkers(✉)

Adam Smith Business School, University of Glasgow, Glasgow G12 8QQ, UK
rob.dekkers@glasgow.ac.uk

This is a modified version of the keynote that was delivered virtually.

Dear participants of the conference,

My name is Rob Dekkers. I hope you are enjoying the conference.

My talk will be about 'why business and management studies do not get design and engineering.' I thank the organisers —Daniela Popescu and Mihai Dragomir— for the opportunity to talk about a topic that is close to my heart.

Looking at the theme of the conference: 'building resilience into production', there are considerable contemporary challenges such as climate change, natural disasters, sustainability, corporate social responsibility, social justice and geopolitical tensions; these affect how supply and production systems are designed and managed. As a starting point, we would consider business and management studies as one perspective of looking at the impact of these challenges, and design and engineering as another perspective. This would be called multidisciplinary research in terms of Aboelela et al.'s [1] classification, i.e. research informed by several disciplines. You could also expect that both disciplines work together to create a shared view on a topic, which is called interdisciplinary research in terms of the same classification. Or perhaps that they even create a new array of subject terms and language as a result of these interactions, becoming truly transdisciplinary research. Looking at the theme of the conference, one would expect that there would be interdisciplinary, and perhaps, transdisciplinary approaches to research on tackling the impact of the challenges faced by those establishing and managing supply and production systems.

So where do these disciplines meet in strands of research? At plain sight, these strands of research seem to be vastly different, with business and management studies focusing on getting things done through people, and design and engineering concentrating on artefacts. Their intersection seems to be limited to a few domains, such as industrial engineering, which may be called operations research in business and management studies, and new product development, that also covers innovation and technology management in this presentation. Inherently, the two disciplines are quite different in their way of thinking. This is illustrated by Schein's [2] essay, in which he distinguishes between three different cultures in firms: operators, engineers and executives. And, his writing brings to the fore that these cultures have different ways of thinking and working. These different ways of thinking and working are necessary to be integrated in an organisation for it to be effective; at the same time, they may cause conflict and misunderstanding

M. Dragomir et al. (Eds.): ICPR 2023, LNPE, pp. 78–92, 2025.
https://doi.org/10.1007/978-3-031-92082-0_9

between these three cultures. Thus, the starting point of this keynote is that the disciplines of business and management studies, and design and engineering actually have a strong impact on practice, whereas it is more likely that engineers acquire business and management skills than managers get the hang of design and engineering.

This stance also raises the question why business and management studies do not get design and engineering. The two meanings behind the phrasing of this question direct the content of the presentation. It will be addressed by considering their thinking and approaches to research and teaching, and the inter- and transdisciplinary thinking within both disciplines.

The structure of the keynote address is as follows. First, what is so specific for design and engineering will be discussed. This will be related to and compared with business and management studies. After highlighting similarities and differences between the two disciplines, the question what do the deliberations add up to, will be answered.

1 What Is So Specific About 'Design and Engineering'?

Starting with the perspective of design and engineering, the discipline is concerned with the design and creation of artefacts. Making these artefacts work is informed by scholarly knowledge, but typically, hardly in one go as will be demonstrated. Some key points will conclude this section.

1.1 It Is About Creating and Designing Artefacts

With the essence of design and engineering being about designing and creating artefacts, it not only extends to the creation of products and services, but also production and supply systems to support these products and services in the context of the presentation. A specific example of an artefact is the recently built Learning and Teaching Hub that that was created at the University of Glasgow. It serves as a central space for teaching activities of different kinds, such as lectures for large groups of students, and as a meeting space for students; its multifunctionality extends to hosting conferences and events. Creating this building, establishing its function and how to fulfil this function required engagement with stakeholders, among them students and staff. Even those living in adjacent residential areas were consulted, particularly for the construction phase as these activities disrupted the quality of living, and thus, restricted the hours of construction. In brief, design and engineering is about the future state of systems, which can extend to optimisation and improvement.

1.2 And, Making Them Work

However, you cannot know what the actual effects of artefacts are unless they have been created. If you take again the Learning and Teaching Hub at the University of Glasgow, during its design and construction phase there still was uncertainty whether the function and requirements will be fully met. Only when the building was opened, and students and staff started to use it, then it became clear to what extent the original objectives were met. In the case of the hub one objective was to create a natural flow of students, staff and other visitors for using its teaching and learning spaces; when the first period of teaching in the hub started, it could be noticed that this was successful. During the design, there can also be experimentation with a proof of concept such as a mock-up or prototype. The prototype can be a simulation model, for instance, you can virtually walk through the building to assess its functionality and practicality. The experimentation and evaluation of proof of concepts is an essential phase and allows to look at subsystems and aspectsystems [3: 27–32]. It also allows the integration of both subsystems and aspectsystems, with the ventilation of a building being a subsystem, and health and safety an aspect. Finally, experimentation in any form aims at exploring the conditions under which a prototype functions according to set requirements. The emphasis of design and engineering is on reducing uncertainties over the course of time; it starts with broad specifications, and at the end it is about nuts and bolts.

1.3 Informed by Scholarly Knowledge

The process of design and engineering is supported by the active use of scholarly knowledge. This concerns knowledge from distinct disciplines on aspects relevant to the artefact that is created and produced. Technological advances using recent understanding, or maybe even insight into the possibilities of new technologies are among them. Typically, insight is reflected in laws of observed regularities, technological rules and modelling in which quantification plays an important role. Optimisation of new or existing artefacts does not mean that the design meets all criteria. Think about the cost and fitness for purpose of the building. Perhaps, the costing turns out to be above budget and then you have to rethink how you can achieve both criteria, or otherwise make trade-offs. Trade-offs and evaluating whether concepts work or not imply an open systems perspective for elicitation of requirements and their utilisation by end users. To this purpose, it also calls on engaging with stakeholders. Once objectives and (functional) requirements have been set design and engineering is associated more with a closed systems perspective. Thus, there are two simultaneous processes at play: one about usefulness of products and services for eliciting (functional) requirements with the product and service provision as open system, and another one directed at designing and delivering a product and service as a closed system.

1.4 But Hardly in One Go

Once you have set the objectives and requirements for the design and engineering of artefacts, it is essentially a decision-making process. Notwithstanding clarity about objectives and requirements, decision making is iterative as it is based on progressive insight as depicted in Fig. 1 for the design and engineering of products [4: 321]; in it there are different stages for moving from the elicitation of requirements and using technological insight to detail instructions for supply and manufacturing (and, not to forget the use of products), but is also depicts a secondary process on the right-hand side, where evaluation and progressive insight mean you may have to go back to reconsider solutions, make trade-offs and optimise against criteria. Even considering what objectives and (functional) requirements are, might be on the table in higher echelons of the figure. Incomplete and incommensurate information throughout is a key characteristic of this

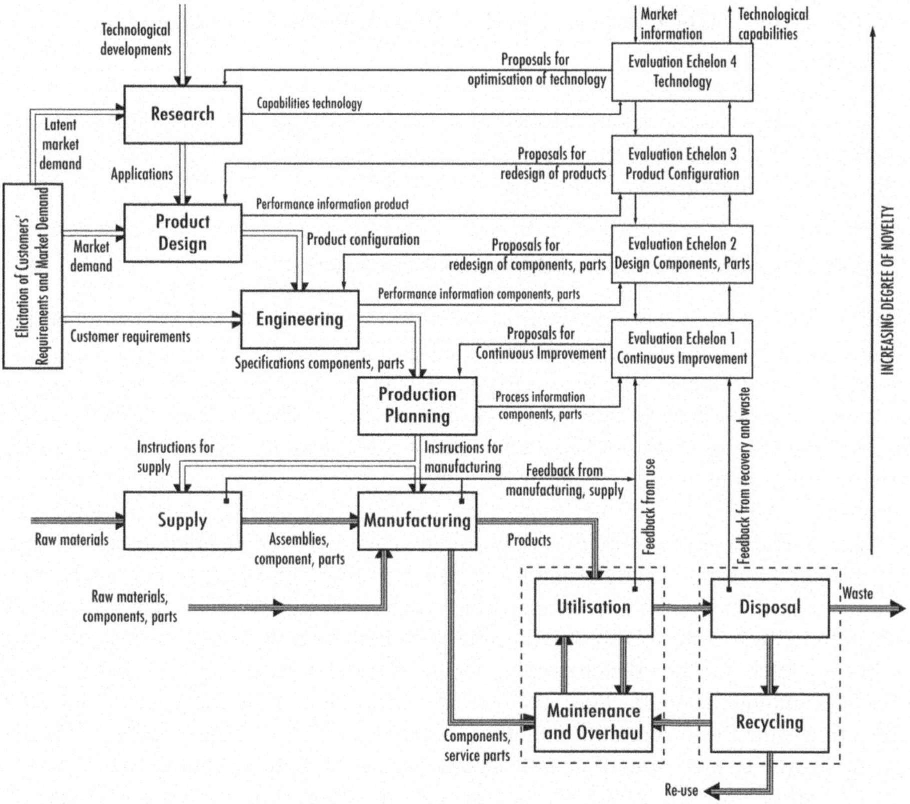

Fig. 1. Generic model for processes of design and engineering artefacts [4: 321]

process and leads to iterations in decision making [5: 319]. And again, it is informed by the application of scholarly knowledge but also the outcomes of experimentation, simulation, proof of concept and prototyping. Formal methods, such as Pugh's [6] controlled convergent method, as it is known in design and engineering but typically called set-based concurrent engineering in business and management studies [e.g., 7], support the decision making in a stage-wise fashion iterating between evaluation of requirements and feasibility of solutions.

Key Points

Summarising the points raised in this section, specific to design and engineering is its teleological and iterative nature. It is about shifting technological boundaries, and about studying artefacts as open systems for implementation and use. In addition, it is application-oriented while being informed by scientific advances with regard to its theoretical foundations, laws of observed regularities, modelling and simulation.

2 How Does This Compare with 'Business and Management Studies'?

Before comparing business and management studies with design and engineering, it is important to understand its nature. The comparison that follows will consider aspects of research, taught matter and engagement with practice.

2.1 Understanding 'Business and Management Studies'

The roots of business school go back to accounting, economics and also engineering. The first business school was called the Aula do Comércio (Lisbon, 1759–1844) [8]. The degree that students could obtain from it was in accounting. The second one that is the 'École Superieure de Commerce de Paris' founded in 1819; it has evolved into the ESCP-EAP European School of Management, its current name. Notably, in its original curriculum also courses into engineering were found. Some attribute the development of business schools to this École Superieure in Paris. The current curricula have been informed by teaching in economics, scientific management (now called industrial engineering and operations research) and social sciences. Further examples are teachings in human source management, marketing, operations and supply chain management. As stated before, there are only few domains that are directly related to design and engineering. For instance, product development, also captured by the domain of innovation of technology and management. And, the second one is a domain spanning from industrial engineering, operations research to mathematical optimisation. In this respect, it is perhaps interesting to know that the well-known journal MIT Sloan Management Review was originally called Industrial Management Review publishing articles on industrial engineering and operations research. Business and management studies have evolved from courses on accounting, economics and engineering to the inclusion of a broad range of domains, whereas today traces of its early developments are still found.

What has remained is that often the focus of business and management studies is directed towards financial performance of firms. It includes profitability and other measures that capture this overarching performance indicator. However, thinking about performance also is about capabilities, particularly for production and supply systems. Consequently, following from the search for financial performance, there is an emphasis on objectives and strategy in business and management studies. The topics being investigated are subject to two main approaches. Quantitative studies are typically undertaken by using questionnaires and data sets; such can be in the form of economic data sets available in databases. A second strand of research is conducting qualitative studies, normally interviews with stakeholders, managers and employees. A popular method is case studies, though care must be taken of its design and analysis of data; see [9] for some indicative thoughts on this matter with regard to the number of case studies and its implications for trustworthiness of a study. More recently there has been attention for combining these methods in research which is mixed-methods research. So, the approach to research and knowledge in business and management studies differs from the discipline of design and engineering.

Table 1. Comparing the Disciplines As Research

	Business and management	Design and engineering
Focus	• Performance of firms • Perspectives of managers, employees, stakeholders, etc • Technologies as enablers for change	• Elicited requirements • Shifting technological boundaries • Technologies as determinant of change
Research	• Correlation (factors)	• Effect size
Unit of analysis	• Organisations as open systems (interactions, stakeholders, behaviour) • Or aspectsystem	• Proof of concepts • Experimentation • Integration (solutions)
Knowledge	• Mainly domains within discipline (e.g., marketing, human resources, supply chain management)	• Domains within disciplines • But also, multiple disciplines (inter- and transdisciplinary research)
Captured by	• Conceptualisations (e.g., matrices, models for causal relationships between factors) • Mid-level theories • But commonly, retrospective	• Theories and laws of observed regularities • Modelling (simulation) • Methods • Technological rules • Aiming for predictive
Design research methodology	• Nascent	• For artefacts: induction
Hypothetico-deductive research methodologies	• Falsification cumbersome • Often relying on surveys • Ignores difference in continuous and discrete data • Cross-sectional studies dominant	• Hypothetico-deductive research methodologies to test and falsify theories, laws of observed regularities, technological rules
Inductive research methodologies	• Mostly as qualitative studies • Less attention to abduction	• Inductive research methodologies for implementation and use
Prevalent research paradigms	• (Post-)positivism • Interpretivism • Constructivism • Pragmatism • Participatory research	• Mostly, positivism • For design and implementation of artefacts post-positivism, interpretivism, pragmatism

2.2 Comparing the Disciplines as Research

Table 1 captures the differences between the two disciplines from the perspective of approaches to research. It shows that the focus of business and management studies is on performance of firms, whereas in design and engineering it is on requirements that have been elicited from users or consumers. Both disciplines hold a different view on technological advancements. Notably, business and management studies see technological progress as enabler for change in organisations and their performance. For design and engineering, this progress provides an opportunity to shift technological boundaries and performance requirements. It is commensurate with an emphasis on correlation in research related to business and management studies, meaning a scale-free measure of change, and effect size in design and engineering as an absolute measure of performance. The units of analysis also differ across both disciplines. Organisations can be seen as an artefact, mostly approached as open systems in terms of their interactions, stakeholders and behaviour, and in studies there may be a focus on aspectsystems (for the later, see [3: 29–32]), represented by domains such as marketing and supply chain management. However, there is emphasis on how work organisation and structure facilitate managing resources and employees. The discipline of design and engineering has distinct domains, too, like mechanical engineering but draws heavier on other disciplines with mathematics for optimisation, physics and acoustics among them. Moreover, a mainstay of design and engineering is about proof of concepts, experimentation and integration of solutions.

The distinctions are reflected in how scholarly knowledge is captured. Business and management studies tend to focus on conceptualisations such as matrices, models building on causal relationships between for factors. In this regard, matrices typically have two dimensions to reflect characterisations or choices made by managers but they are limited what they incorporate. There is also attention to mid-level theories, which are developed for a specific domain. A point of attention is that sometimes these conceptualisations and mid-level theories are oversimplified. Examples are the three linear phases for dynamic capabilities: sensing, seizing and transforming [10: 1342], open innovation [11] criticised for being a substitute for collaboration [12: 216] and the resource-based view [13] seen as an undeveloped concept for competitive heterogeneity [14]. Somehow, this implies that business and management studies may follow a reductionist approach, perhaps related to their 'audience' being managers. In design and engineering there is also a reductionist approach, but this is taken as looking for theories, laws of observed regularities, technological rules, modelling, methods, etc. that predict the future state of systems, products and services in more fundamental terms. This is captured by methods and methodologies for designing and engineering, while in business and management studies design approaches to are still nascent, even after calls [e.g., 15] to include this in management research. Also, in terms of research paradigms a dominant approach is the inductive research methodology in business and management studies alongside questionnaires in the form of cross-sectional studies (while noting that impact of abduction is relatively less considered and that many published studies with questionnaires ignore the difference between continuous and discrete data). Falsification of theoretical foundations often is cumbersome because of the complexity of social interactions. Largely, the discipline of design and engineering follows a hypothetico-deductive research methodology to test and extend theories, laws of observed regularities and technological rules (including methods). In addition, the inductive research methodology can be associated with the implementation and use of artefacts arising from theoretical foundations and technological advances. Consequently, there are also differences in which research paradigms dominate. In the domain of design and engineering, positivism is associated with the design and implementation of artefacts, and post-positivism for the engagement with users and stakeholders about their use and implementation. Since there is a focus on design and solutions, design and engineering some of its research may also be classified as pragmatism. In the domain of business and management studies, commonly there is a broader range of research paradigms in use: (post)positivism, interpretivism, constructivism, pragmatism and participatory research. However, the use of research methodologies and association with research paradigms is strongly related to the content of both disciplines and how they view objects or subjects of study.

Table 2. Comparing the Disciplines As Taught Matter

Business and management	Design and engineering
• Segregated domains, e.g. • Accounting and finance • Entrepreneurship • Innovation and technology management • Logistics and supply chain management • Marketing • Operations and service management • Integrative courses and tutorials less common • In-depth limited to specific domains but influenced by 'strategising'	• Typically informed by multiple disciplines, among them • Mathematics • Physical sciences • Courses (lectures) building on each other • Prerequisites • Integrative courses and tutorials • In-depth investigation and application • Design • Experiments • Simulation • Proof-of-concept

2.3 Comparing the Disciplines as Taught Matter

Taught matter also differs across the two disciplines; see Table 2. The typical, underpinning domains of programmes and degrees in business and management studies are strongly related to the development of business schools to which this keynote only alluded. Except for economics, though related to the founding of business schools, and some domains in social sciences such as psychology, there is less looking outside the discipline. Also, integrative courses, tutorials, simulations and exercises are less found. Furthermore, teaching is influenced by 'strategising', another word for focusing on strategy, with the book by Slack et al. [16] on being an instance as a dressed down version of a more extensive textbook, which often leads to less attention to quantitative methods (which is somewhat odd for a domain such as operations management). Contrastingly, teaching in design and engineering is informed by multiple disciplines. Mathematics and physical sciences are among them. Another feature of the curricula for design and engineering that successive courses in their topics build on previous ones. For instance, you need to complete certain courses in mathematics before you can make calculations for fluid dynamics. Typically, in design and engineering there typically are integrative courses and tutorials. Going back to the case of the teaching hub at the University of

Glasgow, construction or architecture degrees will include where students design the building as a whole rather than just looking at specific aspects. This leads to in-depth investigation and application during design, experiment, simulation proof of concept to understand better how aspects such as ventilation, electrical wiring, etc. need to be integrated in a building. Such a holistic approach tends to be missing for business and management studies. Furthermore, the argument in business and management studies is that self-study and tutorials replace experiments. Another tendency in teaching business of management is avoiding quantitative approaches, except for topics such as operations and supply chain management and research methods; but even then, other than statistical methods for research methods support dissertations, these are often simple formulae.

Table 3. Comparing the Disciplines As Engaging With Practice

Business and management	Design and engineering
• Object of study	• Application embedded in its approaches
• Commonly known as field studies and action research in business and management studies	• Though varies in effectiveness, see Dekkers [2010]
• Less prevalent methods (though found in ranked journals such as Journal of Operations Management)	• Long recognised variety in outputs and merits [KNAW, 2005]
• Emphasises idiographic stance • Needs evaluation of contexts • (e.g., contingency theory)	• More collaboration, industrial placements (uncommon in business and management studies

2.4 Comparing the Disciplines' Engagement with Practice

The final topic for comparing both disciplines is their engagement with practice and practitioners. In both disciplines practice is object of study; see Table 3. Looking at business and management studies, engagement with practice commonly takes place in the form of case studies and field studies, with the latter advocated for operations management [e.g., 17]. It has less prevalence compared to other canonical methods such as econometric modelling, quantitative research in the form of statistical correlation and questionnaires. The use of case and field studies allows better considering the context in which a study takes place and its influence on outcomes, for instance [9]. Consequently, the engagement with practice implies an idiographic stance. Somewhat similar, design and engineering also engages with practice. Last but not least by industrial placements, which are more common than in business and management studies. It could also be the outcome of a scholarly setting; for example, industrial placements in business and management studies are found more in the Netherlands than in the United Kingdom. Scholarly institutions such as the white paper by the Royal Netherlands Academy of Arts and Sciences [18] draw attention to measures beyond publications in scholarly journals and their citation rates; it includes the relevance of field work and case studies. The increasing relevance of engagement with practice is also found in the funding guidelines for the Innovative Manufacturing Programme by the Engineering and Physical Sciences Research Council in the United Kingdom, though its effectiveness for this engagement varies across the research institutes involved [19].

A point of attention for case studies and field work is the involvement of researcher. Related to the case study methodology is action research in which there is more collaboration between researcher and industry; for the domain of operations management, see Westbrook's [20] description. Walsham [21] sees the participation of the research as a continuum. This brief reflection means that even when case studies and field work are conducted, the involvement of the researcher eventually determines the degree of engagement (which is valid for both disciplines).

Key Points

As you can see, there are quite substantial differences between the two disciplines in terms of what is investigated, to what purpose methods are being used and how it is being taught even though both have a strong orientation towards practice. Noting so, the engagement with practice appears to be more intense in design and engineering but good practices are also found in business and management studies.

3 What Does It All Add up to?

The deliberations so far on how the disciplines compare raises the question what the observations and conjectures imply. This will be related to the so-called T-profile, the focus of the presentation and the theme of the conference: building resilience into production.

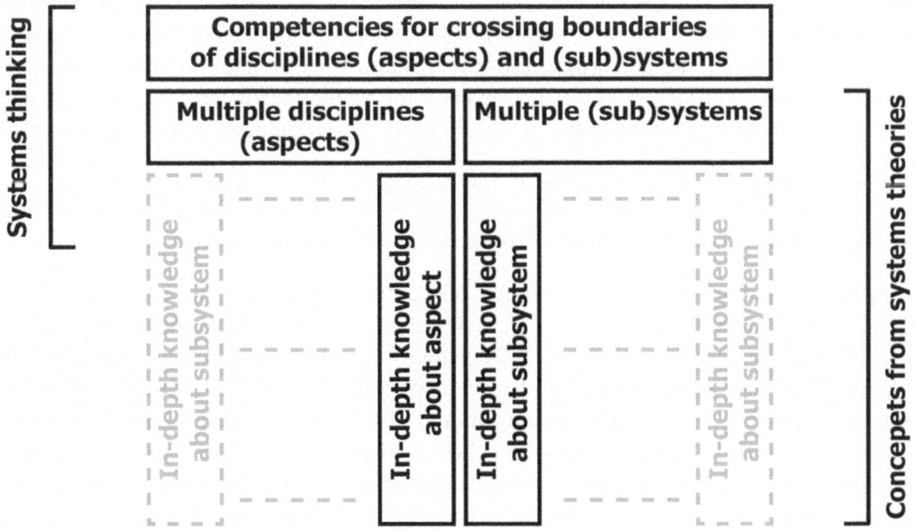

Fig. 2. T-profile including systems theories and thinking

3.1 How Do the Disciplines Come Together in the T-Profile?

A first point to look at is how these disciplines can be mapped in the so-called T-profile for competencies. This profile has become popular in practice. There are strong links to education, particularly in the context of degrees and acquisition of skills after a design and engineering degree [22, 23]. The T-profile of competencies has two axes; see Fig. 2. The vertical one is about in-depth knowledge on a specific aspect [see 3: 29–32] or subsystem. The horizontal represents boundary-crossing competencies for multiple aspects or disciplines, and multiple systems. The tenet of the profile is that both axes are necessary to make well-informed decisions that are integrative and at the same time consider the practicalities for making aspects and subsystems work. In this respect, graduates from business and management studies often possess broad knowledge but have less likely been exposed to skills for analytical thinking (critical thinking seems to be preferred), problem solving and design associated with in-depth knowledge. This contrasts somewhat with graduates from design and engineering studies where these skills are integral to the curricula. However, these graduates may have gaps in a broader orientation towards disciplines. The studies mentioned in this paragraph outline the advantages of exposing and educating engineers with business skills, etc. So, both disciplines have to be brought together for addressing many of the challenges that we face, including building resilience into production and supply systems; for which it may be easier for design and engineering graduates to acquire skills associated with business and management studies than vice versa.

Systems thinking may provide this bridge between the two disciplines. There is interest into systems thinking in business and management studies, but it is seen as being of a bygone era. In the 1970s, 1980s and 1990s there was attention to systems thinking in business and management studies, but after these years interest in it subsided. The terminology such as holistic thinking is used. However, in reality, it often reflects econometric modelling where the interaction of a system with an environment is reduced to factors that influence the performance of the system; for instance, the model for operational processes in Slack et al. [16]. However, the challenges faced by ecological and industrial systems, e.g., resilience and sustainability, are mounting, implying a stronger need for systems thinking.

3.2 How to Build Resilience into Production and Supply Systems?

Going back to the theme of the conference, all points raised do not imply that we should give up bringing the two disciplines together. In business and management studies there should be more attention to principles and epistemology of design and engineering, which does not only benefit operations but also the application of scholarly knowledge in this domain to practice. These opportunities should come along with in-depth application of knowledge rather than the current trend of curricula with a broad range of courses. In this respect, interdisciplinary research could serve as starter. Systems thinking and system theories may provide a foundation for such inter- and transdisciplinary teaching and research, with applied systems theory [3], soft systems methodology [24], system dynamics [25] and the viable system model [26]. I hope that the arguments provided in the keynote makes it clear why there should be a renewed emphasis on system thinking, and attention to design and integration in business and management studies; a mechanism to achieve would be the disciplines working together more intensely.

This approach would benefit building resilience into production and supply systems. Bhamra et al. [27: 5380] observe three dimensions for looking at resilience (slightly reformulated here for consistency):

- Readiness and preparedness.
- Responsiveness and capability for adaptation.
- Recoverability and accommodation.

All three dimensions require decision making and well-designed systems, thus calling on both insight from business and management studies, and design and engineering approaches. Consequently, all three are a product of human interactions, information and communication technologies to provide adequate information, and production and supply systems. One cannot be seen disconnected from another. Therefore, the key will be how both disciplines benefit from each other's scholarly knowledge and collaborate to achieve preparedness and the capability to deal with adversity and adapt to changing circumstances.

3.3 Why 'Business and Management Studies' Do not Get 'Design and Engineering'?

I am now taking the opportunity to recap the observations in this keynote and bringing it back to the title of this presentation. Generally speaking, the discipline of design and engineering is characterised by the integration of in-depth knowledge. And, by nature it is interdisciplinary and often has specific domains for which integration plays a role, too. In the discipline, there is a dominance of systems thinking, even though it is often in the restricted form of systems engineering. Thus, it combines the development of in-depth knowledge and related competencies, and the integration of knowledge from domains and disciplines. Contrastingly, in business and management studies, there is hardly recognition of necessity of integrated activities in the curriculum. There are some exceptions to this rule, such as running a business simulation, for example [28], or developing a (detailed) business plan. Furthermore, classifications and causal relationships dominate teaching and research, rather than seeking laws of observed regularities, methods and models as supporting predictive approaches to performance of systems; this puts the emphasis on selection rather the design. It is this subtle difference (and others) that sets the two disciplines apart instead of bringing them together.

Despite this perhaps somewhat gloomy look, there are also rays of hope. There are postgraduate taught degrees that combine courses in engineering and management. Also, in the discipline of design and engineering more provision of courses on business and management is seen as beneficial. Furthermore, there is attention to design science in business and management studies, though still in its infancy stages. Finally, there are increasingly funding opportunities, opportunities to work together for both disciplines, though its effectiveness varies.

4 Some Reflexivity on the Opinions Expressed Here

A point here is that the viewpoint expressed in the keynote is likely influenced by observations in my career trajectory. This includes the multidisciplinary approach instilled during studying for a master's degree in mechanical engineering, though veering off towards operations management in its final stages, the industrial academic career holding managerial positions, and then embarking on an academic career in which interdisciplinary and transdisciplinary approaches played a large role. Particularly, the observations and experiences in business schools have led to reflections on practices for both education and research. Attending and involvement in interdisciplinary conferences such as the International Conferences on Production Research have kept this spirit of thinking.

And, another word of caution, the observations and reflections here must not be taken as absolute statements. There is a considerable scholarly body of knowledge on the intersection of the two disciplines. Moreover, there is a broad variety of approaches to teaching students, ranging from multidisciplinary takes to transdisciplinary pedagogy. Not all have been observed and not to all justice has been done here. Furthermore, for the sake of argumentation generalisations have been used and they are not doing necessarily capturing all provisions in business and management studies, and in design and engineering. Setting these points aside, what I hope is that this keynote may serve as inspiration for advancing inter- and transdisciplinary research, and more integrative pedagogy teaching so that both domains can benefit constructively from each other's insights.

All this should be placed in the context of the internationalisation of education. Although we see this a trend it constitutes the foundation of universities more than a millenium ago in the first place (i.e. The University of Al Qarawiynn in 859 AD). A worry coming along with the current trend is not only the position of degrees, but also the shift towards making degrees less intensive as observed for business and management studies. A lessened intensity does not coincide with the intensity that is needed for teaching design and engineering. However, there are also some studies pointing out that this may be necessary and that alternative pedagogies, particularly in business and management studies, may lead to students acquiring in-depth skills [e.g., 28]. Nevertheless, current approach to business and management studies becoming less intensive is at odds with what is required for integrative studies.

5 Bringing the Keynote to an End

To conclude, business and management studies do not get design and engineering, and they don't get it!

I wish you a very enjoyable conference and that you have fruitful interactions with our community for production research. Thank you for your attention.

Acknowledgement. An earlier opportunity to share thoughts on this matter in a webinar for the Department of Materials and Production at Aalborg University (5th April 2022) was provided by Yang Cheng and Verena Stingl.

References

1. Aboelela, S.W., et al.: Defining interdisciplinary research: conclusions from a critical review of the literature. Health Serv. Res. **42**(1, Part 1), 329–346 (2006). https://doi.org/10.1111/j.1475-6773.2006.00621.x
2. Schein, E.H.: Three cultures of management: the key to organizational learning. Sloan Manage. Rev. **38**(1), 9–20 (1996)
3. Dekkers, R.: Applied Systems Theory, 2nd edn. Springer, Cham (2017)
4. Dekkers, R., Chang, C.M., Kreutzfeldt, J.: The interface between "product design and engineering" and manufacturing: a review of the literature and empirical evidence. Int. J. Prod. Econ. **144**(1), 316–333 (2013). https://doi.org/10.1016/j.ijpe.2013.02.020
5. Shishank, S., Dekkers, R.: Outsourcing: decision-making methods and criteria during design and engineering. Prod. Plann. Contr. Manage. Oper. **24**(4–5), 318–336 (2013). https://doi.org/10.1080/09537287.2011.648544
6. Pugh, S.: Concept selection: a method that works. In: International Conference on Engineering Design, pp. 497–506, 9–13 March Rome (1981)
7. Ward, A., Liker, J.K., Cristiano, J.J., Sobek, D.K., II.: The second toyota paradox: how delaying decisions can make better cars faster. Sloan Manage. Rev. **36**(3), 43–61 (1995)
8. Rodrigues, L.L., Gomes, D., Craig, R.: The portuguese school of commerce, 1759–1844: a reflection of the "enlightenment." Account. Hist. **9**(3), 53–71 (2004). https://doi.org/10.1177/103237320400900304
9. Dekkers, R., Hicks, C.: Saturation and the case study methodology: how many cases do you need?. In: 25th International Conference on Production Research, Chicago (2019)
10. Teece, D.J.: Explicating dynamic capabilities: the nature and microfoundations of (sustainable) enterprise performance. Strateg. Manag. J.. Manag. J. **28**(13), 1319–1350 (2007)
11. Chesbrough, H.W.: Open innovation: the new imperative for creating and profiting from technology. Harvard Business School Press, Boston (2003)
12. Dekkers, R., Koukou, M.I., Mitchell, S., Sinclair, S.: Engaging with open innovation: a scottish perspective on its opportunities, challenges and risks. J. Innovation Econ. Manage. **28**(1), 193–226 (2019). https://doi.org/10.3917/jie.028.0187
13. Barney, J.B.: Firm resources and sustained competitive advantage. J. Manag.Manag. **17**(1), 99–120 (1991). https://doi.org/10.1177/014920639101700108
14. Priem, R.L., Butler, J.E.: Is the resource-based "view" a useful perspective for strategic management research? Acad. Manag. Rev.Manag. Rev. **26**(1), 22–40 (2001). https://doi.org/10.5465/amr.2001.4011928
15. Aken, J.E., van.: Management research as a design science: articulating the research products of mode 2 knowledge production in management. Br. J. Manag.Manag. **16**(1), 19–36 (2005). https://doi.org/10.1111/j.1467-8551.2005.00437.x
16. Slack, N., Lewis, M.: Operations Strategy. Harlow: Pearson (2010)
17. Meredith, J.: Building operations management theory through case and field research. J. Oper. Manag.Oper. Manag. **16**(4), 441–454 (1998). https://doi.org/10.1016/S0272-6963(98)00023-0
18. KNAW. Judging research on its merits. Amsterdam (2005). http://72.14.207.104/search?q=cache:F_y8gUHmrKEJ:www.knaw.nl/publicaties/pdf/20051029.pdf+judging+research+on+its+merits&hl=en&ct=clnk&cd=1
19. Dekkers, R.: Research at the innovative manufacturing research centres: analysis from the mid-term review of the innovative manufacturing programme. J. Manuf. Technol. Manag.Manag. **21**(1), 63–99 (2010). https://doi.org/10.1108/17410381011011498
20. Westbrook, R.: Action research: a new paradigm for research in production and operations management. Int. J. Oper. Prod. Manag.Oper. Prod. Manag. **15**(12), 6–20 (1995). https://doi.org/10.1108/01443579510104466

21. Walsham, G.: Doing interpretive research. Eur. J. Inf. Syst. **15**(3), 320–330 (2006). https://doi.org/10.1057/palgrave.ejis.3000589
22. Karjalainen, T.-M., Koria, M., Salimäki, M.: Educating T-shaped Design, Business and Engineering Professionals. 19th CIRP Design Conference, Cranfield: Cranfield University Press (2009)
23. Cederberg, L., Wallén Axehill, J., Herzog, E.: Experience from a Program for Accelerating the Creation of T-shaped Technical Leaders 29th INCOSE International Symposium, Orlando, FL (2019)
24. Checkland, P.: Systems Thinking, Systems Practice. John Wiley & Sons, Chichester (1981)
25. Forrester, J.W.: Industrial Dynamics. M.I.T. Press, Cambridge, MA (1961)
26. Beer, S.: Brain of the Firm - the Managerial Cybernetics of Organization. John Wiley & Sons, Chichester (1972)
27. Bhamra, R., Dani, S., Burnard, K.: Resilience: the concept, a literature review and future directions. Int. J. Prod. Res. **49**(18), 5375–5393 (2011). https://doi.org/10.1080/00207543.2011.563826
28. Dekkers, R., Koukou, M., Zhou, Q., Salgado, E.G.: Simulation games for engaging students when teaching operations and supply chain management to students: Using the QpQ simulation game as example. In: Zhang, Y., (ed.), Handbook on Teaching and Learning in Operations Management, pp. 192–212. Cheltenham: Edwar Elgar Publishing (2024). https://doi.org/10.4337/9781802201949.00017
29. Dekkers, R.: The paradox of the Chinese learner in the context of the research-teaching nexus: a matter of module design? Int. J. Inf. Oper. Manag. Educ.Oper. Manag. Educ. **7**(1), 66–92 (2021). https://doi.org/10.1504/IJIOME.2021.114739

Unravelling Eight Myths for Publishing Literature Reviews in the Domains of Operations and Supply Chain Management

Rob Dekkers[1(✉)], Pauline Found[2], and Yang Cheng[3]

[1] Adam Smith Business School, University of Glasgow, Glasgow G12 8QQ, UK
rob.dekkers@glasgow.ac.uk
[2] Cardiff Business School, Cardiff University, Cardiff CF10 3EU, UK
foundp@cardiff.ac.uk
[3] Department of Materials and Production, Aalborg University, DK 9220 Aalborg, Denmark
cy@mp.aau.dk

Abstract. With literature reviews, particularly, protocol-driven ones, becoming more accepted in the domains of operations and supply chain management, eight myths (and a ninth one) that influence authors in their approach have been extracted from submissions to a special issue in the Journal of Manufacturing Technology Management. What to do about these myths, reflected in the comments by reviewers and the guest editors, is at the heart of this paper. Some guidance here complements existing advice found in textbooks and publications in academic journals, whilst the discussion of other points leads to suggestions hardly addressed so far, even though they are essential to getting literature reviews published. Thus, the purpose of this paper is to encourage scholars to reflect on their approaches to literature reviews to increase chances of publication.

Keywords: publishing · systematic literature reviews · writing

1 Writing Literature Reviews Challenging Despite Guidance

The engagement with existing scholarly knowledge in the form of appraising publications on their merits for informing other studies is the foundation for academic practice; given its long-standing tradition one would think writing and publishing literature reviews would pose few challenges for scholars. This presumption was tested during the consideration of 39 manuscripts initially submitted to a special issue in the *Journal of Manufacturing Technology Management* (this excludes 10 desk rejects and two manuscripts that followed a different route). As guest editors we were only accepting literature reviews, though we did not specify which types, and the only direction provided was the scope being manufacturing management and its related domains. The submissions received to the special issue demonstrated that writing literature reviews for publications poses considerable challenges for authors.

This is not to say that there is no guidance. For example, Maier [1] instructs authors to think about the holes in Swiss cheese as metaphor for gaps to be identified from studies

M. Dragomir et al. (Eds.): ICPR 2023, LNPE, pp. 93–106, 2025.
https://doi.org/10.1007/978-3-031-92082-0_10

when conducting a literature review. Beyond such a simplified approach to literature reviews, perhaps oversimplified, there is some general guidance on how to conduct literature reviews, such as Cronin et al. [2], Green et al. [3] and Tranfield et al. [4]; this type of guidance mostly concerns protocol-driven literature reviews. Some are directed at operations management, for instance, Thomé et al. [5], albeit that the latter is a reiteration of techniques for systematic reviews derived from known practices in the domains of medicine and nursing, and some at supply chain management, for example, Durach et al. [6] and Hochrein et al. [7: 255 ff.]. A majority of these works provide directions for so-called protocol-driven literature reviews. It also implies that authors should be able to let their literature reviews be informed by best practices.

Given the abundance of advice, particularly for protocol-driven literature reviews, this paper explores whether submissions to the special issue followed best practice using eight myths. These myths arose during conversations when guest editing the special issue. Before going into more detail on the myths, first, a few notes follow on what literature reviews ought to be. Then the eight myths are presented. The eight myths are followed by a ninth one, which did not directly follow from the initial 39 submissions. Afterthoughts on our own views on literature reviews for the domains and the contribution of this chapter conclude.

2 Keeping in Mind Some Notes

One tricky challenge for authors is what a literature review is about. As mentioned in the introductory section, Maier [1] reduces it to searching for gaps. Also, Hochrein et al. [7: 241] have a limited view: 'the aim of stand-alone literature reviews [ed.] is to identify and summarize primary studies on a specific topic.' In healthcare, summarising studies, particularly using their findings only, is called meta-aggregation, to which Bergdahl [8] agitates; she [9: 391] later denounces these writing as 'topical surveys', following Sandelowski and Barroso's [10: 83] designation. However, literature reviews can be seen as synthesis of scholarly knowledge and evaluation of the suitability of scholarly knowledge for topics. The latter implies that literature reviews have a strong emphasis on analysis of evidence collected in a set of works.

This is also reflected in the users (or readership, for that matter) of published literature reviews. The first group constitutes scholars, covering a broad audience, that can use literature reviews to inform their own studies. This also means that reviews should set out points how the analysis leads to how theoretical foundations can be used but not to specific propositions, frameworks or alike that only suit one or a limited number of studies. The second group of readers are practitioners. This group of readers, depending on the implications of the topic, should be able to take away points that they can use for their own practice; in the case of operations and supply chain management, this encompasses practices, strategies, structuring of operations and supply systems and operational guidance. In this spirit, for instance, passed over by many citing Tranfield et al. [4] is that they concentrate on evidence-based management, akin to best-evidence synthesis in health care, and less on how reviews inform scholarly studies. A third group is policy makers, which are less addressed in studies into operations and supply chain management; however, contemporary changes in these domains, such as the application of artificial intelligence, achieving sustainability across product life-cycles and designing resilient manufacturing systems, have also implications for regulatory and legislative options in policies, which could be covered by literature reviews. These three groups —scholars, practitioners and policy makers— require paying attention to readability and setting out clearly what the implications (perhaps, recommendations), are and how they are underpinned by the quality of evidence found in retrieved studies.

3 Discovering Eight Myths

Taking these notes about literature reviews in mind, we came across eight myths that authors of literature review seem to implicitly hold. These myths can be seen as abduction in the spirit of Peirce's [11] thinking, although he is infamously known for changing his mind on this way of reasoning. However, as Dekkers [12: 61–2] states facts have to be sought that either confirm or refute conjectures reached by abduction, in our case the myths. We did so by reviewing evidence in the 39 publications (see Table 1). Let us not dwell further on the methodology but look at the eight myths we found, how they are evidenced and how they inform suggestions to prevent them for occurring when conducting and writing literature reviews.

3.1 Myth 1: I (We) Can Do This, Too

Whereas other works may inspire academics, such does not mean that their approaches should be copied without considering appropriate guidance; this observation leads tp to the first myth: **I (we) can do this, too**. In the case of our special issue, some submissions were obviously taking both topics and approaches from already published reviews. In this vein, for methodological issues they referred to these published reviews rather than consulting guidance and considering best practice; sometimes, this was evident from similar phrasing, graphs and tabulations. And, some others did not mention at all guidance for the production of literature reviews. It is not so much that authors get inspired by others or adapt approaches deployed by others, but it is like any other scholarly undertaking, works and approaches should be considered on their merits to the topic of a review; particularly on methodological issues in other published reviews caution should be exerted.

Based on our experiences, this myth can be avoided by authors forming analytical review questions and designing an appropriate review methodology based on guidance (the latter opposed to mimicking other published reviews). First, it implies is that descriptive reviews should be avoided; these are sometimes called 'no-findings reports' and 'topical surveys' [9: 395]. Often this relates to how review questions are formed. Guidance for reviews on topics related to operations and supply chain management can be taken from formats in healthcare, such as population-intervention-outcome (PIO), its variants, and other formats; see Dekkers et al. [13: 124–4]. Also, some specific formats have been formulated, for example, context-intervention-mechanism-outcome [14: 4]; for instance, this can be applied to reviews into aspects of implementing the concept of lean production. Second, search strategies need to follow guidance with regard to databases used. Normally, there should be at least two, according to Green et al. [3:

107], and these should not be specific to publisher; see Dekkers et al. [13: 537–40] for an overview of suitable (generic) search engines and databases. Moreover, all studies should be included that are relevant to the topic. The inclusion of journal rankings or citation rates as indication of quality is not advised since they do not reflect the quality of an individual publication or why it is cited. Third, there are many approaches to the analysis and syntheses of literature reviews. Particularly, in the domains of operations and supply chain management, content analysis, e.g., [5: 409 ff.] [15], and meta-analysis (for instance, [5: 409 ff.][16]) has been propagated for protocol-driven literature reviews, but beyond these indicated there are many approaches to analysis and synthesis. By way of illustration, Kidd et al. [17: 362] provide an overview of 20 methods for qualitative analysis and synthesis, of which content analysis is one. Thus, conducting, writing and publishing a protocol-driven literature review requires setting analytical review questions, adopting an appropriate search strategy and using an adequate method for analysis, to which for all guidance is available.

3.2 Myth 2: If I (We) Conduct a Literature Review for a Study, then It Can be Published

A second myth is that if for a graduate, doctoral study or research project a literature review was undertaken before an empirical study, then it is worth to publish; resulting in the myth: **if I (we) write a literature review for a study, then it can be published, too**. Often, such reviews either add little to scholarly knowledge or are propositional in nature or do not always connect a model, framework, etc. to detailed analysis of literature; this can also mean that literature is not investigated sufficiently in-depth but that a review merely provides an overview for the justification of the propositional nature of the submissions. Such manuscripts derived from literature reviews for specific studies or research projects are less likely to be used by others.

Going back to why a literature review is worth to be published in a journal, there are multiple reasons, from which three are discussed here. One is to demonstrate advances in scholarly thought or its application. A second reason is that the review contains information that reaches beyond using it for a single study. A third reason could be synthesising scholarly knowledge for both academics and practitioners. However, the manuscripts that gravitate towards being propositional did not pass the litmus test. This is commonly called the 'so what' test for manuscripts, see Dekkers et al. [13: 504–5]. This also leads to guidance on this matter. There should be focus to review questions and analysis followed by synthesis into frameworks, models, etc. that organise constructs, factors at levels of aggregation and their relationships.

Thus, when conducting and writing literature reviews for submissions to journals, not only appropriate review questions should be considered, but also the analytical approach so that other studies can build on them. It is often the selection of an appropriate method that seems to be lacking in protocol-driven literature reviews. Setting aside manuscripts using bibliometric analysis, content analysis and meta-analysis, authors place more emphasis on the protocol than the actual analysis and synthesis.

3.3 Myth 3: The Protocol is Everything

This brings about the third myth: **the protocol is everything**, which implies that there is an overemphasis on the protocol relatively to the content. This is more likely to happen in case of bibliometric analysis of any kind (bibliometric analysis is a tool for scientometrics as a discipline). However, none of the submissions containing bibliometric analysis managed to establish a meaningful synthesis of scholarly knowledge that would meaningfully inform other studies on the topic. An instance is the popular topic of Industry 4.0, where such analysis leads to how attention for specific technologies shifted over the years; but it does not provide guidance for scholars in terms of content that should be investigated in further studies. Another example is a submission that devoted 6 out of 18 pages of text to the description of the methodology; with two pages introducing the review, it only leaves just over half for the presentation of results, drawing of inferences and findings, and their discussion and implications. This concern —methodology takes over from interpretation of scholarly knowledge— is what MacLure [18] voices, albeit her writing is directed at educational research. To some degree valid for the domains of operations and supply chain management, too, is that the evaluation and synthesis of scholarly knowledge from the viewpoint of a specific topic is the onus of a literature review; the protocol for systematically identifying and selecting studies is only part of the approach to systematic literature reviews but insufficient to ensure a contribution to knowledge.

To steer clear of this myth, three points should be considered. First, in the writing of the review it should be clear that the focus is on analysis and synthesis and that a protocol is only a way for identifying relevant studies. Second, when reporting common descriptive statistics can be omitted. Often these descriptive do not lead to conjectures that inform or support later findings. To give an example, in most published reviews there is reporting of the journals in which studies are found, but stating journals that contain the most studies for the review does not lead to any meaningful insight; separate from the fact that a ranking of journals does not indicate the quality of individual studies. A specific illustration how to make use of which journals studies are published in is found in Dekkers et al. [19: 9], when they demonstrate that the make-or-buy decision got a more strategic focus considering a 47-year period. This leads to the third point; it is almost always possible to submit a supplementary file that not only contains the extracted data but also descriptive statistics that are of less interest for the story in the manuscript. An example of such a data file is Dekkers and Salgado [20] used for Salgado and Dekkers [21]. In such supplementary files more detailed information about the search strategy and inclusion criteria can be provided. In sum, even though a protocol may contribute to transparency it should not lead to overtaking analysis and evaluation of studies considered in a review; the suggested supplementary file with descriptive statistics could be an efficient way, in the sense of using text in a manuscript, and effective way, for the purposes of transparency and accessibility.

Despite the emphasis being on protocol-driven literature reviews, that is only two narrative reviews were submitted, all systematic (literature) reviews did not consider the quality of evidence. Some did consider ranking of the journals they were published in but as deliberated on before it cannot be considered a reliably proxy for the quality of evidence. In healthcare, it is common to use the hierarchy of evidence pyramid, e.g., [13:

211–3]), which uses the study design as point of departure. However, in operations and supply chain management such an acceptable measure does not exist, perhaps related to the methodological variety, diversity in units of analysis and bias by sampling. Notwithstanding this variety, the credibility or trustworthiness of individuals studies, including how evidence informs findings should be assessed. To this purpose, Dekkers et al. [13: 220–3] provide some suggestions. Sometimes, it is possible to make remarks about the quality of evidence in the writing; an example would be: 'author names assert that..., though weakly related to the interpretation of evidence in their study.' Such remarks also help readers to interpret writings about other studies and whether they should take these points at face value. Since reviews aim to direct others in research or practice, the expression of the quality of evidence in and across studies is an inherent feature of writing them.

3.4 Myth 4: Identifying Gaps is its Only Purpose

There were also authors who held in mind the Swiss cheese model, i.e. they were mostly focused on identifying gaps in the literature, leading to the fourth myth: **identifying gaps is its only purpose**. Particularly, this can be associated with reviews that aim at providing overviews rather than seeking analytical rigour and considering trustworthiness for synthesis. There is a catch here from two perspectives. The first one is that identifying a gap does not necessarily mean that studies to cover it will yield insight. By way of illustration, say that a gap is identified but there are numerous other studies. If the scholarly knowledge embedded in those other studies could predict the outcome of a study into the identified gap, then outcomes of such a study would be self-evident. A second perspective is that from how a gap can be identified if there is no overarching conceptualisation or model; guidance to this matter can be found in [13: 126–32]. This means that gaps can only arise from either an overarching model or inconsistencies and weak evidence for conjectures from a review.

3.5 Myth 5: Quantification Yields Insight

Observing a notable number of submissions resorting to quantitative analysis, ranging from bibliometric analysis to content analysis to meta-analysis, yielded the fifth myth: **quantification yields insight**. The three types of manuscripts that were submitted had their own peculiarities and will be dealt with separately. However, a common theme is that they seem to poorly capture that drawing inferences differs from obtaining insight. In this sense, drawing inferences is about statements directly related to results; as an illustration serves the increased number of publications by year on a topic, which generally is noted by authors. To turn it into insight it needs to be coupled with increasing numbers of journals, issues by year, etc.; this is done for outsourcing [19: 7] and results in the notion that a 9-fold increase in available publication outlets over 47 years mostly explains the increased interest by researchers. Thus, in-depth analysis and further investigating the causes of conjectures may lead to insight that is informative.

A popular category of protocol-driven literature reviews is systematic reviews with quantitative content analysis, given the submissions to the special issue; however, using this approach comes along with some particular points to be considered. A first concern is that simply noting frequency does not yield insight. For example, combinations of principles for lean thinking [20: 910] investigated in studies into lean product development yield more insight than looking at the individual frequencies of the principles. As a second point it is worthwhile to connect the individual classifications to other classifications. For instance, the type of literature reviews connected to types of studies [19: 19] shows that narrative overviews, i.e. inexhaustive literature reviews, dominate for studies on outsourcing and propositional papers tend to have poor literature reviews. As a third consideration, these studies tend to have poor writing styles, see also the sixth and seventh myths. Even though (quantitative) content analysis may be an attractive option as propagated by some such as [5], it only yields worthy outcomes when conducting in-depth analysis and narrative of the retrieved studies.

When reviews opted for bibliometric analysis, they tended to remain descriptive. This is due to the nature of data used in this type of analysis, i.e. bibliographical data and keywords. Particularly, the use of keywords requires caution. First, not all keywords are author-generated as many seem to assume. Some journals have preset keywords authors need to select from, other journals allow authors to define their keywords and some other journals allow a mix of these. And, when authors of retrieved studies filled them out they could have opted for keywords that cover methodologies, theories and topics; such diversity complicates bibliometric analysis. Furthermore, the number of keywords may vary, thus not weighting individual publications equally. Hence, for this type of analysis either the ones undertaking a review must assign keywords in a coherent manner while doing justice to individua publications or keywords need to be cleared before any analysis takes place. Second, to be meaningful to synthesising scholarly knowledge and advancing research agendas there should be some kind of qualitative analysis. Only with the correct data and an in-depth analysis, these bibliometric studies will make sense for those active in operations and supply chain management.

Looking at systematic reviews with meta-analysis, authors had not only mimicked other studies but poorly understood what it takes to interpret outcomes. A particular weak point, often not addressed, are the implications of statistical between-study heterogeneity. When these show a high degree of heterogeneity, the outcomes of a meta-analysis are unreliable and subgroup analysis should be explored; but when the subgroup analysis also yields high degrees of statistical between-study heterogeneity, then the only options are further to refine subgroups or define other sets of subgroups. Most likely though, there are also variances in sampling and methodologies, particularly how constructs are determined, that could play a role or factors that were not included in the initial analysis have a greater than expected influence. A published study in which this poor understanding of statistical between-study heterogeneity is the systematic review with meta-analysis on sustainability, supply chains and firm performance by Govindan et al. [22]. They find I^2 as indicator for heterogeneity to be between 94% and 96% [22: 101923/10]. Even their plot on the same page indicates a high degree of variety, with studies having a high value of statistical precision but yielding quite different results. This makes it impossible to investigate their formulated hypotheses, something they seem to be unaware of, despite abundant guidance on this matter. Rather they should have focused on variance in sampling, methodological variance and factors not considered in their further analysis and discussion of findings instead of misleading the reader that results are trustworthy. This points to the necessity to consult guidance and understand this well before embarking on this type of literature review.

3.6 Myth 6: Collating and Cherry-Picking Reflects Close Reading

Often in manuscripts that report literature reviews, there is a quick succession of citations-in-text from different sources, which are called collations of citations-in-text when they are just taken from retrieved studies and cherry-picking when authors select citations-in-text that are favourable to arguments made in the literature review; this observation in introductory sections and analyses of retrieved studies within the submitted manuscripts to the special issue gives way to the sixth myth: **collating and cherry-picking reflects close reading**. Perhaps, authors want to demonstrate that they have read literature well and are reluctant to add their own reasoning and evaluation. Repeatedly, collating and cherry-picking are accompanied with citations-in-text not including page numbers for quoting and referencing, which makes the traceability for readers and reviewers cumbersome. Thus, collating of citations-in-text and cherry-picking should be avoided at any time.

There are three points that authors can consider when writing text for reviews. A first point is that traditional advice for structuring paragraphs should be followed. Normally, a paragraph (following the UK English meaning) has an introductory sentence and ends with a transition (or termination) sentence; normally these do not contain citations-in-text. A transition or termination sentence contains a recap what could be learned from reading the entire paragraph. A second point is that sentences with or about citations-in-text should contain an evaluation from the viewpoint of the topic of the review, an indication of the quality of evidence in a study and the context of the study (e.g., country or region, industrial sector, order entry points [make-to-stock, assemble-to-order, make-to-order, engineer-to-order]). A third point is that the text should indicate why it is relevant to mention a next citation-in-text. This will create logic to the reader why citations-in-text and interpretations to the topic at hand follow each other. These three points —structuring paragraphs, context and evaluation, signposting transitions— alleviate most points that we noted about writing.

3.7 Myth 7: Reporting the Analysis (and Evaluation) of Literature is About Listing Studies Read

Related to the sixth myth, another observation is that authors desperately try to mention all studies that they looked at, giving the impression that some had a seventh myth in mind: **reporting the analysis (and evaluation) of literature is about listing studies read**. Enumeration results in summary citations-in-text (aka as bracketing and reference stacking), see Steenhuis et al. [23: 96] and not considering sources used on their merits and credibility for the topic at hand. Also, this hardly leads to a coherent argumentation and logic flow of thought, which is a necessity for making reviews accessible to a diversity of readers.

In addition to clear indicating quality of evidence for conjectures, findings and recommendations, reviews can avoid listing studies read by two approaches. Certainly, when the number of retrieved studies a supplementary file as earlier suggested can be used. This allows the use summary data from the supplementary file, such as tabulations and visualisations, instead of addressing each study. Notable points from selected, illustrative studies can then be used for the narrative, whereas readers can consult the supplementary file if they want more detail or form their own opinion (it also enhances transparency). A second point for guidance is to identify logic themes for the analysis, evaluation and synthesis of results, conjectures and findings. Such not only improves the readability but also increases the trustworthiness of reviews.

3.8 Myth 8: Writing Reviews is Easy

A final abduction is that submitted manuscripts lacked a finishing touch in terms of how the matter of their review is presented to audiences; this resulted in our eighth myth: **writing reviews is easy**. This was evidenced from poor structuring of the text to grammar mistakes and typos. Other particular points of interest noted are multiple but different review questions, sometimes found across manuscripts, poorly written introductions, use of collations and cherry-picking, poorly articulated findings derived from the analysing, raising points in the later parts of manuscripts that could not be related to the methodology, results, conjectures and findings, and poorly written concluding sections. The lack of evidence for appropriate writing skills is also apparent in a content analysis of reviews and decision letters reported at the EurOMA conference [24]. Advice presented so far already clarifies that appropriate conducting a literature review requires attention to detail, with some points mentioned here getting less attention than they should, and that writing and publishing needs structuring and attentiveness; when less experienced authors might consider working together with colleagues that are more proficient, consult proofreaders and in final stages employ copy-editors (electronic copy-editing services have to be treated with caution since they also introduce potential mistakes or propose rewording that is not appropriate to the scholarly writing). Thus, paying attention to coherent, accurate and efficient writing is necessary for a manuscript on a review to become publishable; together with consulting colleagues and peers, and use of professional copy-editors, taking the writing as a process should overcome most of the points noted here.

4 And, a Ninth One

Not only did we observe the eight myths in the way manuscripts were submitted, but also how revisions were dealt with.

4.1 Marginal Changes as Response to a Review of a Manuscript Will Get It Published

In this vein, we observed that revised manuscripts did initiate a ninth myth: **marginal changes as response to a review of a manuscript will get it published**. Even when fundamental concerns were raised during a review, authors would respond by changing or adding sentences here and there. Or, they would add a number of paragraphs rather than considering more fundamental flaws for rewriting a manuscript that were noted.

Moreover, some authors were defiant towards to the serious comments raised during the review process. An example of the latter is standing by review questions that are descriptive or not informative to the audience of the journal. Another instance was not incorporating advice for the writing of the text, ultimately leading to rejection. Certainly, major revisions indicate that some serious thought by authors is necessary how they can substantially revise the manuscripts based on its reviews.

4.2 How to Respond to Reviews and Decision Letters for Manuscripts

A golden rule for revision is that it should be made as easy as possible for (guest) editors and reviewers to see what has been done. However, authors should avoid saying that they made substantial changes when these were only marginal. And, best is to address each comment made by (guest) editors and reviewers. When doing so, the detailed changes made should be listed in a response letter in addition to highlighting or marking it in the revised manuscript. This supports reviewers in a next round how and where comments were taken onboard. In the case of rebutting, it still may be worthwhile to amend text because somehow the text triggered a point raised by reviewers and could be indicative of more serious concerns that are not explicitly addressed. A few times, authors simply said: 'thank you for the comment, we have addressed this point in the revised text'; such a brief explanation is not very helpful for reviewers to get an idea what has been done. So, at all times, clarify changes in a revision to the (guest) editors and reviewers in sufficient detail and in an accessible manner.

5 With Some Afterthoughts

After reviewing the evidence for the nine myths, there are three main concerns for those that intend to conduct, write and publish a literature review. The first is that conducting a literature review should be based on a single. Interesting review question, perhaps with corollaries, but broader than suitable for one study or a limited number of specific studies. The second is that there is sufficient guidance available which should be consulted when the designing the methodology of the review. Even when inspired by other works this guidance should be looked at since other works have not necessarily followed best practice. Perhaps a detailed comment but nonetheless evidence of engagement with guidance is that PRISMA is not a method, unlike what many write, but a reporting standard; see Liberati et al. [26]. A third concern is that writing of reviews should be done in a coherent way following practices about writing that have been well-known for a long time. This may be a skill that needs to be acquired, although support by more proficient authors, proofreaders and copy-editors could accelerate this path.

5.1 Taking a Step Back

Notwithstanding these comments and critical thoughts about conducting and writing up literature reviews, there were some pearls in the submissions. There was a submission with a relatively new approach to literature reviews, i.e. morphological analysis as a form of content analysis, but also an attempt to synthesise case studies because of their rich information; the latter does justice to Yin and Heald's [25] call for the case survey method (although they seemed to have had a positivist approach in mind). Other submissions took a more canonical approach to the literature reviews, but some presented really interesting topics. Notwithstanding, in early stages we noted the generic points across the submissions that led us to formulating the nine myths.

5.2 A Dose of Reflexivity

Perhaps, our thoughts on the myths presented here were shaped during the regular meetings we held to discuss submitted manuscripts we received and decisions about progressing manuscripts. When starting to receive submissions to the special issue, we started to have meetings every two to three weeks. During these meetings there was extensive discussion on whether topics of submissions were of potential interest, which manuscripts should be taken into review and which ones would undergo an editorial review. Since we sought to seek consensus, this way of considering submitted manuscripts may have shaped our views expressed in this paper.

Related to the myths, we started to a more developmental view on the submissions, too, across the stages of the special issue. Early submissions to the special issue already made it clear to us that there were topics of interest, where both the protocol and analysis lacked depth, and the writing needed considerable improvement. This was evidenced that after the first round of reviews, only rejections and major revisions were the decisions we took. In subsequent interactions with authors —some asked for clarification and some were offered a virtual meeting— there were also indications of what they saw as hurdles and challenges to producing manuscripts. Mostly, these interactions with authors aiming at ensuring that comments by reviews were understood and aiming at reducing the cycles of revision confirmed our myths.

Another influence was the manuscript of a book that became available during the process of considering manuscripts for the special issue. In advance of submissions to the special issue Dekkers et al. [13] was made available as draft to the guest editors in 2021, while in August 2022 it became available as published book. The book offers a few departures from other publications pertinent to literature reviews; for example, it stresses the need for more analytical review questions, the importance of evidencing saturation achieved through the search, the quality of evidence and adequate writing. Based on the book, webinars were provided to universities in which the other guest editors participated, too; instances are a presentation to doctoral and postgraduate students at Cardiff University in March 2022, and to academics in the Newcastle's universities in January 2023 (the latter included our observations as guest editors). Furthermore, the book was informed by writings and guidance drawn from a broad range of disciplines, including contributors to sections and chapters. Correspondence and exchange of thoughts found their way not only in the book but also in the way of thinking about approaches and methods for literature reviews. Undoubtfully, these interactions with contributors and scholars during webinars about the book also found their way to our thoughts, implicitly and explicitly.

5.3 And, What Does It all Add up to?

The myths and the related advice presented here to avoid them brings to the fore that publishing literature reviews should not be taken lightly; it does not mean that we want to discourage authors. First, there is abundant guidance to conducting and writing literature reviews from which only a selection has been presented here. Reading multiple sources on this guidance will certainly help to get a better grip on what should and could be done. Second, in this paper there are recommendations for both conducting and writing literature reviews that are not or poorly addressed in existing guidance for reviews into operations and supply chain management. An example is considering the quality of evidence, which is not mentioned in common writings on literature reviews such as [4–6]. Also, the necessity of appropriately writing text here is mentioned. The latter should not be equated with proficiency in English. For example, this fragment of a sentence does not make logically sense: '... With particular attention to two main new players that have entered the global arena in the recent years, namely digital platforms and mobile applications.' This means that authors should consider what they intend to write and the literal meaning of what they wrote down. Creating manuscripts on literature reviews means from getting it right for the topic and methodology to putting it on paper, keeping readers in mind.

5.4 A Final Thought

In this vein, it was once said that writing a poor manuscript takes as much time as writing a good one. However, some of our myths indicate that insufficient time was taken to map out how a literature review should be conducted and that authors should pay attention to crafting manuscripts. Setting aside this point, our case of the special issue points out that writing a poor manuscript takes more time, somewhat contradicting the saying. When submitted a poor manuscript may be rejected or granted a major revision, hopefully leading to authors rewriting their manuscripts, which will consume additional time. And poorly written manuscripts are likely to be subjected to more revision cycles than well-written ones. Consequently, this also impacts editors and reviewers, since they have to spend more time on administering and reviewing manuscripts. Thus, taking in guidance provided here to present more coherent, analytical literature reviews that are informative and pleasant to read is not only beneficial for authors but also a service to efforts by editors of journals and reviewers when considering publication of manuscripts.

Acknowledgements. We are grateful for the support we received from Harm-Jan Steenhuis, the editor-in-chief for the *Journal of Manufacturing Technology Management*, and a broad panel of reviewers that devoted time to reviewing manuscripts.

References

1. Maier, H.R.: What constitutes a good literature review and why does its quality matter?. Environ. Model. Softw. 43, 3–4 (2013). https://doi.org/10.1016/j.envsoft.2013.02.004
2. Cronin, P., Ryan, F., Coughlan, M.: Undertaking a literature review: a step-by-step approach. Br. J. Nurs. **17**(1), 38–43 (2008). https://doi.org/10.12968/bjon.2008.17.1.28059

3. Green, B.N., Johnson, C.D., Adams, A.: Writing narrative literature reviews for peer-reviewed journals: secrets of the trade. J. Chiropr. Med. **5**(3), 101–117 (2006). https://doi.org/10.1016/S0899-3467(07)60142-6
4. Tranfield, D., Denyer, D., Smart, P.: Towards a methodology for developing evidence-informed management knowledge by means of systematic review. Br. J. Manag. **14**(3), 207–222 (2003). https://doi.org/10.1111/1467-8551.00375
5. Thomé, A.M.T., Scavarda, L.F., Scavarda, A.J.: Conducting systematic literature review in operations management. Prod. Plann. Contr. **27**(5), 408–420 (2016). https://doi.org/10.1080/09537287.2015.1129464
6. Durach, C.F., Kembro, J., Wieland, A.: A new paradigm for systematic literature reviews in supply chain management. J. Supply Chain Manag. **53**(4), 67–85 (2017). https://doi.org/10.1111/jscm.12145
7. Hochrein, S., Glock, C.H., Bogaschewsky, R., Heider, M.: Literature reviews in supply chain management: a tertiary study. Manage. Rev. Q. **65**(4), 239–280 (2015). https://doi.org/10.1007/s11301-015-0113-4
8. Bergdahl, E.: Is meta-synthesis turning rich descriptions into thin reductions? a criticism of meta-aggregation as a form of qualitative synthesis. Nurs. Inq. **26**(1), e12273 (2019). https://doi.org/10.1111/nin.12273
9. Bergdahl, E.: Meta-Summary: Aggregating Findings and Preparing for Meta-Synthesis. In: Dekkers, R., Carey, L., Langhorne, P. (eds.) Making Literature Reviews Work: A Multidisciplinary Guide to Systematic Approaches, pp. 390–396. Springer, Cham (2022)
10. Sandelowski, M., Barroso, J.: Handbook for Synthesizing Qualitative Research. Springer, New York (2006)
11. Peirce, C.S.: On the natural classification of arguments. Proc. Am. Academy Arts Sci. **7**, 261–287 (1867)
12. Dekkers, R.: Applied Systems Theory, 2nd, edition Springer, Cham (2017)
13. Dekkers, R., Carey, L., Langhorne, P.: Making Literature Reviews Work: A Multidisciplinary Guide to Systematic Approaches. Springer, Cham (2022)
14. Denyer, D., Tranfield, D., van Aken, J.E.: Developing design propositions through research synthesis. Organ. Stud. **29**(3), 393–413 (2008). https://doi.org/10.1177/0170840607088020
15. Seuring, S., Gold, S.: Conducting content-analysis based literature reviews in supply chain management. Supply Chain Manage. Int. J. **17**(5), 544–555 (2012). https://doi.org/10.1108/13598541211258609
16. Forza, C., Di Nuzzo, F.: Meta-analysis applied to operations management: summarizing the results of empirical research. Int. J. Prod. Res. **36**(3), 837–861 (1998). https://doi.org/10.1080/002075498193714
17. Kidd, L., Dekkers, R., Carey, L., Langhorne, P.: Principles of Qualitative Synthesis. In: Dekkers, R., Carey, L., Langhorne, P. (eds.) Making Literature Reviews Work: A Multidisciplinary Guide to Systematic Approaches, pp. 353–387. Springer, Cham (2022)
18. MacLure, M.: 'Clarity bordering on stupidity': where's the quality in systematic review? J. Educ. Policy **20**(4), 393–416 (2005). https://doi.org/10.1080/02680930500131801
19. Dekkers, R., Barlow, A., Chaudhuri, A., Saranga, H.: theory informing decision-making on outsourcing: a review of four 'five- year' snapshots Spanning 47 Years [Online]. Glasgow: Adam Smith Business School (2020). https://ssrn.com/abstract=3691983
20. Dekkers, R., Salgado, E.G.: Data for Systematic Literature Review 'Lean Product Development.' University of Glasgow, Glasgow (2017)
21. Salgado, E.G., Dekkers, R.: Lean product development: nothing new under the sun? Int. J. Manag. Rev. **20**(4), 903–933 (2018). https://doi.org/10.1111/ijmr.12169
22. Govindan, K., Rajeev, A., Padhi, S.S., Pati, R.K.: Supply chain sustainability and performance of firms: a meta-analysis of the literature. Trans. Res. Part E: Logistics Transp. Rev. **137**, 101923 (2020). https://doi.org/10.1016/j.tre.2020.101923

23. Steenhuis, H.-J., Dekkers, R., Carey, L., Langhorne, P.: Quality of Literature Reviews. In: Dekkers, R., Carey, L., Langhorne, P. (eds.) Making Literature Reviews Work: A Multidisciplinary Guide to Systematic Approaches, pp. 57–105. Springer, Cham (2022)
24. Dekkers, R., Found, P., Cheng, Y.: How to Publish Literature Reviews in Supply Chain and Operations Management: Lessons for Authors from a Special Issue in JMTM. Paper presented at 30th EurOMA Conference, Leuven, 3–6 July (2023)
25. Yin, R.K., Heald, K.A.: Using the case survey method to analyze policy studies. Adm. Sci. Q. **20**(3), 371–381 (1975). https://doi.org/10.2307/2391997
26. Liberati, A. et al. The PRISMA statement for reporting systematic reviews and meta-analyses of studies that evaluate health care interventions: explanation and elaboration. J. Clin. Epidemiolo.**62**(10), e1–e34 (2009). https://doi.org/10.1016/j.jclinepi.2009.06.006

A Study on On-The-Job Training (OJT) Operator Allocation Considering Fatigue in Cell Manufacturing System

Moe Endo and Harumi Haraguchi(✉)

Graduate School of Science and Technology, Ibaraki University, 4-12-1, Nakanarusawa, Hitachi, Ibaraki, Japan
harumi.haraguchi.ie@vc.ibaraki.ac.jp

Abstract. In labour-intensive cellular production systems, it is important to train operators efficiently because productivity depends on operator skills. In our previous study, we proposed a 'Skill Index' to classify operator skills based on the time required for each task and used this method to allocate operators with a primary focus on training. The 'Skill Index' uses a learning curve to estimate the time required. However, in actual workplaces, operators are expected to accumulate fatigue caused by repetitive tasks, which affects their execution time. Therefore, in this study, a fatigue-learning curve was employed. We proposed an operator allocation that includes fatigue and compared it with computer experiments in our previous study. As a result, it is possible to represent the decrease in proficiency due to fatigue from repetitive tasks and to include fatigue in the design phase. To validate the proposed method, we conducted an experiment to assemble LEGO robots with the operator allocation from the computer experiment and analysed the assembly time for each operator. The results showed that the Skill Index Values were the same for more than half of the tasks in the computer experiment and the assembly experiment, indicating that the proposed method is effective. Efficient operator training is critical in labour-intensive cellular production systems. Our previous study proposed a 'skill index' to classify operator skills based on the time required for each task and used this method to allocate operators for focused training. However, fatigue resulting from repetitive tasks affects operator execution time. Therefore, we proposed an operator allocation method that considers fatigue using a new fatigue learning model and compared it with computer experiment in a previous study. As a result, we proposed a new proficiency curve to capture how fatigue from repetitive tasks reduces operator performance. We conducted an assembly experiment using LEGO robots with the operator allocation obtained from the computer experiment. The results showed that the skill index values were consistent for more than half of the tasks in both experiments, indicating the effectiveness of our proposed method.

Keywords: Human factor · Operator training · Cellular manufacturing system · Operator allocation · fatigue-dependent learning rate

M. Dragomir et al. (Eds.): ICPR 2023, LNPE, pp. 107–124, 2025.
https://doi.org/10.1007/978-3-031-92082-0_11

1 Introduction

In labor-intensive cellular production system, it is necessary to understand the operator skills and provide effective skills training accordingly because productivity and quality are highly dependent on the operator skills.

In manufacturing plants, use ‘Skill Map’, which is skill management charts, to clarify the skills required of operators and confirm and visualize their skills at the present time. Haraguchi et al. (2015) applied the Skill Map and proposed a ‘Skill Index’ to classify operators’ skills based on the execution time for each task. As operators perform the same task repeatedly, they become more proficient, and the time required decreases. The Skill Index estimates the time required to complete a task by simulating proficiency, thereby updating the skill level.

However, in actual workplaces, it is found that the factors affecting the execution time are not only related to proficiency, but also to other factors related to the operators’ ability and the work environment (Abubakar and Wang 2018; Karwowski et al. 2019). These factors include age, experience, work design, and the effects of forgetting due to differences in the length of breaks separating work cycles (Abubakar and Wang 2019; Neumann et al. 2016; Sikström and Jaber 2012). Furthermore, human factors such as fatigue have been found to affect the speed of learning (Daria et al. 2017). From the above, it can be said that not only proficiency, but also human factors need to be considered to accurately predict the operators’ skills. Asadayoobi et al. (2021), Mályusz and Pém (2014) and Digiesi et al. (2004) proposed a learning model that considered fatigue.

In this study, we perform a simulation that more closely resembles actual workplaces on task-divided multi-operator cellular production systems. Therefore, we focus on the influence of human factors. Among various human factors such as forgetting and fatigue, this study focuses on the influence of fatigue. The model proposed by Mályusz and Pém (2014) is incorporated into the conventional method (Haraguchi et al. 2015) to propose an operator allocation that considered the influence of fatigue. Then, the simulation results of the conventional method (Haraguchi et al. 2015) and the proposed method are compared. We also verify the effectiveness of the proposed method by conducting experiments to assemble LEGO blocks with the operator allocation obtained by the proposed method and comparing the assembly times. In the following, we present in Sect. 2 a literature review of the learning curve and the fatigue-learning curve. In Sect. 3, we present the Skill Index, which is a measure of operators’ skills, and the fatigue-learning model used. In Sect. 4, we present the conditions and formulations of the target model, and in Sect. 5, we perform computer simulations to obtain the operator allocation. In Sect. 6, we show the practicality of the proposed method by means of assembly experiments using LEGO blocks. Finally, Sect. 7 contains the conclusion and future study.

2 Literature Review

Manual assembly operators remain a necessity in many manufacturing industries (Fantini et al. 2016). They provide flexibility to the assembly line that is not possible with a self-directed system (Hoedt et al. 2020). Diversifying consumer needs are driving manufacturers to shift from high-volume, low-volume production to high-mix, low-volume production (Hasegawa et al. 2009; Gan et al. 2023; Abu-Samah et al. 2017; Alduaij et al. 2020). Therefore, operators are expected to perform a wide range of tasks, and it is important to understand the operators' skills and to train them.

One of the most important indicators of an operators' experience is the time required for a task (Hoedt et al. 2020). In general, as the same task is repeated, the time required for that task decreases. This phenomenon is called 'learning' (Morooka 1968). The learning curve is a graphical representation of an individual's performance (time or cost) per repetitive task. This form of learning is called 'autonomous learning' or 'learning by doing' (Adler and Clark 1991; Glock et al. 2019; Asadayoobi et al. 2021). Learning curves can be used to predict the time required for an operators' task and are typically used to optimize production planning and operator allocation.

2.1 Learning Curve

The learning curve has been studied considerably since it was first proposed by Wright (1936). The model proposed by Wright (1936), also called the "power model," represents a constant percentage decrease in the execution time of a task for each doubling of its cumulative output (Globerson and Gold 1997). The behaviour of the model is shown in Eq. (1).

$$T_n = T_1/n^{\alpha} \tag{1}$$

where T_n is the execution time of the nth unit, T_1 is the execution time of the first unit, α is the learning exponent, and β is the learning rate, with the relationship $\alpha = -\log_2\beta$. The learning curve proposed by Wright has become a well-known model due to its simplicity and ability to fit a wide range of training data well (Yelle 1979; Badiru 1992; Jaber 2006). Haraguchi et al. (2015) estimated Skill Index values by using this Wright learning curve. Since the studies of Wright (1936) and Hovland (1951), various learning curves have been developed, including exponential, hyperbolic, and S-shaped curves (Jaber et al. 2013; Anzanello and Fogliatto 2011). Several two- and three-parameter exponential function models have been developed (Knecht 1974; Mazur and Hastie 1978; Towill 1990; Naim and Towill 1990). The exponential model provides a more accurate estimate of the production rate than the log-linear model (Anzanello and Fogliatto 2011; Nembhard and Uzumeri 2000a). Anzanello and Fogliatto (2007) also showed that hyperbolic models are more robust than three-parameter exponential and time constant models (Anzanello and Fogliatto 2011). The S-curve reflects three behaviors of performance. The learning curve is slow at the beginning, steadily improves in the middle, and slows again when proficiency is reached (Peltokorpi and Jaber 2020). A review of learning curves can be found in Yelle (1979), Dutton et al. (1984) and Dar-El (2000).

2.2 Fatigue-Learning Curve

When repetitive tasks become extremely large, Wright's learning curve indicates zero performance (Anzanello and Fogliatto 2011; Hurley 1996; Eden et al. 1998). To solve this problem, a plateau model has been developed that includes a constant term in Wright's model (Anzanello and Fogliatto 2011). This new model reaches a steady state when the iterations become very large. In practice, however, the decrease in physical and cognitive resources that accompanies an increase in the number of tasks results in fatigue, which has been shown to negatively affect the progress of learning (Daria et al. 2017). Accumulated fatigue can lead to a decline in operators' judgment, decision-making, and reaction time (Hartzler 2014; Langner et al. 2010). Furthermore, it has been shown that low levels of fatigue improve operator performance, while high levels of fatigue negatively affect operator performance (Bahdur et al. 2019; Hillman et al. 2003). Examples of tasks with high levels of fatigue include heavy-load tasks and tasks with long cycles (Chang et al. 2012). Winkelhaus et al. (2018) found that the performance of most operators, affected by effects of learning and fatigue, improved up to a certain threshold but deteriorated beyond it. This pattern of performance suggests a U-shaped learning curve (Asadayoobi et al. 2021). Therefore, the model which plateaus as iterations become larger does not accurately imitate the actual workplaces.

Available learning models that considered fatigue have been proposed by Asadayoobi et al. (2021), Mályusz and Pém (2014), and Digiesi et al. (2009). Asadayoobi et al. (2021) proposed a learning exponent that considered fatigue to investigate the combined effects of fatigue and learning on individual performance. This model's methodology incorporates the influence of fatigue accumulation in the learning exponent by setting a fatigue exponent for each task. Mályusz and Pém (2014) proposed a learning model with fatigue coefficients for repetitive tasks in the construction industry. This model's methodology estimates the time required for each task by means of a fatigue coefficient and the number of iterations. Also, this model is U-shaped. Digiesi et al. (2004) proposed a model that considered the combined effects of both learning and fatigue phenomenon to evaluate human execution time in repetitive manual tasks. This model's methodology includes two more parameters than the two learning models that considered fatigue described above, and these parameters are difficult to estimate (Asadayoobi et al. 2021).

In this study, we estimate Skill Index values using the fatigue-learning curve proposed by Mályusz and Pém (2014).

3 Operator Skill Indicators

3.1 Skill Index

In this study, the 'Skill Index' proposed by Haraguchi et al. (2015) is used to describe the skills possessed by operators. If the standard execution time for each task k of product j is ST_{jk} and the actual execution time for each operator i is ET_{ijk}, the Skill Index value S_{ijk} is obtained by Eq. (2).

$$S_{ijk} = ST_{jk}/ET_{ijk} \tag{2}$$

For example, if the standard execution time ST_{13} of task 3 of product 1 is 90 and the actual execution time ET_{213} of operator 2 is 100, the Skill Index value S_{213} is 0.9. A smaller value of the skill index indicates that the operator is not good at the task, while a larger value indicates that the operator is good at the task. Using the Skills Index, classify the operators' task-specific skills as newcomer, standard, or expert. Newcomers have $0.6 \leq S_{ijk} < 0.9$, standards have $0.9 \leq S_{ijk} < 1.3$, and experts have $1.3 \leq S_{ijk} \leq 1.5$.

The skill level of the operator is expressed as the average value of the Skill Index for all tasks for which the operator is responsible. While the reduction in execution time due to learning is theoretically unlimited, actual human skills are limited. Therefore, in this study, the maximum Skill Index is set at 1.5 and the minimum Skill Index is set at 0.6.

3.2 Fatigue-Learning Model

In this study, we adopted the fatigue-learning model proposed by Mályusz and Pém (2014). The model of Mályusz and Pém (2014) is shown in Eq. (3).

$$T_n = T_1 \lambda n^{-\alpha} + T_1(1-\lambda)(k+1-n)^{-\alpha} \tag{3}$$

where k is the total number of cycles and λ is a real number between 0 and 1. If $\lambda = 1$, the original learning curve is obtained. A larger value of the fatigue coefficient λ indicates a slower accumulation of fatigue, while a smaller value of λ indicates a faster accumulation of fatigue. The fatigue-learning curve represented by Eq. (3) and the learning curve represented by Eq. (1) are shown in Fig. 1. In Fig. 1, the learning rate is 90%, and the fatigue coefficient λ in Eq. (3) is 0.80.

Figure 1 shows that Wright's model in Eq. (1) approaches zero as the number of iterations increases. On the other hand, Mályusz and Pém (2014) model in Eq. (3) properly represents the effect of increased execution time due to fatigue at the end of the number of repetitions.

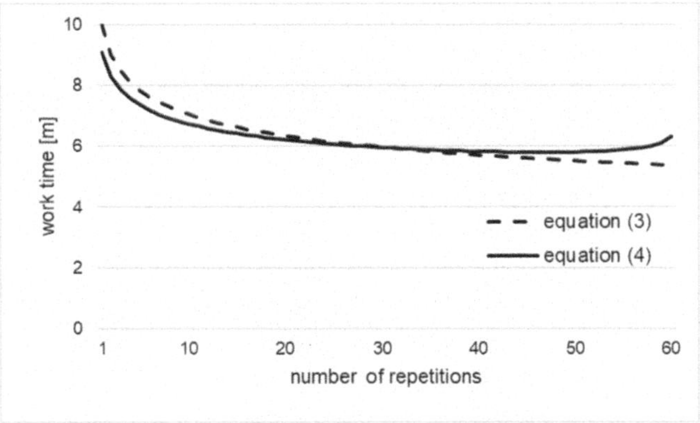

Fig. 1. Comparison of fatigue and learning curves

4 The Target Model

4.1 Conditions

The target model is a labor-intensive cellular production system in which operators' skills play a major role in productivity. The features are as follows.

(1) The same order consisting of multiple products is received repeatedly.
(2) Products are assembled in order of arrival.
(3) Each product is completed assembled in a single cell.
(4) The cell facilities can handle both single-operator and multi-operator cells, and operators are always allocated to one cell.
(5) In multi-operator cells, a task-divided cell configuration is adopted, and the task is allocated to one dedicated operator, meaning only one operator is allocated to one task.
(6) If an operator is responsible for multiple tasks, the tasks should be performed consecutively.
(7) This is a job shop type cell where the processing order of tasks varies for each product.

4.2 Symbols

Table 1 shows the symbols used to formulate the conditions in Sect. 4.1.

Table 1. Symbols

Variables	
i	The operator number (1, 2, …, I)
c	The cell number (1, 2, …, C)
j	The product number (1, 2, …, J)
k	The task number (1, 2, …, K)
t	The order number (1, 2, …, T)
S_{ijk}	The Skill Index values for each task of each operator
x_{itc}	The decision variable represents who to assign to which cell
y_{itjk}	The decision variable represents who to assign to which task
C_{max}	The maximum number of operators that can be allocated to a single cell
K_{max}	The maximum number of tasks that can be allocated to a single operator
F_{itjk}	The first task number allocated to the operator
E_{itjk}	The last task number allocated to the operator
Q_{tjall}	The order quantity of product j in order t
Q_{tcj}	The order quantity in cell c of product j in order t
$TaktTime_{tcj}$	The takt time in cell c of product j in order t

4.3 Formularizations

The purpose of this study is to achieve efficient operator allocation in On-the-Job Training (OJT). To achieve this, tasks that operators are not good at will be allocated to them preferentially, aiming to improve their skill proficiency. Therefore, the objective function is to minimize the sum of Skill Index for each task allocated to operators. The objective function is shown in Eq. (4), where S_{ijk} is the Skill Index values for each task of each operator, the decision variable x_{itc} represents who is assigned to which cell, and y_{itjk} represents who is assigned to which task.

$$min \sum_{i=1}^{I} \sum_{c=1}^{C} \sum_{j=1}^{J} \sum_{k=1}^{K} S_{ijk} x_{itc} y_{itjk} \ (\forall_t) \tag{4}$$

The constraints are defined by Eqs. (5) to (10). Equations (5) and (6) are constraints on the allocation of operators to cells.

$$\sum_{c=1}^{C} x_{itc} = 1 \ (\forall_{t,i}) \tag{5}$$

$$\sum_{i=1}^{I} x_{itc} \leq C_{max} (\forall_{t,c}) \tag{6}$$

Equation (7) represents the constraint on task continuity.

$$\sum_{k=F_{ijk}}^{E_{ijk}} y_{itjk} = E_{itjk} - F_{itjk} + 1 \leq K_{jmax} (\forall_{i,t,j}) \tag{7}$$

In Eq. (8), the number of operators who can be assigned to the same task within a cell is set to be one, achieving a task-dedicated system.

$$\sum_{i=1}^{I} x_{itc} y_{itjk} = 1 \ (\forall_{t,c,j,k}) \tag{8}$$

The takt time for each cell is calculated by Eq. (9). Based on the takt time, the number of products to be assigned to each cell is determined by Eq. (10).

$$TaktTime_{tcj} = max \left\{ \sum_{k=1}^{K} \frac{ST_{jk}}{S_{1jk} x_{1tc} y_{1tjk}}, \sum_{k=1}^{K} \frac{ST_{jk}}{S_{2jk} x_{2tc} y_{2tjk}}, \ldots \right\} (\forall_{t,c,j}) \tag{9}$$

$$Q_{tcj} = Q_{tjall} / TactTime_{tcj} \sum_{c=1}^{C} \frac{1}{TactTime_{tcj}} \ (\forall_{t,c,j}) \tag{10}$$

5 Computational Experiments

5.1 Solving Procedure

For each order, determine the operator allocation and the quantity of products to be allocated to each cell. Then, update the Skill Index values based on the allocated tasks and quantity of products, and to determine the operator allocation and quantity of products for the next order. Use the Simulated Annealing (SA) algorithm to find a solution for each order. The solution process is illustrated in Fig. 2.

First, set the Skill Index values in < 2 >, and in < 4 >, judge the terminal criterion for the number of orders t. After that, enter the SA algorithm. After initializing the temperature $Temp_r$ and the number of steps r in < 6 > (r=1), the operators are allocated to each cell and task in < 7 >. After the number of products in each cell is determined in < 8 >, the solution acceptance judgement is performed in < 9 >. The acceptance

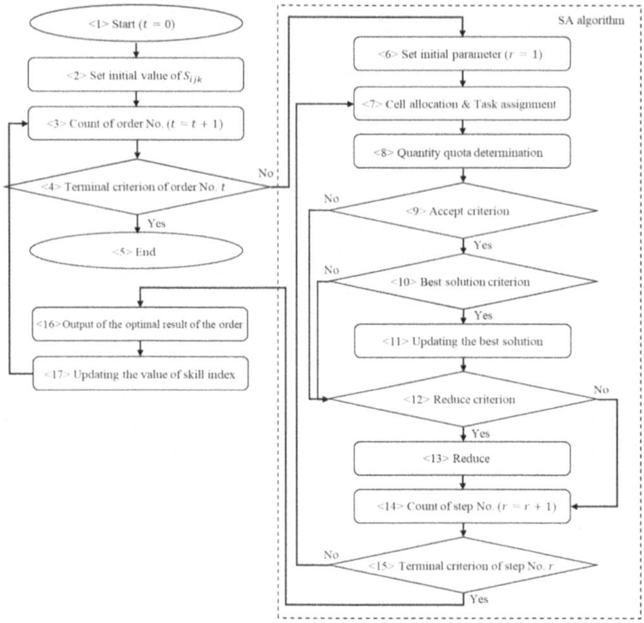

Fig. 2. Solution process

judgment is determined by Eq. (11), where x is the current solution, $x\prime$ is the next solution, and Δx is the value of $x - x\prime$.

$$(acceptx) = \begin{cases} \Delta x < 0 \ldots 1 \\ otherwise \ldots \exp\left(-\frac{\Delta x}{Temp_c}\right) \end{cases} \tag{11}$$

If$\Delta x < 0$, update the best solution in $< 11 >$. If $\Delta x > 0$ and x is accepted, do not update the best solution, and proceed to the cooling judgement in $< 12 >$. $Temp_r$ represents the temperature at step$r$. In $< 13 >$, the next temperature $Temp_{next}$ is set using Eq. (12) with cooling rateμ.

$$Temp_{next} = \mu Temp_r (0.8 \leq \mu \leq 1) \tag{12}$$

The terminal criterion of the SA algorithm is determined by the number of steps$r$. If the terminal criterion is not reached, return to $< 7 >$ operator allocation and task assignment. After completing the step and terminating the SA algorithm, the best solution is output as the order t solution in $< 16 >$. Then, in $< 17 >$, based on the task assigned and the number of products assigned to each cellQ_{tcj}, the execution time is determined, and the Skill Index values are updated using Eq. (2). To do this, the conventional method uses the learning curve proposed by Wright (1936) (Eq. (1)) to determine the execution time. The proposed method uses.

the fatigue-learning curve (Eq. (3)) proposed by Mályusz and Pém (2014). The execution time of the proposed method is obtained by Eq. (13).

$$ET_{ijk} = T_1 \lambda / {Q_{tcj}}^{\alpha} + T_1 (1 - \lambda) / \left(Q_{tcj} + 1 - Q_{tcj}\right)^{\alpha} \tag{13}$$

where $T_1 = ST_{jk}/S_{ijk}x_{itc}y_{itjk}$.

The above is repeated until the value of the number of orders t reaches the terminal criterion in $< 4 >$.

5.2 Conditions

The setting conditions assumed 3 cells (C1 ~ C3) with equivalent facilities and 8 operators (O1 ~ O8). The number of operators allocated to each cell ranged from 1 to 4. 3 types of products (G1 ~ G3) were set, with 4 tasks (K1 ~ K4) for each product. Additionally, 50 units of each product were set for every order, and the total number of orders was set to 5 (T1 ~ T5). The breakdown of operator skills was 2 expert operators, 4 standard operators, and 2 newcomer operators. The Skill Index values for each operator are randomly set within the ranges: 1.1 to 1.5 for expert operators, 0.8 to 1.2 for standard operators, and 0.6 to 1.0 for newcomer operators. The learning rate was randomly assigned between 90% and 95% for each operator and task. The fatigue coefficients used in the proposed method are Patterns I ~ X, where all tasks are unified to 0.05, 0.10, 0.15, 0.20, 0.25, 0.75, 0.80, 0.85, 0.90, and 0.95. The SA algorithm involves randomly determining the initial value and then proceeding with the search using an exchange neighbourhood. Consider orders T1 ~ T5 as one experiment and initialize the number of steps and temperature at the end of each order. The experiment will be conducted for both the proposed method using ten patterns of fatigue coefficients and the conventional method, with each method being tested 20 times.

5.3 Results

Tables 2 and 3 shows the basic statistics of the minimum value of the sum of Skill Index for each order for 20 trials. Table 2 covers for both the conventional method and the patterns I ~ V of the proposed method. Meanwhile, Table 3 focuses on for both the conventional method and the patterns VI ~ X of the proposed method.

5.4 Discussion

From Tables 2 and 3, for the proposed method, the average value of Pattern I is the smallest, and conversely, the average value of Pattern X is the largest. In T1, all methods had about the same average value. In T2 to T5, for all patterns of the proposed method, the average value was smaller than that of the conventional method. This is because the upward adjustment of actual execution time after each order completion became smaller because of influence of fatigue. For pattern X of the proposed method, the average value was almost the same as that of the conventional method. This is because the value of the fatigue coefficient λ is close to 1 and approximates the original learning curve.

In conclusion, using the model of Mályusz and Pém (2014), we were able to reflect the negative impact of fatigue on execution time in addition to the reduction in execution time due to learning.

Table 2. Basic statistics for the conventional method and the patterns I ~ V of the proposed method.

		T1	T2	T3	T4	T5
Previous study	Avg.	32.540	36.421	40.528	45.930	49.552
	Max	32.900	37.222	41.345	46.744	50.265
	Min	32.000	35.556	39.314	45.330	48.878
	SD	0.205	0.432	0.523	0.376	0.399
Pattern I	Avg.	32.610	32.859	33.027	33.237	33.428
	Max	32.900	33.177	33.312	33.507	34.046
	Min	32.200	32.443	32.728	32.861	33.098
	SD	0.189	0.180	0.137	0.159	0.208
Pattern II	Avg.	32.540	33.000	33.366	33.843	34.359
	Max	32.900	33.355	33.581	34.201	34.680
	Min	32.100	32.347	33.101	33.422	33.776
	Avg.	0.233	0.234	0.141	0.207	0.196
Pattern III	Max	32.540	33.239	33.873	34.428	35.143
	Min	32.900	33.492	34.134	34.729	35.510
	SD	32.200	32.757	33.579	33.996	34.777
	Avg.	0.162	0.171	0.145	0.204	0.181
Pattern IV	Avg.	32.585	33.390	34.265	35.164	36.014
	Max	32.800	33.782	34.578	35.488	36.266
	Min	32.000	32.982	33.710	34.975	35.618
	SD	0.185	0.225	0.184	0.150	0.166
Pattern V	Avg.	32.560	33.735	34.598	35.718	36.837
	Max	32.900	34.121	34.980	36.234	37.124
	Min	32.200	33.475	34.165	35.272	36.314
	SD	0.180	0.143	0.216	0.200	0.208

6 Assembly Experiment

We verify the effectiveness of the proposed method by comparing the estimated execution time using the proposed method with the actual execution time of assembling LEGO blocks.

6.1 Conditions

Two types of products were used in this experiment, as shown in Fig. 3. The entire process of each product was divided into 4 tasks (K1 to K4), and standard execution time was set for each. The Table 4 shows the number of processes, parts, and standard execution time for each task for each product. The target cell for the assembly experiment consisted of 2 cells with equivalent facilities and 8 operators, with 4 operators allocated to each cell. 4 tasks are set for each product, and 20 units of each product type are assembled. The number of orders is 1. The operators are undergraduates or graduate students in their early 20s. Operators assembled the products in advance and calculated their Skill Index values and the learning rate based on the assembly time. Table 5 shows the operators' Skill Index. For Product 1, there were two expert operators, five standard operators, and one newcomer operator. For Product 2, there were two expert operators, four standard operators, and two newcomer operators. The fatigue coefficient λ of all tasks was set at 0.80. The reason for this is that according to Mályusz and Pém (2014), when the fatigue coefficient λ is 0.80, the Mályusz and Pém (2014) model gives better estimates than the original learning curve.

Table 3. Basic statistics for the conventional method and the patterns VI ~ X of the proposed method.

		T1	T2	T3	T4	T5
Previous study	Avg.	32.540	36.421	40.528	45.930	49.552
	Max	32.900	37.222	41.345	46.744	50.265
	Min	32.000	35.556	39.314	45.330	48.878
	SD	0.205	0.432	0.523	0.376	0.399
Pattern VI	Avg.	32.645	35.477	38.476	42.370	46.137
	Max	33.100	35.854	39.007	43.255	46.773
	Min	32.100	34.974	37.570	41.632	45.301
	SD	0.222	0.243	0.438	0.476	0.454
Pattern VII	Avg.	32.595	35.645	39.128	43.129	46.988
	Max	32.900	36.281	40.110	44.150	47.767
	Min	32.200	34.856	38.470	41.936	46.372
	SD	0.220	0.371	0.368	0.538	0.404
Pattern VIII	Avg.	32.635	35.900	39.566	44.010	47.736
	Max	33.100	36.570	40.187	44.761	48.488
	Min	32.200	35.392	38.982	43.493	46.711
	SD	0.231	0.259	0.328	0.338	0.419
Pattern IX	Avg.	32.635	36.091	39.844	44.540	48.293
	Max	32.800	36.762	40.935	45.267	48.912
	Min	32.500	35.415	38.941	42.975	47.349
	SD	0.128	0.323	0.570	0.568	0.402
Pattern X	Avg.	32.610	36.220	40.498	45.278	49.037
	Max	32.900	36.785	41.142	46.142	49.598
	Min	32.100	35.349	39.694	44.332	48.116
	SD	0.202	0.390	0.421	0.481	0.401

6.2 Results and Discussion

Under the conditions described in Sect. 6.1, we performed computer experiments under the same SA algorithm conditions as in Sect. 5.1. As a result, the best solution for 20 trials was 16.93. Table 6 shows the operator allocation when the best solution is obtained.

Fig. 3. The LEGO blocks used in the assembly experiment.

Table 4. The number of processes, parts, and standard execution time for each task for each product.

Task	Product 1			Product 2		
	Steps	parts	standard execution time	Steps	parts	standard execution time
1	23	50	407	32	102	569
2	16	41	323	19	41	233
3	23	32	209	21	41	299
4	24	40	239	36	52	355
Total	86	163	1178	108	236	1455

The number of products assigned to each cell was 13units of Product 1 and 11units of Product 2 in Cell 1, and 7 units of Product 1 and 9units of Product 2 in Cell 2.

Based on the operator allocation and number of products obtained from computer experiments, assembly experiments were conducted in a cellular production system to measure the assembly time for each task. The assembly manuals ware provided for each task. The assembly manuals are illustrated with parts and procedures for each process, and these are compiled in a booklet. In addition, there are typically parts supply teams in actual workplaces, creating an environment where operators can focus on assembly work. Therefore, one supporter was assigned to each operator, and the supporter supplied the parts. Operators received and assembled the parts supplied by the supporter in process order. Figure 4 shows the experimental design.

Simple dispatching rules determined the order in which the two types of products were assembled. Figure 5 shows the order of assembly. In Fig. 5, assembly time was determined using Mályusz and Pém (2014) model. In Cell 1, most operators assembled Product 1 before assembling Product 2. On the other hand, in Cell 2, all operators assembled the Product 1 and Product 2 in mixed order.

6.3 Results and Comparisons

Tables 7 and 8 show the Skill Index values at the end of the task on the computational experiment and the assembly experiment, respectively. Only the tasks that each operator was assigned are shown.

Table 5. The operators' Skill Index.

Operator	Skill Level	Product 1				
		K1	K2	K3	K4	Average
O1	Expert	1.50	1.44	1.23	1.26	1.36
O2	Expert	1.50	1.48	1.07	1.10	1.29
O3	Standard	1.50	1.16	1.12	0.89	1.17
O4	Standard	1.50	1.42	1.01	1.04	1.24
O5	Standard	1.30	1.28	0.97	1.02	1.14
O6	Standard	1.50	1.50	1.34	1.45	1.45
O7	Newcomer	0.78	0.79	0.76	0.73	0.76
O8	Newcomer	1.10	0.98	0.84	0.94	0.97
Operator	Skill Level	Product 2				
		K1	K2	K3	K4	Average
O1	Expert	1.45	1.47	1.44	1.34	1.43
O2	Expert	1.30	1.28	1.42	1.21	1.30
O3	Standard	1.26	1.29	1.34	1.17	1.27
O4	Standard	1.20	1.25	1.22	1.35	1.26
O5	Standard	1.08	1.06	1.14	1.07	1.09
O6	Standard	0.93	0.92	0.94	1.02	0.95
O7	Newcomer	0.71	0.62	0.65	0.60	0.65
O8	Newcomer	1.10	0.98	0.84	0.94	0.97

Table 6. Operator allocation

	Product 1				Product 2			
	K1	K2	K3	K4	K1	K2	K3	K4
C1	O6	O1	O4	O3	O4	O6	O1	O3
C2	O7	O8	O5	O2	O5	O7	O8	O2

Comparing the Skill Index values of operators' assigned tasks in Tables 8, 9, and 5, the Skill Index values after the computer experiment and assembly experiment have increased compared to before the experiments. The Skill Index values of expert operators reached the maximum value of 1.50, with standard operators improving their skill level to expert, and newcomer operators improving their skill level to either standard or expert. Compare Tables 8 and 9. The difference in Skill Index values was within 0.10 for 10 of the 16 tasks. Among them, 9 tasks matched the Skill Index values. On the other hand, the task with the largest difference was Product 1 Task 1 for Operator O7 in Cell 2, with a difference of 0.40. T-test was conducted with the Skill Index values for each product at a significance level of $\alpha = 0.05$, and the P-values were 0.07 and 0.44, respectively, showing no significant difference. In addition, T-test was also conducted on the execution time after the experiment for each product at a significance level of $\alpha = 0.05$. The results

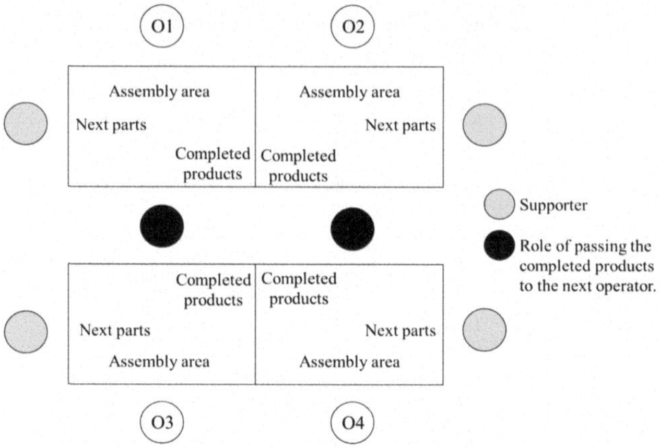

Fig. 4. Experimental design

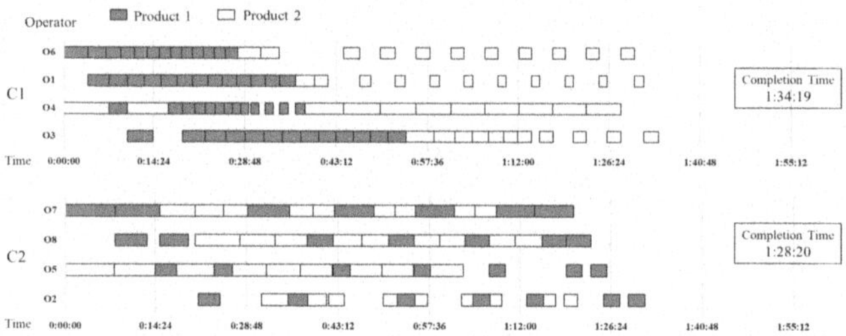

Fig. 5. Computational Scheduling

Table 7. Skill Index Values at the end of the task on the computational experiment.

Operator	Product 1				Product 2			
	K1	K2	K3	K4	K1	K2	K3	K4
O1	–	1.50	–	–	–	–	1.50	–
O2	–	–	–	1.50	–	–	–	1.50
O3	–	–	–	1.39	–	–	–	1.50
O4	–	–	1.50	–	1.50	–	–	–
O5	–	–	1.37	–	1.50	–	–	–
O6	1.50	–	–	–	–	1.50	–	–
O7	1.10	–	–	–	–	1.13	–	–
O8	–	1.39	–	–	–	–	1.50	–

showed no significant difference with P values of 0.32 and 0.50, respectively. Therefore, there is no significant difference between the two sets of data (Fig. 6).

Table 8. Skill Index Values at the end of the task on the Assembly experiment.

Operator	Product 1				Product 2			
	K1	K2	K3	K4	K1	K2	K3	K4
O1	–	1.50	–	–	–	–	1.50	–
O2	–	–	–	1.50	–	–	–	1.50
O3	–	–	–	1.50	–	–	–	1.50
O4	–	–	1.50	–	1.50	–	–	–
O5	–	–	1.50	–	1.23	–	–	–
O6	1.50	–	–	–	–	1.50	–	–
O7	1.50	–	–	–	–	1.11	–	–
O8	–	1.50	–	–	–	–	1.33	–

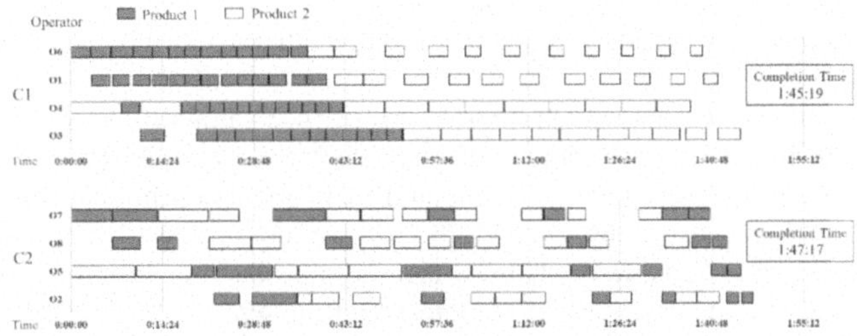

Fig. 6. Actual scheduling

7 Conclusion

In this study, we proposed an operator allocation method that considers the influence of fatigue by introducing a fatigue learning model into the conventional method to achieve more realistic operator allocation in the actual workplace. 10 patterns of proposed method with different values of the fatigue coefficient and a conventional method that does not consider fatigue were used to conduct a computer experiment, and the results were compared. As a result, the average of the minimum values of the sum of Skill Index of all patterns of proposed method was lower than that of the conventional method that did not consider fatigue. From this, it was found that the proposed method can consider not only the reduction of execution time due to learning but also the negative influence of fatigue on execution time. To validate the proposed method, we also conducted an assembly experiment using LEGO robots with the operator allocation obtained from the computer experiment and analyzed the assembly time for each operator. The results showed that the Skill Index values were the same for more than half of the tasks in the computer experiment and the assembly experiment. In addition, T-tests were conducted on the Skill Index values and execution time at the end of each task, respectively, and no significant differences were found. Therefore, the proposed method was shown to be effective.

Our future studies should consider not only fatigue but also other human factors such as age and forgetting. Also, the influence on the learning curve when multiple products are assembled with mixed scheduling needs to be studied. In assembly experiments, we aim to construct experimental designs that are more like workspaces. In the current assembly experiment, only a total of 40 products were assembled, and the experiment took less than 2 h per cell. However, in actual workplaces, assembly work may require longer working hours. Therefore, in our future studies, we must focus on constructing assembly experiments that reflect the actual workplace conditions. By doing so, the effectiveness of the proposed method can be more appropriately evaluated.

References

Harumi, H., Toshiya, K., Nobutada, F.: A study on operator allocation aiming at the skill improvement for cell production system. Transa. JSME (in Japanese) **81**(825), 14–00646 (2015). https://doi.org/10.1299/transjsme.14-00646

Abubakar, M.I., Wang, Q.: Prioritizing key human factors on effect of human centred assembly performance using the extent analysis method. J. Ind. Intell. Inform. **6**(1), 23–30 (2018). https://doi.org/10.18178/jiii.6.1.23-30

Karwowski, W., et al.: The complexity of human performance variability on watch standing task. Appl. Ergon. **79**, 169–177 (2019). https://doi.org/10.1016/j.apergo.2018.07.006

Abubakar, M.I., Wang, Q.: Key human factors and their effects on human centered assembly performance. Int. J. Ind. Ergon. **69**, 48–57 (2019). https://doi.org/10.1016/j.ergon.2018.09.009

Neumann, W., Kolus, A., Wells, R.W.: Human factors in production system design and quality performance – a systematic review. IFAC-PapersOnLine **49**(12), 1721–1724 (2016). https://doi.org/10.1016/j.ifacol.2016.07.830

Sikström, S., Jaber, M.Y.: The depletion–power–integration–latency (DPIL) model of spaced and massed repetition. Comput. Ind. Eng. **63**(1), 323–337 (2012). https://doi.org/10.1016/j.cie.2012.03.005

Daria, B., Martina, C., Alessandro, P., Fabio, S., Valentina, V.: Fatigue and recovery: research opportunities in order picking systems. IFAC – PapersOnLine **50**(1), 6882–6887 (2017). https://doi.org/10.1016/j.ifacol.2017.08.1211

Asadayoobi, N., Jaber, M.Y., Taghipour, S.: A new learning curve with fatigue-dependent learning rate. Appl. Math. Model. **93**, 644–656 (2021). https://doi.org/10.1016/j.apm.2020.12.005

Mályusz, L., Pém, A.: "Model for "Bath Tub" effect in construction." Creative Construction Conference 2014 (2014)

Digiesia, S., Kock, A.A.A., Mummolo, G., Rooda, J.E.: The effect of dynamic worker behavior on flow line performance. Int. J. Prod. Econ. **120**(2), 368–377 (2009). https://doi.org/10.1016/j.ijpe.2008.12.012

Fantini, P., et al.: Exploring the integration of the human as a flexibility factor in CPS enabled manufacturing environments: methodology and results. In: IECON 2016 - 42nd Annual Conference of the IEEE Industrial Electronics Society, pp. 5711–5716 (2016). https://doi.org/10.1109/IECON.2016.7793579

S. Hoedt, A., Claeys, E.H., Aghezzaf, Cottyn, J.: Real time implementation of learning-forgetting models for cycle time predictions of manual assembly tasks after a break. Sustainability 2020 **12**(14), 5543 (2020). https://doi.org/10.3390/su12145543

Katsuhisa, H., Yasuaki, F., Shin, S.: A Study on the Workload in Cell Production. Japanese J. Ergonomics **45**(4), 219–225 (2009). https://doi.org/10.5100/jje.45.219

Adler, P.S., Clark, K.B.: Behind the learning curve: a sketch of the learning process. Manage. Sci. **37**(3), 267–281 (1991). https://doi.org/10.1287/mnsc.37.3.267

Glock, C.H., Grosse, E.H., Jaber, M.Y., Smunt, T.L.: Applications of learning curves in production and operations management: a systematic literature review. Comput. Ind. Eng. **131**, 422–441 (2019). https://doi.org/10.1016/j.cie.2018.10.030

Yelle, L.E.: The learning curves: historical review and comprehensive survey. Decis. Sci. **10**(2), 302–328 (1979). https://doi.org/10.1111/j.1540-5915.1979.tb00026.x

Badiru, A.B.: Computational survey of univariate and multivariate learning curve models. IEEE Trans. Eng. Manage. **39**(2), 176–188 (1992). https://doi.org/10.1109/17.141275

Jaber, M.Y.: Learning and Forgetting Models and Their Applications. Handbook of Industrial and Systems Engineering (2006)

Hovland, C.I.: Human learning and retention. S.S. Stevens (Ed.), Handbook of Experimental Psychology, Wiley, Oxford, England. 613–689 (1951)

Jaber, M.Y., Givi, Z.S., Neumann, W.P.: Incorporating human fatigue and recovery into the learning–forgetting process. Appl. Math. Model. **37**(12–13), 7287–7299 (2013). https://doi.org/10.1016/j.apm.2013.02.028

Dutton, J.M., Thomas, A., Butler, J.E.: The history of progress functions as a managerial technology. Bus. History Rev. **58**(2), 204–233 (1984). https://doi.org/10.2307/3115048

Dar-Ei, E.M.: Human Learning: From Learning Curves to Learning Organizations. Kluwer Academic Publishers, Boston (2000)

Hartzler, B.M.: Fatigue on the flight deck: the consequences of sleep loss and the benefits of napping. Accid. Anal. Prev. **62**, 309–318 (2014). https://doi.org/10.1016/j.aap.2013.10.010

Langner, R., Steinborn, M.B., Chatterjee, A., Sturm, W., Willmes, K.: Mental fatigue and temporal preparation in simple reaction-time performance. Acta Physiol (Oxf.) **133**(1), 64–72 (2010). https://doi.org/10.1016/j.actpsy.2009.10.001

Bahdur, K., Gilchrist, R., Park, G., Nina, L., Pruna, R.: Effect of HIIT on cognitive and physical performance. Apunts. Medicina de l'Esport. **54**(204), 113–117 (2019). https://doi.org/10.1016/j.apunts.2019.07.001

Hillman, C.H., Snook, E.M., Jerome, G.J.: Acute cardiovascular exercise and executive control function. Int. J. Psychophysiol. **48**(3), 307-314 (2003). https://doi.org/10.1016/S0167-8760(03)00080-1

Chang, Y.K., Labban, J.D., Gapin, J.I., Etnier, J.L.: The effects of acute exercise on cognitive performance: a meta-analysis. Brain Res. **1453**, 87–101 (2012). https://doi.org/10.1016/j.brainres.2012.02.068

Winkelhaus, S., Sgarbossa, F., Calzavara, M., Grosse, E.H.: The effects of human fatigue on learning in order picking: an explorative experimental investigation. IFAC-PapersOnLine. **51**(11), 832–837 (2018). https://doi.org/10.1016/j.ifacol.2018.08.442

Anzanello, M.J., Fogliatto, F.S.: Learning curve models and applications: literature review and research directions. Int. J. Ind. Ergon. **41**(5), 573–583 (2011). https://doi.org/10.1016/j.ergon.2011.05.001

Globerson, S., Gold, D.: Statistical attributes of the power learning curve model. Int. J. Prod. Res. **35**(3), 699–711 (1997). https://doi.org/10.1080/002075497195669

Hurley, W.J.: When are we going to change the learning curve lecture? Comput. Oper. Res. **23**(5), 509–511 (1996). https://doi.org/10.1016/0305-0548(95)00031-3

Eden, C., Willians, T., Ackermann, F.: Dismantling the learning curve: the role of disruptions on the planning of development projects. Int. J. Project Manage. **16**(3), 131–138 (1998). https://doi.org/10.1016/S0263-7863(97)00053-7

Nembhard, D.A., Uzumeri, M.V.: An individual-based description of learning within an organization. IEEE Trans. Eng. Manage. **47**(3), 370–378 (2000). https://doi.org/10.1109/17.865905

Knecht, G.R.: Costing, technological growth and generalized learning curves. J. Oper. Res. Soc. **25**(3), 487–491 (1974). https://doi.org/10.1057/jors.1974.82

Mazur, J.E., Hastie, R.: Learning as accumulation: a reexamination of the learning curve. Psychol. Bull. **85**(6), 1256–1274 (1978). https://doi.org/10.1037/0033-2909.85.6.1256

Towill, D.R.: Forecasting learning curves. Int. J. Forecast. **6**(1), 25–38 (1990). https://doi.org/10.1016/0169-2070(90)90095-S

Naim, M.M., Towill, D.R.: An engineering approach to LSE modelling of experience curves in the electricity supply industry. Int. J. Forecast. **6**(4), 549–556 (1990). https://doi.org/10.1016/0169-2070(90)90033-8

Anzanello, M.J.: Learning curve modelling of work assignment in mass customized assembly lines. **45**(13), 2919–2938 (2007). https://doi.org/10.1080/00207540600725010

Peltokorpi, J., Jaber, M.Y.: Potential models of group learning in production. Adv. Transdisciplinary Eng. **13**, 205–216 (2020). https://doi.org/10.3233/ATDE200158

Gan, Z.L., Musa, S.N., Yap, H.J.: A review of the high-mix, low-volume manufacturing industry. Appl. Sci. **13**(3), 1687 (2023). https://doi.org/10.3390/app13031687

Abu-Samah, A., Shahzad, M.K., Zamai, E.: Bayesian based methodology for the extraction and validation of time bound failure signatures for online failure prediction. Reliab. Eng. Syst. Saf. **167**, 616–628 (2017). https://doi.org/10.1016/j.ress.2017.04.016

Alduaij, A., Hassan, N.M.: Adopting a circular open-field layout in designing flexible manufacturing systems. Int. J. Comput. Integr. Manuf.Integr. Manuf. **33**(6), 572–589 (2020). https://doi.org/10.1080/0951192X.2020.1775300

Developing a Framework for Value Creation in the Context of Digital Transformation

R. A. Fayet[1(✉)], E. Pinheiro de Lima[2], F. Deschamps[1], and S. Gouvea da Costa[2]

[1] PPGEPS, Pontifícia Universidade Federal do Paraná, 1155 Imaculada Conceição, Curitiba, Paraná, Brazil
rafael@gestaointeligente.com

[2] PPGEPS, Universidade Tecnológica Federal do Paraná, 3165 Sete de Setembro, Curitiba, Paraná, Brazil

Abstract. Although many organizations are "digitized," they generally need to adopt a more holistic approach to digital transformation. In this context, the literature lacks theory on how digital transformation should be considered in operations strategy to create value for customers. Based on Pettigrew's work (1987), content analysis was conducted, focusing on value dimensions, competitive priorities, and value characteristics that guide operations strategy in the context of digital transformation. Correlations were established to identify variables related to value. The content analysis identified the need for greater attention to defining digital transformation operations strategies and creating value that meets market needs, while also exploring internal resources and capabilities. The proposed conceptual framework establishes two main flows of value-related dimensions (proposition, creation, delivery, and capture of value): the value innovation flow and the value co-creation flow.

Keywords: digital transformation · operations strategy · value proposition · value creation · value co-creation · value innovation

1 Introduction

The global business scenario witnesses the evolution of a relatively cheap and increasingly accessible global digital infrastructure with new digital technologies cloud computing, Internet of Things (IoT), Big Data and analytics, 3D printing, and intelligent autonomous systems. These digital technologies are transforming the strategic context of organizations and generating the need for rapid adaptation to maintain competitiveness [5, 8]. In this sense, although many organizations are "digitized" in some processes of the corporate value chain, in general, organizations need to adopt a more holistic approach to digital transformation [5].

This article presents a contextualist analysis, based on the approach of [17], on the dimensions and characteristics of value for the direction of the operations strategy, in the context of the digital transformation, that is, how the digital transformation must be considered in the strategy of operations to create value for customers. In the end, to

M. Dragomir et al. (Eds.): ICPR 2023, LNPE, pp. 125–132, 2025.
https://doi.org/10.1007/978-3-031-92082-0_12

outline a conceptual model that addresses a strategic vision of digital transformation, a framework that operationalizes the proposition, creation, delivery, and value capture flow is proposed.

2 Theoretical background

2.1 Digital Transformation and Digital Transformation Strategy

Digital transformation is a technology-driven multidimensional phenomenon that has a revolutionary impact on businesses, individuals, and society. In a more practical approach, digital transformation refers to the use of digital technology for a more efficient and effective economic value creation process [16]. It can generate broad organizational implications by changing the organization's core business model using digital technology [22].

[22] identifies three phases of digital transformation: digitization, digitalization, and digital transformation. While digitization describes a technology, or system of technologies, digitalization answers why that technology is relevant to a specific process or organization. In this way, digitalization is the sociotechnical process of leveraging digitized products or systems to develop new organizational procedures, business models, or commercial offerings. Digital transformation, on the other hand, is defined as a socio-cultural process of companies adapting to new organizational forms and adapting to the sets of skills necessary to remain viable and relevant in a digital scenario [16].

Thus, a digital transformation strategy should not treat technology in isolation, as a separate component of business operations, but should consider how digital technology will be used to reshape the company's processes and relationships with customers, markets, employees, partners, and others [3].

For [5], a digital transformation requires the development of several organizational capabilities allied to technology, not just considering technology in isolation. Ultimately, an organization-wide approach is needed that integrates the impacted operational and functional business areas enabling enhanced, expanded, or redefined customer value propositions [3].

2.2 Operations Strategy and Value Proposition

For [18], an operations strategy defines how the production function can help achieve business objectives to ensure the effectiveness of operations, through the adequate provision of structural elements (facilities, equipment, etc.) and appropriate infrastructure (human resources, organizational structure, control policies, etc.).

Thus, operations strategy is the total pattern of decisions that shape the long-term capabilities of any type of operation and their contribution to the overall strategy by reconciling market requirements with operations resources. They include not only the production processes but also the processes that contribute to the creation and delivery of the product, such as purchasing processes, physical distribution, and after-sales service, among others [21].

The composition of the operations strategy includes the definition of decision areas and competitive priorities. Decision areas represent key areas where appropriate strategic decisions must be made to adjust the strategy. These decisions affect the organization's ability to achieve its competitive objectives [11, 21]. Operations strategy decisions, or decision areas, are grouped according to capability strategies, supply network strategies, process technology strategies, and development and organization strategies [21].

The performance through a pattern of actions in the decision areas generates results in the competitive priorities, based on the particular objectives of the organization [11, 21]. In other terms, the corporate strategy requires a compelling understanding of competitive priorities (or performance objectives) and strategic choices that provide for an appropriate operations strategy [11]. The authors identify five performance objectives: quality, speed, reliability, flexibility, cost [21], and innovativeness [11].

The definition of operations performance objectives must meet the market requirements by determining its competitive factors, that is, things that a customer can see or experience [21]. Despite different nomenclatures, the concepts of "competitive factors" and "value propositions" [9] are compatible.

The value proposition establishes a strategic focus through the unique combination of attributes offered to customers, through products and services, to generate loyalty and satisfaction in target segments. These attributes can be divided into three categories: product/service attributes, customer relationships, and image and reputation [9].

Developing a sustainable operations strategy consists of ensuring alignment between the organization's internal activities and the value proposition. Thus, a clear value proposition provides the fundamental target on which the strategic themes, internal business processes and infrastructures of the organization should focus [9].

3 Development of the Conceptual Model: Context, Content and Process

3.1 Methodology

For strategic changes or organizational transformations, [17] proposes that a contextualist investigation should involve questions about the content, context, and process of change, along with the interconnections between these three broad analytical categories. The analysis of the organizational transformation must start from the understanding that the formulation of a new strategy content inevitably involves the management of its context and process [17].

The context, derived from the analysis of the internal and external context, represents the "why" of the change. While the external context refers to the competitive environment in which the company operates, the internal context refers to the company's structure and corporate culture. Content refers to "what" is intended to change, such as technology, labor, products, or even corporate culture. The process refers to the "how" of the change, that is, the actions, reactions, and interactions of the various stakeholders in the change process [17].

This article presents the contextualist investigation, according to [17], based on the researched articles, from a systematic literature review (SLR) in the SCOPUS and Web of Science databases, including search terms related to three category axes: the context of digital transformation (axis 1), concepts related to value (axis 2) and business and operations strategy (axis 3). The search in the bases, with subsequent application of the exclusion criteria and the snowballing technique, resulted in 24 documents for the final portfolio.

Content analysis was developed through the coding process focused on value dimensions, competitive priorities, and value characteristics considering the most recent digital transformation approaches. Correlations were established to identify the variables related to the value used in the proposed framework.

3.2 Results and Discussion

In this section, the content analysis of the SLR articles will be presented, according to [17], to understand the context (why), content (what), and process (how) for defining the operations strategy, which reconciles a value proposition and organizational resources, in the context of digital transformation.

Context

In the context of the competitive environment, there is agreement that digital transformation generates changes in the value creation process through new configurations of resources, that is, how resources are exchanged, combined, and integrated [2]. For [20], this requires new strategies from companies to remain focused on their customers.

In the internal context of companies, the authors identify multiple gains that can be achieved by digital transformation, such as increased sales, productivity, efficiency, innovations in value creation, new ways of interacting with customers, and business expansion, among others [6, 13]. In terms of barriers to adoption digital technologies, for [10] difficulty most faced by organizations is of a managerial and cultural nature, and not related to data and technology.

Content

The content analysis focused on identifying the definitions of the concept of value and the characteristics of value in the context of digital transformation. Regarding the conceptual approach to value in this work, different nomenclatures and approaches were found in the research.

In agreement with researched articles [3, 13], despite the growing amount of literature on value creation in the digitized world, there is still little guidance for a common scientific understanding, perhaps due to the complexity and diversity of the subject. This finding is supported by the varied understanding of the dimensions of value (creation, proposition, delivery, co-creation, innovation, and capture) found in the research. The most common terms were: value creation (15 out of 24 articles), value proposition (7 out of 24 articles), value co-creation (7 out of 24 articles), and value capture (5 out of 24 articles). Some authors argue that the concept of value creation already includes the concept of the value proposition. Other value-related terms were: value innovation (1 of 24 articles), value offering (1 of 24 articles), and value delivery (1 of 24 articles).

Moreover, none of the articles significantly addresses the characteristics of customer value as drivers of competitive priorities and decision areas, key concepts in operations strategy. The main approaches are:

- [1] addresses, superficially, the need for companies to transform the company's competitive priorities and restructure operational capabilities, such as quality, flexibility, delivery time, services, and costs;
- [4] addresses the flexibility performance dimension as a performance/quality, time/deadline, and cost alignment factor, and identifies types of flexibility;
- [20] addresses the value characteristics according to three dimensions of the perceived quality of the service (reliability, guarantee, and responsiveness);
- [19] addresses the value characteristics according to the set: price, delivery/deadline/time, availability, place of delivery, flexibility, and assistance/support.

Process

Despite the different perspectives of the models identified in this research, none presents a process approach that operationalizes the steps of the value creation flow through an operations strategy in the context of digital transformation.

The identified approaches include the "Manufacturing Value Modeling Methodology – MVMM" [4], the conceptual model based on the Resource-Based Theory [15], the strategic roadmap for the transition to Industry 4.0 [6, 7], the value innovation stages [14] and the "Customer-Product-Process-Resource - CPPR 4.0" model [12].

3.3 Conceptual Model

Based on the references of the models identified in the Systematic Literature Review [4, 7, 12, 15], to establish a model in the context of digital transformation, some organizational elements are needed as input information, such as purpose, environmental context (external and internal), customer needs, key partners (including customers), key resources, organizational capabilities, and competencies.

Thus, to establish the content and process for the model that involves the concept of value, in the context of digital transformation, this work suggests adopting the approach and nomenclature of value domains based on the proposal of [12], which establishes interconnected elements related to the value of a business model.

This work proposes changes in this approach that allow greater flow integration and domains in the model. The proposal establishes two main flows of interconnected elements related to value: the value innovation flow and the value co-creation flow.

The value innovation flow emphasizes the development and commercialization of future-oriented value concepts [14], which means creating fundamentally new and superior value for the customer. The value co-creation flow involves customers or interested parties in all processes related to value creation, enhancing the ability to convert joint knowledge into a sustainable competitive advantage [1, 19, 20].

The value innovation and value co-creation flows are composed of the following value domains (Fig. 1, below): value proposition, value creation, value delivery, and value capture. In general, the main difference between the flows is the involvement of other actors in the value co-creation flow, which can result in distinct processes within the value domains, while still allowing for innovations to be developed in the value co-creation flow.

The value proposition domain is the way to communicate how the company aims to provide value to customers or what it offers as a superior value package to target customers [15]. This value proposition should justify why customers choose one company over another [12]. Value propositions should include product/service attributes, customer relationships, image, and reputation that are representative of target customers and associated with the competitive priorities of the operations strategy for digital transformation.

The value creation domain represents the processes (business, support, and management), resources, and capabilities needed to deliver the value proposition [12]. The main factors that create value are related to the capabilities and organizational resources of the company and, in the co-creation flow, also of the customer and the partner network. Thus, the creation of value represents the digital transformation plan that establishes how the organization intends to meet the competitive priorities and fulfill the value proposition, defining the operations strategies in the areas of decision for the digital transformation.

The delivery of value domain is the ability of the company, or together with the client and partner network in the co-creation flow, to carry out digital transformation operations strategies through actions and projects in the decision-making areas. This domain precedes value capture and is related to the organization's capacity for effective action.

Finally, the capture of value domain represents the capture of sustainable results from the delivery of value [12]. In this way, value capture can include monetary, relational, and learning aspects.

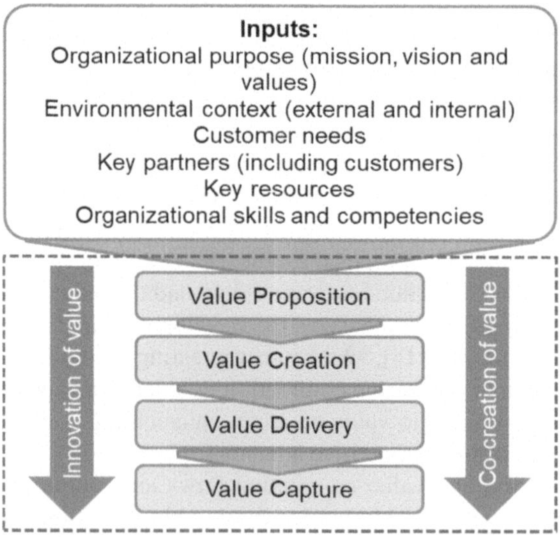

Fig. 1. Innovation flow and Co-creation flow of value

4 Conclusion

Organizations often find it difficult to create value in today's dynamic, globalized, interconnected, and increasingly complex technological environments. While current technology trends potentially offer organizations great value, emerging technology implementations and transformations often fail to achieve the desired outcomes of value-creation.

This article helped identify the importance of proposing, creating, and delivering sustainable value to the customer through differentiated value characteristics that align with customer expectations. In this sense, it was noticed that there is a need for more attention to how value is created and how it guides the development of operations strategies to respond to market opportunities, leverage resources and internal capabilities, and possess organizational skills, among others.

While the basic principles of digital transformation strategies have been established, researchers agree that there is a need for a better understanding of the phenomenon, particularly regarding how to formulate, implement, and evaluate digital transformation strategies. However, understanding how digital transformation will impact the definition of the operations strategy requires a stronger academic research foundation to clarify, make it more rigorous, and consolidate it.

Finally, the theoretical framework developed in this research helped identify conceptual gaps in the terms studied, taking into account the context of digital transformation. Similarly, regarding digital transformation models, there is no comprehensive overview that organizes the field of study of digital transformation due to its multidisciplinary nature.

The proposed conceptual model correlates and integrates the value dimensions (creation, proposition, delivery, co-creation, innovation, and capture) by defining the steps in the flow of innovation and co-creation of value. Thus, the importance of determining the strategic focus in the context of digital transformation, related to value creation through digital technologies, can be identified as a means to gain a competitive advantage in the market segments in which an organization operates.

References

1. Agrifoglio, R., et al.: How emerging digital technologies affect operations management through co-creation. Prod. Plann. Control Napoli **28**(16), 1298–1306 (2017)
2. Amit, R., Han, X.: Value creation through novel resource configurations in a digitally enabled world. Strateg. Entrepreneurship J., New Jersey **11**(3), 228–242 (2017)
3. Brown, N., Brown, I.: From digital business strategy to digital transformation - how. Proc. South African Inst. Comput. Sci. Inform. Technol. 2019 on ZZZ - SAICSIT '19, Skukuza, South Africa (2019)
4. Burger, N., et al.: Investigating flexibility as a performance dimension of a manufacturing value modeling methodology (MVMM): a framework for identifying flexibility types in manufacturing systems. Procedia CIRP, Amsterdã **63**, 33–38 (2017)
5. Carcary, M., Doherty, E., Conway, G.: A dynamic capability approach to digital transformation: a focus on key foundational themes. In: The European Conference on Information Systems Management. Academic Conferences International Limited, Évora, p. 20 (2016)

6. Ghobakhloo, M.: The future of manufacturing industry: a strategic roadmap toward Industry 4.0. J. Manufact. Technol. Manage. **29**(6), 910–936 (2018)
7. Ghobakhloo, M., Fathi, M.: Corporate survival in industry 4.0 era: the enabling role of lean-digitized manufacturing. J. Manuf. Technol. Manage. **31**(1), 1–30 (2020)
8. Kaltenbach, F., et al.: Smart services maturity level in Germany. In: 2018 IEEE International Conference on Engineering, Technology and Innovation (ICE/ITMC), Stuttgart (2018)
9. Kaplan, R.S., Norton, D.P.: The strategy-focused organization: How balanced scorecard companies thrive in the new business environment. 1th edition. Harvard Business Review Press. Boston, MA Rio de Janeiro: Campus (2001)
10. Lavalle, S., et al.: Big data, analytics and the path from insights to value. MIT Sloan Manage. Rev. **52**(2), 21 (2011)
11. Leong, G.K., Snyder, D.L., Ward, P.T.: Research in the process and content of manufacturing strategy. Omega **18**(2), 109–122 (1990)
12. Martínez-olvera, C., Mora-vargas, J.: A comprehensive framework for the analysis of industry 4.0 value domains. Sustainability **11**(10), 2960 (2019)
13. Matt, C., Hess, T., Benlian, A.: Digital transformation strategies. Bus. Inf. Syst. Eng. **57**(5), 339–343 (2015)
14. Matthyssens, P.: Reconceptualizing value innovation for industry 4.0 and the industrial internet of things. J. Bus. Ind. Mark. **34**(6), 1203–1209 (2019)
15. Payne, A., Frow, P., Eggert, A.: The customer value proposition: evolution, development, and application in marketing. J. Acad. Mark. Sci. **45**(4), 467–489 (2017)
16. Peter, M.K., Kraft, C., Lindeque, J.: Strategic action fields of digital transformation. J. Strateg. Manag. **13**(1), 160–180 (2020)
17. Pettigrew, A.M.: Context and action in the transformation of the firm. J. Manage. Stud. **24**(6), 649–670 (1987)
18. Platts, K.W., Gregory, M.J.: Manufacturing audit in the process of strategy formulation. Int. J. Oper. Prod. Manag. **10**(9), 5–26 (1990)
19. Rantala, T., et al.: Creating sustainable customer value through digitality. World J. Entrepreneurship Manage. Sustainable Dev. **15**(4), 325–340 (2019)
20. Saunila, M., Rantala, T., Ukko, J.: Characteristics of customer value creation in digital services. J. Serv. Sci. Res. **9**(2), 239–258 (2017)
21. Slack, N., Lewis, M.: Operations strategy. 3. ed. Harlow, England: Pearson (2011)
22. Verhoef, P.C., et al.: Digital transformation: a multidisciplinary reflection and research agenda. J. Bus. Res.Res. **122**, 889–901 (2021)

Advanced Energy Management: Guidelines to the Industrial Management

C. C. Ferreira[1], N. J. Monteiro[2(✉)], S. E. Gouvea da Costa[3], F. Deschamps[1,4], and E. P. de Lima[3]

[1] Production and Systems Engineering Graduate Program, Pontifícia Universidade Católica do Paraná, Rua Imaculada Conceição 1155, Curitiba, Paraná, Brazil
[2] Department of Engineering, Universidade do Estado do Pará, Av. Hiléia s/n, Marabá, Pará, Brazil
nathalia.monteiro@uepa.br
[3] Production and Systems Engineering Graduate Program, Universidade Tecnológica Federal do Paraná, Av. Sete de setembro, 3165 Curitiba, Paraná, Brazil
[4] Department of Mechanical Engineering, Federal University of Paraná, R. XV de Novembro, 1299 Curitiba, Brazil

Abstract. Effective energy management has become increasingly crucial for companies due to the need to enhance competitiveness in the face of resource scarcity. By investing in energy management (EM), companies can reduce energy consumption and costs. One way to achieve this is by implementing an energy management system (EnMS), which can serve as a blueprint for companies looking to incorporate EM practices. However, integrating EnMS with digital technologies can be challenging. To help organizations structure their EnMS with the support of digital technologies, this paper proposes guidelines based on a thorough content analysis of literature and expert interviews. In total, 23 guidelines were proposed across four dimensions, and they were refined by specialists with at least five years of experience in the industry, consulting, or academia. These guidelines can help organizations of any size or sector optimize their EM practices by effectively integrating EnMS and digital technologies. The guidelines were successfully applied in three departments of an automotive company and related to the company's current practices. By implementing them, organizations can reduce their energy consumption and costs while enhancing their competitiveness.

Keywords: Energy management · Advanced energy management · Industry 4.0 · Digital technologies

1 Introduction

The issue of EM is a significant concern for organizations due to various external pressures. These pressures include the need to reduce environmental impacts, media coverage of climate change, legal requirements, and the scarcity and rising prices of energy sources [1, 2]. Additionally, studying energy can also provide a competitive advantage [3, 4]. Companies can benefit not only from reducing their resource consumption but

M. Dragomir et al. (Eds.): ICPR 2023, LNPE, pp. 133–140, 2025.
https://doi.org/10.1007/978-3-031-92082-0_13

also from improving their image as environmentally friendly. Therefore, it is essential for organizations to prioritize EM and consider it a valuable aspect of their operations.

As stated by Schulze et al. [5], EM encompasses all the activities, procedures, and systematic routines of an organization, including strategy, planning, implementation, operation, control, organization, and culture. By incorporating all these processes, an organization can effectively reduce energy consumption and its associated costs.

As companies strive to increase their production while reducing their energy consumption, the concept of energy efficiency (EE) comes into play. Essentially, this means finding ways to produce the same output but with less energy input [6]. To achieve this goal, businesses can utilize EnMS, which provide a structured and systematic approach to controlling and optimizing their energy usage [7].

One way to enhance EnMS is to increase participation and use of advanced technology information systems, as May et al. [8] have suggested. Wang et al. [9] further support this notion, stating that the Internet of Things (IoT) can be a valuable tool in developing decision systems that aim to improve energy efficiency. The use of autonomous technologies can also lead to better real-time monitoring of production systems, which can have a positive impact on the availability of data related to energy consumption. This, in turn, can help organizations to identify opportunities for improvement and implement more effective energy efficiency practices [10].

Despite the benefits, there are few studies that combine information and communication technologies (ICT) and sustainable systems [11]. This indicates that there is still a lot of untapped potential in this area. Considering this scenario of optimizing energy management through digital technologies, the present work aims to identify the key factors that are essential for this integration. In doing so, it seeks to answer the following question: What are the guidelines that companies need to follow to adopt energy management systems that are supported by digital technologies?

2 Methodology

This work defines guidelines as universal recommendations that organizations can use to adopt and apply EMS supported by digital technologies of information and communication. This definition is based on the work of Deschamps et al. [12]. This work has two significant steps, the literature review and the refinement of the guidelines, explained in Sects. 2.1 and 2.2, respectively.

2.1 Literature Review

The research that was conducted followed the systematic literature review (SLR) steps recommended by Tranfield et al. [13]. This included (i) defining the research purpose, (ii) constructing the research protocol, (iii) conducting research in relevant databases, (iv) selecting interesting works, and (v) reading and analysing the selected works to answer the research question. The aim of the SLR was to identify the major themes of this work, which were energy management and industry 4.0.

The research protocol for energy management and industry 4.0 are in Table 1 and 2, respectively.

Table 1. Research protocol for energy management

Characteristics	Description
Search terms (Title, abstract, keywords)	Group 1: ("industrial energy efficiency" OR "industrial energy management" OR "energy efficiency manufacturing"); Group 2: ("energy efficiency" AND "manufacturing"); Group 3: ("energy management" AND "industry")
Research strategy	"OR" between groups
Database	Scopus
Delimitation of studies areas	Energy, Engineering, Environment, Businesses, Economy and Decision Sciences
Language	English
Type of publication	Literature reviews
Date of publication	Not specified

Table 2. Research protocol for industry 4.0

Characteristics	Description
Search terms (Title, abstract, keywords)	("industr* 4.0" OR "connect* manufactur*" OR "smart manufactur*" OR "industrial internet of things" OR "21st manufactur*" OR "4th industrial revolution")
Database	Scopus
Delimitation of studies areas	Energy, Engineering, Environment, Businesses, Economy and Decision Sciences
Language	English
Type of publication	Literature reviews
Date of publication	Not specified

After applying the research protocol, 50 articles were selected for analysis. The guidelines were extracted from these works and refined in Sect. 2.2.

2.2 Refinement of the Guidelines

The guidelines selected were refined with a framework based on the Cambridge Design Process Approach of Platts and Gregory [14].

To choose the specialists, two perspectives were considered. The first was their expertise on the topics of energy management and/or digital technologies The second perspective was their field of work, which included manufacturing, consulting, and academia. Were selected individuals who worked in one or more of these domains, depending on their experience and knowledge.

Five specialists with diverse functions were selected to refine the energy management guidelines. The ones from the industry have experience with the automotive and telecommunications sectors.

Six experts were interviewed to refine the guidelines from digital technologies. They have diverse experience with industrial automation, digital platforms, and software development.

The interviews were done in person or by an online platform for about one and a half hours. The researchers presented the work and its goal, followed by the presentation of the guidelines and the specialist's opinion about implementing energy management and/or digital technologies.

The specialist was questioned if he agreed with the guideline. If not, he could eliminate it or change its concept. Even when he agreed, he was questioned about the definition of the guideline and if it could be combined with another one.

The guidelines after the refinement were presented in Sect. 3.

3 Case Study

3.1 Guidelines

After refinement, the specialists were left with 23 guidelines that are divided into four dimensions. These dimensions are part of the maturity model for Industry 4.0 proposed by Schuh et al. [15]. The dimensions of the model are explained below.

- **Resources:** Refer to the tangible assets of the organization. The resources should have an interface between the physical and the digital world.
- **Information systems:** These are socio-technical systems where the information is provided based on economic criteria. They prepare, process, store, and transfer data and information.
- **Organizational structure:** Refers to the company's internal organization (structure and operational processes) and its position in the value network.
- **Culture:** It is the value system within the company, also describing the soft factors of collaboration.

The guidelines for advanced energy management and their description are in Table 3.

Table 3. Proposed guidelines for advanced energy management

Number	Guideline	Description	Dimension
1	Sharing goals	The company needs to communicate its goal of enhancing energy efficiency to its employees, stakeholders, and other interested parties	Culture
2	Organizational commitment	To ensure effective energy management, it is important for the organization to establish an energy policy that reinforces its dedication to this cause and provides the necessary resources for managing energy	Culture
3	Risk management	To successfully implement energy management in an organization, it's important to understand the factors that can enable or hinder this process	Culture
4	Awareness and engagement	To promote participatory energy management, it is essential to provide training and encouragement to individuals. This will help them understand their roles and ways of contributing to the cause	Culture
5	Stock risk management	To effectively manage energy, it is important to understand how it will impact conventional indicators such as quality, cost, and flexibility. Having a clear vision of these impacts is necessary	Culture
6	Compliance with regulations	The organization needs to show that it is dedicated to following all current regulations	Organizational Structure
7	Prioritization of energy efficient actions	Employees should understand that the company values and prioritizes actions, attitudes, and technologies that promote energy efficiency	Organizational Structure
8	Setting goals and action plans	To achieve success, energy goals and objectives should align with the company's overall objectives, and clear action plans must be established	Organizational Structure
9	Financial analysis	It is important for individuals with expertise in both technology and finance to assess the financial analysis of energy efficiency projects	Organizational Structure
10	Definition of relevant KPIs	To effectively monitor an organization's objectives and goals, it is important to create and implement relevant and suitable Key Performance Indicators (KPIs)	Organizational Structure
11	Action tracking	It's important to consistently monitor the efficiency and effectiveness of any improvement actions taken	Organizational Structure
12	Knowledge consolidation	It is important to establish a comprehensive knowledge base on energy to support future applications and to identify potential discrepancies and opportunities	Organizational Structure

(*continued*)

Table 3. (*continued*)

Number	Guideline	Description	Dimension
13	Monitoring the energy profile	Regularly recording energy baseline results aids in monitoring and analyzing the system, identifying opportunities to reduce consumption and prioritize improvements	Information Systems
14	Active management	To better control energy consumption and expenses, the organization should adopt technologies that enable a shift from reactive to proactive measures	Information Systems
15	Proper definition of energy data	It's important to collect energy data that aligns with the company's objectives to ensure it supports monitoring activities effectively	Information Systems
16	Data reliability	To ensure accurate analysis results, it is important to guarantee the reliability and repeatability of the collected data	Information Systems
17	Interoperable systems	To ensure smooth exchange of information between systems (interoperability), it is important to establish reliable data and communication standards	Information Systems
18	Statistical analysis	The production system should be structured in a way that allows for statistical analysis and data mining in order to optimize its performance	Information Systems
19	Simulation of scenarios	To prevent issues during application, it's important to use simulation tools to predict the effects of actions and decisions regarding investments before implementing them	Information Systems
20	Visualization tools	To effectively monitor, understand, and manage energy consumption indicators, it is important to utilize visualization tools such as dashboards	Information Systems
21	Management support	Top management must demonstrate their support and commitment to energy management	Resources
22	Definition of responsibility	To ensure effective management of energy and access to accurate information, it is important to designate an energy manager who is directly connected to top management. This person will be responsible for overseeing and improving energy management procedures	Resources
23	Allocation of resources	To effectively carry out energy management actions, it's important to determine the available resources such as human, financial, and physical assets	Resources

3.2 Application of the Guidelines as a Diagnostic Tool

The guidelines were used to diagnose issues in a car manufacturing company that has multiple sites across the world. The study was conducted in Brazil, where the site has been operating for over 20 years, making it a significant contributor to the region's economy.

Two central departments were studied (industrial maintenance and fluid management). The maintenance team is responsible for the energy management on the production site. The fluid management team takes care of the energy consumption and the site's global indicators.

Three employees from the operational level were interviewed to evaluate the alignment of the guidelines with the EMS of the organization. Employee 1 works on the maintenance team of the production site, Employee 2 works on the fluid management team, and Employee 3 works for the maintenance team, but in another production area of the enterprise, with intensive energy consumption. All the employees have more than two years in the industrial energy management of the site.

After analyzing each guideline and considering employee feedback, it is evident that the organization values the importance of energy consumption in its processes. However, employees are primarily focused on short-term goals and there is no specific energy policy in place despite the high management's support for energy efficiency projects. While the teams are aware of their roles, there is no regular monitoring of energy indicators. Additionally, the company is concerned about regulations, but this information is not shared with employees.

The organization and advanced energy management guidelines can be better aligned by following these suggested propositions:

- Enhance communication and make the medium and long-term energy goals more visible.
- Improve the tracking of energy usage targets and progress in areas with lower energy consumption.
- To promote ongoing improvement and increase employee awareness, it's advisable to implement established regulations such as ISO 50001.

4 Conclusions

Organizations are increasingly concerned about energy, but there is still room for improvement in terms of understanding and spreading awareness throughout the entire company.

The EMS can reach its full potential in the organization with the help of new information and communication technologies.

In this scenario, the aim is to assist organizations in structuring their advanced EMS. To achieve this, 23 guidelines have been proposed, which have been selected through an SLR and refined by specialists from manufacturing, consulting, and academia.

A car manufacturer company used them as a diagnostic tool to analyze their EMS. Three employees provided their opinions, and it was observed that there was an issue with unclear communication among all employees despite the organization's concern with energy consumption. The company's short-term goals were prioritized. Recommendations were made to enhance energy management systems with digital technologies.

This work has limitations due to the subjective nature of its analysis. It utilized an SLR, which is constrained by the research protocol, and the company's analysis was derived from the opinions of only a few employees.

In order to overcome these limitations, some propositions of future works are recommended. These include expanding the research protocol for the SLR and applying the guidelines to multiple case studies. This will help in developing a tool to assess the maturity of companies.

References

1. Thollander, P., Backlund, S., Trianni, A., Cagno, E.: Beyond barriers - a case study on driving forces for improved energy efficiency in the foundry industries in Finland France, Germany, Italy, Poland, Spain, and Sweden. Appl. Energy **111**, 636–643 (2013)
2. Perroni, M.G., Gouvea da Costa, S.E., Pinheiro de Lima, E., Vieira da Silva, W., Tortato, U., Measuring energy performance: a process based approach. Appl. Energy **222**, 540–553 (2018)
3. Moreno, B., García-Álvarez, M.T., Ramos, C., Fernández-Vázquez, E.: A general maximum entropy econometric approach to model industrial electricity prices in Spain: a challenge for the competitiveness. Appl. Energy **135**, 815–824 (2014)
4. Roth, J., Brown IV, H.A., Jain, R.K., Harnessing smart meter data for a multitiered energy management performance indicators (MEMPI) framework: a facility manager informed approach, Appl. Energy **276** (2020)
5. Schulze, M., Nehler, H., Ottosson, M., Thollander, P., Energy management in industry - a systematic review of previous findings and an integrative conceptual framework **112**, 3692–3708 (2016)
6. Perroni, M.G., Gouvea da Costa, S.E., Pinheiro de Lima, E., Vieira da Silva, W.: The relationship between enterprise efficiency in resource use and energy efficiency practices adoption. Int. J. Prod. Econ. **190**, 108–119 (2017)
7. Fiedler, T., Mircea, P.M., Energy management systems according to the ISO 50001 standard - challenges and benefits. In: 2012 International Conference on Applied and Theoretical Electricity, ICATE 2012 – Proceedings (2012)
8. May, G., Taisch, M., Kelly, D., Enhanced energy management in manufacturing through systems integration. In: IECON 2013 - 39th Annual Conference of the IEEE Industrial Electronics Society, pp. 7525–7530 (2013)
9. Wang, W., Yang, H., Zhang, Y., Xu, J.: IoT-enabled real-time energy efficiency optimisation method for energy-intensive manufacturing enterprises. Int. J. Comput. Integr. Manuf. **31**, 362–379 (2018)
10. Shrouf, F., Miragliotta, G.: Energy management based on internet of things: practices and framework for adoption in production management. J. Clean. Prod. **100**, 235–246 (2015)
11. May, G., Stahl, B., Taisch, M., Kiritsis, D.: Energy management in manufacturing: from literature review to a conceptual framework (2017)
12. Deschamps, F., Pinheiro De Lima, E., Gouvea Da Costa, S.E., Alves, E., Santos, P., Van Aken, E.: Development of enterprise engineering guidelines for enterprise diagnosis and design (2013)
13. Tranfield, D., Denyer, D., Smart, P., Towards a methodology for developing evidence-informed management knowledge by means of systematic review * (2003)
14. Platts, K.W., Gregory, M.J.: Manufacturing audit in the process of strategy formulation. Int. J. Oper. Prod. Manag. **10**, 5–26 (1990)
15. Schuh, G., Anderl, R., Dumitrescu, R., Krüger, A.: Industrie 4.0 Maturity Index (2020)

A Method of Description the Production Process for Production Planning and Control Purposes

M. Fertsch(✉)

Poznan University of Technology, 5 M. Skłodowska-Curie Square, 60-965 Poznan, Poland
michal.fertsch@put.poznan.pl

Abstract. The aim of the study was to develop and verify a method for describing the structure of the production process for the purposes of production planning and control, responding to the identified challenges occurring in these areas in production companies. The description of the structure of the production process is understood as a set of information about the longevity of the production process and its stages, resources necessary to implement the stages of the production process and the relationships between individual stages of the production process The development of a method for describing the production process is designed to enable the construction of tools using the description of the structure of the production process, the use of which will be possible in enterprises.

Keywords: Production process description · production planning · production control

1 Introduction

The aim of the study was to develop and verify a method for describing the structure of the production process for the purposes of production planning and control, responding to the identified challenges occurring in these areas in production companies The description of the structure of the production process is understood as a set of information about the longevity of the production process and its stages, resources necessary to implement the stages of the production process and the relationships between individual stages of the production process The development of a method for describing the production process is designed to enable the construction of tools using the description of the structure of the production process, the use of which will be possible in enterprises..

2 A Method of Describing the Production Process for Production Planning and Control Purposes

2.1 Principles of Functioning of the Proposed Method

The challenges posed to the method of describing the production process result from the increase in customer requirements. They are widely described in the literature and are characterized, among others, by pressure on companies to shorten the time of order

M. Dragomir et al. (Eds.): ICPR 2023, LNPE, pp. 141–147, 2025.
https://doi.org/10.1007/978-3-031-92082-0_14

execution or the willingness of customers to influence the final shape of finished products. Such a challenge required the search for a universal solution that could be used in a large group of companies of various sizes, operating on different markets and in various industries. The requirement for the proposed solution is to enable the analysis of data concerning: quantities and production dates, demand forecasts, stocks of finished products, load on production resources. In the area of production control, the proposed solution should use data on: production programs and dates, work in progress, warehouse stock, production time, production cycles, individual quantities or collective time advance and allow for the analysis of these data.

2.2 Rules for Developing a Description of the Production Process

The process of developing the production process runs in few steps. The process of developing the structure of the production process for the purposes of production planning and control is shown in Fig. 1.

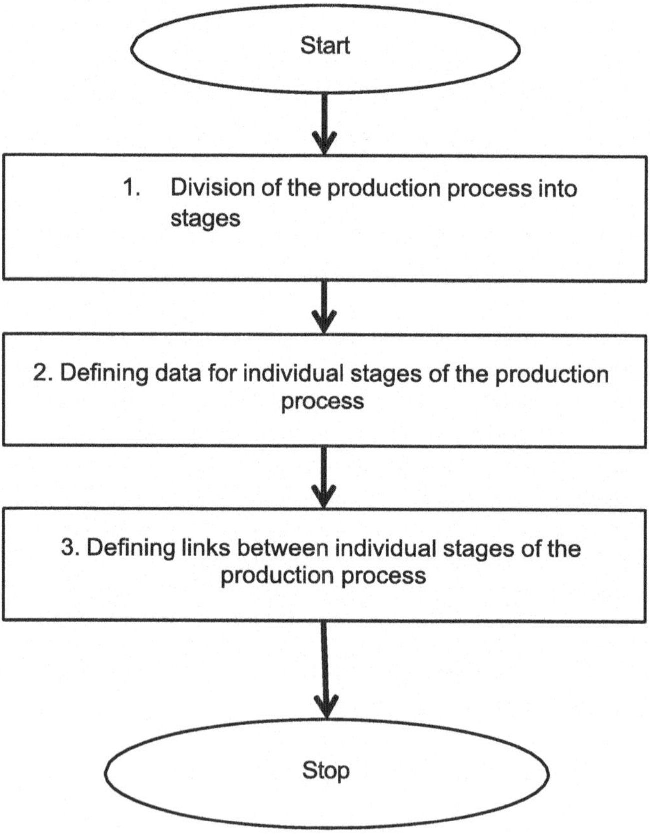

Fig. 1. The process of developing the structure of the production process.

The first step in defining a production process, for the purposes of production planning and control, is to divide the entire production process into stages. In order to divide the production process into stages, it is necessary to refer to the technology functioning in the enterprise. The division of the production process into stages is closely related to the resources that are used to carry out production. One stage can only include successive technological operations carried out on one production station. Changing the production station entails the need to define the next stage of the production process. If several operations are carried out at one station, but they are not carried out one after the other, a larger number of stages should be designated so that all operations to be carried out at that post are taken into account, while maintaining the continuity of the operations performed. The rules of division the production process into stages is shown on Fig. 2.

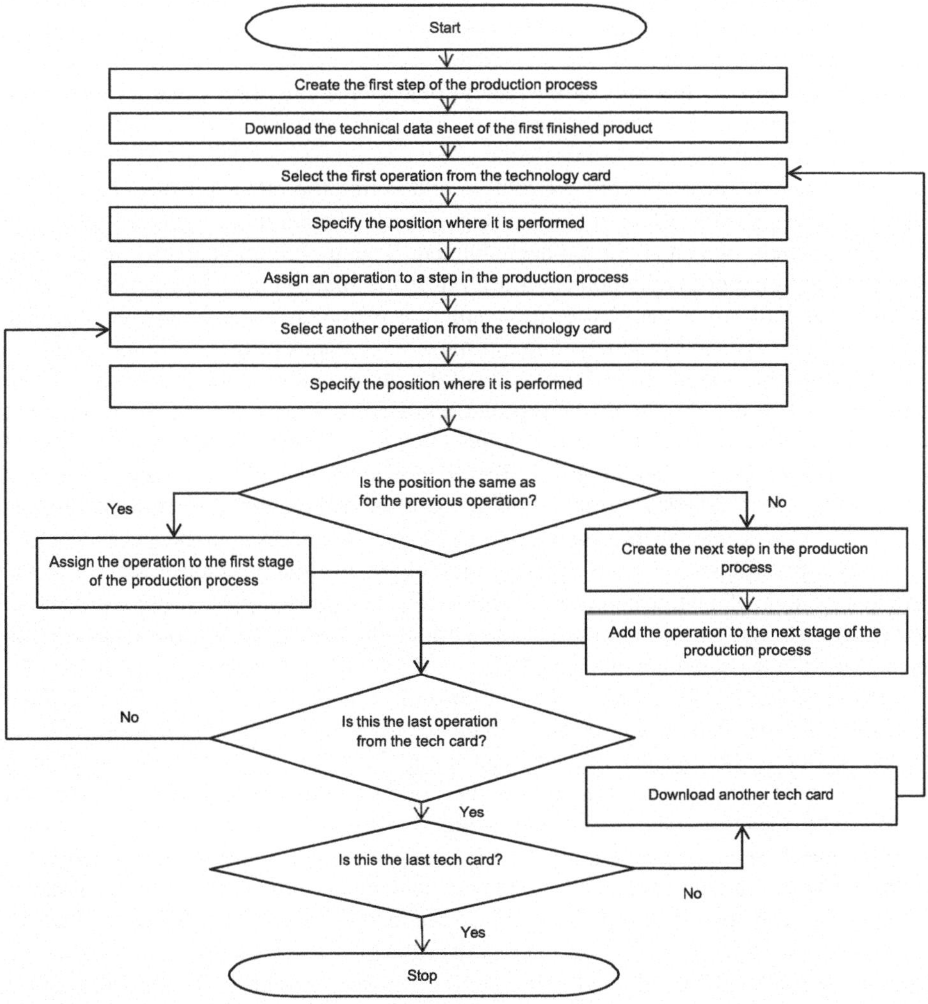

Fig. 2. Defining the stages of the production process

After completing the development of the division of the entire production process into stages, you should start defining data for each of the stages of the production process. The first information to be developed is the effect of the implementation of a given production stage. In order to ensure the functionality of the proposed solution, some strictly defined rules for determining the effect of the production process stage have been developed. For each stage, its effect must be determined so that it can function as a single piece. To do this, it is necessary to analyse the technical conditions in the enterprise. If, for example, an enterprise produces small products and they are sold or stored on transport pallets, in this situation, the result of the last stage of production should be one pallet of finished products. Defined as one piece, without isolating smaller elements within it. If we define the effect, for the production of elements included in the finished product, we have two options to choose from. In a situation where at further stages of production is used, for example, a set of parts, as a result of a given stage of production we can define: one part or the whole set. However, this will have consequences. If we define this effect as one set of parts, then it will not be possible to plan individual parts as products for sale or as spare parts. It will be necessary to plan whole sets. In a situation where the parts are not intended for sale as spare parts, it is simpler to define the effect as one set of parts.

Then go to the next stage of production and define the effect for this stage. After determining what is the effect of a given stage, check whether there is already a stage in the list for which such an effect has been defined. If such a situation arises, such a stage of the production process should be removed from the analysed list. The fact that several stages occur with the same effect may result, for example, from the use of the same parts in several finished products. Separating several stages of the production process with the same effect from the point of view of the functionality of the proposed solution is pointless. These steps should be repeated until all defined stages of the production process have been exhausted. Another element that must be determined for each stage of the production process is its durability. Technological data should be used to determine the longevity. It is necessary to take into account the technological operations that the stage covers and determine the number of their repetitions necessary to produce a specific effect. The quantitative relationship between the effect of the production stage and the structure element of the finished product should be taken into account. Using the example described earlier, if we define one part as an effect, the number of repetitions of a given operation will correspond to one part. If the result of the stage of the production process is to be a set of parts, the number of repetitions of individual operations should tell the assembly of parts. Analysing various variants of the stages of the production process, we can encounter three groups of them. The first of these are the stages of the production process that do not generate demand for the effects of other stages of the production process. This means that they are supplied directly from the supply area. Only materials from the purchase are necessary for their implementation. The second group of stages of the production process are stages that generate demand for the effects of other stages of the production process, but their effects do not feed other stages of the defined stages of the production process. This means that the effects of these stages of the production process are intended for sale and correspond to finished products. The third group of stages of the production process are stages that require both the effects of other stages of the production process and feed other stages of the defined stages of the production process with their effects. Defining the links between the stages of the production process should start with determining what effects of other stages of the production process are necessary to feed a given stage of the production process for the stages corresponding to the finished products. In a situation where the supply effect is determined, the relationship between the amount of demand generated by the powered stage of the production process and the size of the powered stage effect should be established. The developed stages of the production process will not constitute a coherent structure of the production process. Depending on the decisions made in the production planning and control process, the structure of the production process will adapt to the decisions made by people involved in production planning and control. In a situation where these people decide to use the stock of one of the effects of the stage of the production process, they may resign from its production. This will automatically exclude from the structure of the production process all power stages necessary to implement a given stage.

3 Functioning of Production Planning and Control Based on the Proposed Solution

The production planning process using the proposed production description method for production planning and control purposes should start with the transcription of the plan or sales forecast from the units used to develop them into the corresponding results of the stages of the production process. In order to make this transcription of production, the planner must have two pieces of information for each product item covered by the sales plan or forecast. This information is: the planned sales volume in the period in which we plan production and the relationship between the effect determined for the final stage of production and the unit used when planning the sale of the finished product. The simplest relationship that can be defined here is a one-to-one relationship. This is possible if sales and production are defined at the level of one piece, or when sales are carried out in bulk packaging in accordance with the size of the effect assigned to the stage of the structure of the production process. In another case, e.g. when the effect of the production stage is determined at the level of one pallet of the finished product, and sales are determined at the level of individual pieces of the final product, in order to plan production, it is necessary to determine how many pieces of the finished product are on one pallet. In order to plan the determination of multiples of repetitions of the final stage of the production process, divide the planned sales volume by the "n" factor, then round up to the whole. The date of completion of production is the latest possible date of completion of production, giving the possibility of timely sales. By further implementing tasks related to planning for the level of finished products, we can determine the dates of commencement of the planned implementation of the entire production as well as individual stages of production In a situation where the implementation of all production stages corresponding to finished products has been planned, it is necessary to proceed to the planning of the production of parts This is possible because we have information about the deadlines for the implementation of individual stages of the production process and thanks to the description of these stages, we know what amounts of power effects are necessary for implementation in order to be able to implement the powered stages. The deadlines for the implementation of the remaining supply stages should be set in the same way. Both if they supply the stage whose effect corresponds to the finished product as well as other feeding stages. Assuming that at the input, the date of completion will be the one that corresponds to the appropriate stage of the power supply. By planning production in this way, after the planning stage, it is possible to obtain information about the entire demand for production potential, assigned to specific production stations, in the corresponding time units. It should be noted, however, that in the above example no inventory analysis was carried out, either at the level of finished products, the corresponding effects or the effects feeding any stage of the production process. The situation when we want to take into account inventory levels at the production planning stage is presented below. When planning the implementation of stages corresponding to finished products, the desire to increase or decrease stocks should be taken into account at the level of the finished product. The number of effects corresponding to the finished product should not be manipulated. In a situation where the supply effects reserve is taken into account, the use of the proposed solution at the production planning stage assumes a decrease in these stocks to the level specified by the person planning production. The implementation of the supply stages corresponding to the planned reduction in the amount of inventory will not be planned. If one of the supply stages is not planned, no other stages of the production process will be planned to feed it. In this way, depending on the stock levels of the various feed effects, a number of different structures of the production process and corresponding production plans will be created. Thanks to the proposed method of describing the production process, this will be reflected in the information on the demand for production potential.

4 Verification of the Proposed Solution

The analysis of the obtained results showed that it can be used in a wide group of companies. The use of the method made it possible to:

- Facilitate the translation of long-term plans by accurately determining the demand for production capacity and its assignment to the right production stations. Access to this data was obtained at the MPS level,
- Enabling the analysis of production potential by accurately determining the demand for production potential and its assignment to the appropriate production stations. Access to this data was made available in the early planning stages.
- Simplify the process of compiling the data necessary for the implementation of production planning by using the existing defined stages of the production process.
- Enable material consumption analysis by accurately determining material requirements.
- The effects of the stages of the production process were defined as the smallest technologically possible units. It was possible to analyze the inventory of all the effects of the stages of the production process. The exact relationship between the supply effect and the demand generated by the fed stage was established.
- It was possible to define internal transport and tools of special resources as necessary for the implementation of the stage of the production process.

References

1. Fertsch, M.: Metoda planowania zapotrzebowania materiałowego w planowaniu produkcji i sterowaniu jej przebiegiem. Poznań: Wydawnictwo Politechniki Poznańskiej (2013)
2. Hadaś, Ł., Fertsch, M., Cyplik, P.: Planowanie i sterowanie produkcją. Poznań: Wydawnictwo Politechniki Poznańskiej (2012)
3. Fox, T.Q., Dog, T.L.: The quick brown fox jumps over the lazy dog, 5th edition, Animals eds., The Forest (2022)
4. Ohno, T.: Toyota Production System: Beyond Largescale Production. Productivity Press, Cambridge MA (1988)
5. Singh, N., Falkenburg, D.R.: Kanban systems. W R. C. Dorf Handbook of Design Manufacturing and Automation, pp. 567–585. John Wiley, New York (1994)
6. Thürer, M., Stevenson, M.: Bottleneck- oriented order release with shifting bottlenecks: an assessment by simulation. Int. J. Prod. Econ. **197**, 275–282 (2018)

Model of Selection of the Trajectory of the Transformation of the Production Cell

M. Fertsch(✉)

Poznan University of Technology, 5 M. Skłodowska-Curie Square, 60-965 Poznan, Poland
marek.fertsch@put.poznan.pl

Abstract. Let us assume the continuity of the development of production systems, from the beginning of their foundation to Industry 4.0 model. Let us formulate recommendations on how to start from the current state, which can correspond to each generation of production system models, and move in the shortest possible time and at the lowest possible cost to the optimal level, which currently corresponds to Industry 4.0 model. The analysis is carried out for the conditions of the machine industry, which supplies the other manufacturing industries with machines and is characterized by the most diverse operating conditions. The analysis does not apply to the conditions of high-tech manufacturing or the automotive industry as these industries have developed their own practices in the organization of production. The analysis also does not apply to the conditions of the food, chemical or clothing industries, in which the specificity of the processed materials and the technologies used determine the organization of production.

Keywords: Production cell · Design of production cell · Industry 4.0

1 Introduction

Currently, there are two models of the design of production systems. These are the American-European model and the Japanese model. The existing differences result from differences in understanding the essence of the production cell operation. In the American-European model, a separate production cell is designed with a predefined location, separate resources and tasks assigned to it. In the Japanese model, the flow of production between lined up workstations is designed.

The condition presented above refers to production environment characterized by a moderate saturation of production processes with automation and robotics. With regard to the conditions in which the saturation with automation elements increases until the production is fully automated, the models that emphasize the technical aspects of production processes apply. These are, in chronological order, flexible manufacturing systems, computer-integrated manufacturing (CIM) and reconfigurable manufacturing systems. The contemporary model of the production system is presented by the concept of Industry 4.0.

M. Dragomir et al. (Eds.): ICPR 2023, LNPE, pp. 148–151, 2025.
https://doi.org/10.1007/978-3-031-92082-0_15

2 State of the Art

The following overview of production system models is presented in chronological order. Individual models differ in the tangible and intangible resources used. In the American-European elementary model, a production unit is a set of resources organized in the form of a line, cell or workshop. Higher-order production units (departments) are created by grouping elementary cells and joining separate sets of the organizational and management subsystem. A variation of the American-European model is the QR - quick response manufacturing model [Suri 2020]. In the quick response manufacturing model, first-stage production units are small machining or assembly cells. In the Japanese model, the production process is organized in a linear form. The production line is divided into segments and as a whole constitutes a production cell. The line segments are separated by warehouses - buffers, where inventories of works in progress are stored.

The relatively clear difference between the American- European model and the Japanese model is obscured by two specific approaches to production systems. One is called lean manufacturing (LM) and the other is agile manufacturing (AM). Both present a specific comprehensive vision of the functioning of the production cell presented in the form of principles and techniques of its operation. The range of principles and techniques that recommend both approaches partially overlaps. It seems that the vision presented by lean production is closer to the Japanese model, and the vision presented by agile production corresponds to the American-European model. In the conditions of lean production, the organization of the production system is modeled on the basis of the Japanese model. In the conditions of agile production, the production units are short u-shaped lines. When grouped they form higher-order production units.

In a flexible manufacturing system, the production cell is a system configured of machining centers and tool exchange, transport and storage systems [Lis S., 1994]. In addition, in a flexible manufacturing system there are subsystems for supplying lubricants and coolants as well as subsystems for waste disposal. They do not differ significantly from those used in machines and conventional production cells. A machining center is a single numerically controlled machine tool equipped with devices for changing tools and workpieces. Often supplemented with an input/output warehouse for short-term storage. It occurs alone or is part of higher-level flexible production systems. The structure of the higher-level flexible production system is complemented by the automated storage and retrieval systems ASRS and the automated guided vehicles systems AGVS.

The CIM model assumes implementation of an automated production system - a production plant. The first level production units in such a system are flexible machining and assembly systems [Sch., 1994]. The automated production system is controlled by an IT system. The entire automated production system is controlled by an IT system. This system consists of a number of CaX subsystems. Individual subsystems cooperate with each other through a common database.

The CIM concept introduced a new element into the structure of the production system, an IT system that controls the operation of the production cell. The introduction of this element resulted in the emergence of new models of production systems, in which the focus was placed on the functioning of the IT system. These models are the intelligent production system proposed by C. Dagli [Dagli C. H., (ed), 1994] and the swarm-based system [Dagli C. H., Meyyappan L., 2005]. It seems that both of these models have remained in the theoretical sphere. There is no information in the literature on the implementation of both models in practice.

The model of the reconfigurable production system assumes that the design of the production system should enable the introduction of quick changes in the structure both in the mechanical part (machines - machining centers) and in the software in order to adapt to rapid changes in the demand for potential and the produced assortment within a specific class of parts [Dashchenko A. I. (ed), 2006]. It can be assumed that a reconfigurable production system is a flexible production system or a CIM class system equipped with the ability to change the structure of both the technical subsystem and the control subsystem.

Industry 4.0 is a continuation of the trend of emphasis on the functioning of the IT system as a subsystem of the production cell. It proposes the structure of the IT system consisting of five levels [Lee, 2015]:

smart connection level,
data to information conversion level, cyber level,
cognition level,
configuration level.

There is no general model of production cell optimization in the literature. The design process is iterative and consists in successive application of partial criteria. These are maximizing the similarity of the produced assortment, minimizing the size of the set of assortment items produced in a given production cell, minimizing the diversity of machines and devices in a given production cell. The last partial criterion is limiting cooperation in the execution of the set of assortment items assigned to a given cell and striving to perform it in its entirety in a given production cell.

3 Problem Formulation

The historical development of production systems meant that each of the models created during this development is present not only in the literature of the subject, but also in the form of solutions functioning in industrial practice. As a result, people responsible for production management face the problem of choosing a path (trajectory) to transform the currently existing production system. A rational choice is not to repeat the whole path of systems development, but to go to the highest level available in the existing conditions and limitations.

The proposed method of formulating recommendations for the selection of the production system transformation trajectory is based on comparing the structure of the existing production system with the structure of the target system and taking into account the existing limitations.

4 Heading 1

The developed method of problem solution requires solving tfe following equation:

Objective function:x i al + xi bk = max

Constraints:Σ x i caj + Σ xi cbk $\leq$ C

> xiτal+ Σ xi τbk $\leq$ T

where:

xi - decision variable, xi = {0, 1}, i = {1, …., n+m},

al - material resources set (machines, software), l ={ 1,…,n}

caj – cost of resource aquisition ai, l = {1,…,n}, τal – cycle of resource aquisition ai, l = {1,…,n},

bk - intangible resources set (knowledge, competence), k= {1,..,m}

cbk - cost of resource aquisition bk, k = {1,…,m}, τ

bk – cycle of resource aquisition bk, k = {1,…,m},

C - assumed value of outlays for the transformation of the production system,

T - the period adopted for the transformation of the production system.

The list of considered tangible and intangible assets is determined by comparing the current model of the production system and the target model. The value of outlays and the length of the period adopted for the implementation of the system transformation are determined by the persons managing the production. They use the project management method. The task is solved by the review method.

References

Dagli, C.H., (ed.): Artificial Neural Networks for Intelligent Manufacturing. Chapman & Hall (1994)

Dagli, C.H., Meyyappan, L.: Swarm based system of system behavior simulation for network centric enterprise. In: Proceedings of 18th International Conference on Production Research, Salerno, Italy (2005)

Dashchenko, A.I. (ed.): Reconfigurable Manufacturing Systems and Transformable Factories. Springer – Verlag, Berlin Heidelberg (2006)

Lee, Y., Bagheri, B., Kao, H.: A cyber – physical systems architecture for industry 4.0 – based manufacturing systems. Manuf. Lett. **3** (2015)

Lis, S., Santarek, K., Strzelcza, S.: Organizacja Elastycznych Systemów Produkcyjnych, Wydawnictwo Naukowe PWN, Warszawa (1994)

Scheer, A.-W.: CIM: Computer Integrated Manufacturing: Toward the Factory of the Future. Springer – Verlag, Berlin – Heidelberg, 3th edition (1994)

Suri, R.: Quick Response Manufacturing. CRC Press (2020)

Designing the Information Architecture of an OSS-Based IIOT Application for a Weld-And-Polish Robotic Cell

M. Fulea[1(✉)], C. Ioaneş[2], B.Mocan[1], and M. Murar[1]

[1] Department of Design Engineering and Robotics, Technical University of Cluj-Napoca, Cluj-Napoca, Romania
michal.fertsch@put.poznan.pl
[2] Inno Robotics SRL, Cluj-Napoca, Romania

Abstract. The ubiquitous connectivity between machines, sensors and actuators, which is the IIoT, is significantly changing how goods are manufactured. It has a great potential to reduce costs, improve production efficiency, and enhance manufacturing and service quality. It also raises challenges for companies, especially for SMEs, like data management or investment costs.

In this work we undertake a case study, within a robotic solution integrator SME, which addresses these two issues. We report on designing the information architecture of an OSS-based, IIoT operating and control software application for a weld-and-polish robotic cell, for which we relied on several taxonomies, identified in the literature, related to manufacturing processes. We also report the technical and managerial implications of this technical endeavour, the present case study aiming to serve as a good practice for other SMEs interested in adopting OSS as a backbone of their IIoT applications.

Keywords: IIoT · information architecture · manufacturing processes · OSS

1 Introduction

The Industrial Internet of Things (IIoT) offers manufacturing businesses huge financial prospects and challenges [19]. The IIoT is significantly changing how goods are manufactured. It has a great potential to reduce costs, improve production efficiency, and enhance manufacturing and service quality. Nevertheless, it challenges companies, especially SMEs, on data management, IT infrastructure, environmental requirements, or investment costs [23].

The IIoT-based solutions are rather software-first technology, as compared to traditional hardware solutions, being much more efficient, flexible, and able to absorb changes without greatly impacting operating costs [6]. However, as manufacturers increasingly rely on software to increase their operations' effectiveness [14], all components of costs associated to software become important issues of concern. As such, open source software, which is typically made available at no cost, for anyone to use, become increasingly

M. Dragomir et al. (Eds.): ICPR 2023, LNPE, pp. 152–163, 2025.
https://doi.org/10.1007/978-3-031-92082-0_16

appealing, reportedly bringing over £40 billion of potential value to businesses in UK only [31].

To SMEs, both the IIoT and open source software raise several challenges, apart from their great potential. Possible unfavourable effects of IIoT adoption could be related to data security and data chaos, complexity and user acceptance of the new technologies, among others [19]. Regarding OSS adoption, decision makers in SMEs consider technical support, lack of drivers, functionality or innovativeness as potential problematic issues [5]. As, to date, established paths or frameworks of OSS adoption cannot be identified [1], the above identified challenges and issues still have to be addressed by any such endeavour, and managers have to carefully balance costs with desired production flexibility.

Our focus in this work is on manufacturing data management in custom-built IIoT solutions based on open-source software. We undertake a case study, within a robotic solution integrator SME, by which we report on designing the informational architecture of an OSS-based, IIoT operating and control software application for a weld-and-polish robotic cell. To design the process information architecture, we relied on several taxonomies, identified in the literature, related to manufacturing processes. We also report the technical and managerial implications of this technical endeavour, the present case study aiming to serve as a good practice for other SMEs interested in adopting OSS as a backbone of their IIoT applications.

The remaining of this article is organized as follows. Section 2 discusses OSS adoption in manufacturing SMEs and taxonomies for manufacturing processes data with respect to the existing literature. Section 3 is dedicated to the case report, and Sect. 4 discusses the results and their implications.

2 Literature Support

2.1 Open Source Software Adoption in Manufacturing SMEs

Manufacturing processes become increasingly complex, under the pressure of product complexity, energy costs, mass customization trends, shorter delivery times, shortage of workers, etc. [14, 15, 22]. As such, manufacturers increasingly have to rely on software to increase their operations' effectiveness [14].

But it's not just effectiveness that counts. Business agility is also important, and, in the manufacturing industry, open source software (OSS) is considered to be critical in managing it [10, 25]. Several trials can be done with different OSS components to see how things work. And instead of building or requesting new (proprietary) solutions, organizations may better act like integrators, choosing the right (existing) software component for each problem, applications being built on top of open-source blocks [25].

OSS is far from being a cost free alternative to proprietary solutions [1]. To use OSS, businesses need to invest a significant effort, and several factors have to be considered when adopting such a solution.

Although several formalized processes to adopt OSS components have been proposed ([27], for instance), organizations developed their own approaches in this respect [1]. Choosing the adequate OSS component is not just a matter of functional requirements

[26]. Instead, multiple additional factors have to be considered, the selection process being rather complex [20].

These factors include, apart from quality, maturity and reliability: availability of support from either community or commercial sources, security, flexibility and customisation possibilities, documentation quality, software licence type, cost of maintenance, integration effort, and lock-in risk.

The open-source nature is reported to strongly contribute to the development of local capacity [10], i.e. the ability of (local) people, organizations and society as a whole to manage their affairs successfully [17]. This allows decreasing development and maintenance costs over time, contributing to business sustainability [3, 10, 31].

Regarding figures, OSS is reported to have brought £46.5 billon of "potential value" to businesses in the UK in 2020. It contributes $59 billion each year in UK's GDP, this country being the top contributor to OSS in Europe and the fifth biggest contributor worldwide [31].

2.2 Taxonomies for Manufacturing Processes Data

With IIoT architectures, large volumes of structured, semi-structured, and unstructured data, acquired from a variety of heterogeneous sources - the so-called big data - has to be processed. This may be difficult not due to the large datasets or the need for improved performance or efficiency, but to the changes in the architecture that is needed to manage this data [11].

Taxonomies are what information-science professionals call controlled vocabularies – planned, prescriptive ways of adding descriptive metadata to content so that it can be retrieved effectively. The idea is that the taxonomy defines a limited set of terms for describing our content in the background; content creators must attach them to any new piece of content. Put simple, taxonomies illustrate what can or cannot be done in a process or organization [12].

A taxonomy ensures that data is properly categorized and classified. Information can be shared across multiple platforms without worry that the data may be misinterpreted or misunderstood. Taxonomies can be therefore of great help in designing the process information architecture. Several taxonomies related to manufacturing processes have been discussed in the literature, whether referring to process productivity, cybersecurity, resource consumption, environment, control variables, manufacturing business models, or maintenance.

Manufacturing processes are core value generators in organizations, many times determining their competitiveness and sustainability. Taxonomies covering manufacturing processes (e.g. [2, 7]) can be of great help in determining what process data to manage. For instance, in [2] the following key values are important in assessing operations efficiency: control variables (data from sensors to regulate these control variables), productivity variables (e.g. quality of material produced, scrap), energy and service variables (e.g. monitoring energy consumption in the process, temperature or pressure in-s and out-s), and emission variables (emissions generated by the process).

Taxonomies covering manufacturing strategies and business models [8, 13, 30] can help in managing process data in case of, for instance, industrial distributed manufacturing, where a network of geographically dispersed manufacturing facilities is coordinated using information technology [7].

As manufacturing systems became interconnected and data-driven, they also became subject to at least the same cyber threats as compared to only IT systems [24]. Taxonomies related to security threats and their associated risks are discussed in several works, for instance [29] and [18], and cover attack methods, targets and location, consequences, and risk models. Countermeasures to attack methods are also proposed [24], like fingerprinting for false data injection (at sensor or actuator level), or side channel monitoring for physical tampering, at the physical process level.

Maintenance management (MM) plays a primary role for competitiveness in manufacturing [21], impacting on system availability, quality, productivity, and operational costs. Human factors and information flows are critical factors in maintenance management [28]. Decision making in MM has to be thoroughly supported through proper information, and taxonomies and ontologies are reported to be of great help in this respect [4, 16, 21].

3 The Case Report

We report on designing the information architecture of a weld-and-polish robotized line, producing parts of metal furniture. Before automating it, the process was carried out exclusively by hand with welders and mechanical lathe operators, being problematic in terms of part quality, productivity, work conditions (a lot of dust and noxious substances are produced in the process), and availability of personnel. A robotized line to automate the process was therefore designed and built. The key hardware components are a robot, two welding tables, a MIG/MAG welding equipment, a polishing tool, and an automatic tool changer. The production data is to be taken from the PLC by a dedicated software component (the Broker) and transferred over VPN to a local server for further processing. The robotized line and its architecture can be seen in Fig. 1. The PLC and broker, and the local server, represent the layers 1 and 2 of an IIoT architecture.

The initial requirement for data management was to collect production (context information) data consisting of cycle time per piece, length of weld cords on each piece, welding time per piece, length of cord to be polished, polishing time, cell status (manual/auto), preparation time for a new batch of parts, and maintenance time. Reports on production data should reflect (at least) productivity and OEE (overall equipment effectiveness = availability x performance x quality).

Following the prototype, an improved version of the control software application was required, hence a new information architecture had to be designed. Regarding this, the main concern of the company was to have an effective and efficient data management.

3.1 General Considerations

The company considers OSS to be a key enabler of business agility, to commoditize automation, and to successfully tackle integration issues in complex projects, so they

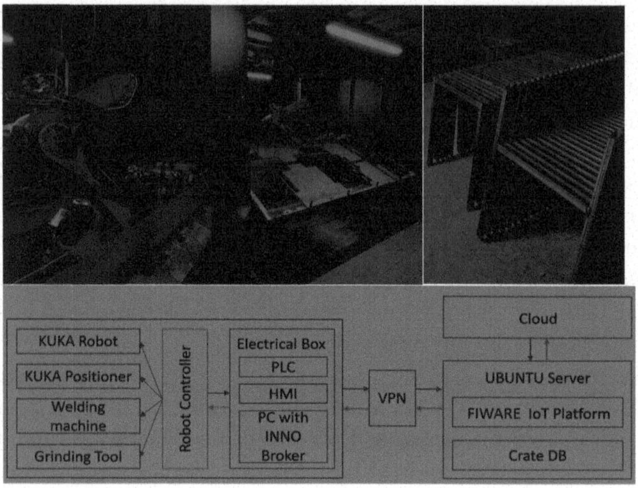

Fig. 1. (a) robot arm and tool stand, (b) table with fixture and parts, (c) welded parts, (d) the system architecture.

decided to rely on OSS in this project, specifically on the FIWARE software framework (fiware.org). By doing so, this technical endeavour should attain cost-effectiveness, flexibility, maintainability, and upgradability.

Fiware is a cloud-based framework of open source components which help manage the so-called context information (i.e. process data, in our case). Fiware's core component, the Orion Context Broker, integrates all the components, and allows other applications to update and / or consume the context information. Fiware also provides a component for backend device management, used to connect IoT devices to the platform.

3.2 Asset Configuration

To model our robotic line, we set up a simplified plant model, having one site, one area, one workstation, and two units (one for welding and one for polishing). Physically we have one robotic cell performing part welding and polishing.

We installed and configured the Fiware stack on a virtual machine running Ubuntu. After building and configuring it, and checking that it runs as needed, we deployed it into production, on a virtual machine server, and assigned it its own IP address in the organization's network.

A script containing cUrl commands was developed to manage the assets within the context broker. The workstation (the robotized line) and its two units (the welding and the polishing unit) do generate time series data, which is managed by Fiware's context broker, and can be viewed and interpreted locally, using Grafana charts and graphs, and remote, using the data exploration and visualization platform Apache Superset.

3.3 The Time Series Data Structure

As with any process automation, a significant volume of data is now generated in the process, its sources being the manufacturing equipment, the automation system, and the work pieces. This is the information context of the process, and it has to be sent to the context broker. We discuss here the time series data structure that were designed, inspired by the taxonomies discussed above, aiming to have both an as-complete-as-possible and easily-manageable context information.

While developing and testing the robotized line prototype, two time series data types were employed, according to the existing experience and practices: one for managing the workstation states, and one for managing the part production.

However, although it was suitable for hardware prototype development and testing & integration, we decided to completely redesign the data management, using the above taxonomies as a reference.

To effectively manage the information related to the weld and polish operations, we started from the operational business artifacts that represent the context of the technical process [9]. The two relevant artifact classes, in our case, are the *equipment* performing the operations (the two units of the robotic line) and the *part* (the unfinished product the robotized line is working on).

To manage the workstation (i.e. equipment – the two units of the robotized line) data, we kept the initial artifact class, which we further extended; the new data structure is summarized in Table 1. Additions and changes to the initial data structure are in italic; the taxonomies which inspired these changes are mentioned in square brackets.

For instance, "booting" the workstation would generate a record like {ws_state:starting, op_type:initializing, part_state:null, ...}; once "booting" completed, the corresponding record would be like {ws_state:waiting-for-user, op_type:null, part_state:null, ...} because the interface waits for an operator to authenticate.

To manage the actual operations, two additional record types were set up: one to store operations, and one to store details of a specific operation, like the electrode wire advance speed, the welding seam geometry (thickness, height), voltage, current, etc. This second record type was inspired by taxonomies covering operations efficiency. The data structures are summarized in Tables 2 and 3.

To persist the time series data, three subscriptions to Fiware's REST service for storing, querying and retrieving spatial-temporal data had to be created, one for each data structure described above. Then, data instances for each data type could be persisted in CrateDB, Fiware's time-series database, using HTTP requests.

3.4 Data Visualization

Production data should be reported under the form of production figures and OEE (overall equipment effectiveness). This data was computed using the query engine provided by Apache Superset, one of the software components included in the used open source stack on the highest architectural layer. Data visualisation was implemented using the Dashboards feature of Superset (sample graphs are presented in Fig. 2).

The OEE was computed by first computing availability, performance and quality. Availability was computed from the workstation table, by determining the time-auto

Table 1. The equipment time series data structure.

Key	Remarks
ws_state	*workstation current state (possible values: "starting", "waiting-for-user", "available", "auto", "changing-setup", "manual", "stopping", "stopped")* [manufacturing-operations, security]
op_type	operation currently being performed (possible values: "initializing", "*authenticating*", "*changing-setup*", "auto-processing", "manual-processing", "stopping", "shutting-down") [manufacturing-operations, security]
op_reference	an identifier of the operation in which the part was produced, to be used in relation with the operations tables [manufacturing-operations]
part_state	state of the produced part, for auto- and manual- processing (possible values: "good", "non-compliant", "not tested")
ws_msg	*an optional additional text message with details regarding the current operation* [manufacturing-operations]
user_id	*the currently authenticated user id (if applicable)* [security]
order_reference	*an identifier of the customer order or maintenance work order, to be linked to other information systems (e.g. the ERP)* [manufacturing-operations, maintenance]
idx_meter_ electric	*the current index of the electric meter (to count the consumed energy per operation) [manufacturing-operations, maintenance, efficiency]*
idx_meter_ emissions	*similar to the above, but counting gas emissions* [manufacturing-operations, maintenance, efficiency]
fingerprint	*a hash value of the data of one record, to check the record authenticity* [security]

percent of the total robotized line functioning time. Performance was computed by comparing (weld_len / weld_time) and (polish_len / polish_time) against the desired values. Quality was determined based on the number of compliant and non-compliant produced parts.

4 Discussion on Results

We discuss in what follows how the case report results, in particular the lessons learnt, support the SME in attaining superior responsiveness to customers and strategic partners, as a key competitive advantage. The discussion is organized around a couple of competitiveness factors that the company aims to foster, related to its operations. These factors are technical- and managerial- related. In doing so, the present case study aims to serve as support and good practice in information architecture design, for other SMEs interested in adopting OSS as a backbone of their IIoT applications.

Table 2. The operation time series data structure.

Key	Remarks
part_id	the part identifier
op_reference	*in relation to the equipment data record [manufacturing-operations]*
part_status	if the part is compliant or not
op_time	operation time
weld_len	length (mm) of welding
weld_time	time (s) needed for the welding
polish_len	length (mm) of polishing
polish_time	time (s) needed for the polishing
part_msg	*a customizable string message related to the operation outcome on a part [manufacturing-operations]*
fingerprint	*a hash value of the data of one record, to check the record authenticity* [security]

Table 3. The operation details time series data structure.

Key	Remarks
op_reference	*in relation to the equipment data record [manufacturing-operations]*
part_id	*in relation to the operation data record* [manufacturing-operations]
welding_seam_ thickness, _height	*welding seam geometry* [manufacturing-operations]
electrode_wire_ advance_speed	*m / s* [manufacturing-operations]
voltage	*V* [manufacturing-operations]
current	*A* [manufacturing-operations]
protective gas flow	*liters / min* [manufacturing-operations]
robot_motion_speed	*TCP speed* [manufacturing-operations]
torch_orientation	*orientation of the welding torch* [manufacturing-operations]
noise_level	*noise instant level when polishing, dB* [manufacturing-operations]
fingerprint	*a hash value of the data of one record, to check the record authenticity* [security]

4.1 Obtained Results

The direct result of the technical endeavour presented in the case study is an effective and efficient data structure for information management within the robotized line. By

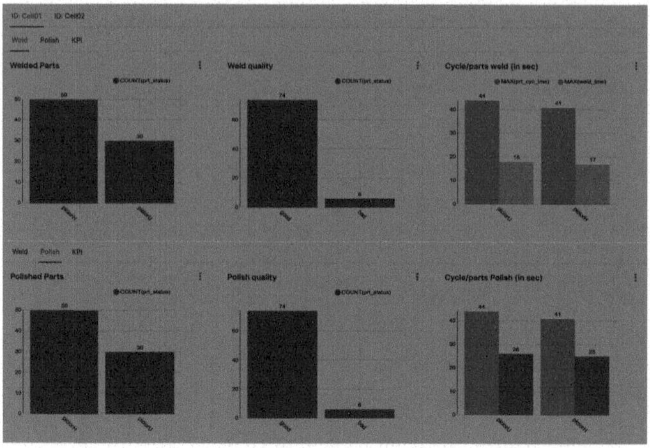

Fig. 2. Sample graphs of production figures.

considering previous technical projects as a reference, the company concluded that the information management in this endeavour better covers data safety and security, predictive maintenance, resource efficiency, and traceability, while also allowing a streamlined implementation of the operating and control software application for a weld-and-polish robotic cell.

Nevertheless, the most important result is an indirect one – the newly achieved local capacity, within the SME, of OSS integration and software development. While not necessarily being technically mature in it's current form, the developed software application and its future variants will help the engineering team grow and become more skilled in providing customized I4.0 (and above) solutions. It will also help, on the long term, with reducing costs with proprietary licences; money invested in such costs will be better invested in human resources, thus contributing to the sustainability of its business.

4.2 Implications for Practice

Open source platforms adoption. For many SMEs, access to state-of-the-art technologies in terms of functionality is very important, and rarely do they afford to invest in proprietary software, especially in R&D projects or when experimenting new ideas. OSS fits well for many of the automation and data management features SMEs wish to develop. Adopting OSS in SMEs is also a worldwide trend, and many such companies aim to align to it.

Avoidance of vendor lock-in situations. On the long term, relying on OSS for feature implementation will allow an organization to avoid such traps, which are difficult to foresee or avoid in particularly challenging projects.

Experimentation of niche technologies for R&D or challenging or highly context-tailored projects. Innovative SMEs do get involved in R&D projects and actively participate in proposing such projects with partners both in industry and academia, and investing in proprietary software for such purposes is not feasible and sometimes not ethical (especially with grants). On the other hand, SMEs are many times asked by clients to provide one-of-a-kind technical solutions. In such a case, proprietary software is not economically and technically suited, even with the technical support it may come.

4.3 Managerial Implications

Cost reduction. OSS components and frameworks are a cost-free alternative to investments that SMEs many times do not afford. And, rather than investing in technical solutions that soon become deprecated without further investment, one may consider better investing in in-house technical knowledge on OSS.

Response speed towards clients / future project partners. In short, it's better for an SME to use open-source software than to create original software, for which it many times lacks adequate capabilities. By building a „local" competence of combining and using the right (open) technologies for a particular problem, an SME can significantly reduce the time in which it deploys projects (or R&D project phases).

5 Conclusions

In this work we focused on manufacturing data management in custom-built IIoT solutions based on open-source software. Via a case study within an SME integrating robotic solutions, we reported on designing the informational architecture of an OSS-based, IIoT operating and control software application for a weld-and-polish robotic cell. To design the process information architecture, we relied on several taxonomies, identified in the literature, related to manufacturing processes. The obtained result was an effective and efficient data structure for information management within the robotized line. By considering previous technical projects as a reference, the company concluded that it better covers data safety and security, predictive maintenance, resource efficiency, and traceability, while also allowing a streamlined implementation of the operating and control software application for a weld-and-polish robotic cell.

We also reported the technical and managerial implications of this technical endeavour, the present case study aiming to serve as a good practice for other SMEs interested in adopting OSS as a backbone of their IIoT applications.

Future work will focus on extending the data visualisation features and on further strengthening the newly developed local capacity of OSS integration.

References

1. Butler, S., et al.: Considerations and challenges for the adoption of open source components in software-intensive businesses. J. Syst. Softw. **186**, 111152 (2022)
2. Cortés, D., Ramírez, J., Ponce, P., Molina, A.: S3 manufacturing process taxonomy. J. Manuf. Process. **67**, 579–610 (2021)
3. Crichton, D.: Open source sustainability | TechCrunch (2018). https://techcrunch.com/2018/06/23/open-source-sustainability/. Accessed May 2023
4. D'Amico, R.D., Erkoyuncu, J.A., Addepalli, S., Penver, S.: Cognitive digital twin: an approach to improve the maintenance management. CIRP J. Manuf. Sci. Technol. **38**, 613–630 (2022)
5. D'MacRedie, R.D., Mijinyawa, K.: A theory-grounded framework of Open Source Software adoption in SMEs **20**(2), 237–250 (2017).https://doi.org/10.1057/ejis.2010.60

6. Dobrasevic, S.: Navigating the Costs of Automated Technology. https://www.sme.org/technologies/articles/2022/october/navigating-the-costs-of-automated-technology/. Accessed May 2023
7. Fox, S., Alptekin, B.: A taxonomy of manufacturing distributions and their comparative relations to sustainability. J. Clean. Prod. **172**, 1823–1834 (2018)
8. Frohlich, M.T., Dixon, J.R.: A taxonomy of manufacturing strategies revisited. J. Oper. Manag. **19**(5), 541–558 (2001)
9. Fulea, M., Kis, M., Blagu, D., Mocan, B.: Artifact-based approach to improve internal process quality using interaction design principles | FULEA | ACTA technica napocensis - series: applied mathematics, mechanics, and engineering. ACTA Technica Napocensis-Ser. Appl. Math. Mech. Eng. **64**(4s), 697–706 (2021)
10. He, Y., et al.: Development and national scale implementation of an open-source electronic laboratory information system (OpenELIS) in Côte d'Ivoire: sustainability lessons from the first 13 years. Int. J. Med. Informatics **170**, 104977 (2023)
11. Ismail, A., Truong, H.L., Kastner, W.: Manufacturing process data analysis pipelines: a requirements analysis and survey. J. Big Data **6**(1), 1–26 (2019)
12. Kapetaniou, C., Rieple, A., Pilkington, A., Frandsen, T., Pisano, P.: Building the layers of a new manufacturing taxonomy: how 3D printing is creating a new landscape of production eco-systems and competitive dynamics. Technol. Forecast. Soc. Chang. **128**, 22–35 (2018)
13. Kathuria, R.: Competitive priorities and managerial performance: a taxonomy of small manufacturers. J. Oper. Manag. **18**(6), 627–641 (2000)
14. Van Leeckwyck, F.: Using Open Software in Manufacturing | ManufacturingTomorrow. https://www.manufacturingtomorrow.com/article/2019/01/using-open-software-in-manufacturing/12736. Accessed May 2023
15. Mocan, B., Fulea, M., Brad, E., Brad, S.: State-of-the-art and proposals on reducing energy consumption in the case of industrial robotic systems. In: 2014 International Conference on Production Research - Regional Conference Africa, Europe and The Middle East and 3rd International Conference on Quality and Innovation in Engineering and Management (ICPR-AEM 2014), pp. 328–333 (2014)
16. Ochoa, W., Larrinaga, F., Pérez, A.: Context-aware workflow management for smart manufacturing: a literature review of semantic web-based approaches. Futur. Gener. Comput. Syst. **145**, 38–55 (2023)
17. OECD. DAC Guidelines and Reference Series The Challenge of Capacity Development Working Towards Good Practice (2006)
18. Pan, Y., et al.: Taxonomies for reasoning about cyber-physical attacks in iot-based manufacturing systems. Int. J. Interact. Multimedia Artif. Intell. 4, Special Issue on Adv. Appl. Internet Things Cloud Comput. 45–54 (2017)
19. Peter, O., Pradhan, A., Mbohwa, C.: Industrial internet of things (IIoT): opportunities, challenges, and requirements in manufacturing businesses in emerging economies. Procedia Comput. Sci. **217**, 856–865 (2023)
20. Petersen, K., et al.: Choosing component origins for software intensive systems: in-house, COTS, OSS or outsourcing? - a case survey. IEEE Trans. Software Eng. **44**(3), 237–261 (2018)
21. Polenghi, A., Roda, I., Macchi, M., Pozzetti, A.: Data taxonomy to manage information and data in maintenance management. IFAC-PapersOnLine **53**(3), 245–250 (2020)
22. Pop, E., Campean, E., Braga, I.C., Ispas, D.: New product development of a robotic soldering cell using lean manufacturing methodology. Sustainability **14**, 14057 (2022)
23. Popescu, D., Dragomir, M., Popescu, S., Dragomir, D.: Building better digital twins for production systems by incorporating environmental related functions—literature analysis and determining alternatives. Appl. Sci. **12**, 8657 (2022)

24. Rahman, M.H., Wuest, T., Shafae, M.: Manufacturing cybersecurity threat attributes and countermeasures: review, meta-taxonomy, and use cases of cyberattack taxonomies. J. Manuf. Syst. **68**, 196–208 (2023)
25. Spiegel, R.: Open-Source Software Moves into Industry 4.0. (2020). https://www.designnews.com/automation/open-source-software-moves-industry-40
26. Spinellis, D.: How to select open source components. Computer **52**(12), 103–106 (2019)
27. Stol, K.J., Ali Babar, M.: A comparison framework for Open Source Software evaluation methods. In: IFIP Advances in Information and Communication Technology, 319 AICT, pp. 389–394 (2010)
28. Tsang, A.H.: Strategic dimensions of maintenance management. J. Qual. Maintenance Eng. **8**(1), 7–39 (2002)
29. Yampolskiy, M., et al.: Security of additive manufacturing: attack taxonomy and survey. Addit. Manuf. **21**, 431–457 (2018)
30. Zhao, X., Sum, C.C., Qi, Y., Zhang, H., Lee, T.S.: A taxonomy of manufacturing strategies in China. J. Oper. Manag. **24**(5), 621–636 (2006)
31. State of Open: The UK in 2021 Phase Three The Values of Open (2021)

Smart and Resilient Manufacturing: Machine Learning Approach to Identify Patterns of Employee Skill Levels for Cognitive Assistance Within Inspection Processes

Erdem Gelec[1], Christian Saba[2(✉)], and Johannes Wimmer[2]

[1] Fraunhofer Institute for Industrial Engineering IAO, 70569 Stuttgart, Germany
[2] University of Stuttgart, Institute of Human Factors and Technology Management (IAT), 70569 Stuttgart, Germany
christian-oswaldo.saba-gayoso@iat.uni-stuttgart.de

Abstract. Towards a resilient manufacturing using digital twins and cognitive assistant systems, human-focused and data-based approaches are essential. Work skill gaps and differences among employees still belong to the most common challenges that have a high impact in manual manufacturing processes. For measuring and inspecting of parts manually in between manufacturing process to secure perfect quality, it is essential that those measurements are performed in the correct way. Especially in manual processes, lack of data collection and usage for analysis purposes result in general in loss of competitiveness, whereas extracting data knowledge and reacting to skill level gaps immediately via algorithm predictions allows continuous improvement of measuring and inspection processes and employee skills.

Hence, the main purpose of this study is to empirically explore and predict skill related patterns that affect process performance and the quality of 3D hand-based measurements for quality inspection tasks in the automotive industry context using machine learning approaches. The setting is a manual measuring and inspection station in the ARENA2036 in Stuttgart, Germany using a battery case as an example part for the performed measurements. The algorithms are selected following the human-in-the-loop-approach, where understanding of the statistical data and the involvement of the process experts to validate results should be addressed. In this sense, principal component analysis (PCA) and clustering algorithms are used in the first exploration stage of linear separability. Subsequently, predictions are done using multiple classification algorithms as K-Nearest Neighbours (KNN), Random Forest, Support Vector Machines (SVM) and Naive-Bayes to evaluate accuracy. Thus, the novelty of this paper lies in validating predictions for qualification levels within the manual inspection context and promptly developing modules and guidelines for the use of data to enhance cognitive digital assistance systems for resilient factories.

Keywords: cognitive assistant systems · cognitive digital twin · resilient production · machine learning · AI-based inspection · upskilling

M. Dragomir et al. (Eds.): ICPR 2023, LNPE, pp. 164–174, 2025.
https://doi.org/10.1007/978-3-031-92082-0_17

1 Introduction

In the course of advancing digitalization in factories and production, data is assuming an increasingly important position in the production process. It can be used to analyse production processes and, by means of artificial intelligence, increase efficiency and productivity or minimize rejects and errors. In addition, processes can be planned and optimized virtually at an early stage using digital twins. The sensor technology required for this is already supplied with the production systems or can be applied externally.

These technologies all aim at the target state of a resilient factory, which controls itself and can react independently to deviations.

However, the role of humans is non-existent in many automated and digital processes. At the same time, there will still be manual activities in a resilient factory in the future and humans will continue to play an existential role in production.

In the following, we will therefore present a concept that uses AI to depict the user behaviour of workers in the resilient factory and show how cognitive assistance systems can support employees.

The specific research question on which this publication is grounded is how the employee differences in skill level can be systematically explored based on process data in manual tasks. It is envisaged that in the next step the cognitive assistance system for the work tasks will integrate machine learning in order to perform more adaptive, user-oriented, context-sensitive, and higher quality assistance by capturing and processing data continuously from a quality inspection work station.

2 State of the Art

2.1 Cognitive Assistance System for Production

Humans will remain an important part of the factory with their flexibility, intuition and creativity [1]. Cognitive assistance systems are technical systems that process information and support humans in performing their tasks [2, 3], enhancing the worker's capabilities [4]. The expected benefits according to industrial companies are mainly higher productivity, better process control and higher product quality [5]. Following a human-centred design approach, cognitive assistance systems consider the identification of the user's context, the specification of user requirements, the creation of design solutions, and their evaluation [6]. The adaptability to the user's qualification has been identified as one of the relevant requirements of assistance systems, while it is not necessary to adapt the worker assistance system to every individual, but to define different qualification levels like, beginner, intermediate and expert [7]. The difference regarding the assistance to be provided depends on the use case as well as the user skills and can affect the amount of information and the information content.

2.2 Artificial Intelligence in Production

Currently, many companies in different industries are using AI and can generate added value with the respective applications. Examples of applications in production and quality assurance include AI-based maintenance of equipment using acoustic signals, optimization of inspection processes using thermal imaging technology, and AI-assisted quality assurance in electroplating [8].

One example from research is quality improvement with the aim of identifying and eliminating defective products before customer delivery [9]. AI deployment aims to reduce reclamation costs while maintaining productivity, which reduces inspection workload, and increases customer satisfaction [9].

2.3 Clustering and Classification

In many use cases in production, the two AI types of supervised (classification) and unsupervised learning (clustering) are used, which were also used in the approach underlying this paper.

In the context of classification, objects are to be assigned based on their properties (input) either to none, to one class or to several classes from a given class structure (output). The model created by learning is also called a classifier. Classification is the most frequently considered problem in the field of machine learning [10].

Cluster analysis or clustering combines similar objects within a data set into groups. The groups formed are referred to as clusters, classes or segments. In contrast to manual segmentation by humans, the classification in clustering is performed by an algorithm. The data-driven grouping of similar objects requires that the similarity or the distance between objects can be defined meaningfully within the framework of an application [10].

Cluster analysis and classification are closely related. In classification, the classification of objects into an existing class structure is learned, whereas in clustering, a class structure is to be found first. Such a cluster analysis is called descriptive clustering. If the model is capable of assigning new data points to the classes of the newly identified class structure, then it is predictive clustering [10].

3 Approach and Data Analysis

3.1 Technical System

The technical system used consisted of the FARO Quantums Max, the FARO CAM2 v2023.0 measurement software and a Lenovo workstation as in Fig. 1.

Fig. 1. Technical Systems used

3.2 Measurement Setup

The measurement setup consisted of the FARO Quantum^{S} Max, which has 7 axis and an 8th rotary axis, which was mounted on the measurement table and on which the component to be measured lay. By means of a 3D-printed holder, the component could always be brought into the same position and fixed in such a way that measurements could be carried out correctly (see Figs. 2 and 3).

The measuring software FARO CAM2 v2023.0 used during the measuring process was displayed on a screen placed behind the measuring table, the exact steps of the measuring process could be displayed on the screen of the workstation used, which was placed sideways on the measuring table as shown in Fig. 1.

Fig. 2. Measuring system and axes

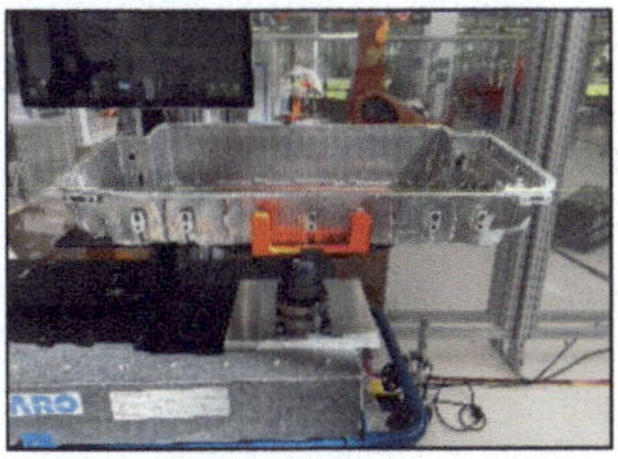

Fig. 3. Clamping system

3.3 Data Interface

As a data interface, the measuring arm was integrated into the Asset Administration Shell of ARENA2036 using the software development kit (SDK) of the software used for the measuring arm and thus a connection of the arm software to the integrated MQTT broker of ARENA2036 was created. This made it possible to dock the data of the measuring arm or the tracked joints of the arm to a Grafana database in order to temporarily store the data here. Finally, the stored data could be extracted and analysed by means of CSV files (Fig. 4).

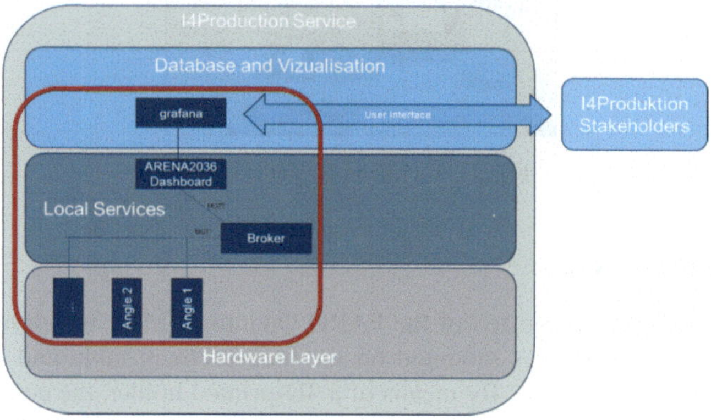

Fig. 4. Data interface

3.4 Measurement Process

The measurement process provided for individual surfaces on the outside and holes on the inside to be recorded at the beginning for both surface quality and internal diameter during the measurements by means of a tactile measurement. In addition, at the end of the measurement process, a surface scan of the entire lateral surface on the outside and inside of the component was performed. In total, there were 22 tactile measurement points and 2 scans during the defined process. Process times differed between 6 to 15 min.

3.5 Data Analysis and Results

All measurements were generated by random subjects with different skill levels in handling the FARO Quantum Max, from beginner level to intermediate and expert, over a 3-month period. From all readings taken, only those with proper quality (39), meaning not significantly interrupted processes by hardware or software issues, were selected for the subsequent data analysis.

Every reading (CSV-file) in the Grafana database consisted of tabular time-series records of the 11 different axis values, mentioned previously. All 39 files could then be imported to a Jupyter-Notebook environment for machine learning analysis. A statistical analysis was first performed to evaluate the linear separability of all data in a first attempt to group measurements with similar properties and recognize patterns. Features analysed in this context were variance, integral, range, mean, maximum, minimum as well as derivation values of each curve. By sorting all processes according to their variance values within the halftime reading zone a good matching percentage to the expertise level could be achieved. This could be explained by the fact that beginners tended to fluctuate more when handling the FARO measurement arm than experts, what could also be graphically visualized and analysed in the time-series plot (Fig. 5).

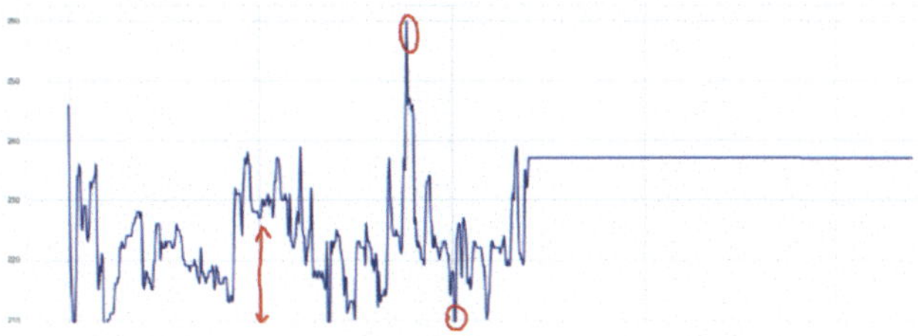

Fig. 5. Time-series record of rotation angle 1 (a1)

Secondly, a clustering analysis was performed using K-means algorithm and n = 3 clusters to compare cluster predictions to the annotated expertise level of the measurements. All the time-series steps (second cycle time) were the same for every reading and thus selected as attributes (dimensions) for this analysis, including the extracted statistical features. Principal component analysis (PCA) was then used to reduce the number of dimensions to 2 be able to visualize the built clusters as in Fig. 6, representing results for first rotation angle.

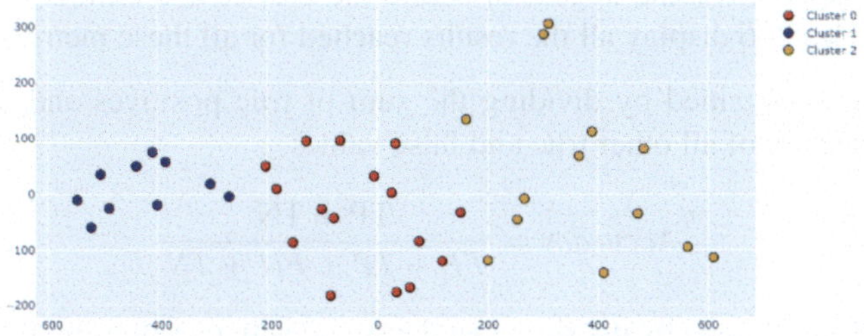

Fig. 6. Clustering results for rotation angle 1

Comparison of clustering results (predicted clusters) with annotated expertise levels over all 11 axis values revealed that rotation angle 1 delivered the best accuracy rate with 91%. Table 1 resumes all results for all axis values:

In the third and last step a classification analysis was performed based on different machine learning algorithms such as Support Vector Machines, Naïve-Bayes, K-nearest neighbour, Decision Trees, and Random Forest. The data was shuffled and split into 67% training and 33% testing set. 10-fold cross-validation was applied. Trained classes were the three expertise levels of the subjects.

1) Support Vector Machines: It performs the classification test by drawing a hyperplane.
2) Naïve-Bayes: It works on the principles of conditional probability as given by Bayes theorem. We used Gaussian.

Table 1. Prediction rate results for every position value

Position value	Accuracy
x	0.78
y	0.35
z	0.75
a1	0.91
a2	0.64
a3	0.66
a4	0.54
a5	0.67
a6	0.63
a7	0.61
a8	0.59

3) K-nearest neighbour: It works by calculating distance between data points through a function.
4) Decision Tree: It breaks the data into smaller subsets using decision nodes.
5) Random Forest: It combines different decision trees for more statistical power.

The metrics used to measure the performance of the classification models were the accuracy, precision, recall and f1-score values. They are based on rate calculations about the true and false positive and negatives within the classes and building an average on all of them. Tables 2–6 display all the results reached for all these metrics.

1) Accuracy: It is obtained by dividing the sum of true positives and true negatives values by the sum of all other true and false values.

$$Accuracy = \frac{\mathrm{TP} + \mathrm{TN}}{FP + TP + FN + TN} \tag{1}$$

2) Precision: It is obtained by dividing just the true positives values by the sum of false positives and true positives.

$$Precision = \frac{\mathrm{TP}}{FP + TP} \tag{2}$$

3) Recall: It is obtained by dividing just the true positives values by the sum of true positives and false negatives.

$$Recall = \frac{\mathrm{TP}}{FN + TP} \tag{3}$$

4) F1-score: It combines precision and recall into a single score.

Table 2. Classification results with Support Vector Machine algorithm

Training	Precision	Recall	f1-score
Beginner	0.86	1.00	0.92
Expert	0.91	0.91	0.91
Intermediate	0.88	0.78	0.82
Accuracy			0.88
Test	Precision	Recall	f1-score
Beginner	0.67	0.67	0.67
Expert	0.57	1.00	0.73
Intermediate	0.67	0.33	0.44
Accuracy			0.62

Table 3. Classification results with Naïve-Bayes algorithm (Gaussian)

Training	Precision	Recall	f1-score
Beginner	0.86	1.00	0.92
Expert	0.91	0.91	0.91
Intermediate	0.88	0.78	0.82
Accuracy			0.88
Test	Precision	Recall	f1-score
Beginner	0.50	0.67	0.57
Expert	0.57	1.00	0.73
Intermediate	0.50	0.17	0.25
Accuracy			0.54

Table 4. Classification results with K-nearest neighbour algorithm

Training	Precision	Recall	f1-score
Beginner	0.83	0.83	0.83
Expert	0.82	0.82	0.82
Intermediate	0.78	0.78	0.78
Accuracy			0.81
Test	Precision	Recall	f1-score
Beginner	0.50	0.67	0.57
Expert	0.80	1.00	0.89
Intermediate	0.75	0.50	0.60
Accuracy			0.69

Table 5. Classification results with Decision-Trees algorithm

Training	Precision	Recall	f1-score
Beginner	1.00	1.00	1.00
Expert	1.00	1.00	1.00
Intermediate	1.00	1.00	1.00
Accuracy			1.00
Test	Precision	Recall	f1-score
Beginner	1.00	0.67	0.80
Expert	0.50	0.75	0.60
Intermediate	0.80	0.67	0.73
Accuracy			0.69

Table 6. Classification results with Random-Forest algorithm

Training	Precision	Recall	f1-score
Beginner	1.00	1.00	1.00
Expert	1.00	1.00	1.00
Intermediate	1.00	1.00	1.00
Accuracy			1.00
Test	Precision	Recall	f1-score
Beginner	0.50	0.67	0.57
Expert	0.80	1.00	0.89
Intermediate	0.75	0.50	0.60
Accuracy			0.69

4 Discussion and Limitations

Across all classification algorithms, the Decision-Tree algorithm as well as the Random-Forest and the K-nearest-neighbour algorithm showed the best accuracy rates when averaging results of all classes with an accuracy score of 0.69 in all three cases. This score is acceptable [11], but also lower than the 0.91 achieved through the K-means clustering algorithm. Nonetheless, given the novelty of this approach using machine learning for identifying and classifying expertise levels of subjects, it represents a first success.

The results show big potential for the data-driven prediction of qualification level of the employees in manual assembly context. Sufficient data collection and clean processing before the analysis with machine learning algorithms were here key factors for this purpose. The possibility of building an assistance system that is adaptive to the skill level of the workers is therefore given. Processes can this way be constantly optimized and redesigned and also be adapted to the user behaviour and new services according to his or her needs following the human-in the-loop approach [12] (see Fig. 7).

The clustering and the classification algorithms employed were all open-source and the invested time for the machine-learning analysis was manageable. On the other hand,

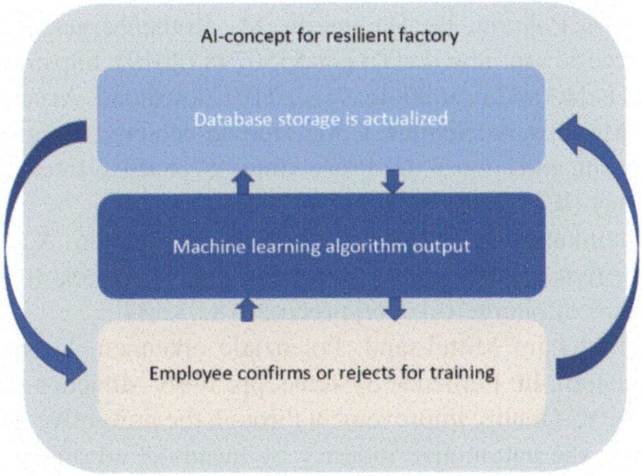

Fig. 7. Final AI-concept to improve productivity of resilient manufacturing.

the limitation of this study relies in the relative low number of taken measurements (39), so that the obtained results should still be tested with far more measurements for universality check.

5 Conclusions and Further Work

In general, the use of machine learning in production with the goal of improving processes and assisting the employee to perform diverse tasks more effectively should be carried on in future studies after these first results. In particular, the introduction of domain knowledge for conducting these analyses, not only in manual assembly processes, but at nearly every stage of a manufacturing company, turns into an essential success factor.

Additionally, it is important to seek for, apply and combine different technologies that can collect data of different type. In this sense, eye-tracking emerges as a potential technology to complement the monitoring process of employee conditions and skills and can be combined to the available process data. In this way, with more data, process stability as well as domain knowledge, new service offerings and business models can emerge, whereas employee and user satisfaction within production context can improve.

References

1. Munir, S., Stankovic, J.A., Liang, C.J.M., Lin, S.: Cyber physical system challenges for human-in-the-loop control. In: Proceedings of the Presented as part of the 8th International Workshop on Feedback Computing, USENIX, 24–28 2013, vol. 4, pp. 1–4 (2023)
2. Longo, F., Nicoletti, L., Padovano, A.: Smart operators in industry 4.0: a human-centered approach to enhance operators' capabilities and competencies within the new smart factory context. Comput. Ind. Eng. **113**, 144–159 (2017)
3. Bellgran, M., Säfsten, K.: Production Development—Design and Operation of Production Systems, 1st edn. Springer, London, UK (2010)
4. Lusic, M., Fischer, C., Bonig, J., Hornfeck, R., Franke, J.: Worker information systems: state of the art and guideline for selection under consideration of company specific boundary conditions. Procedia CIRP **2016**(44), 1113–1118 (2016)

5. Klapper, J., Gelec, E., Pokorni, B., Hämmerle, M., Rothenberger, R.: Potenziale digitaler assistenzsysteme access, urn: nbn:de:0011-n-5550798 (2019). https://publica.fraunhofer.de/bitstreams/e6dc1241-d400-42af-a690-4ea91eeb71b1/download. Accessed 25 Jun 2023
6. Nelles, J., Kuz, S., Mertens, A., Schlick, C.M.: Human-centered design of assistance systems for production planning and control. In: Proceedings 2016 IEEE International Conference on Industrial Technology (ICIT), pp. 2099–2104 (2016)
7. Müller, R., Vette-Steinkamp, M., Hörauf, L., Speicher, C., Bashir, A.: Worker centered cognitive assistance for dynamically created repairing jobs in rework area. Procedia CIRP **72**, 141–146 (2018). https://doi.org/10.1016/j.procir.2018.03.137
8. Dowling, M., et al.: KI im Mittelstand: Potenziale erkennen, Voraussetzungen schaffen, Transformation meistern. In: Lernende Systeme, pp. 1–96, München (2021)
9. Vannucci, M., Colla, V.: Quality improvement through the preventive detection of potentially defective products in the automotive industry by means of advanced artificial intelligence techniques. In: Ireneusz Czarnowski (Hrsg.) u.a.: Intelligent Decision Technologies 2019, pp. 3–12. Springer, Singapore (2019)
10. Lanquillon, C.: Blockchain und maschinelles Lernen: Wie das maschinelle Lernen und die Distributed-Ledger-Technologie voneinander profitieren. 1st ed. 2019. Berlin, Heidelberg (2019)
11. Lara, F., Lara-Cueva, R., Larco, J.C., Carrera, E.V., León, R.: A deep learning approach for automatic recognition of seismo-volcanic events at the Cotopaxi Volcano (2021)
12. Wu, X., Xiao, L., Sun, Y., Zhang, J., Ma, T., He, L.: A survey of human-in-the-loop for machine learning. Futur. Gener. Comput. Syst. **135**, 364–381 (2022)

Patterns for Resilient Workforce Structures and Strategies in Manufacturing

S. Gerlach(✉) and M. Hämmerle

Fraunhofer Institute for Industrial Engineering, IAO, Nobelstr. 12, 70569 Stuttgart, Germany
stefan.gerlach@iao.fraunhofer.de

Abstract. A resilient setup of manufacturing facilities covers all factors: buildings, equipment, processes, supply chains and human labour. Especially the utilization of human labour and resources is a key success factor for manufacturing companies to maintain their business responsiveness and competitiveness in volatile market environments. To build resilience in human labour is a strategic task which has to reflect different criteria, e.g. cost of labour and releasing of staff, developing and securing of skills and training, shortage of labour market, time for onboarding. Therefore, "firing and hiring" of employees may be a measure in case of disturbed markets or supply chains. But it must not be the solely and appropriate answer in every case. This paper highlights the requirements for resilient human resource allocation and describes rules and patterns to analyze and build resilience into human labour in manufacturing companies. The patterns give operational advice to experts how to analyze the requirements and to compare them with specific solutions. As a result, companies are able to develop and establish long term strategies for human resource allocation. The rationale is based on selected HR-instruments, which needs a longer time up to several years for preparation to utilize them. Therefore, adjusted strategies for human resource allocation are needed.

Keywords: Human resource allocation · personnel capacity planning · strategies for manufacturing resilience

1 Introduction

There are many influences on the actual output and workload of manufacturing companies. The main external drivers are fluctuations in customer demands as a result of marketing campaigns, large, singular orders, individualized products, and shortened delivery times. Combined with the globalisation of markets and emerging international competitors, it is difficult to estimate the actual workload for a given period [1]. Figure 1 shows an example of the workload of a production unit in a machine building company. The workload is calculated in terms of the working hours needed to assemble the customer orders with delivery due date in each week.

This short-term flexibility is superimposed by long-term effects such as seasonal demands, product life cycles or the global economic activity [3]. In particular, it seems

M. Dragomir et al. (Eds.): ICPR 2023, LNPE, pp. 175–182, 2025.
https://doi.org/10.1007/978-3-031-92082-0_18

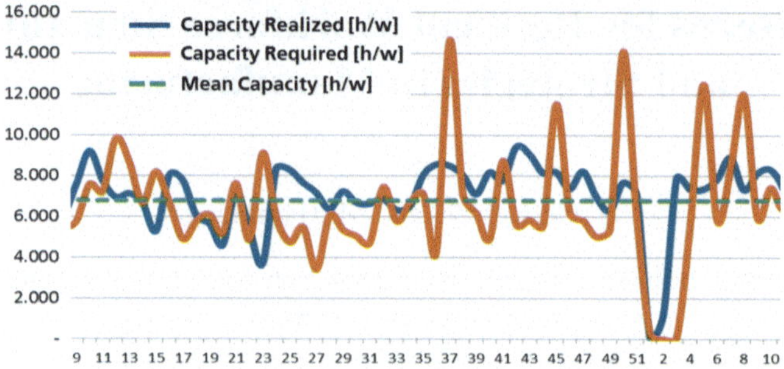

Fig. 1. Workload machine building company [2]

impossible to reliably predict the global economic activity [4]. Figure 2 illustrates the major global economic crisis since the late 1990 s, which have occurred approximately every 10 years. In these crises the decline of customer orders and thus the workload in many companies can be dramatically higher as the GNP graph in Fig. 2 indicates. Especially the machine building industry is very sensitive to the global economy. Figure 3 shows the fall in new orders and workload in 2009 for the same machine building company mentioned above. The decline of minus 41% over more than 12 months gives an idea of the big difference between short-term flexibility and so-called "volatility" long-term effects, which in many cases have a duration of at least several months.

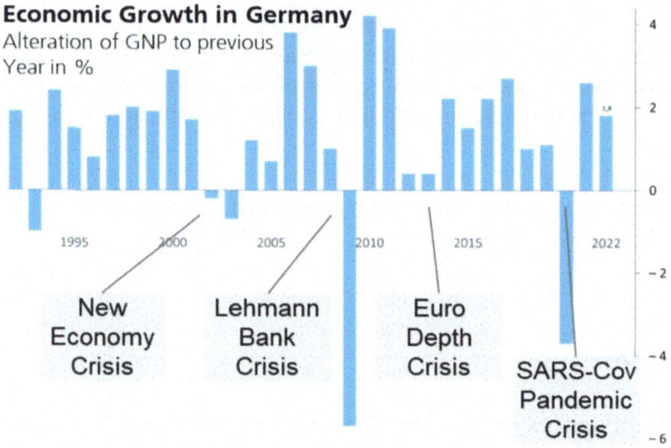

Fig. 2. Global Economic Crisis [5]

Nevertheless, manufacturing companies have to meet high demands for short delivery times, low stock levels and competitive costs [6, 7]. They must meet customer demands for products of high quality faster and more reliably than their competitors [8]. Centralised planning and scheduling of production demands and orders [9] doesn't meet the needs of dynamic and volatile markets. Therefore, companies are forced to utilise flexible forms of production activities, schedules, and labour [10]. To reflect the demands of long-term volatility in particular, labour utilisation must follow a strategically oriented planning of personnel in terms of quantity and quality, i.e. volume and skills [11]. The following paper introduces the concepts of flexible labour utilization. Afterwards the basic patterns for resilient workforce structures will be described and explained.

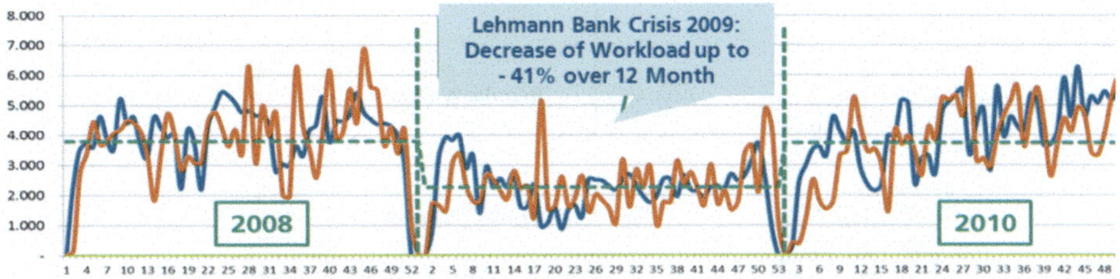

Fig. 3. Workload machine building company

2 Human Resource Flexibility

The flexibility of the workforces in a company depends on specific measures for the utilization of working hours, often referred to in this paper as flexibility instruments. A comprehensive discussion of the common flexibility instruments can be found e.g. in [12]. A brief overview is given below.

In general, there are two ways for flexible labour utilisation. First, the working time of a group of employees can be reduced or increased, perhaps on a daily basis. Second, the group of employees can be reduced or increased [13]. Figure 4 gives an idea of the two ways using the example of a so-called "U-shaped assembly system". In this system one worker can assemble an entire product by walking around all the workstations in one walking circle. According to German law can work a maximum of ten hours a day. There is no law on minimum working hours. However, for economic reasons, there is usually a minimum daily working time. If a higher workload is needed, an additional worker can be added to the assembly system. The two workers share the assembly tasks between them. Obviously, the maximum number of workers within the assembly system is given by the number of workstations.

The example shows that a flexible labour utilisation can be achieved by working with more or fewer workers, or by working longer or shorter hours, or by both measures. While each of these measures can meet the additional demand, they differ in terms of cost, time to utilize and coverable amount of capacity. Further constraints may result e.g. from long training periods or high skill requirements. Therefore, the mix and size of the flexibility instruments must be determined carefully based on a calculation of the total benefit and the financial budget required [14]. In addition, a short-term control of flexibility instruments and staff utilisation is necessary [15].

Various recent studies display the widespread distribution of instruments of personnel flexibility in the Germany. The project FlexPro surveyed more than 1200 companies on the application of different flexibility instruments in Germany. More than 2/3 of these companies do business in the manufacturing industry [16]. The results demonstrate the wide range of flexibility instruments, which are regularly used in manufacturing area

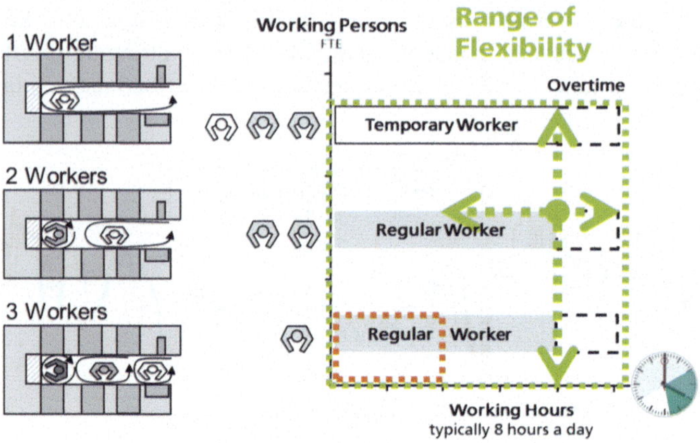

Fig. 4. Flexible Working Hours [13]

(Fig. 4). Instruments that are quickly available, such as overtime and extra work or internal personnel shifts are highly prioritized by the companies surveyed. Furthermore, part-time and temporary workers are used strongly. The working time of these employees can be adjusted flexibly to the short-term changes in capacity requirements of volatile markets.

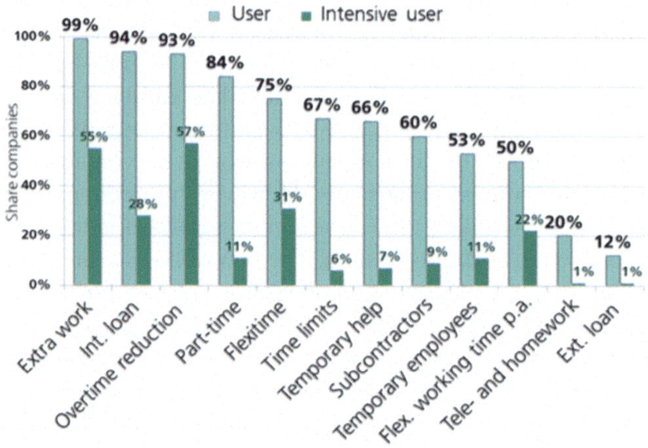

Fig. 5. Share of flexibility instruments [16]

3 Resilient Workforce Structures

Over the past fifteen years, the authors of this paper have developed resilient workforce structures in a variety of companies. The following patterns represent the general experience of the authors in this task.

3.1 Phrase Your Risks

Like any strategy, resilient workforce structures need to be oriented on a clear and comprehensive strategic target. With respect to the topic of resilience a target may be expressed more clearly as a "risk" and the most basic pattern for building resilient workforce strategies is "risk awareness". Any risk can be described in terms of.

- Volume of workload decline,
- Duration of the long-term effect causing the risk,
- Time to react after the effect occurs,
- Frequency or expected distance between two effects.

Figure 5 shows four general risk scenarios, each described with common values in the four dimensions and a short phrase to name and characterise the scenario.

While in general every risk follows more or less one of these scenarios, not every company is exposed to all of them. Each company faces with its own and specific set of risks. For example, automotive suppliers have to deal with long-term fluctuations in market demand, product lifecycles and economic crisis, while the food and beverage industry typically to deal with massive seasonal demand, but is only less exposed to global economic activity.

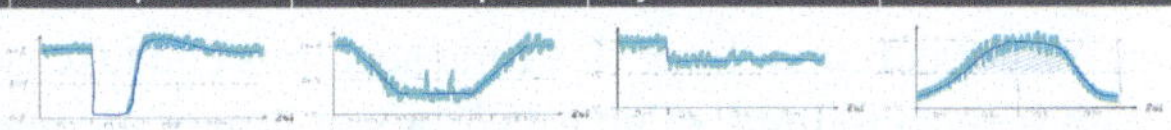

	Catastrophe	Market Slump	Key Account Loss	Season
Volume, Capacity	Up to 100%	30-80% Decrease	15-30 % Decrease	Min +- 30%, sometimes +300 to – 80%
Duration	Weeks to Months	Severals Months up to two Years	Permanent	Months to one Year
Time to React	Immediatly	1 – 3 Months	Immediatly up to several Weeks	Season Cycle, one Year manytimes
Frequency	Very rare, hopefully never	Business Cycles every 7 – 10 Years	Business dependent	1 x every Season
Examples	Cyber-Attack, Natural Hazard	Lehman-Bank 2008, Flop of new Produkt	Food Industry, Contract Manufact.	Retail Sale, Product Life Cycle

Fig. 6. General Risk Scenarios

Every company therefore needs to have a clear understanding and vision about the risks it is exposed. A "phrase" that describes each risk and gives it a short name, is very helpful for the stakeholders to be aware of the risks, talk about the vision and set up a resilience strategy. This is the rationale behind the name "Phrase you Risks" of this pattern.

Beyond the names, a vision must be based on a thorough analysis of the risks. A collection of patterns for the analysis of human resource flexibility can be found in [2].

3.2 Cocktail of Instruments

Once the risks are phrased and the strategy is known, a resilient workforce structure needs to be developed. Figure 8 lists flexibility instruments, that are appropriate for making workforce structures resilient to volatility. The flexibility instruments can be described by the same set of dimensions as the risks:

- Volume of possible workload adjustment,
- Duration of the possible use,
- Period of latency after implementation of the instrument,
- Frequency of potential use or time to rebuild the instrument.

With these dimensions the coverage of instruments over a risk scenario can be determined. Figure 6 shows how the typical values of each dimension fit with the common values of the risks. + + or + indicates a good fit, 0 indicates applicability but to limited extent, -- or - indicates, that this dimension prohibits the use of the instrument.

In almost all cases, there is no single flexibility instrument, that covers all the dimensions of the risk scenarios. Rather, a specific mix, a "cocktail of instruments", is needed to protect the workforce structure against volatility risks.

3.3 On-the-Job Preparation

Most of the flexibility instruments require a preparation before they can be used:

- Large working time accounts for employment protection, capable of storing several hundred working hours must be collected,
- Part-time, fixed-term or temporary workers need to be part of a company's daily workforce during periods of normal workload.

In particular, the collection of large flex time accounts can take considerably amount of time to fill, with respect to the size up to 5 Years or longer are necessary. Companies often fail to prepare the instruments. A big mistake is, that they rely on order and workload peaks to prepare these instruments, which is very problematic. In practice, there are no order peaks over 5 years to fill a time account. Nor are workers usually willing to spend 500 h of overtime without being paid for it. And hiring temporary workers only in the peak season means that in mid-season at average workload there are no temporary workers who can be dismissed in risk scenarios.

Therefore, the preparation of most of the flexibility instruments must be part of a regular workload and the daily work, i.e. "on-the-job". In order to fill large time accounts for employment protection a fixed amount of the daily or weekly contract hours is not paid, but this time is spent by the employees to fill the time account. A typical amount might be 2 h per week. A time account of 500 h, for example, would take about 5 years to fill.

3.4 Preparation Guidance

Once the right cocktail of instruments has been "mixed" to mitigate the phrased risks, it is necessary to decide if and when the on-the-job preparation is complete and in a good shape. As part of a "preparation guidance" the level of preparation needs to be monitored and assessed against the risks.

Figure 7 shows an example of a company, that in the past agreed to records of up to 400 h. The aim was to establish a volatility reserve in the range from 250 to 350 h for each employee. The graph shows the flextime records of each of the 254 employees sorted by the size of the record five years after implementation.

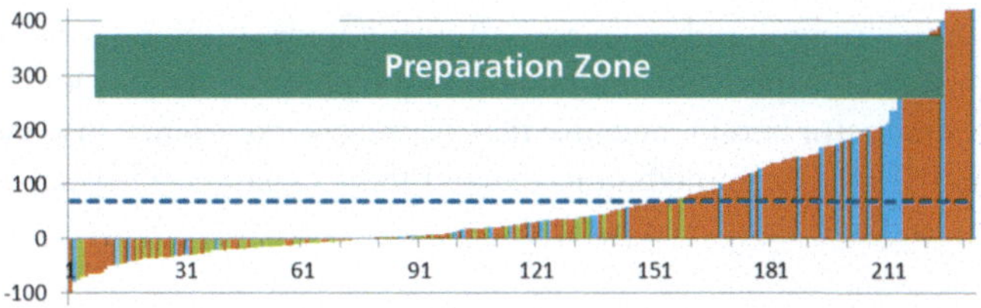

Fig. 7. Example of an Employee Protection Account

	Catastrophe				Market Slup				Key Account Loss				Season			
	V: 100%	**R:** 1 D	**D:** 2-4 W	**F:** >1 Y	**V:** 50%	**R:** 1-3 M	**D:** 0,5-2 Y	**F:** 10 Y	**V:** 25 %	**R:** 1-4 W	**D:** > 2 Y	**F:** >1 J	**V:** 300%	**R:** 1 Y	**D:** 1-6 M	**F:** 1 Y
Flextime Account	++	++	0	+	++	++	-	+	++	++	--	+	0	+	+	+
Employ. protection Account	++	++	++	+	++	++	0	+	++	++	-	+				--
Dissmiss part-time Employ.		--			-	+	-	+	0	++	++	+				--
Plant holidays		--			0	+	-	+	-	--			0	++	++	++
Stop fixed-term Employees		--			0	-	++	+	0	0	++	+	++	++	++	++
Stop temporary Employees	0	++	-	+	0	++	++	+	0	++	++	+	++	++	++	++
Reduced contract hours	-	--			0	++		+			--		--			
Reduced working hours	+	--	-		++	+	0	+			--		--			
Dissmiss regular staff		--	--		-	-	++	+	+	-	++	+				--

Fig. 8. Flexibility Instruments and Risk Scenarios

It is easy to see, that the company is missing its target. It was expected that most of the employees has reached the preparation zone, the diagram graph should have a big "belly" with an asymptotic expansion. One reason for missing the target was the lack of a preparation guidance. The other reason was, that no on-the-job preparation was agreed and organised. The company had hoped that the time account will be filled by fortune of a good economic situation with a lot of voluntary overtime hours of the employees. But about a third of the employees didn't work any overtime. So the company has agreed on instruments for long-term risk, but there is no chance of using this instrument.

4 Conclusion

The idea of describing general rules and experiences as "patterns" was originally invented by the architect Christopher Alexander, who published a comprehensive collection of patterns for designing cities and buildings [17]. Later his idea was successfully adopted to software engineering [18] and the analysis of business processes [19] in a more formal notation of patterns.

In the chapter above, patterns for resilient workforce structures and strategies are presented in a natural and more expressive style. Patterns are experience-based, their source is not formal theory or calculation. Nevertheless, they provide a helpful guidance for the analysis process in real situation and a specific company. The patterns presented are the essence of round about thirty projects with German companies to develop resilient and flexible workforce structures. Also, the patterns look more or less "obviously clear and evident" to experts, many of the responsible stakeholders in the companies are impressed and gain strong impulses in reconfiguring their solutions for human resource flexibility. In this case, especially the "obviously clearness and evidence" of patterns may be one success factor herein, because they represent experience in a natural scheme. Experts are usually able to immediately expand their existing experience, to applicate the new patterns and add their own patterns.

References

1. Tolio, T.: Design of Flexible Production Systems: Methodologies and Tools. Springer (2009)
2. Gerlach, S., Hämmerle, M., Schuler, S.: Patterns for analysis of human resource flexibility in manufacturing. In: Proceedings of the 25th ICPR, pp. 9–14. Elsevier, Chicago (2019)
3. Spath, D., Hämmerle, M.: Analysis of sales rhythm as a basis for human resource capacity planning in manual assembly systems. In: Proceedings of the 21th ICPR. Fraunhofer Verlag, Stuttgart (2011)
4. Bullinger, H.J., Spath, D., Warnecke, H.J., West-kämper, E.: Handbuch Unternehmensorganisati-on – Strategien, Planung, Umsetzung. Springer, Berlin (2009)
5. Nachhaltige Entwicklung in Deutschland, Indikatorenbericht 2022. Statistisches Bundesamt (Destatis), Berlni (2023)
6. Salvendy, G.: Handbook of Industrial Engineering: Technology and Operations Management. Wiley, NewYork (2001)
7. Hopp, W.J., Spearman, M.L.: Factory physics, 3rd edn. McGraw-Hill, Boston (2008)
8. Spath, D.: Global Challenges and the Need for Enhanced Performance. Presentation, Professional Training Facts, Stuttgart (2008)
9. Artiba, A., Elmaghraby, S.: The Planning and Scheduling of Production Systems: Methodologies and applications. Springer (1996)
10. Oechsler, W.A.: Personal und Arbeit – Grund-lagen des Human Resource Management und der Arbeitgeber – Arbeitnehmer-Beziehungen. Verlag Oldenbourg, Munich (2011)
11. Rump, J., Eilers, S. (eds.): Strategische Personalplanung. Springer Gabler, Berlin (2020)
12. Reilly, P.: Balancing Flexibility – Meeting the Interests of Employer an Employee, Burlington (2001)
13. Bauer, W., Gerlach, S., Hämmerle, M., Strölin, T.: Rule base for operative planning and control of flexible labour hours. In: Proceedings of the19th IFAC World Congress, Cape Town (2014)
14. Witte, K.-W., Vielhaber, W. (eds.): Flexible und wirtschaftliche Serienmontage – Wege zu zukunftsstabilen Montagesystemen (MORATIO). Shaker Verlag, Aachen (2003)
15. Hämmerle, M., Spath, D., Bauer, W.: Require-ments for an evaluation model for the proactive management of human resource capacity in volatile markets. In: Proceedings of the 22nd ICPR, Iguasso Falls (2013)
16. Sende, C., et al.: Flexibler Mitarbeitereinsatz – Ergebnisse einer deutschlandweiten Unternehmensbefragung. Industrie Management **27**(4) (2011)
17. Alexander, C., et al.: A Pattern Language: Towns, Buildings, Construction. Oxford University Press, New York (1977)
18. Buschmann, F., Henney, K., Schmidt, D.C.: Pattern-Oriented Software Architecture: On Patterns and Pattern Languages. John Wiley & Sons Ltd., Chichester (2007)
19. Fowler, M.: Analysis Patterns: Reusable Object Models. Addison-Wesley, Menlo-Park (1997)

Integrating Lot-Sizing Environmental Factors Estimation for a Green Reorder-Level Inventory Control Policy

M. Giacomelli(✉), F. Pilati, and M. Brunelli

Department of Industrial Engineering, University of Trento, Via Sommarive 9, Trento, Italy
marco.giacomelli-2@unitn.it

Abstract. With the aim of reducing pollution as firms start to consider environmental issues, this work focuses on the modeling and optimization of a sustainable reorder-level lot-sizing policy that integrates a tailored estimation of environmental factors for comprehensive green inventory control. In particular, the goal is to minimize total greenhouse gas emissions by modeling the impact of warehousing and resupply of each order quantity. Other than operational aspects that include energy usage for lighting, and thermal control inside a warehouse, building characterization and the related embodied carbon are other often-neglected significant sources of greenhouse gas emissions. This integrated approach allows, in a lost-sales setting constrained by minimal service level, to model the direct relation between emission components and the environmental impact of managing the inventory of a product, which carries clear valuable managerial insights.

Keywords: Inventory control · reorder-level policy · green supply chain management · embodied carbon · lost sales

1 Introduction

With the growing sensibility of both customers and legislators, businesses are pressured to make their products and operations greener [1]. In general, environmental sustainability encompasses many aspects, such as raw material extraction, manufacturing, transportation, and product disposal. Within this context, inventory control is responsible for greenhouse gas (GHG) emissions, mainly due to warehousing and reordering [2]. Moreover, warehousing is related to the energy demand of the building during occupation and in other life cycle phases, the latter of which can be measured through the embodied carbon measure [3]. Embodied carbon captures the carbon, or carbon equivalent, footprint related to direct and indirect energy usage due to construction [4].

By explicitly optimizing for environmental performance, businesses can contribute to the necessary global effort to reduce carbon footprints. This can be enabled by developing performance measures that drive decisions toward green operations [5]. In particular, sustainability in inventory logistics can be achieved by relating inventory decision variables and parameters to the environmental objective [2].

M. Dragomir et al. (Eds.): ICPR 2023, LNPE, pp. 183–191, 2025.
https://doi.org/10.1007/978-3-031-92082-0_19

Even though several efforts have been made to integrate environmental considerations in inventory control models, a detailed integrated investigation on the estimation GHG emission factors seems to be missing. To this end, this paper presents a reorder-level inventory control policy aimed at minimizing the expected total GHG emissions over the planning horizon. The goal is to identify the optimal reorder level and quantity, r and Q, given a lost-sale context where customer orders that cannot be fulfilled with on-hand inventory are lost. This scenario may also be representative whenever backorders are fulfilled with emergency orders, that are out of the scope of the standard stock system [6]. By introducing a comprehensive formulation for the environmental performance of the inventory, the estimation of GHG emissions is integrated within a service level constrained optimization model for the (r, Q) policy.

Subsequently, Sect. 2 provides a comprehensive overview of key contributions that integrate environmental sustainability measures in continuous-review policies. Section 3 then presents the formulation for estimating the considered environmental drivers, followed by the formulation of the constrained optimization model in Sect. 4. Section 5 explores realistic sample results driven by a case study aimed at showcasing the model's capabilities. Finally, Sect. 6 concludes the paper by summarizing the key findings and contributions, while also discussing potential areas for future research and development for green inventory policies.

2 Literature Review

Different approaches have been developed to integrate environmental factors into reorder policies. Andriolo et al. [7] presented a survey on economic order quantity models that also focuses on environmental considerations in lot sizing. In this regard, Battini et al. [8] considered emissions for a sustainable reordering model through additional cost factors for transport and inventory holding. In particular, analogously to Kazemi et al. [9], they model storage carbon emission costs with respect to the average space occupied in the warehouse. Other contributions focus instead on explicit environmental modelling. For example, van der Veen and Venugopal [10] and Bouchery et al. [11] presented an emission function that considers holding and transport coefficients for a deterministic order-quantity policy. Moreover, as environmental impact is directly connected to energy usage, Zavanella et al. [12] included energy considerations in lot sizing for production.

From the point of view of warehousing, the energy profile of the building can be used to derive overall building carbon footprint [13]. In this regard, Fichtinger et al. [14] employed simulation to identify how order quantities are related to inventory and warehousing emissions. They identified, while underlining the lack of attention to warehouse emissions in sustainable inventory management, heating and lighting as the main carbon-emission drivers. Moreover, their results identify how smaller reorder quantities can result in lower storing emissions. Additionally, Ries et al. [13] point out how the literature often considers lighting as the main energy driver in warehouses. Moreover, they use the degree days technique for estimating energy requirements of thermal control inside a warehouse, which also depends on building transmission coefficient, warehouse size, and thermal system efficiency. CIBSE [15] defined the degree days measure as the cumulative temperature difference between outdoor and setpoint temperatures over a period of time. By quantifying degree days, energy needs and the related emissions can then be estimated.

A further component for energy considerations in buildings is embodied energy. Whilst operational emissions are related to the energy needed to keep a warehouse running, embodied emissions are the result of energy usage for raw materials, their transportation, and construction activities during the first portion of a building life cycle [16], and can be measured as embodied carbon [17]. In particular, Zajac [18] identified warehouses among the lowest energy-intensive facilities, while Yohanis and Norton [16] highlighted the relevance of embodied energy especially for low-energy buildings. Materials are by far the greatest drivers for embodied carbon [4], with steel and concrete that can account, according to Rai et al. [19], for more than 80% of total embodied energy for a distributional warehouse. Additionally, the authors focus also on operational emissions for carbon footprinting of a warehouse building via the estimation of heating demand through simulation.

3 Environmental Sustainability Drivers

The baseline for the optimization model of the reorder-level policy is given by the estimation of its environmental sustainability drivers. As in most (r, Q) formulations, the considered inventory components relate reordering frequency and average inventory to total long-run performance measures. Consequently, the goal for the considered inventory components is to estimate the GHG emissions due to order issuing and storing. The notation used in the remainder of this section, dedicated to presenting the formulation for the sustainability drivers, is described in Table 1.

3.1 Inventory Storage

Maintaining a warehouse operational requires energy usage and is thus responsible for a portion of the environmental carbon footprint related to inventory control. Additionally, embodied carbon quantifies the emissions incurred for warehouse building. Total emission per unit volume per unit time h are thus modelled as the sum of lighting (h_l), thermal control (h_c), and embodied carbon components (h_e). In particular, emissions due to lighting requirements are related to the luminous efficacy F of the lighting system [20] and are estimated as:

$$h_l = \frac{tE_v A}{VF} c_l \tag{1}$$

where E_v is the required illuminance, dependent on warehouse operational specifications, A is the floor area, t the daily operating time, and V the total warehouse storage volume. Quantifying the required energy for lighting allows for the estimation of emissions via the emission coefficient c_l, which depends on energy sources.

The other component for operational emissions is given by the need to provide thermal comfort in the warehouse. Accordingly, the emissions related to this driver depend on the indoor set temperature, but also on the outdoor one. Using the technique of degree days, it is possible to estimate the energy requirements in relation to the

Table 1. Notation for estimation of sustainability drivers.

Storage		
h	Daily emissions per storage volume	$\left[kgCO_2/m^3\right]$
V	Useful warehouse volume	$\left[m^3\right]$
E_V	Required illuminance	$[lx]$
F	Luminous efficacy	$[lm/kW]$
A	Useful floor area	$\left[m^2\right]$
t	Daily operating time	$[h]$
c_l	Lighting energy emission factor	$[kgCO_2/kWh]$
D_H	Average daily heating degree days	$[K/day]$
$U\prime$	Overall building heat loss coefficient	$[kW/K]$
η	Heating system efficiency	
c_c	Heating energy emission factor	$[kgCO_2/kWh]$
EC	Overall building embodied carbon	$[kgCO_2]$
p	Building useful lifespan	$[days]$
Order issue		
k	Emissions per reorder	$[kgCO_2]$
e_t	Emission factor for transport	$[kgCO_2/km]$
k_0	Fixed emissions per reorder	$[kgCO_2]$
l	Supply transport distance	$[km]$

cumulative temperature differences for a building. The resulting storage emissions due to this environmental driver for each stored unit results as proportional to the average heating degree days D_H per day and the overall building loss coefficient $U\prime$ as per Eq. (2). This coefficient depends mainly on thermal transmittance of each building component [15].

$$h_c = \frac{tD_H U'}{V\eta} c_c \tag{2}$$

In particular, η is the efficiency of the heating system and c_c the emission coefficient.

In order to estimate thoroughly emissions for inventory storing, the embodied carbon component needs to be considered in addition to the previous drivers. Specifically, the total embodied carbon of a warehouse EC depends on the materials of each building component, and on the embodied carbon of each material. The resulting factor h_e of this portion storing emissions is obtained, through the useful storing volume V as done previously, as the ratio between EC and total building useful lifespan p.

3.2 Order Issue

Each time a reorder is issued to the supplier, the reorder quantity needs to be transported. The environmental impact of reordering is thus related to a fixed component k_0 due to the organizational effort of order issue, and a variable one given by transport, assumed to be by truck. By adapting the formulation of Bouchery et al. [11] for the transport component, the resulting resupply emission coefficient is given by:

$$k = k_0 + e_r 2l \lceil \frac{Q}{Q_{max}} \rceil \tag{3}$$

In particular, e_r is the emission rate per traveled distance, the capacity of a single vehicle is Q_{max}, and a round trip of distance $2l$ is considered from supplier to the warehouse.

4 Optimization Model

Given the estimation of the emission drivers of storing and order issuing, it is now possible to define the objective function for the reorder-level policy aimed at minimizing carbon-equivalent emissions in the considered time horizon of G days. Customer demand during the lead time is considered as stochastic, with a probability density function $f(x)$, while the expected total demand during the planning horizon is D. Additionally, lost sales are assumed whenever the inventory system is in stockout.

The resulting objective function for long-run total inventory emissions, for product of volume v per unit, is given by:

$$\begin{aligned} Z(r, Q) = {} & k \frac{D}{Q + \int_r^{\infty} (x - r) f(x) \mathrm{d}x} \\ & + hvG\left(\frac{Q}{2} + \int_0^{\infty} (r - x) f(x) \mathrm{d}x\right) \end{aligned} \tag{4}$$

Unlike a backlogging assumption, where the expected number of reorders is unaffected by the stockout size, for lost sales the order issue component needs to be estimated by considering the amount of lost sales per cycle. The expected value of such quantity is given by all the demand that exceeds the reorder level during the lead time, when the reorder quantity is outstanding. On the other hand, average inventory is also affected by the demand distribution via the safety stock, formulated as the remaining inventory after the lead time.

The optimization problem is constrained by the assumption of a single outstanding order per cycle, and additionally by customer service level $\mathrm{SL_a}$, or ready rate. In particular, the ready rate is the minimal service level measure of probability of avoiding a stockout at each reorder cycle [21] and can be formulated, following the definition of Silver et al. [22], as the cumulated distribution of demand during lead time in r:

$$F(r) = \int_0^r f(x) dx \geq \mathrm{SL_a} \tag{5}$$

The service level constraint restricts the decision space for the variable r due to the required ready rate α. Since this constraint is independent on Q, it can be solved to find a lower bound r_{LB} that satisfies the service level (Eq. 7). The resulting constrained minimization problem is the following:

$$\min_{r,Q} \quad Z(r, Q) \tag{6}$$

$$\text{subject to} \quad r \geq r_{LB}(\text{SL}_\text{a}) \tag{7}$$

$$r \leq Q \tag{8}$$

where Eq. (8) identifies the solutions for a single outstanding order per cycle [23]. It is assumed that r or SL_a are large enough to avoid negative values of safety stock.

5 Results

The model integrating the estimation of emission drivers of Sect. 3 and the optimization problem of Sect. 4 has been applied for a case study of an Italian multi-regional distributor that operates from a central warehouse. Customer demand for the considered product has been estimated from daily demand and lead time data as gamma distributed with shape and scale parameters of $\boldsymbol{\alpha} = 3$ and $\boldsymbol{\beta} = 5$ respectively. The other case study input parameters used for the optimization, listed in Table 2, are estimated both from case study, and from technical and literature contributions. Transport distance is estimated from geographical data of a representative supplier, which supplies via a truck with 33 $\mathbf{m}^3$ of transport volume.

Table 2. Parameters of environmental drivers.

V	8448 m^3	η	0.8
E_V	300 lx	c_c	0.184 $kgCO_2/kWh$
F	50000 lm/kW	EC	381807 $kgCO_2$
A	3600 m^2	p	7300 days
t	12 h	e_t	0.5462 $kgCO_2/km$
c_l	0.161 $kgCO_2/kWh$	k_0	6.33 $kgCO_2$
D_H	7.65 K/day	l	42.8 km
U'	1.79 kW/K		

The goal is to optimize the environmental performance of the inventory system of the considered product for a yearly time horizon, considering a service level of at least 90%. Firstly, the estimation of the emission drivers modeled in sect. 3 with the input data results in daily emissions for storing of $\boldsymbol{h} = 0.0144$ per day per unit volume and emissions per reorder of $\boldsymbol{k} = 53.1$ for a single resupply shipment.

By integrating emission estimates and demand parameters, it is possible to approach the global minimization problem of Eq. (6, 7, 8). In particular, each unit of product occupies 0.05 $\mathbf{m}^3$ of storage space, and the company expects a demand $\boldsymbol{D}$ over the planning period of 730 units. The service level constraint of Eq. (5) results in a lower bound for $\boldsymbol{r}$ of 27 units, which restricts the selection of both decision variables.

Due to the non-convexity of the problem, the constrained minimization problem is approached with the numerical global optimization technique of simulated annealing (SA) for avoiding local optima. The resulting values of $\boldsymbol{r}$, $\boldsymbol{Q}$ after 100 iterations of the SA algorithm are $\boldsymbol{r}^* = 27$, $\boldsymbol{Q}^* = 543$ with total emissions performance of 145.6 $\mathbf{kgCO_2}$ in the planning horizon.

Further results analyze the effects of demand variance and emission parameters on optimal total environmental performance. In particular, differences in demand standard deviation are tested, starting from the base case variance σ_0^2, along with transport capacity and storing emissions. The latter are estimated by varying the base case scenario by assuming an increase and a decrease in the building thermal efficiency, meaning that thermal transmission coefficients of each building element are assumed to be one half or double of the base- case values. Results of Table 3 illustrate the relevance of the resupply capacity on the selection of the reorder quantity. Moreover, the optimal reorder level is influenced only by the demand parameters due to the service level constraint. For these scenarios, optimal emissions are achieved always with a single supply trip, where for lower transport volume the optimal reorder quantity is equal to the available capacity. Additionally, the change in storing emissions due to building efficiency is related to relevant changes in the selection of the reorder quantity. In particular, higher storing emissions favor lower values of Q.

Table 3. Inventory optimization results for varying values of transport capacity, storing emissions, and demand parameters.

$Q_{max} = 660$	h	$Z(r^*, Q^*)$	Q^*	r^*	$Q_{max} = 240$	h	$Z(r^*, Q^*)$	Q^*	r^*
$\sigma = \sigma_0$	0.0188	167.3	473	27	$\sigma = \sigma_0$	0.0188	206.4	240	27
	0.0144	145.6	543	27		0.0144	195.7	240	27
	0.0121	133.6	592	27		0.0121	190.3	240	27
$\sigma = \sigma_0/2$	0.0188	165.3	476	21	$\sigma = \sigma_0/2$	0.0188	204.6	240	21
	0.0144	144.1	542	21		0.0144	194.3	240	21
	0.0121	132.3	592	21		0.0121	189.2	240	21
$\sigma = \sigma_0/4$	0.0188	164.3	476	18	$\sigma = \sigma_0/4$	0.0188	203.7	240	18
	0.0144	143.3	542	18		0.0144	193.7	240	18
	0.0121	131.6	592	18		0.0121	188.6	240	18

6 Conclusion

To conclude, this paper has presented a reorder-level inventory policy tailored to address the problem of environmental sustainability of logistics operations. By modeling the presented environmental drivers, inventory decisions can take into account quantifiable effects of the selection of r and Q on total emissions. Moreover, results show how emission driver parameters affect the optimal selection of decision variables.

Future research can focus on further analysis of the different influence of emission parameters on inventory environmental performance. In addition, an explicit integration of the environmental perspective with the classic economic one within a multi-objective optimization framework should be investigated.

Acknowledgments. This research activity is co-funded by the Doctoral Programme "Dottorati su tematiche green" of the National Operative Program (NOP) on Research and Innovation 2014–2020 and by the "Bando piattaforma e-commerce Trentino" call of the Trentino province (Italy) under the Provincial Law 13 May 2020, n° 3, with resolution 933 of July 2020.

References

1. Ansari, Z.N., Kant, R.: A state-of-art literature review reflecting 15 years of focus on sustainable supply chain management. J. Clean. Prod. **142**, 2524–2543 (2017)
2. Arikan, E., Fichtinger, J., Ries, J.M.: Impact of transportation lead-time variability on the economic and environmental performance of inventory systems. Int. J. Prod. Econ. 279–288 (2014)
3. Adalberth, K.: Energy use during the life cycle of buildings: a method. Build. Environ. **32**, 317–320 (1997)
4. Davies, P.J., Emmitt, S., Firth, S.K.: Challenges for capturing and assessing initial embodied energy: a contractor's perspective. Constr. Manag. Econ. **32**, 290–308 (2014)
5. Bonney, M., Jaber, M.Y.: Environmentally responsible inventory models: non-classical models for a non-classical era. Int. J. Prod. Econ. 43–53 (2011)
6. Johansen, S.G., Thorstenson, A.: Optimal and approximate (Q, r) inventory policies with lost sales and gamma-distributed lead time. Int. J. Prod. Econ. **30–31**, 179–194 (1993)
7. Andriolo, A., Battini, D., Grubbström, R.W., Persona, A., Sgarbossa, F.: A century of evolution from Harris's basic lot size model: survey and research agenda. Int. J. Prod. Econ. **155**, 16–38 (2014)
8. Battini, D., Persona, A., Sgarbossa, F.: A sustainable EOQ model: theoretical formulation and applications. Int. J. Prod. Econ. **149**, 145–153 (2014)
9. Kazemi, N., Abdul-Rashid, S.H., Ghazilla, R.A.R., Shekarian, E., Zanoni, S.: Economic order quantity models for items with imperfect quality and emission considerations. Int. J. Syst. Sci.: Oper. Logist. **5**, 99–115 (2018)
10. van der Veen, J.A.A., Venugopal, V.: Economic and environmental performance of the firm: synergy or trade-off? Insights from the EOQ model. In: Choi, T.M. (ed.) Handbook of EOQ Inventory Problems: Stochastic and Deterministic Models and Applications, pp. 121–137. Springer, Boston (2014)
11. Bouchery, Y., Ghaffari, A., Zied, J., Fransoo, J.: Sustainable transportation and order quantity: insights from multiobjective optimization. Flex. Serv. Manuf. J. **28**, 367–396 (2016)
12. Zavanella, L.E., Marchi, B., Zanoni, S., Ferretti, I.: Energy considerations for the economic production quantity and the joint economic lot sizing. J. Bus. Econ. **89**, 845–865 (2019)
13. Ries, J.M., Grosse, E.H., Fichtinger, J.: Environmental impact of warehousing: a scenario analysis for the United States. Int. J. Prod. Res. **55**, 6485–6499 (2017)
14. Fichtinger, J., Ries, J.M., Grosse, E.H., Baker, P.: Assessing the environmental impact of integrated inventory and warehouse management. Int. J. Prod. Econ. **170**, 717–729 (2015)
15. CIBSE, Degree-Days: Theory and Application (2006)
16. Yohanis, Y.G., Norton, B.: Including embodied energy considerations at the conceptual stage of building design. Proc. Inst. Mech. Eng. Part A: J. Power Energy **220**, 271–288 (2006)

17. Hammond, G., Jones, C., Lowrie, F., Tse, P.: Embodied carbon: the inventory of carbon and energy (ICE). Building Services Research and Information Association, University of Bath (2011)
18. Zajac, P.: Evaluation Method of Energy Consumption in Logistic Warehouse Systems. Springer, Cham (2015)
19. Rai, D., Sodagar, B., Fieldson, R., Hu, X.: Assessment of CO2 emissions reduction in a distribution warehouse. Energy **36**, 2271–2277 (2011)
20. Szokolay, S.: Introduction to Architectural Science: The Basis of Sustainable Design. Routledge (2014)
21. Chen, F.Y., Krass, D.: Inventory models with minimal service level constraints. Eur. J. Oper. Res. **134**, 120–140 (2001)
22. Silver, E.A., Pyk, D.F., Thomas, D.J.: Inventory and Production Management in Supply Chains. CRC Press (2017)
23. Bijvank, M., Vis, I.F.A.: Lost-sales inventory theory: a review. Eur. J. Oper. Res. **215**, 1–13 (2011)

The Neglected Nexus: Trade-Offs and Overlaps Between Resilience and Working Capital in Supply Chains

C. A. J. de Goeij[1] and L. M. Gelsomino[1,2(✉)]

[1] Faculty of Business, Media and Law, Windesheim University of Applied Sciences, Campus 2, 8017 CA Zwolle, The Netherlands
l.m.gelsomino@rug.nl

[2] Faculty of Economics and Business, University of Groningen, Netelbosje 2, 9747 AE Groningen, The Netherlands

Abstract. Covid-19 and geopolitical tensions in Ukraine show that there are several working capital disruptions impacting supply chains, which are to a large degree neglected in supply chain resilience literature. Resilience and working capital strategies are both investigated in supply chain literature, however they are not looked at in conjunction. Based on focus group research with 17 participants representing 14 different companies, we study trade-offs and overlaps between resilience and working capital strategies in supply chains. We identify three key themes at the intersection between both: 1) inter-departmental inventory and payments management, 2) inter-company collaborative working capital management and 3) financial process improvement.

Keywords: Supply chain resilience · working capital management · focus group · financial supply chain management

1 Introduction

The recent global events, from the Covid-19 pandemic to geopolitical tensions, have brought an unprecedented wave of attention towards the topic of supply chain resilience (SCRes). In the last three years we have seen supply chains facing disruptions caused by personnel shortages, unpredictability in container prices, inflation and shortages of raw materials and components like chips, copper, iron ore, steel, corn, coffee, wheat, soy beans, plastic and card board [1]. Recent contributions in the field of SCRes have given much attention to issues related to manufacturing and logistics, typically analyzing how companies deal with them [e.g. 2].

At the same time supply chains suffer from working capital disruptions, such as the extension of payment terms and the negative liquidity effects of increased inventory levels [3]. Intrum Justitia [4] shows that payment terms in the EU have been extended since Covid-19, and in addition, that the gap between the contractual payment terms and actual payment terms is growing. In B2B the average contractual payment term

M. Dragomir et al. (Eds.): ICPR 2023, LNPE, pp. 192–201, 2025.
https://doi.org/10.1007/978-3-031-92082-0_20

was 39 days in 2022, while actual payment terms where 52 days: a gap of 13 days in comparison to an 11 day gap in 2021. In the public sector payment terms are even longer: the average contractual payment term is 49 days, while the actual payment term is 62 days (also a 13 days gap, in comparison to an 11 day gap in 2021). Especially SME suppliers could be victims of extended payment terms, due to a weaker negotiation position [5]. Both in supply chain resilience and working capital research there is little empirical research on the SME perspective [6, 7].

According to industry reports the most popular risk response strategy was increasing the amount of inventory [8]. Oftentimes buyers require suppliers to hold on to more inventory to have a buffer in times of scarcity, and thereby passing on working capital pressure to suppliers [9]. Just like in the aftermath of the financial crisis of 2008/2009 these are signs of aggressive buyer behaviour, which can result in further supply chain disruptions. For example, it could lead to worsened relationships between buyers and suppliers and even bankruptcies for suppliers [10]. After the financial crisis, many scholars in financial supply chain management (FSCM) suggested a collaborative perspective on payment terms and inventories between supply chain partners, to avoid such problems [e.g. 11].

Both resilience and working capital strategies are investigated in supply chain literature, however they are not looked at in conjunction. A redundancy strategy in resilience, which has been applied very often in recent years [9], is in its core focused on more availability of inventory, and therefore directly negatively influences the working capital strategy. A visibility approach in resilience is typically focused on creating transparency in flows of goods, but could also entail more transparency in the sending, sharing and approving of orders and invoices to achieve faster payment and better working capital management [12]. Such examples show us many intersections and overlaps, but also tensions and trade-offs could exist between resilience and working capital strategies in supply chains. In this paper we aim to provide insights in these intersections, overlaps, tensions and trade-offs.

2 Literature Review

2.1 Supply Chain Resilience

Supply chain resilience (SCRes) can be defined as "the adaptive capability of an organisation to prepare for, respond to and recover from any type of supply chain disruption" [13], p. 31]. SCRes could not only entail returning to a previous state before a disruption, but also moving to a more desirable state [14]. There are many different categorizations for these adaptive capabilities. In this research we look at the four most used resilience capabilities in literature: redundancy, collaboration, flexibility and agility [15, 16].

Redundancy in its core is about mitigating negative effects of supply chain risks by increasing the amount of inventory [17]. To realize this strategy companies can simply choose to increase their own inventory levels, or for example choose to increase the amount of suppliers for certain services or components, search for more diversity in the regions of suppliers or request suppliers to hold on to more inventory [18].

Collaboration is about developing synergy between supply chain partners by sharing information and jointly thinking about the strategy for continuity. Companies can do this, for exampling, by investing in long term relationship, do joint planning or thinking along with supply chain partners about fair contract terms [18].

Flexibility enables supply chains to deal with higher levels of uncertainty, to adapt with minimal time and effort to changing environmental demands [19]. Companies can influence their level of flexibility for example by requesting flexibility in INCO terms, being able to quickly change routings or modes of transportation and modular desing of goods [20].

Agility can be divided into velocity and visibility. Velocity is about creating short lead times in the supply chain [21]. Companies can for example focus on frequent and fast restocking, frequent production runs and quick decision taking. Visibility is about creating transparency in the flows of goods, money and information in supply chains [22]. This means regular exchange of information between supply chain partners, for example by sharing and assessing real time data on the status and location of products or sharing forecasting information [18, 20].

2.2 Working Capital in Supply Chains

Working capital measures are important performance metrics for checking the ability of companies to manage capital. Such measures also show business dynamics beyond the own company, since they include payment terms from customers and towards suppliers. One of the most crucial working capital metrics is the cash conversion cycle, defined by the length of time between payment for products, materials or components and the sale of (end) products, which therefore "measures the time a firm has funds invested in working capital" [23, p. 25].

Most finance literature explains companies should try to have Days Sales Outstanding (average number of days to collect payment for a sale) and Days Inventory Outstanding (average number of days inventory is held before being sold) as short as possible, but the Days Payables Outstanding (average number of days it takes to pay suppliers) as long as possible [24]. Often larger, focal companies in the supply chain dictate payment terms and inventory requirements towards smaller suppliers [11]. This could lead to increased supply risk due to more financial pressure for such suppliers [10]. Furthermore, such practices could be seen as unethical buyer behaviour [25].

Therefore, in financial supply chain management (FSCM) a collaborative view on working capital was introduced [11]. To achieve a better supply chain performance on working capital, FSCM literature suggests there should be more attention for efficiency in order-to-cash (O2C), purchase-to-pay (P2P) and forecast-fo-fulfill (F2F) processes, for faster invoicing and order management, faster payments and decreasing average inventory levels in supply chains [12]. Furthermore, FSCM literature describes the potential of instruments such as reverse factoring, dynamic discounting or inventory financing for faster payments or optimizing inventory management in supply chains [26, 27]. Such collaborative practices to working capital management are claimed to reduce supply chain risk and increase SCRes [10, 28].

2.3 Intersections Between Resilience and Working Capital in Supply Chains

In SCM literature there is very limited attention for the impact of working capital management on resilience. 'Financial strength' is in some contributions described as a resilience capability, but is operationalized mainly with generic indicators such as profitability or access to financial reserves, with limited or no attention for working capital [20].

There are some exceptions. Pettit et al. [18, p. 73] are slightly more specific and describe financial strength as the 'capacity to absorb fluctuations in cash flow'. Han et al. [29] are one of the few that mention the cash conversion cycle in relation to SCRes. Quang & Hara [30] and Ghadge et al. [31] describe delayed payments and potential bankruptcy of suppliers as risk factors influencing resilience. Bak et al. [32] link the ability of companies to obtain bank financing to resilience. Chen et al. [33 p. 2435] explain next to inventory bullwhip effects there could be 'financial bullwhip effects' in supply chains, when 'liquidity risk effects become greater upwardly along the supply chain'. Van Hoek [9] mentions buyers directly after Covid-19 increased financial pressure for suppliers by more strict inventory requirements and extended payment terms.

In the specific operationalization of SCRes capabilities attention for working capital is even more scarce. For example, there seems to be no attention for working capital effects of increased inventory levels from scholars describing the redundancy approach [20]. In addition, authors reporting on velocity elaborate on reducing lead times in delivery and production, but not on reducing lead times in the O2C process [e.g. 29], while in the operationalization of visibility there seems to be no attention at all for transparency in sharing order and invoice information [18].

While FSCM literature shows there could be many potential effects of working capital elements on SCRes and vice versa, there is a scarcity of empirical research focusing on the intersections, in the form of overlaps and trade-offs. Therefore, the research question in this paper is:

Which overlaps and trade-offs exist between working capital and resilience strategies in supply chains?

3 Methodology

In this first stage of the research, we used focus group methodology to identify the most important themes at the intersection of working capital and resilience strategies in supply chains. A second stage of the research will follow, wherein we will built on the focus group results in a multiple case study.

In focus group research the interaction between research participants is used to generate data [34]. These group dynamics can result in extracting "deeper and richer data" which is based on the synergy of the group interaction, in comparison to one-to-one interviews [35, p. 656]. We aim to understand how intersections between supply chain resilience and working capital strategies influence the decision making of supply chain partners. Focus groups seem to suit this complex and multidisciplinary themes well, since group interactions can help participants explain and interpret different points of view.

3.1 Design

Before the focus group started we asked the participants to share with us descriptive information about their companies and the supply chains they are part of, and asked them basic information about the disruptions they have faced in the previous years and the most important strategies to deal with these disruptions. We started the focus group discussions with a 30-min presentation wherein the key concepts in supply chain resilience and working capital were explained. This presentation was done to get respondents familiar with the most important terminology and provide an elemental structure of supply chain resilience and working capital management. We divided the group into three subgroups with three different themes [34, 36]. The first group was focused on financial supply chain flows, the second group on physical flows of goods and the third group on flows of information.

In every subgroup there was a moderator and an observer. Each subgroup lasted for 30 min, after which respondents were rotated among groups to ensure group dynamics and to reduce dominance of single group members for each theme. Altogether, three rounds of discussion for each time, led to a total of 90 min of group discussions per theme.

3.2 Sample

17 respondents (representing 14 different companies) participated in the focus group. These respondents allow for a sample with a good mix of homogeneity and heterogeneity. The majority of the involved companies were SMEs (10), while 4 large companies participated. Some of the companies are owned by companies in other countries, but all of them have operations in The Netherlands. 11 production companies participated while 3 companies were service providers (logistics, recruitment and IT). The production companies represent a variety of industries: metal industry (2x), hardware (2x), construction, energy, cleaning products, trailers, production machinery and optical lenses. We incorporated relevant functional perspectives in our focus group: the respondents were purchasing managers (6x), supply chain or operation managers (6x), CEOs (2x), sales managers (2x) and one combined CPO/CFO role. In addition, 2 CFOs, 3 sales managers and 2 purchasing managers from 7 of the 14 companies assisted in gathering additional data after the focus group. From 5 out of the 14 companies an extensive amount of extra data (interview data and secondary data like financial reports and extraction from the companies' ERPs with invoice data) on resilience and working capital was collected, due to having research assistants inside those companies for five months each. These research assistants reported observations weekly to the members of the research team.

3.3 Data Collection and Analysis

The discussions in all three tables of the focus group were recorded and transcribed at a later stage. Immediately after the focus group sessions, the involved researchers collected all notes and flip charts, and compared the most important results per theme, thereby improving reliability and internal consistency (Kidd and Parshall, 2000). Codes used for data analysis were based on the literature review in Sect. 2. For supply chain resilience the 4 resilience capabilities were used as codes. The working capital codes were also based on literature, the codes used were "payment terms to suppliers", "payment terms from buyers", "Inventory Size", "O2C Process", "P2P Process", "F2F Process", "Credit Limits", "External financing needs" and "SCF Instruments". Three core themes on the intersection of supply chain resilience and working capital were identified by carefully analyzing similarities and differences.

4 Preliminary Results

4.1 Supply Chain Resilience: Disruptions and Strategies

The most frequently mentioned disruption among the focus group respondents is inflation. Higher prices and unpredictability in prices seem to affect all companies, both service providers and production companies. Delayed deliveries and unpredictability in delivery time are the second most mentioned disruption, all of the production companies were affected by this as a result of lockdowns, container shortages and capacity problems of suppliers. The capacity problems are a result of mainly scarcity of materials and personnel shortages. This in turn, led to delays in the own production planning.

The most popular strategy among the participants to deal with such disruptions was redundancy. Companies have built extra inventories, sometimes because larger buyers demanded them to increase inventories, and in addition often requested their own suppliers to improve inventory levels. The CEO of a company in construction mentions: "*Because of these extra inventories we were able to continue production in times of scarcity and unreliable delivery times and finish our projects.*" In addition, the majority of production companies have put considerable effort in finding additional suppliers for the most important raw materials and components. Becoming more agile by improving visibility, for example by investing in forecasting and monitoring systems, was mentioned by almost all companies as a strategy they want to pursue. While plans on increasing inventory seem to be directly realized because of immediate necessity, plans on visibility are more vaguely described in terms of future ambitions. Surprisingly the other factor in agility, velocity, does not receive the same attention from the involved companies.

The majority of respondents explains collaboration with supply chain partners is seen as more important in recent years. The procurement director of a construction company explains: "*We receive information about delays much sooner nowadays from our suppliers, which helps us explain adjustments in production schedules towards customers.*" Surprisingly, flexibility is among the four capabilities the least popular in our group of respondents. This is not necessarily because of unwillingness but could be the result of regulations or other limitations in the own industry or market. For example, the procurement manager of an optical lenses company explains: "*We can not just switch suppliers or switch components easily. In our industry everything is regulated and we would need external approval for such decisions which could take months.*"

4.2 Working Capital Management in Supply Chains: Disruptions and Strategies

Investing in extra inventories could have big effect on the working capital of involved companies. It seems for most companies that production and turnover come first, before working capital: "*production must always go on*", as explained by the operations manager of a construction company. The CFO of a company in the metal industry explains however that the effects of such an approach can be harmful: "*Despite our biggest turnover ever, our results were wiped out because of negative liquidity effects*".

In addition, multiple companies report on the extension of payment terms by buyers leading to further working capital problems. A procurement manager of a large IT company mentions: "*after Covid-19 we immediately checked if we could get working capital benefits by extending payment terms to suppliers.*" Not all companies share this vision on payment terms, as explained by a procurement manager in production machinery: "*helping suppliers by paying earlier could really improve our relationship with suppliers.*"

Some companies tried to protect themselves from aggressive buyer behaviour by introducing credit limits or strictening terms on already existing credit limits. A procurement manager of an office furniture company explains they notice operational constraints as a result of credit limits: "*we try to buy more at once, while at the same time credit limits are more strict which urges us to pay faster*". Furthermore, one of the companies mentions in recent years they do more creditworthiness checks on (new) customers.

Several companies explain there was a need for more bank financing as a result of working capital pressure caused by more inventories and/or delayed payments. While most companies rely on traditional bank financing, multiple companies mention they started to use instruments such as factoring, reverse factoring and dynamic discounting for liquidity benefits in recent years.

Several of the involved companies explain that a smooth O2C-process is important for them. The sales manager of a company in cleaning products explains inflation contributes to this: "*the invoice values are bigger nowadays, so it's even more important to chase these invoices for fast payment*". Furthermore, the increased liquidity pressure as a result of extended payment terms and more inventories, makes O2C-process optimization even more crucial.

5 Discussion: Overlaps and Trade-Offs

Based on our results we identified three important themes at the intersection between supply chain resilience and working capital management in supply chains, which receive limited attention in supply chain resilience literature.

First of all, inter-departmental inventory and payments management is a key theme. With a one-sided short-term operational view on inventory management, the working capital perspective is not always seen as a priority. While having lots of extra inventory could lead to outproducing competitors, our results show this could also seriously harm the financial position of companies. A multi-dimensional view with operational, procurement and finance managers involved, can result in a more balanced perspective on inventory. Likewise, an inter-departmental perspective on payment terms seems to be necessary. As explained by Dekkers et al. [36], financial managers often want to extend payment terms to suppliers for working capital benefits, while at the same there are purchasing managers in our research that would like to pay early to create good relationships with suppliers. Such inter-departmental trade-offs should be considered in supply chain resilience strategies.

Second, collaborative working capital management is a crucial theme for achieving supply chain resilience. Buyers that extend payment terms and request suppliers to invest in inventories, do not seem to have financial stability of suppliers on top of mind. Such an approach could backfire, De Goeij et al. [6] already explain it could lead to suppliers increasing prices. In our results we also see suppliers introducing or tightening credit limits, which could limit the order quantity of buyers. Chen et al. [33] report on a 'financial bullwhip effect', which was also seen in the supply chains of multiple participating companies in our research. These companies had to pass on financial pressure to suppliers by extending payment terms, since buyers did the same to them. While multiple articles in FSCM literature report on the importance of a collaborative perspective on working capital management [e.g. 11], there seems to be limited to no attention for this in supply chain resilience literature, even though it could have a major impact on both the 'physical' and financial supply chain.

Third, while supply chain resilience literature on agility is mainly focused on optimizing physical processes of production and delivery, our results show the necessity of also focusing on optimizing financial processes such as the O2C-process. Multiple companies in our sample explain this could partly offset the negative liquidity effects of payment term extensions, inflation and extra inventories. They invest in changes related to the organization and the digitalization of the process to send and approve orders and invoices faster.

While in some occasions there are clear trade-offs, mainly between financial and operational goals linked to inventories and payment terms, our research shows there can also be many overlaps between working capital management and resilience strategies in supply chains. A collaboration approach could include faster payments to suppliers and an agility approach could also include O2C-process improvements and better visibility in the flow of orders an invoices.

6 Preliminary Conclusions

The management of financial flows is included in many supply chain definitions nowadays. However, within the specific field of supply chain resilience, which is rapidly growing in terms of corporate attention and papers published, this financial flow is underrepresented. Our results show the importance of including this financial perspective, and identify three key themes at the intersection of working capital and resilience strategies in supply chains: 1) inter-departmental inventory and payments management, 2) inter-company collaborative working capital management and 3) financial process improvement.

By not taking these themes into account supply chain resilience literature falls short in providing advice for many of the supply chain problems arising since Covid-19 and the geopolitical tensions in Ukraine, such as working capital problems and 'financial bullwhip effects' caused by inventory investments and payment terms extensions, inflation and unpredictability in prices and delivery times.

More empirical research and exploratory research is still needed at the intersection of WCM and SCRes research. Our focus group research will be followed up by case study research to further investigate the overlaps and trade-offs between resilience and working capital strategies in supply chains. Especially, more attention seems to be needed for the perspective of SMEs in supply chains, since most research is only focused on how large companies respond to disruptions.

Acknowledgements. This research was conducted in the context of the NWO SIA RAAK project "Supply Chain Risk Management for SMEs", under grant number "RAAK.MKB14.021".

References

1. Bouckaert, L.: De wereldeconomie heeft plotseling een tekort aan alles en er is 'geen einde in zicht'. Business AM (2021)
2. Kähkönen, A.K., Patrucco, A.: A purchasing and supply management view of supply resilience for better crisis response. J. Purchas. Supply Manage. 100803 (2022)
3. Rice Jr., J.B., Klibi, W., Trepte, K.: Overcoming the financial barriers to building resilient supply chains. Harv. Bus. Rev. (2022)
4. Intrum Justitia. European Payment Report 2022 (2022)
5. Caniato, F., Moretto, A., Rice Jr., J.B.: A financial crisis is looming for smaller suppliers. Harv. Bus. Rev. (2020)
6. de Goeij, C., Gelsomino, L.M., Caniato, F., Moretto, A.M., Steeman, M.: Understanding SME suppliers' response to supply chain finance: a transaction cost economics perspective. Int. J. Phys. Distrib. Logist. Manag. **51**(8), 813–836 (2021)
7. Fan, Y., Stevenson, M.: A review of supply chain risk management: definition, theory, and research agenda. Int. J. Phys. Distrib. Logist. Manage. (2018)
8. McKinsey. It's quicker to build inventories than factories (2022)
9. van Hoek, R.: Research opportunities for a more resilient post-COVID-19 supply chain–closing the gap between research findings and industry practice. Int. J. Oper. Prod. Manag. **40**(4), 341–355 (2020)
10. de Boer, R., Steeman, M., van Bergen, M.: Supply chain finance, its practical relevance and strategic value: the supply chain finance essential knowledge series. Hogeschool Windesheim (2015)
11. Hofmann, E., Kotzab, H.: A supply chain-oriented approach of working capital management. J. Bus. Logist. **31**(2), 305–330 (2010)
12. Hofmann, E., Strewe, U.M., Bosia, N.: Supply Chain Finance and blockchain Technology: The Case of Reverse Securitisation. Springer, Cham (2017)
13. Scholten, K., Scott, P.S., Fynes, B.: Building routines for non-routine events: supply chain resilience learning mechanisms and their antecedents. Supply Chain Manage.: Int. J. **24**(3), 430–442 (2019)
14. Waters, D.: Supply Chain Risk Management. Vulnerability and Resilience in Logistics. Kogan Page, London and Philadelphia (2007)
15. Tukamuhabwa, B.R., Stevenson, M., Busby, J., Zorzini, M.: Supply chain resilience: definition, review and theoretical foundations for further study. Int. J. Prod. Res. **53**(18), 5592–5623 (2015)
16. Sawyerr, E., Harrison, C.: Developing resilient supply chains: lessons from high-reliability organisations. Supply Chain Manage.: Int. J. **25**(1), 77–100 (2020)
17. Sheffi, Y., Rice Jr, J.B.: A supply chain view of the resilient enterprise. MIT Sloan Manage. Rev. **47**(1) (2005)
18. Pettit, T.J., Croxton, K.L., Fiksel, J.: Ensuring supply chain resilience: development and implementation of an assessment tool. J. Bus. Logist. **34**(1), 46–76 (2013)

19. Stevenson, M., Spring, M.: Flexibility from a supply chain perspective: definition and review. Int. J. Oper. Prod. Manag. **27**(7), 685–713 (2007)
20. Kaviani, M.A., Tavana, M., Kowsari, F., Rezapour, R.: Supply chain resilience: a benchmarking model for vulnerability and capability assessment in the automotive industry. Benchmarking: Int. J. **27**(6), 1929–1949 (2020)
21. Christopher, M., Peck, H.: Building the resilient supply chain. Int. J. Logist. Manage. **15**(2) (2004)
22. Jüttner, U., Maklan, S.: Supply chain resilience in the global financial crisis: an empirical study. Supply Chain Manage.: Int. J. **16**(4), 246–259 (2011)
23. Moss, J.D., Stine, B.: Cash conversion cycle and firm size: a study of retail firms. Manag. Financ. **19**(8), 25–34 (1993)
24. Farris, M.T., Hutchison, P.D.: Cash-to-cash: the new supply chain management metric. Int. J. Phys. Distrib. Logist. Manage. (2002)
25. Cowton, C.J., San-Jose, L.: On the ethics of trade credit: understanding good payment practice in the supply chain. J. Bus. Ethics **140**, 673–685 (2017)
26. Gelsomino, L.M., de Boer, R., Steeman, M., Perego, A.: An optimisation strategy for concurrent supply chain finance schemes. J. Purch. Supply Manag. **25**(2), 185–196 (2019)
27. Elliot, V.H., De Goeij, C., Gelsomino, L.M., Woxenius, J.: Supply chain finance is not for everyone. Int. J. Phys. Distrib. Logist. Manag. **50**(9/10), 775–807 (2020)
28. Popa, V.: The financial supply chain management: a new solution for supply chain resilience. Amfiteatru Econ. J. **15**(33), 140–153 (2013)
29. Han, Y., Chong, W.K., Li, D.: A systematic literature review of the capabilities and performance metrics of supply chain resilience. Int. J. Prod. Res. **58**(15), 4541–4566 (2020)
30. Quang, H.T., Hara, Y.: Risks and performance in supply chain: the push effect. Int. J. Prod. Res. **56**(4), 1369–1388 (2017)
31. Ghadge, A., Jena, S.K., Kamble, S., Misra, D., Tiwari, M.K.: Impact of financial risk on supply chains: a manufacturer-supplier relational perspective. Int. J. Prod. Res. **59**(23), 7090–7105 (2021)
32. Bak, O., Shaw, S., Colicchia, C., Kumar, V.: A systematic literature review of supply chain resilience in small–medium enterprises (SMEs): A call for further research. IEEE Trans. Eng. Manage. (2020)
33. Chen, T.K., Liao, H.H., Kuo, H.J.: Internal liquidity risk, financial bullwhip effects, and corporate bond yield spreads: supply chain perspectives. J. Bank. Financ. **37**(7), 2434–2456 (2013)
34. Kitzinger, J.: Focus group research: using group dynamics. Qual. Res. Health Care **56**, 70 (2005)
35. Rabiee, F.: Focus-group interview and data analysis. Proc. Nutr. Soc. **63**(4), 655–660 (2004)
36. Dekkers, R., et al.: Evaluating theoretical conceptualisations for supply chain and finance integration: a Scottish focus group. Int. J. Prod. Econ. **220**, 107451 (2020)

Three-Stage Data-Driven Approach to Fast and Accurate Job Dispatching Using Learning-to-Rank Techniques

Young-Suk Han[1] and Jae-Yoon Jung[1,2](✉)

[1] Department of Big Data Analytics, Kyung Hee University, 1732 Deogyeong-daero, Giheung-gu, Yongin-si, Gyeonggi-do, Republic of Korea
jyjung@khu.ac.kr

[2] Department of Industrial and Management Systems Engineering, Kyung Hee University, 1732 Deogyeong-daero, Giheung-gu, Yongin-si, Gyeonggi-do, Republic of Korea

Abstract. Recently, a few studies have been conducted to construct data-driven job dispatching methods for hybrid flow shop such as semiconductor and display manufacturing systems. The data-driven job dispatching models can be used to simulate the production scheduling or dispatching without knowing the original dispatching rules. To learn historical job dispatching data, they adopt machine learning and deep learning methods for classification, which specifically aim at choosing the dispatched job among candidate jobs. However, such classification-based job dispatching engines often take a long inference time despite high accuracy because they require a lot of comparison between among jobs. This paper proposes a three-stage modelling approach that filters prioritized jobs using a learning-to-rank technique before performing a pairwise comparison. The method was evaluated using two major semiconductor process datasets obtained from a commercial simulation-based scheduling engine. The experimental results demonstrate that the proposed method outperforms the traditional pairwise dispatching model in terms of speed while maintaining high dispatching accuracy.

Keywords: Job scheduling and dispatching · Data-driven job dispatching method · Machine learning · Learning-to-rank

1 Introduction

Discrete event simulation (DES) is an essential tool in high-tech manufacturing industries such as semiconductors and displays [1]. It allows manufacturers to model and optimize complex production processes, which are characterized by many dependent steps. This can be challenging to identify and address bottlenecks and other issues. Simulation with different scenarios helps managers to identify opportunities for improvement, reduce cost, and increase productivity.

In high-tech manufacturing, real-time schedules are generated by using complicated dispatching rules [2]. However, high-fidelity simulation using such real dispatching rules

M. Dragomir et al. (Eds.): ICPR 2023, LNPE, pp. 202–209, 2025.
https://doi.org/10.1007/978-3-031-92082-0_21

is too time-consuming, and so it is required to develop the dispatching models that are much faster but can generate the similar schedules to the complicated dispatching rules.

To address this issue, automatic model generation (AMG) is adopted to emulate real-world complicated dispatching rules in production lines. Previous studies have leveraged machine learning techniques to automatically create dispatching models from log data obtained real-world scheduling systems [3, 4]. However, a significant limitation of these studies is the time-consuming nature of scheduling, as the model relies on pairwise comparisons among jobs to choose the most relevant job to be dispatched.

This paper proposes a three-stage modelling approach to generate the dispatching models based on machine learning. The multi-stage approach uses a learning-to-rank (LTR) technique to filter prioritized jobs before performing the time-consuming pairwise comparison task. It can result in faster and more accurate dispatching models, which can improve the overall performance of production simulation.

2 Literature Review

2.1 Automatic Model Generation

AMG is a technique for creating simulation models directly from external data sources [3]. AMG has been widely applied in various industries, such as manufacturing, healthcare, and logistics. AMG aims to reduce the time and effort required for simulation model generation. Traditionally, AMG approaches have relied on explicit information about the system being modelled. However, this information is often not readily available. To address this challenge, recent studies have explored the use of machine learning algorithms to train AMG models from historical production data [4].

2.2 Machine Learning of Composite Dispatching Rules

Dispatching rules are used to prioritize jobs in a queue. They can be based on various attributes, such as arrival time, due date, processing time, and queue length. Single dispatching rules use a single attribute and are commonly used for specific performance measures. However, applying single dispatching rules to jobs with diverse characteristics can be challenging and may lead to conflicting results.

To overcome this, composite dispatching rules (CDR) consisting of multiple single dispatching rules are used, allowing for improved decision-making [5]. The composition of dispatching rules can be achieved through weighted combination or hierarchical combination. Weighted combination calculates priority scores for each job by multiplying the rule's value by its weight, while hierarchical combination applies dispatching rules sequentially when previous rules fail to determine the best job.

Machine learning techniques such as decision trees [3] and neural network approaches [4] have been employed to learn dispatching rules. Notably, recent studies have explored LTR for learning dispatching rules using a listwise approach [6]. These studies have been conducted to learn simple dispatching rules such as shortest processing time (SPT) and earliest due date (EDD) and CDR such as the weighted combination of simple ones.

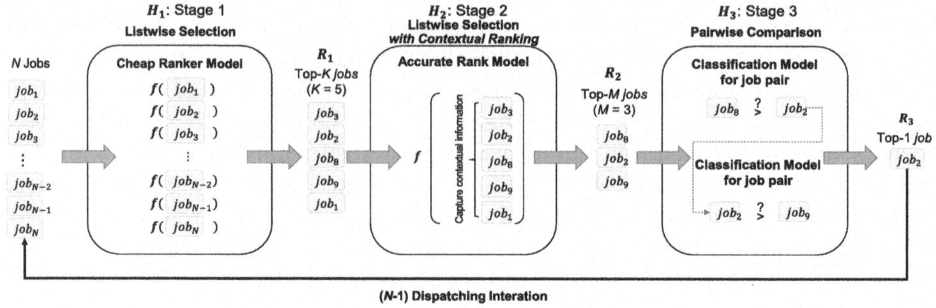

Fig. 1. Structure of Three-Stage Dispatching Model

2.3 Learning-to-Rank

LTR refers to the construction of ranking models for information retrieval systems using machine learning techniques. It has applications in information search, machine translation, and cooperative filtering [7].

There are two categories of LTR algorithms. One is the boosting algorithms such as LambdaMART, XGBoost ranker, and LightGBM ranker, while the other is the neural network algorithms such as ListNet, ListMLE, and ListPL.

In LTR, there is a trade-off and effectiveness and efficiency. Highly effective ranking models tend to be slower due to computationally expensive features, while fast ranking models may sacrifice efficiency. Multi-stage ranking aims to generate a set of candidate documents efficiently and iteratively reorder them using more expensive machine learning techniques [8]. By minimizing the number of candidate documents in the early stages, overall processing time can be significantly reduced.

This study aims to learn CDRs composed of hierarchical combination effectively and efficiently through an LTR-based multi-level ranking structure.

3 Three-Stage Dispatching Method

This paper deals with the creation of AMG models to amulate CDRs for scheduling. A CDR consists of multiple state-dependent rules that are hierarchically combined. A state-dependent rule has a form of 'IF <state>, THEN <dispatching rule>'. The CDRs are used to finally select the most appropriate dispatching rule based on the specific situation.

In this study, jobs are iteratively dispatched to the machines with the lowest load. The goal of this research is to learn historical production data so that the AMG model can generate the most similar schedules to the original CDRs.

The architecture of the proposed dispatching method consists of three stages, H_1, H_2, and H_3, as shown in Fig. 1. The first two stages estimate the scores of given jobs quantifying how relevant each job is to dispatching event. They generate ranked lists, R_1 and R_2, which consist of K and M relevant jobs, respectively. The final stage H_3 provides a ranked list R_3 where we can choose the most relevant job to be assigned.

3.1 Stage 1: Listwise Selection (Fast and Cheap Model)

One can compare the structures between an information retrieval task and a job dispatching task as follows. A query in information retrieval is mapped to a dispatching in job dispatching, a document set to be queried is mapped to a job set to be assigned in dispatching, and the relevance score of a document is mapped to the relevance score of a job.

The first stage receives N jobs as input and returns K candidate jobs for dispatching. For given dispatching iteration d and N jobs, the feature vector of jobs is denoted by $\mathbf{x}^d = (x_1{}^d, \ldots, x_N{}^d)$, where $x_i{}^d$ contains the feature values of job j at dispatching iteration d. And the corresponding labels of the jobs are denoted by $\mathbf{y}^d = (y_1{}^d, y_2{}^d, \ldots, y_N{}^d)$, where $y_i{}^d$ represents the degree of relevance with job j at the dispatching d. If job j was assigned to the machine at the dispatching d, $y_i{}^d = 1$, otherwise, $y_i{}^d = 0$.

In stage H_1, a listwise selection LTR model estimates scores for given jobs and creates a selected job list R_1 consisting of the top K candidate jobs based on their scores. The LTR algorithms described in Sect. 2 can be applied to rank jobs according to their relevance scores.

3.2 Stage 2: Listwise Selection with Contextual Ranking (Accurate Model)

The second stage H_2 receives K candidate jobs in R_1 as input at the dispatching iteration d, because the first stage reduced N jobs to K candidates. The second stage allows to extract the ability to capture contextual information for the entire candidate set. Contextual ranking was adopted in the second stage. Four construction strategies were proposed to capture contextual information called rank-based features [9]. This approach was applied to learning dispatching rules.

For a given feature $f \in F$, the feature Rank_f is the rank of job j in the candidate job set in descending order of f. The feature RvsRank_f is the reverse vector of Rank_f, that is, in ascending order of f. The feature DistMin_f is given by the absolute value of the difference between the value of f and the smallest value of f in the candidate job set. The feature DistMax_f is similar to DistMin_f, but the difference with the largest value of f in candidate job set. These features are expected to improve the scoring ability because they contribute to providing contextual information about the entire set of candidate jobs. In the second stage, listwise with a contextual ranking model estimates scores for each candidate job and create a new candidate list R_1 consisting of top M candidate jobs based on these scores.

3.3 Stage 3: Pairwise Comparison (Very Accurate Model)

The M candidate jobs in R_2, which were returned at the previous stage H_2, is used as input to the pairwise comparison model at this stage H_3. In the pairwise comparison, the model estimates the probability $p_{i,j}$ that the candidate job_i is more relevant than the candidate job_j. Therefore, the listwise selection model receives as input all candidate jobs at the dispatching iteration, while the pairwise comparison model receives as input candidate job_i and job_j pairs at the dispatching iteration d.

In the pairwise comparison model, its feature vector is denoted by $x_{ij}{}^{d} = < x_i{}^{d}, x_j{}^{d}, x_i{}^{d}\text{-}x_j{}^{d} >$, where $x_i{}^{d}$ and $x_j{}^{d}$ represents attributes of jobs i and j at the dispatching d, and $x_i{}^{d}\text{-}x_j{}^{d}$ represents the difference of attributes between jobs i and j at the dispatching d. The corresponding label $y_i{}^{d}$ represents the relative relevance of jobs i and j. If job i is more relevant than job j, $y_i{}^{d} = 1$, otherwise, $y_i{}^{d} = 0$. The pairwise comparison model takes a relatively longer time than the listwise selection model because the pairwise comparison is needed to calculate every pair in the job list. Sufficient candidate filtering was performed in the previous two stages. This stage can finally obtain the accurate ranks of M jobs through the pairwise comparison, which can be expected to lead to good performance of dispatching.

4 Experiment Results

The proposed method was applied to two synthetic dispatching datasets collected from a commercial scheduling system that can employ CDRs in flexible flow shop processes such as semiconductor and display manufacturing. Table 1 shows the description of the CDRs used in the two manufacturing processes. CDR-A and CDR-B consist of 12 and 17 state-dependent rules, respectively. CDR-B has a larger number of rules with more attributes, making it more difficult to learn than CDR-A.

A total of 80 schedules were simulated using CDR-A, while 104 schedules done for CDR-B. For CDR-A, 70 schedules were used for model training, and the remaining 10 schedules were reserved for model evaluation. For CDR-B, 90 schedules were used for model training, and the remaining 14 schedules done for model evaluation.

In the experiment, the performance of three strategies was compared in terms of dispatching accuracy and speed. The first strategy is the one-stage method using only pairwise comparison, the second is the two-stage method that combines listwise selection and pairwise comparison, and the final is the three-stage method that combines a listwise selection model, a listwise selection model with contextual ranking model, and a pairwise comparison model.

The experimental results for CDR-A are presented in Table 2. The results indicate that all three models achieved 100% dispatching accuracy. However, the three-stage strategy exhibited significantly fast performance (5.72 s), compared to the one-stage strategy (124.43 s) and the two-stage strategy (29.99 s).

The difference in speed among the three strategies primarily is caused by the number of pairwise comparisons required in the strategies. The one-stage strategy necessitates pairwise comparisons between every two jobs at (N-1) dispatching iterations, resulting in a slow average speed of 124.43 s. The two-stage strategy employs a listwise approach to reduce the number of candidates requiring pairwise comparisons, resulting in fewer comparisons and a faster average speed of 29.99 s. Finally, the three-stage model utilizes a contextual ranking listwise approach to further minimize the number of candidates requiring pairwise comparisons, resulting in the fewest comparisons and the fastest average speed of 5.72 s.

In the case of CDR-B, the results indicate that the two-stage and three-stage strategies achieve higher dispatching accuracy than the one-stage strategy. Both accuracies of the two-stage and three-stage strategies were 99.13%, slightly higher than the accuracy of

Table 1. Two composite dispatching rules, CDR-A and CDR-B

CDR	Rule	Rule Description
CDR-A	Rule 1	IF Attr1 = True, THEN Prioritize the job with a small value of Attr2
	Rule 2	Prioritize the job with a large value of Attr3
	Rule 3	Prioritize the job with a large value of Attr4
	Rule 4	IF Attr5 <= Attr3 AND Attr5 >= Attr6, THEN Prioritize the job with a large value of Attr7
	Rule 5	IF Attr5 <= Attr3 AND Attr5 >= Attr6, THEN Prioritize the job with a large value of Attr8
	Rule 6	IF Attr9 = True, THEN Prioritize the job with a large value of Attr7
	Rule 7	Prioritize the job with a small value of Attr8
	Rule 8	Prioritize the job with a large value of Attr9
	Rule 9	Prioritize the job with a large value of Attr7
	Rule 10	Prioritize the job with a large value of Attr10
	Rule 11	Prioritize the job with a large value of Attr11
	Rule 12	Prioritize the job with a large value of Attr12
CDR-B	Rule 1	IF Attr1 = True, THEN Prioritize the job with a large value of Attr2
	Rule 2	IF Attr1 = True, THEN Prioritize the job with a small value of Attr3
	Rule 3	Prioritize the job with a small value of Attr3
	Rule 4	IF Attr4 <= 7 AND Attr5 = Attr6, THEN Prioritize the job with a large value of Attr2
	Rule 5	IF Attr4 <= 7 AND Attr5 = Attr6, THEN Prioritize the job with a small value of Attr7
	Rule 6	IF Attr4 <= 7, THEN Prioritize the job with a large value of Attr2
	Rule 7	IF Attr4 <= 7, THEN Prioritize the job with a small value of Attr7
	Rule 8	Prioritize the job with the same values of Attr8 and Attr9
	Rule 9	Prioritize the job with Attr10 values less than 0.1
	Rule 10	Prioritize the job with a large value of Attr2
	Rule 11	Prioritize the job with a large value of Attr11
	Rule 12	Prioritize the job with a large value of Attr12
	Rule 13	Prioritize the job with a large value of Attr13
	Rule 14	Prioritize the job with a large value of Attr14
	Rule 15	Prioritize the job with a large value of Attr15
	Rule 16	Prioritize the job with a large value of Attr16
	Rule 17	Prioritize the job with a large value of Attr17

Table 2. Performance comparison of three ML-based dispatching strategies for two dispatching datasets

Strategy	CDR-A		CDR-B	
	Dispatching accuracy	Dispatching speed (sec.)	Dispatching accuracy	Dispatching speed (sec.)
One-stage method (pairwise comparison)	100%	124.43	99.02%	60.73
Two-stage method (listwise selection + pairwise comparison)	100%	29.99	99.13%	16.80
Three-stage method (listwise selection + listwise selection with contextual ranking + pairwise comparison)	100%	5.72	99.13%	3.68

99.02% achieved by the one-stage strategy. This difference could be attributed to the fact that CDR-B has more rules and a relatively larger number of columns included in the rules, making it more challenging to learn using a one-stage pairwise approach due to its limitation in handling large and complex datasets. In contrast, the listwise approaches used in the two-stage and three-stage strategies are better suited for such datasets as they consider the entire set of rules together and can leverage contextual relationships between them, thereby capturing more relevant information for accurate dispatching. Furthermore, the three-stage strategy achieved the fastest dispatching speed of 3.68 s, followed by the two-stage model (16.80 s) and the one-stage model (60.73 s).

5 Conclusions

This paper proposed a three-stage strategy to accelerate the data-driven job dispatching models for production simulation. The strategy uses LTR models to filter prioritized jobs before performing the time-consuming pairwise comparison among jobs. This results in faster and more accurate dispatching in the experiments.

Experimental results on two dispatching datasets demonstrated the effectiveness of the three-stage strategy. The strategy could achieve 100% dispatching accuracy for CDR-A and 99.13% for CDR-B, respectively. It was also significantly faster than the one-stage and two-stage strategies.

This data-driven job dispatching method can be used to improve the efficiency of simulation for production planning in high-tech manufacturing. It can also be used with any scheduling system that supports CDRs. The approach will be evaluated with additional scheduling datasets in various manufacturing processes.

Acknowledgments. This work was partly supported by Institute of Information & Communications Technology Planning & Evaluation (IITP) grant funded by the Korea government (MSIT) (No. RS-2022-00155911, Artificial Intelligence Convergence Innovation Human Resources Development (Kyung Hee University)).

References

1. Sakr, A.H., Aboelhassan, A., Yacout, S., Bassetto, S.: Simulation and deep reinforcement learning for adaptive dispatching in semiconductor manufacturing systems. J. Intell. Manuf. 1–14 (2021)
2. Lee, J.H., Kim, Y., Kim, Y.B., Kim, B.H., Jung, G.H., Kim, H.J.: A sequential search method of dispatching rules for scheduling of LCD manufacturing systems. IEEE Trans. Semicond. Manuf. **33**(4), 496–503 (2020)
3. Bergmann, S., Feldkamp, N., Strassburger, S.: Emulation of control strategies through machine learning in manufacturing simulations. J. Simul. **11**(1), 38–50 (2017)
4. Li, X., Olafsson, S.: Discovering dispatching rules using data mining. J. Sched. **8**(6), 515–527 (2005)
5. Zhong, H., Liu, M., Bao, L.: A job-priority based soft scheduling approach for uncertain work area scheduling in semiconductor manufacturing. Int. J. Prod. Res. **60**(16), 5012–5028 (2022)
6. Nagahara, S., Sprock, T.A., Helu, M.M.: Toward data-driven production simulation modeling: dispatching rule identification by machine learning techniques. Procedia CIRP 222–227 (2019)
7. Li, H.: Learning to rank for information retrieval and natural language processing. Synthesis Lect. Hum. Lang. Technol. **7**(3), 1–121 (2014)
8. Culpepper, J.S., Clarke, C.L., Lin, J.: Dynamic cutoff prediction in multi-stage retrieval systems. In: Proceedings of the 21st Australasian Document Computing Symposium, pp. 17–24 (2016)
9. Lucchese, C., Nardini, F.M., Orlando, S., Perego, R., Tonellotto, N.: Speeding up document ranking with rank-based features. In: Proceedings of the 38th International ACM SIGIR Conference on Research and Development in Information Retrieval, pp. 895–898 (2015)

Automating Test Cases for OPC UA Information Models

Tonja Heinemann(✉), Oliver Riedel, and Armin Lechler

ISW, University of Stuttgart, Seidenstraße 36, 70174 Stuttgart, Germany
tonja.heinemann@isw.uni-stuttgart.de

Abstract. In a digitized production, devices are interconnected with each other. This is achievable with OPC UA, which is currently increasing in popularity. To accommodate different kinds of machinery, Companion Specifications (CSs) for OPC UA are being defined. To achieve connectivity at the installation of machinery equipment, interoperability and correct implementation of the CS interfaces must be guaranteed. Discovering faults as soon as possible can be achieved through testing. For OPC UA and CSs, a testing framework exists, but the test cases are written by hand. This is a drawback for CS standardization groups, that have to put in the effort to develop test cases. In this work, the prerequisites to automatically generate CS test cases are outlined. An approach for automatic test case generation is sketched, that synthesizes test cases based on these prerequisites. In a comparison to existing test cases, the possibilities and limits of autogenerating test cases are discussed.

Keywords: OPC UA · Testing · Companion Specifications · Digitization · Industrie 4.0

1 Introduction

The production systems making up the manufacturing environments of tomorrow are connected with software systems, allowing to access their information, and monitor and control the production process.

The more uniform the communication between the interconnected production systems, the easier it is to set up and operate them. Such a uniformity can be achieved through standardization. In Open Platform Communications Unified Architecture (OPC UA), one possible means of communication in manufacturing, such standardization is happening not only for the communication mechanisms, but also for the contents transported in the form of data models in OPC UA Companion Specifications (CSs) [1].

As soon as standards are implemented and used, all the benefits of the uniform communication can come to play, unless the implementations are faulty. To mitigate faults in the implementation of the communication standards, testing of the implementations against the standards can be performed. This work outlines how the generation of test cases for such a scenario can be automated to replace the current manual process for CSs.

M. Dragomir et al. (Eds.): ICPR 2023, LNPE, pp. 210–218, 2025.
https://doi.org/10.1007/978-3-031-92082-0_22

The article is structured as follows: In Sect. 2, the benefits of using CSs in OPC UA are outlined. The problems that arise in practice due to implementation faults with CSs are described in Sect. 3. To find and mitigate these faults early, testing practices can be used. Their scope and generation process, along with current work around test cases, are shown in Sect. 4. The approach to automate the test case generation process is described and exemplified in Sect. 5. In Sect. 6, the results of the exemplary test case generation are compared to manually generated test cases. In Sect. 7, the findings of the comparison are summarized and further steps are outlined.

2 Connectivity with OPC UA Information Models

Major requirements towards communication in production are a uniform data format along with services to control the representation physical assets and the technical process [2]. OPC UA, a popular means of industrial data transport, is a candidate for such a uniform communication solution [2–4].

OPC UA combines multiple aspects of communication, among them the means of transporting messages between communication partners. Using the same means of communication for all entities greatly reduces efforts compared to different protocols and message serializations. However, if the transported data are not standardized, major overheads are still needed when aggregating the data.

Figure 1 sketches an example for this. In the top part, a case without standardized data models is pictured. There are two data providers, both send pressure values to a data consumer. However, they have different labels to identify the data: data provider 1 uses the key "Pressure" while data provider 2 uses "P". While easily understandable by human observers, this difference is not easily perceived by software programs when querying the data. In addition, the data providers don't use the same unit for the value. Data provider 1 shows the pressure in pascal, while data provider 2 represents the pressure in millibar. This leads to additional efforts when using the data, which need to be converted and scaled differently. The pseudo code in the box of the data consumer in the upper part of Fig. 1 exemplifies this: different pieces of code are needed for each data provider to account for the different data labels and the scaling.

In the bottom part of Fig. 1, the same example is pictured with both data providers using the same data models. In the example, this means using the same labels for the data and the same units of measurement. This results in an easier application for the data consumer to achieve the same result: No more differentiation when reading the values is needed, as the labels are the same. And due to the units of measurement being the same, scaling of the data is also not dependent on the source.

For this reason, standardizing data models is currently favored among other organizations by the Mechanical Engineering Industry Association (VDMA). Around 40 international groups to develop CSs have been started by the VDMA so far [5]. These CSs define a data model for use with communication methods as stated by OPC UA.

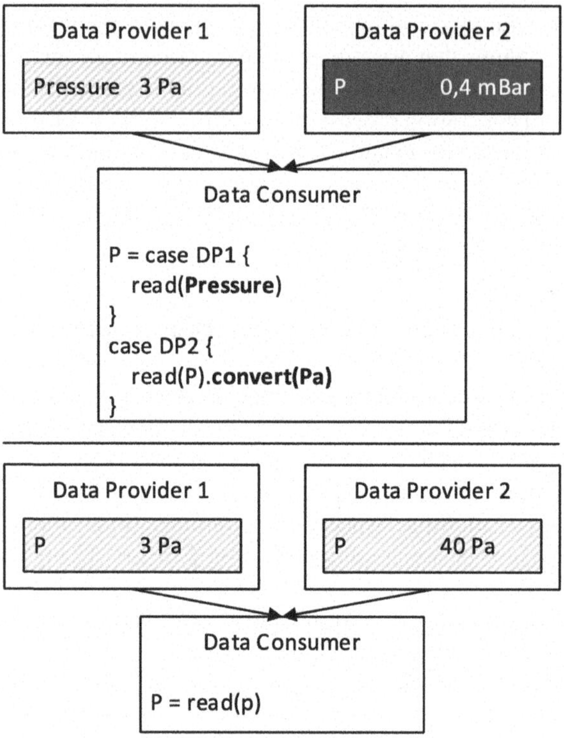

Fig. 1. Usage of standardized data models.

3 Practical Limitations of Connectivity with OPC UA Information Models

To benefit from standardized data models, manufacturers of OPC UA capable devices have to implement the respective standards. However, these are extensive: The OPC UA base specification spans 22 documents, there are 108 CSs defined. [6] Not all of the specification documents are relevant for every application. Nevertheless, reading CSs requires prior knowledge of OPC UA and an overview to choose the relevant specifications requires expertise. This need for education of personnel is a known problem regarding software in combination with production systems [7].

In many instances, an OPC UA interface is added to various kinds of manufacturing equipment. The main task of this equipment is to support the function in manufacturing it is supposed to; a machine tool primarily has to support a production process and a measuring device shall primarily take measurements. OPC UA connectivity is an additional feature to this main functionality, in consequence it plays a minor role in the manufacturing equipment's development compared to the primary manufacturing function. The time and effort that are reserved for the OPC UA interface in development are limited. Likewise, the time and effort the developers can invest in reading specification documents is limited as well.

To reduce the education efforts, implementations can be based on the model libraries and sometimes implementation examples [6, 8]. This practice may however lead to implementation faults, as some aspects of the model behavior are only covered in the documents. In addition, careless mistakes may be introduced in implementation as well. As an example, Fig. 2 shows an abbreviation of element names (A, B, C) and data types (Bool in case of A, Int otherwise) of a specification on the left side. On the right side, an implementation is shown. This implementation introduces the fault, that C is not of data type Int, but of type Bool.

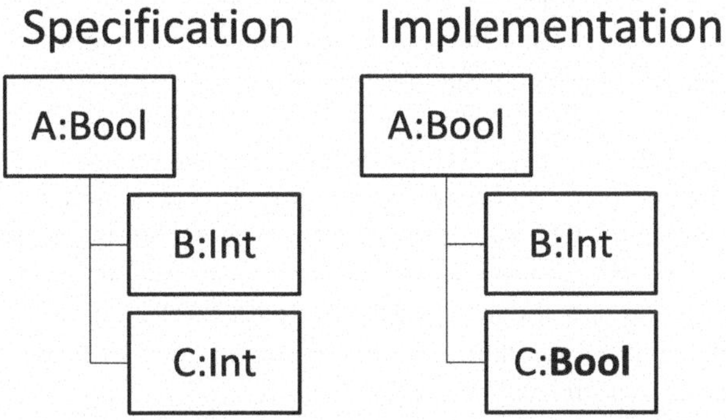

Fig. 2. Implementations may differ from the specification.

Any kind of implementation fault leads to connectivity problems in communication. If the expectations of a communication partner are not met, there may be consequences, e.g. program crashes. In the example in Fig. 2, a receiving communication partner would expect an Int out of element C and handle that data accordingly. If a Bool is transmitted, any results will be incorrect and may lead to further errors, e.g. in type conversions.

The main issue here is not specific to OPC UA: If standards are not implemented correctly, the goal of the standardization cannot be met and additional problems can occur. However, reading all standards is not just cumbersome but not realistic in many implementation projects and careless mistakes may happen. The easiest way to combat implementation faults and their propagation within projects is early and frequent testing of implemented assets.

4 Test Cases in OPC UA and Test Case Generation Today

As testing helps to ensure interconnectivity, the OPC Foundation offers an OPC UA test tool, the OPC UA Compliance Test Tool (UACTT) [9]. This tool can be used by implementors of OPC UA themselves, and it is also used in the official certification process of the OPC Foundation. In combination with the UACTT, test cases for the specifications are offered. These include both automated test cases to be executed by the UACTT and manual test cases to be executed by a manual tester [10]. The test cases are defined by the same working group as the respective specification document, the implementation of the UACTT automated test cases is done in a subsequent step.

For a better overview in compliance tests, the test cases are aggregated in so-called Conformance Units, this aggregation is also done by the working group [11, 12].

Among the test cases to be generated for CS models, two major groups can be differentiated: test cases for the model contents and test cases for the model behavior. The first group contains the names, types, definitions and relations of the CSs being used in the right way. The second group, the behavior tests, contain the knowledge of the specification developers on how to use the CS model [12].

There are first efforts towards automated test generation, especially with the second group of behavior tests as a target [13]. These efforts are focused on extracting the behavior of OPC UA state machines from implemented models and developing test cases to explore all possible states and transitions. Meanwhile, the focus for test creation and execution is testing interface behavior without regard to specification conformance [13].

While all the profiteers of the CSs also benefit from the test cases defined by the CS working groups, the respective test cases are developed purely manually. These test cases concern the elements of the CS. This is sketched in Fig. 3. The specification contains the elements A, B and C with specified data types. Test cases may now check the existence of these elements in the specification (indicated with "A?"), or the correct data types ("A:Bool?"). Additionally, the relations of the elements may be tested, e.g. the correct connection of elements A and B ("A - B?").

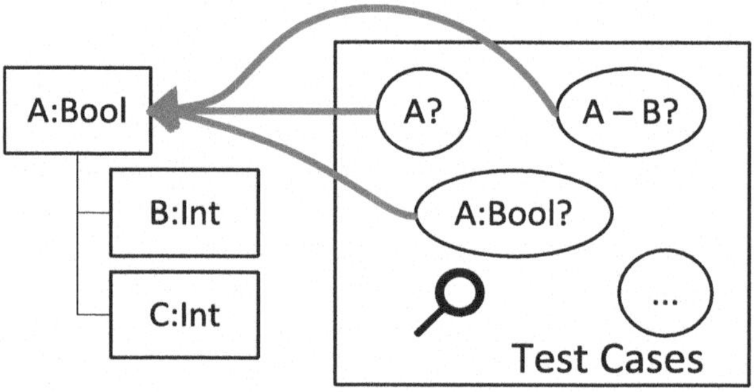

Fig. 3. Test cases can be directly mapped to nodes in the OPC UA model.

These tests can help in managing the complexity of OPC UA systems and reduce cost and time in implementation and commissioning. However, the tests have to be defined and implemented. This step is easily disregarded in specification groups due to the high amount of (repetitive) work and time needed, if test cases are defined manually. An automated generation of test cases for automated execution based on the specification files would allow for minimal efforts in test generation and execution.

Aside from test cases, the CS working groups always provide a textual description of their model. This document also contains the Conformance Units. Due to the template and knowledge exchange between the CS groups, CSs often use similar approaches in describing and aggregating model and test information. The CS Working groups also provide a representation of all model elements in a machine readable format defined by the OPC Foundation, the NodeSet XML [8]. These files can be used in engineering software for OPC UA, e.g. to create OPC UA Servers or Clients.

5 Automated Test Case Generation

In order to enable an automated test case generation for both automated and manually executed test cases in OPC UA, the CS documents have to be analyzed for relevant information. Even if not all test cases desired by standardization groups can be generated automatically, the efforts for test case generation can be reduced by autogenerating some test cases for the CSs. A first step is to focus on the test cases directly related to model elements.

For automated test case generation based on the model elements, the relations that lead from elements in a CS to test cases need to be analyzed and represented in a machine readable way. Figure 4 shows this principle. The elements in the specification are given, the arrows represent the relations needed for automated test case generation. These relations may be found by analyzing existing test cases, that have been developed by working groups so far.

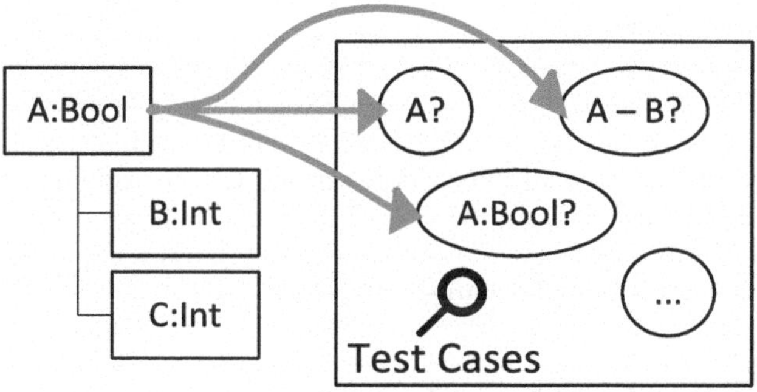

Fig. 4. Test cases are generated from types in the OPC UA model.

The method chosen for this analysis starts with the conformance units. These are topically sorted after the classes suggested by Friedl et al. [12]. These classes are closely related to the different kinds OPC UA model elements. In a subsequent step, all the test cases belonging to the conformance units in one class are compared for similarities and differences. Based on the patterns found by analyzing the test cases, rules for autogenerating test cases can be derived.

As an example, this method is carried out with the CS for weighing technology [14]. Of the 27 conformance units of the CS for weighing technology, 20 are assigned the class "Type Check", three are assigned "Method" and four "Functionality". For the further analysis, only the test cases of the "Type Check" class are regarded, due to the small sample size in the other classes.

In the "Type Check" class, there are 231 test cases defined. All of them are directly related to an OPC UA ObjectType, one kind of OPC UA model element, defined in the CS. When analyzing the test cases, it becomes clear that there are test cases defined for each object type (the ObjectTypes inheritance hierarchy is represented correctly), test cases that are defined only conditionally (e.g. check for child elements if there are child elements defined for the test case) and test cases that are unrelated to the model represented in the NodeSet XML (the requirement that at least one instance of an ObjectType shall be available in an OPC UA Server).

Based on this analysis, similar test cases can be autogenerated. In the NodeSet XML, all ObjectTypes can be identified and the conditions, e.g., if child elements exist, can be checked. On this basis, test case suggestions can be autogenerated. When implementing this simple test case generation and applying it to the CS for weighing technology, 296 possible test cases are found.

Of the original, manually generated test cases, 19 were not found. These were the previously mentioned required instances, that are not contained in the XML as well as checks for child elements that are not direct child elements of the ObjectType but derived in the model.

In addition to the original test cases, the autogenerated test cases suggest seven additional tests for ObjectTypes that are included in the original test cases. These contain some of the conditional test cases. Other than that, 24 types of the specification have not been considered in the original test cases. The obligatory and conditional test cases for these ObjectTypes make up the remaining difference between the number of original and autogenerated test cases.

The same test case generation algorithm is used for another specification, the CS for machine tools [15]. It produces 201 test cases. When comparing these to the manually generated test cases for this specification, the overlap is much lower. The inheritance checks that were defined for all ObjectTypes in the CS for weighing are not among the manually defined test cases in the CS for machine tools for every object type. In comparison, the CS for machine tools contains more test cases requiring special model behavior, these are not among the autogenerated test cases.

6 Comparison to Existing Test Cases

The autogenerated test cases found a few conditional tests the manually generated test cases missed. While working groups can decide what to include or not, autogenerated test cases can prevent unintentionally missing some test cases. This is most likely the reason for the 24 types that were not covered in the original test cases: the working group decided to focus on key parts of the specification and thus reducing the workload in defining test cases.

Some of the original test cases could not be found by the autogenerated test cases. In the case of derived child elements, that can be technically solved in the NodeSet XML parser. Information about the types that are required in OPC UA products however is not contained in the NodeSet XML. To generate such test cases, either the textual specification has to be parsed for such information or it is suggested to the working group for suitable model elements, the test cases are then generated based on this user input.

There are test cases in the CS for weighing technology, that contain additional textual information. This information is also unavailable from the NodeSet XML.

When comparing to a different specification, it becomes apparent that CS working groups use different styles of describing test cases.

7 Conclusion

To make the test case generation for CSs easier, the test case generation process can be automated. As shown in this work, such an automation can be achieved for test cases referring to the model in the NodeSet XML by analyzing existing test cases and deriving rules for test case generation based on the analysis results. For the analysis, more than one specification has to be considered: in the CS for weighing technology, only test cases for ObjectTypes provided a broad enough base for analysis, however there are more model elements in OPC UA. And in the comparison with the CS for machine tools it became apparent, that different styles in test case definition have to be taken into account.

The comparison also showed the potential of test case generation for the model tests, even if they are only part of all possible test cases. For a specification like the CS for machine tools, the autogenerated model tests would add test coverage, as this specification did not define the ObjectType related tests as thoroughly as the OPC UA CS for weighing.

For test cases concerning the model behavior, other sources of information have to be added, e.g. the CS specification text documents or human input by the working groups.

All in all, the working groups have the power to decide what test cases to include and how to structure them into conformance units. An automated test case generation should consider that.

Acknowledgments. This Project is supported by the Federal Ministry for Economic Affairs and Climate Action (BMWK) on the basis of a decision by the German Bundestag.

References

1. OPC Foundation. OPC 10000-1: UA Part 1: Overview and Concepts. Version 1.05.02. (2022)
2. Adolphs, P., et al.: Reference architecture model industrie 4.0 (RAMI4.0): Status Report (2015)
3. VDMA e.V. OPC UA: standardized communication for the world of machines - vdma.org - VDMA. https://www.vdma.org/global-production-language. Accessed 27 Oct 2022
4. VDW—Verein Deutscher Werkzeugmaschinenfabriken e.V. (German Machine Tool Builders' Association). umati - universal machine technology interface – connecting the world of machinery. https://umati.org. Accessed 27 Oct 2022
5. VDMA e.V. Which OPC UA working groups exist in the VDMA? - vdma.org - VDMA. https://www.vdma.org/viewer/-/v2article/render/1248676. Accessed 27 Oct 2022
6. OPC Foundation. Documents - OPC Foundation. https://opcfoundation.org/developer-tools/documents/. Accessed 27 Oct 2022
7. Feichtinger, K., et al.: Industry voices on software engineering challenges in cyber-physical production systems engineering. In: Proceedings of the 27th International Conference on Emerging Technologies and Factory Automation (ETFA) (2022)
8. OPC Foundation. OPC UA NodeSets and Other Supporting Files. https://github.com/OPCFoundation/UA-Nodeset. Accessed: 27 Oct 2022

9. OPC Foundation. OPC UA Compliance Test Tool (UACTT). https://opcfoundation.org/developer-tools/certification-test-tools/opc-ua-compliance-test-tool-uactt/. Accessed: 27 Oct 2022
10. OPC Foundation. Part 1: Certification Process. Standard. OPC Foundation (2018)
11. Heinemann, T., Lechler, A., Riedel, O.: Profiles and testing in OPC UA – current state and future potentials in industry 4.0. In: Intelligent and Transformative Production in Pandemic Times. Proceedings of the 26th International Conference on Production Research (2021)
12. Friedl, S., et al.: Implicit templates for conformance units in OPC UA companion specifications. In: 2021 22nd IEEE International Conference on Industrial Technology (ICIT), vol. 1, pp. 730–735. IEEE (2021)
13. Walchshofer, A., Klammer, C., Fischer, S.: A toolchain for testing OPC UA interfaces. In: IEEE International Conference on Emerging Technologies and Factory Automation, ETFA 2022, Stuttgart, Germany. IEEE (2022)
14. OPC Foundation. OPC 40200: Weighing Technology. Version 1.00 (2020)
15. OPC Foundation. OPC 40501-1: Machine Tools – Monitoring and Job Overview. Version 1.00.0 (2020)

System Dynamic Modelling is Supporting the Urban Production Paradigm

Michael Hertwig(✉)

Digital Engineering, Fraunhofer Institute for Industrial Engineering IAO, Nobelstrasse 12, 70569 Stuttgart, Germany
michael.hertwig@iao.fraunhofer.de

Abstract. Production in the direct proximity of residential areas causes conflicts. On the one hand, they are in competition with residential real estate development; on the other hand, hey are often viewed as undesirable by residents. However, manufacturing companies contribute significantly to the maintenance of prosperity. Cities are therefore called upon not to allow existing value-added structures to migrate. To this end, a correspondingly attractive framework must be created. At the same time, production companies are under increasing pressure in terms of sustainability and workforce development.

To act in a more ecologically and socially responsible manner, manufacturing companies need to be further advanced. In addition to sustainability, companies must not only eliminate negative influences in the future but should also fulfil their local responsibility. Nevertheless, further development is surrounded by economic framework conditions. Therefore, the increase in value for the company in all dimensions of urban production must be presented from a sustainability perspective as well as from a human-factors perspective. By developing a long-term perspective, in which the holistic contribution of further development measures becomes visible, strategic goals can be addressed.

In this article, we will look at how it is possible to combine a holistic approach with a long-term view. For this purpose, different modelling approaches are compared with each other to prioritize a suitable methodology. Since the development of a system dynamic model is costly and time-consuming, a procedural proposal is presented that supports small and medium-sized enterprises in the usability of system dynamic modelling in order to anchor urban production measures in strategic planning. The outlined procedure is intended to support an evaluation of urban production approaches regarding sustainability and environmental compatibility. The outlook will highlight what needs to take place to assess the applicability of the approach. In addition, it will be discussed how the approach can strategically support manufacturing companies in the transformation to near-urban, ecologically safe, and socially compatible production.

Keywords: Urban Production · Urban Manufacturing · sustainable manufacturing · symbiosis with urban environments

M. Dragomir et al. (Eds.): ICPR 2023, LNPE, pp. 219–226, 2025.
https://doi.org/10.1007/978-3-031-92082-0_23

1 Introduction

1.1 Urbanity

Urbanization is a major megatrend whose impact continues unabated, despite a slowdown due to the COVID19 pandemic [1, 2]. The attractiveness of cities is a growth driver due to the high level of economic potential, which is particularly visible in megacities [3].

If the concept of urbanity was always in flux, Lefebvre captures it as a space-tension structure [4]. Siebel makes it more concrete, for whom urbanity is where conflicts due to the spatial proximity of the functions of living, working and recreation are visible, must be fought out and resolved [5].

The situation is particularly challenging in developed countries where, on the one hand, the standard of living is high and, at the same time, urban patterns are shaped by industrialization. In addition, the degree of agglomeration is already very high and growth dynamics are weakened. A transformation of urban structures is associated with major social, economic and ecological challenges [6].

1.2 Transformation of Manufacturing

Production is the creation of material goods, services, and ideal goods, whereby a transformation of input into output happens [7]. Due to changes in external conditions, production has always had to adapt, which has changed established production paradigms [8]. The transformation of production always takes place considering the conflict of targets also minimal cost, minimal time and high-quality results [9].

Current change drivers for production in industrialized countries can be classified as follows:

- Individualization of customer requirements [10];
- Cost and globalization pressures [11];
- Increasing importance of the knowledge economy;
- Demands for work-life balance [12];
- Sustainability and environmental protection [13];
- Demographic change [14];
- Digitalization of production [15];
- Maturity of new production technologies and
- Hybridization of products [16].

Production has been pushed back in favour of a growing service sector in many industrialized nations. This has also been seen as a solution to the employment crisis due to increasing productivity [17]. Despite this, manufacturing value added has been considered an important factor in the creation and preservation of wealth. Thus, in economies with a high share of manufacturing, a stabilizing effect was perceived in the economic crisis of 2008/09 [18]. The acceptance of manufacturing industries by the local population is an important factor for success [19].

2 Urban Production

Production of goods that takes place in densely populated areas, i.e., that meets urbanity, is referred to as urban production in the narrower sense and, in the case of product-service combinations, as urban production in the broader sense (see Fig. 1) [20, 21]. However, this only works if the production is designed sustainably, i.e., negative impacts on ecology and society are reduced [22]. Then, a positive contribution to the urban ecosystem can be created by the urban production paradigm [23]. This is because the mixing of production system with the urban system enables the use of local resources and the establishment of regional value chains with synergy effects and symbiotic interactions between all partners [20, 24].

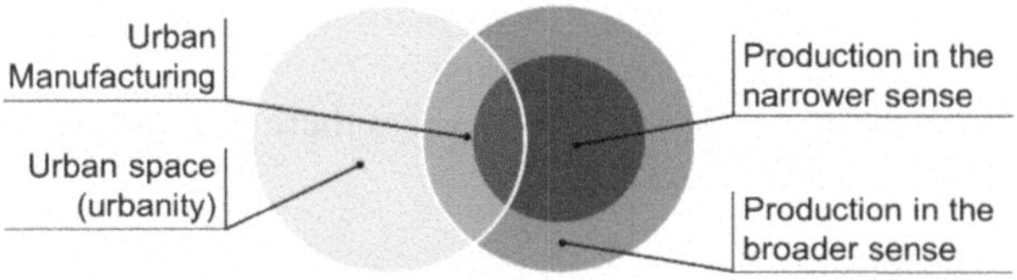

Fig. 1. Urban manufacturing is the overlay of urbanity and production

To leverage the potentials, a transformation of production towards city compatibility must take place [25]. To support this, Stiehm has derived design approaches for city-compatible production that are intended to provide a framework for action [26]. The fields of action from an urban planning perspective differ depending on the type of value creation which can be clustered to urban craftmanship and repair, urban industry or urban agriculture [27]. In the meantime, there are various good target images for urban production including its fields of action, as described for example by Positive-Impact-Factory [28], Ultra-efficiency factory [29] or E^3-factory [30].

However, since the embedding in the strategic further development of the producing companies is missing, an approach is to be outlined hereafter, which uses a model as a basis. The model-based approach is intended to support the shaping of scenarios. By embedding it in the company-specific strategy process, the aim is to provide a tool for the holistic consideration of the elements of urban production.

3 Urban Manufacturing Model

3.1 Framework Conditions and Choice of Modelling Approach

Models have a specific purpose and represent only a subset of reality. It is therefore important to define the scope of analysis and to specify the conditions within which the model will work [31]. For the Urban Manufacturing Model, the level of detail was defined as the factory-environment level, which excludes a detailed consideration of production processes as well as a comprehensive consideration of all ambient parameters. The focus lies on the interfaces between production and the urban surroundings. The study of further development in the sense of urban production is carried out in four perspectives: economic, ecological, social and production. Existing approaches will be used for the modelling. To ensure comparability, all calculations are made based on relative values, and the focus is also on the medium to long term, which is why no individual impulses are examined in detail. All perspectives are considered equally, and the economic perspective is not the focus.

In the investigation of possible modelling approaches, a comparison of system dynamics, discrete event simulation and agent-based systems was carried out. Despite the advantages of each modelling method, system dynamics shows the best fit with the needs created [32]:

- The model has continuous behaviour; thus, time sequences are computable.
- The system dynamics offers a deterministic description of the observation domain.
- The system dynamics models allow interaction analysis and support macroscopic analysis.
- Established modelling software systems exist.
- Due to ongoing research, for more than 50 years there is a variety of models whose components are proven and reusable.

3.2 Qualitative Modelling

Qualitative modelling is used to identify relationships and interactions. The system dynamics modelling approach uses caused-loop diagrams to describe the relationships. As a more fundamental approach, the sustainability dimensions are linked with the human-factors approach (people - technology - organisation) (Fig. 2).

Fig. 2. Framework of relevant perspectives for urban manufacturing (adapted to [33])

The linking of sustainability and socio-economic considerations helps to put the relationship in context by determining whether the situation can be addressed through technical, organisational measures or what changes in the behaviour of those affected need to be motivated. In this context, the economic perspective establishes the reference to competitiveness and financial performance. In the ecological perspective, the fields are emission development, traffic, biodiversity, and surface sealing. The aspects of employees, attractiveness and contributions to the surroundings are summarised in the social perspective. Production, the core of producing companies, is considered separately to take resource flows into account. Through the development of auxiliary relationships, connections that cannot be directly assigned are also included.

3.3 Quantitative Modelling

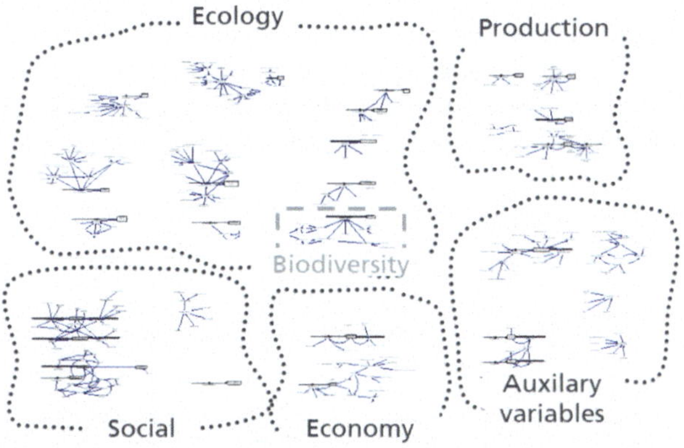

Fig. 3. Quantitative urban manufacturing model as overview of all perspectives

In order to achieve numerical statements, the caused-loop diagrams are converted into stock-flow diagrams. The developments can be quantified through the mathematical description of the relationships. The quantitative model is divided into the different perspectives - ecology, economy, social, production and auxiliary variables - whereby interactions exist between the different perspectives (Fig. 3).

Based on existing sustainability indicators, a selection of relevant analysis parameters was made. For quantification, the Urban Manufacturing Model is subdivided in specific sub models (Fig. 4) in which a standardised parameter is determined.

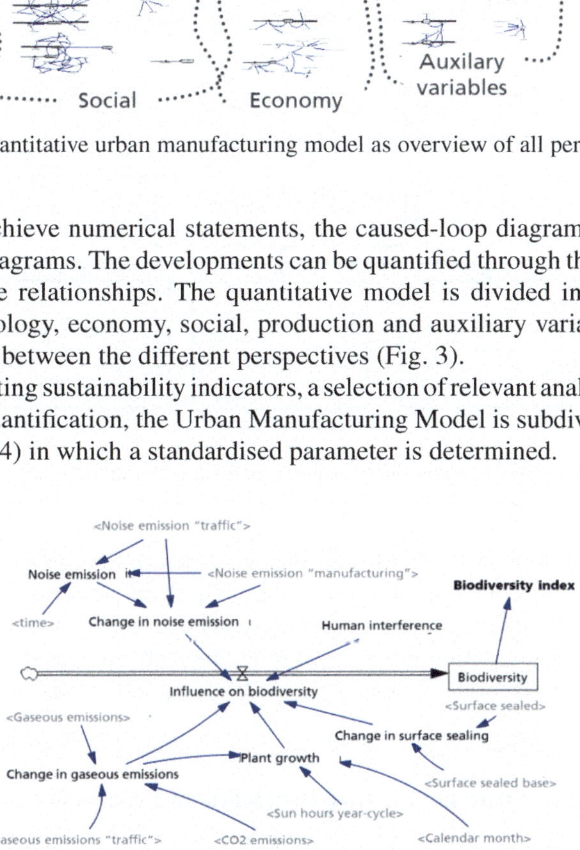

Fig. 4. Sub model for biodiversity in the ecological perspective on the urban manufacturing model

The analysis parameters can be calculated based on a defined analysis period. In the current approach the calculation period will be ten years divided in 524 weeks which allows a resolution of the analysis in weeks. The derived analysis parameters can be visualised in plots with the time steps on the x-axis.

4 Framework for Industrial Application

4.1 Individualization of the Generic System Dynamic Model

A power of system dynamics is the derivation of scenarios by choosing relevant framework conditions in different variations. To support a company-specific transformation, the generic framework must be easily adaptable to the company (Fig. 5). In addition, it must be comprehensible for companies how different perceptions can be expressed in the form of scenarios.

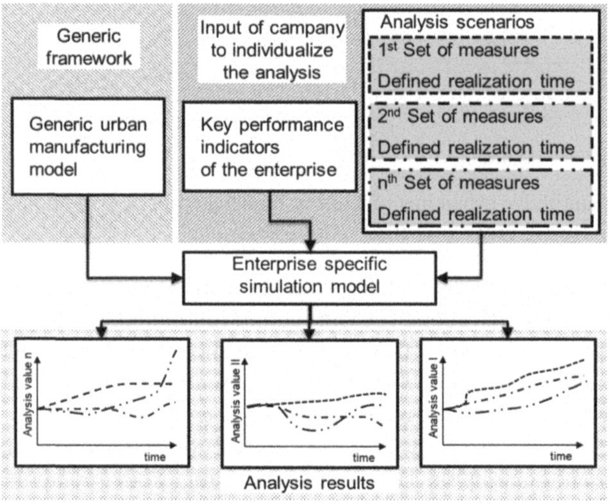

Fig. 5. Approach to derive individual scenario analysis with the urban manufacturing model

For the adaptation to the company, there is a fixed amount of key performance indicators, which the companies have to collect. These are imported and thus adapt the generic urban manufacturing model accordingly. By defining a selection of development measures and their realisation or effective time, the analysis scenarios can be defined. This enables the derivation of company-specific, temporal trends in the results of the analysis.

4.2 Embedding in Strategic Further Development Process for Urban Production

The transformation towards a sustainable and eco-compatible production requires a long-term effort. Therefore, it needs to be integrated into the strategic planning of the company. For this purpose, it is a suitable option to embed the utilization of the urban manufacturing model in a scenario-driven strategy process [34]. This includes the steps of objective definition, scenario-based determination of possible outcomes as well as the translation into specific measures. Since this process includes a feedback loop and is carried out regularly, undesirable developments can be corrected or changed assumptions can be considered.

5 Summary and Further Research Work

In this paper, the scope of the model for urban production was outlined. Based on this, the selection of the modelling method was carried out, with which the interactions and framework conditions can be described in the best possible way. Based on the object of analysis - manufacturing companies in an urban setting - a generic interaction model was developed.

Regarding the application, it is outlined how the generic system dynamic model can be tailored to individual companies. Furthermore, it is shown how scenarios for the evaluation of further development can be derived from it. The embedding in strategic corporate development is described theoretically.

This provides concrete starting points for subsequent research activities. The generic model and the approach to strategy development implementation must be validated and tested using real data. It is also important to verify whether it is possible to embed the model in the strategy process of small and medium-sized enterprises and what effort is required to do so. This is because the effort involved versus the usability of the results can provide an indication of their potential use.

Acknowledgments. Prof. Oliver Riedel of University of Stuttgart, Institute for Control Engineering of Machine tools and Manufacturing Units ISW and of Fraunhofer institute for Industrial Engineering IAO is gratefully thanked for supporting this PhD research.

References

1. Heß, W.: Ein Blick in die Zukunft - acht Megatrends, die Wirtschaft und Gesellschaft verändern: Working Paper, no. 103 (2008)
2. Studtrucker, M., Kalinowski, M., Schneemann, C., Söhnlein, D., Zika, G.: QuBe-Bevölkerungsprojektion für die Kreise und kreisfreien Städte Deutschlands. In: IAB-Discussion Paper, vol. 19 (2022)
3. Dobbs, R., Smit, S., Remes, J., Manyika, J., Roxburgh C., Restrepo, A., Urban world: mapping the economic power of cities. McKinsey Global Institute, Seoul (2011)
4. Lefebvre, H.: Dissolving city planetary metamorphosis. Environ. Plan D **32**(2), 203–205 (2014)
5. Siebel, W.: Urbanität, In: Häußermann, H. (ed.) Großstadt, VS Verlag für Sozialwissenschaften, Wiesbaden (2000)
6. United Nations. Transforming our world: the 2030 Agenda for Sustainable Development, General Assembly, vol. 70, no. 16301, pp. 1–35 (2015)
7. Gutenberg, E.: Grundlagen der Betriebswirtschaftslehre: Die Produktion, 24th edn. Springer, Heidelberg (1971)
8. Boër, C.R., Dulio, S.: Mass Customization and Footwear: Myth, Salvation or Reality? Springer, New York (2007)
9. Ramsauer, C.: Industrie 4.0 - Die Produktion der Zukunft. WINGbusiness, vol. 3, pp. 6–12 (2013)
10. Arnoscht, J., et al.: Individualisierte Produktion, in Integrative Produktionstechnik für Hochlohnländer, pp. 83–255. Springer, Heidelberg (2011)
11. Nicole, E.: Is your business glocal? How to think globally and act locally (2011)

12. Krech, D., Ballachey, E.L., Crutchfield, R.S.: Individual in Society: A Textbook of Social Psychology. McGraw-Hill, New York (1962)
13. Büchele, R., Henzelmann, T., Panizza, P., Wiedemann, A.: GreenTech made in Germany 4.0: Umwelttechnologie-Atlas für Deutschland. Roland Berger Strategy Consultants, Berlin (2014)
14. Fuchs, J., Söhnlein, D., Weber, B.: Projektion des Arbeitskräfteangebots bis 2050: Rückgang und Alterung sind nicht mehr aufzuhalten, Nürnberg (2011)
15. Kletti, J., Schumacher, J.: Die perfekte Produktion: Manufacturing Excellence durch Short Interval Technology, 2nd edn. Springer, Berlin (2014)
16. Barwasser, A., Lentes, J., Riedel, O., Zimmermann, N.: Software-defined product features: reduktion physischer varianz durch virtualisierung. Zeitschrift wirtschaftlichen Fabrikbetrieb **115**(11), 824–828 (2020)
17. Burda, M.: Eine wichtige Säule des amerikanischen Beschäftigungserfolges: der Dienstleistungssektor. Mitt. Arbeitsmarkt Berufsforsch. **30**(4), 819–821 (1997)
18. Reuter, N., Zinn, K.G.: Moderne Gesellschaften brauchen eine aktive Dienstleistungspolitik. WSI Notes **9**, 462–469 (2011)
19. Abele, E., Reinhart, G.: Zukunft der Produktion: Herausforderungen, Forschungsfelder. Carl Hanser Verlag, Chancen (2011)
20. Brandt, M., Gärtner, S., Meyer, K.: Urbane Produktion – ein Versuch einer Begriffsdefinition. Forschung Aktuell **N0**(08), 1–14 (2017)
21. Piegeler, M., Spars, G., Urbane Produktion – Konzept und Messung. In: Schumpeter Discussion Papers, vol. 2019, no. 001, pp. 1–39 (2019)
22. O'Brien, C.: Sustainable production – a new paradigm for a new millennium. Int. J. Prod. Econ. **60–61**, 1–7 (1999)
23. Spath, D., Lentes, J.: Flexibler produzieren in der Stadt. Fertigungstechnisches Kolloquium, FtK2012, pp. 241–259 (2012)
24. Spath, D., Bauer, W.; Lentes, J.: A quick-check to identify potentials by urban manufacturing for industrial enterprises. In: 23rd Conference on Production Research, Manila, pp. 6–11 (2015)
25. Spath, D., Lentes, J., Urban production to advance the competitiveness of industrial Enterprises. In: 22nd International Conference on Production Research, Foz do Iguaçu (2013)
26. Stiehm, S.: Gestaltungsparameter für die (Re-) Integration von Produktion in den urbanen Raum im Kontext von Industrie 4.0 (2017)
27. Brandt, M., Gärtner, S., Meyer, K.: Urbane Produktion, Planungsrecht und dezentrale Finanzsysteme, Forschung Aktuell, vol. 10 (2018)
28. Herrmann, C., Blume, S., Kurle, D., Schmidt, C., Thiede, S.: The positive impact factory-transition from eco-efficiency to eco-effectiveness strategies in manufacturing. Procedia CIRP **29**, 19–27 (2015)
29. Lentes, J., Mandel, J., Schließmann, U., Blach, R., Hertwig, M.: Competitive and sustainable manufacturing by means of ultra-efficient factories in urban surroundings. Int. J. Prod. Res. (2016)
30. Kondziella, H., et al.: Marktdesign, Regulierung und Gesamteffizienz von Flexibilität im Stromsystem – Bestandsaufnahme und Herausforderungen (2019)
31. Affeldt, J.F.: The application of system dynamics (SD) simulation to enterprise management. In: Proceeding of Winter Simulation Conference, Squaw Peak, pp. 1496–1500 (1999)
32. Maidstone, R.: Discrete event simulation, system dynamics and agent based simulation: discussion and comparison. In: Goti, A. (ed.) Discrete Event Simulations, Sciyo (2010)
33. Molamohamadi, Z., Ismail, N.: Developing a new scheme for sustainable manufacturing. Int. J. Mater. Mech. Manuf. **1**(1), 1–5 (2013)
34. Tenaglia, M., Noonan, P.: Scenario-based strategic planning: a process for building top management consensus. In: Planning Review, no. 2, pp. 12–19 (1992)

Enhancing Supply Chain Resilience: Strategic Production Allocation and Eurasian Freight Rail Optimization

Qian Huang[1(✉)] and Ruixue Li[2]

[1] Department of Business Design and Management, Waseda University, Tokyo, Japan
huangqian365@gmail.com
[2] Faculty of Business Administration, Hosei University, Tokyo, Japan

Abstract. This study examines the role of transportation in global supply chain management, focusing on the China Railway Express as an attractive alternative to sea, road, and air freight due to its cost and lead-time advantages. The research proposes a supply chain plan that simultaneously considers production areas and transportation routes and accounts for freight rate risk, aiming to optimize operations. The proposed model addresses uncertainties in fare structures across different routes and multiple carrier partnerships in international rail corridors. By adopting a strategic approach to route selection, businesses can balance production and transportation costs while managing market fluctuations and volatility.

Keywords: China Railway Express · Supply chain optimization · Simultaneous decision model

1 Introduction

Supply chains serve as a link between the upstream and downstream product flows. This is particularly true in the case of global supply chains, where raw materials, parts, and finished products are sourced from various parts of the world, produced in multinational factories, and sold on a global scale. Given the uncertain business environment, it is essential to diversify risks by collaborating with multiple supply chains. As product flows become increasingly complex and interconnected, the supply chain has transformed into a supply chain network. In the context of global supply chain networks, transportation plays a critical role in ensuring smooth and efficient supply chain operations. As a result of the greater distance between the origin and destination of supply activities in the case of global supply chains, the transportation process has become more complex than ever before. Therefore, it is crucial to focus on optimizing the transportation process of global supply chain networks to ensure the smooth flow of materials, parts, and finished products.

Over the past few years, ocean freight rates have experienced a significant surge owing to disruptions in European and intra-Asian routes, port congestion, and other factors, such as ship accidents in the Suez Canal, temporary port closures in response

M. Dragomir et al. (Eds.): ICPR 2023, LNPE, pp. 227–235, 2025.
https://doi.org/10.1007/978-3-031-92082-0_25

to the COVID-19 pandemic, and the suspension of transportation services owing to the Russia–Ukraine War. This has led to a significant increase in transportation costs, forcing Japanese companies to re-evaluate their international transportation routes (Fig. 1).

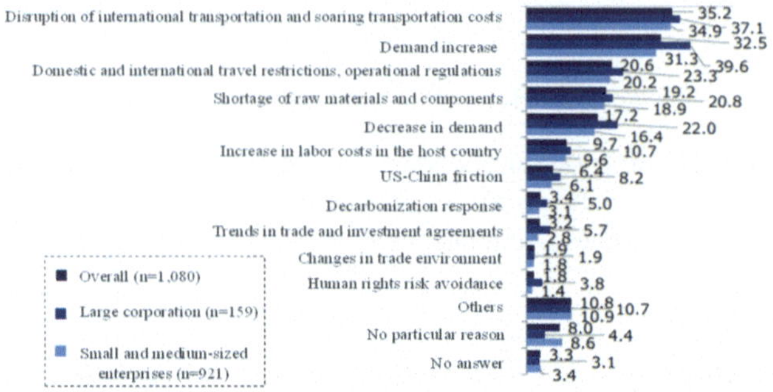

Fig.1. Reasons for redesigning supply chain networks (Source: JETRO, 2022; edited by the authors)

The China Railway Express (CRE) is gaining popularity as a freight railway service due to its advantages in transportation lead time and costs. The lead time is shorter than maritime transportation, taking 20 to 30 days to transport goods from Japan to Europe compared to 40 to 50 days by sea. Additionally, CRE is cheaper than air freight. The volume of goods transported by CRE has significantly increased, with the number of trains rising from 17 in 2011 to over 12,400 in 2020, and the amount of goods transported growing from 40,000 tonnes to over 1.14 million tonnes over the same period. New routes have been added and optimized to better serve customers, including the world's longest railway route linking Yiwu, China, to Madrid, Spain. Railway transportation has the potential to become a primary mode for long-distance goods transport, offering a feasible and cost-effective alternative to sea and air transportation. Moreover, rail transportation is more environmentally friendly, emitting fewer greenhouse gases compared to sea and air transportation.

Choosing the right rail transportation route can be challenging due to the complexity of rail networks. Factors such as cost, distance, transit time, capacity, availability, safety, and security must be considered. Surprisingly, a longer route can sometimes be more cost-effective, as carriers may offer lower rates on routes with excess capacity. Shippers should compare costs, consider additional fees, and explore volume discounts and incentives. However, comparing fares across different routes can be challenging when multiple carriers with varying pricing policies and expertise are involved. To overcome this challenge, shippers should adopt a strategic approach to route selection.

The proposed supply chain plan considers both the production area and transportation route, helping businesses make informed decisions that balance production and transportation costs while considering fluctuating freight rates by rail transportation. Implementing this plan can optimize supply chain operations, reduce lead times, improve delivery performance, and manage exposure to market fluctuations. By addressing fare uncertainty, businesses can reduce costs, increase efficiency, and manage risk, contributing to long-term success in a dynamic market.

2 Relevant Literature

The literature on supply chain planning focuses on the application of mathematical optimization methods to tackle the intricate nature of multi-layer, multi-site, and multi-product networks. The primary objective is to achieve optimal flow of goods and services by minimizing costs and maximizing efficiency. To accomplish this, researchers extensively employ mathematical optimization techniques like linear programming, integer programming, and mixed-integer programming. These methods are employed to construct models and determine the most efficient solutions. Golan et al. (2020) find the need for advanced network analysis and resilience analytics in supply chain literature. It identifies a lack of comprehensive models and integration with transportation and command and control networks. Balancing efficiency and resilience is crucial, and a comprehensive approach is recommended. Chopra (2018) discusses the design and management of supply chain networks, emphasize poor supply planning can damage the organization's performance. Christopher and Lee (2004) realize the uncertainty of supply chain and suggest end to end visibility to enhance the quality of supply chain. Ivanov and Dolgui (2021) propose digital supply chain twins as real-time models for managing disruption risks. It emphasizes their importance in mapping supply networks, ensuring visibility, and enhancing predictive decision-making in global companies. These studies demonstrate the complexity of algorithms and techniques used in supply chain planning and the importance of developing new and innovative methods to address the challenges of uncertainty and complexity in the supply chain network.

Some studies explore the impact of transportation infrastructure on firms' location choices and global value chain structures. Krugman (1991) introduces the "new economic geography" and a core-periphery model, highlighting how transportation infrastructure influences firms' decisions to locate in peripheral regions. Ghiani et al. (2004) discuss logistics network design and its role in long-haul freight transportation planning and management. Saidi and Hammami (2011) emphasize the importance of transportation in attracting foreign direct investment. Meersman and Nazemzadeh (2017) find that transport infrastructure is vital for Belgium's small, open economy, with government investments in highways and ports positively impacting economic growth. Ramya and Devadas (2019) utilize Geographic Information System and decision-making techniques to identify suitable locations for agro-based industries, considering factors such as topography, soil characteristics, transportation networks, resources, and climate. These studies collectively suggest that transportation factors significantly influence decision makings, and improving transportation infrastructure can attract investment and drive economic growth in specific regions.

Eurasian rail transportation is a complex system. To optimize the system, it is necessary to analyse and understand factors such as cargo types, container technology, rail transportation routes, and hub cities from different perspectives. One strand of literature focuses on cargo types and container technology as a means of analysing Eurasian rail transportation. Different types of cargo require different handling and transportation methods, and the choice of container technology can have a significant impact on the efficiency and safety of the transportation process. Vinokurov and Tsukarev (2018) analyze the railway corridors through Eurasia Economic Union. Seven transport corridors are compared by length, capacity, multi-modal, cost and so on to find the highest promising ones. Some researchers conclude that the China Railway Express is a key component of the Belt and Road Initiative. While it has made significant progress in recent years, challenges such as unclear market positioning and infrastructure limitations need to be addressed for its stable and long-term development. Wang et al. (2021) say the EU faces challenges of development disparities, populism, and debt crises. China's One Belt One Road Initiative offers an opportunity for cooperation and economic benefits, particularly for lagging EU countries, through improved connectivity and trade facilitation in Eurasia. Overall, these reviews of the literature provide insights into the various factors that affect Eurasian rail transportation, and demonstrates the importance of analysing the system from multiple perspectives.

In this study, we propose a model that can simultaneously determine the decision-making of production and rail transportation networks, while simultaneously considering the risk of fluctuations in the fare market. In addition, because the fare market varies depending on the transportation route and the logistics company that provides the service, we devise a classification of transportation routes and examine the fluctuation of the fare market in multiple scenarios. This is necessary because transportation costs can have a significant impact on production and supply chain operations. By considering the risk of fare market fluctuations and analyzing the transportation routes of the international rail corridor, manufacturing companies can make more informed decisions about production and transportation.

3 Proposed Model

This study involves a multi-stage, multi-facility, and multi-product network. The locations in this network are divided into production nodes, transportation nodes, and sales nodes, and the transportation routes consist of multiple transportation routes. We use binary variables to represent in which factory each product is produced in each market and which transportation route is used to transport each product. Our objective is to minimize production and transportation costs. The cost of rail container transportation is determined based on the transportation distance, extra fare, and discount fare.

To determine which production area should use which route under different rates. We formulate the mathematical formulations and verify the result from numerical experiments of multiple scenarios.

The supply chain planning model handled in this study is formulated as follows.

Sets

$i, h, j \in \{I \cup H \cup J\}$: Nodes

(I : factory nodes, H : transportation nodes, J: markets nodes)

$k \in K$: Products

$r \in R = \{(i, a), (a, b), \ldots, (t, j)\} \in A$: Routes

$p \in P = (i, a, b, \ldots, t, j) \in A$: Corridors

$A = \{(i, h, j) \ominus i \in I, h \in H, j \in J\}$: Arcs

Parameters

C_i^{ass}: Depreciation cost of production equipment at factory i

C_{ik}^{prod}: Production unit cost of product k at factory i

C_r^{tran}: Standard transportation unit price for route r

C_h^{lport}: Handling unit price of cargo at land port h
$PCap_i$: Production capacity at factory i
$RCap_r$: Booking limit on route r
$Path_{ijp}$: $(i, a, b \ldots t, j)$ Combination of route r from factory i to market j through transportation
s_r: Administrative instability coefficient on route r, $s_r \leq 1$
r_r: Discount rate on route r
ε_r: Loading ratio of required transportation volume for the discount rate s_r
d_{jk}: Demand quantity of product k in market j
Objective function

$$Min\left(\sum_{i\in I} C_i^{ass} + \sum_{j\in J}\sum_{k\in K}\sum_{i\in I} C_{ik}^{prod} d_{jk}y_{jki} + \sum_{p}\sum_{r} RC_{pr}\right) \tag{1}$$

$$if \quad RQ_r = \sum_{j}\sum_{k}\sum_{i}\sum_{p} d_{jk}y_{jki}\, Path_{jip}x_{jkp} \geq RCap_r\varepsilon_r \tag{2}$$

$$then \quad RC_r = C_r^{tran}\left(1 - \frac{r_r}{s_r}\right)RQ_r \tag{3}$$

$$else \quad RC_r = C_r^{tran} RQ_r \tag{4}$$

Decision variables
y_{jki}: Binary variable 0/1 for producing product k in market j at factory i
x_{jkp}: Binary variable 0/1 for transporting product k in market j via route p
RQ_{pr}: Total transportation volume for route r using route p
RC_{pr}: Total transportation cost for route r using route p
Constraints
Restrictions on the number of production areas:

$$\sum_{i\in I} y_{jki} = 1, \quad \forall j \in J, \forall k \in K \tag{5}$$

Restrictions on the number of transportation routes:

$$\sum_{p\in P} x_{jkp} = 1, \quad \forall j \in J, \forall k \in K \tag{6}$$

Production capacity constraint:

$$\sum_{j}\sum_{k} d_{jk}y_{jki} \leq PCap_i, \quad \forall i \in I \tag{7}$$

Transportation capacity constraint:

$$\sum_{j}\sum_{k}\sum_{i}\sum_{p} d_{jk}y_{jki}\, Path_{jip}x_{jkp} \leq RCap_r, \quad \forall r \in R \tag{8}$$

4 Numerical Experiments

In this study, we conduct numerical experiments using data from a manufacturing company, which makes mechanical parts and has factories and markets globally. There are 7 factories and 13 markets around Eurasia continent, and a railway express group that provides rail freight transportation services across the Eurasian continent. There are East, Central, West, South corridors start from China mainland to Europe, which constituted by different major cities in each corridor. (see Tables 1). Some main routes are also list in Table 2. As the name of "Zheng-Ou" means the route start from "Zhengzhou(Zheng)" city to "Europe(Ou)", and Hamburg is the destination city in Europe. The route go though the Frontier station called Alashankou or Horgos or Erenhot, the about transportation time is 15 to 18days, and some major cities are list in the table.

To check the effectiveness of the proposed model in explaining the importance of rail transportation conditions on manufacturing decisions, we define four scenarios with varying manufacturing and rail transportation conditions and examine how they impact on each other. We use the CPLEX optimization software package to conduct this numerical experiment.

Table 1. Four corridors of the railway express network

Corridor	Location of China area	Major Cities
East Corridor	Along China's eastern coast	Beijing, Tianjin, Qingdao, Lianyungang
Central Corridor	In the middle of China	Lanzhou, Zhengzhou, Almaty, Tashkent
West Corridor	In western China	Chengdu, Chongqing, Urumqi, Horgos
South Corridor (Planning)	In southern China	Kunming, Hanoi, Bangkok, Yangon

The four scenarios involve different optimization approaches for transportation and production, and different costs are associated with each scenario, as shown in Table 3. Transportation optimization occurs in Scenarios 1, 2, and 3, but transportation costs vary, whereas the production cost is held constant for all three scenarios, at 234,500. Scenarios 1 and 2 have the same transportation cost of 23,410, whereas Scenario 3 has a lower transportation cost of 21,310. Therefore, Scenario 3 has the lowest total cost at 255,810, whereas Scenarios 1 and 2 have the same total cost of 257,910. Scenario 4 uses both production and transportation optimization, with a production cost of 231,500 and a transportation cost of 21,910. This scenario has the lowest total cost at 253,410. Therefore, on the basis of the information provided, Scenario 4 has the lowest total cost in terms of production and transportation optimization.

In Scenario 4, a discount fare was implemented in the transportation network and its impact on the total cost of the proposed simultaneous determination model was investigated. The results indicate that transportation cost increased, but the total cost decreased from 255,810 to 253,410 owing to a decrease in production cost. This suggests that the simultaneous determination model had a positive impact on the overall cost, despite

Table 2. Example of main routes within corridors (Li, 2018)

Route	Origin	Destination	Frontier Station	Transportation Time (Days)	Major Cities
Zheng-Ou	Zhengzhou	Hamburg	Alashankou/ Horgos/ Erenhot	15–18	Zhengzhou, Erenhot, Warsaw, Hamburg
Yu-Xin-Ou	Chongqing	Duisburg	Alashankou/ Horgos/ Erenhot	15–18	Chongqing, Alashankou, Nur-Sultan, Moscow, Duisburg
Rong-Ou	Chengdu	Duisburg	Alashankou/ Horgos	12–18	Chengdu, Alashankou, Warsaw, Hamburg
…	…	…	…	…	…
Sha-Xin-Ou	Xiamen	Hamburg	Alashankou	16	Xiamen, Alashankou, Minsk, Hamburg

Table 3. Results of the numerical experiment (One million yen)

Comparison	Type	Production cost	Transportation cost	Total costs
Transportation optimization	Scenario 1	234,500	23,410	257,910
	Scenario 2	234,500	23,410	257,910
	Scenario 3	234,500	21,310	255,810
Production and Transportation optimization	Scenario 4	231,500	21,910	253,410

increasing the transportation cost in Scenario 4. By decreasing the production cost, the model was able to achieve a lower total cost, even with the increased transportation cost.

5 Discussion

In Scenario 4, we observe minor changes in the flows of products among railway transportation stations, such as Hamburg–Erenhot, with the most significant changes occurring between the factory location and nearby transportation stations, as shown in Fig. 2 (e.g., Kunshan–Chongqing). This suggests that the decrease in total cost is related to both production and shipping activities. Interestingly, the transportation cost does not decrease but slightly increases, indicating that local production for local markets is not always the best approach. Striking a balance between different production locations and prioritizing global optimization over local optimization leads to greater cost savings.

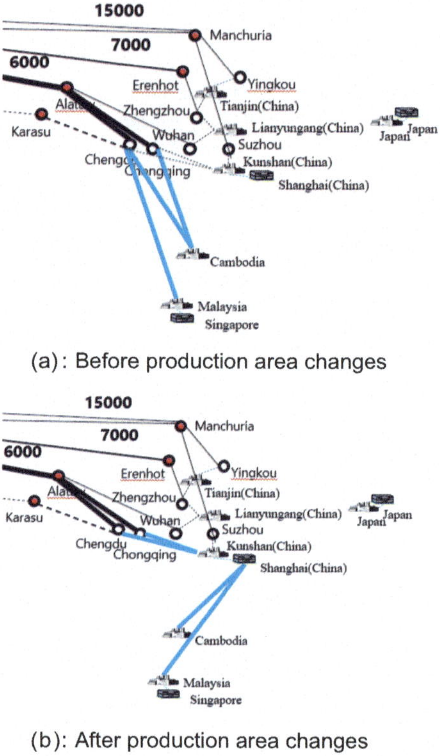

Fig. 2. (a) Before production area changes. (b) After production area changes

6 Conclusion

In this study, we proposed a simultaneous optimization model for production and rail transportation networks that accounted for railway fare fluctuations. The interesting results were acquired from numerical experiments, demonstrating that improved transportation conditions attracted cargo flows and that economies of scale in transportation influenced production location decisions. We drew four conclusions from this study, as follows.

1. Production decisions should be made concurrently with transportation activities, as long-distance global transportation processes significantly influence production decision-making.

2. International rail corridors exhibit unique characteristics, featuring a complex transportation network with multiple participants engaged in transportation activities as trains traverse multiple countries. Therefore, fare market fluctuations should be incorporated into the design of production–transportation networks.
3. Local production for local markets could be an efficient risk mitigation strategy in many cases. However, considering the future, we found that global optimization was generally more advantageous than local optimization.
4. International rail corridor transportation activities are not solely a cost calculation for manufacturers; they also serve as crucial factors in strategic decision-making regarding factory locations and production volumes.

Acknowledgments. This work was supported by JSPS KAKENHI (Grant Number 23K13518), Daiichi Institute of Technology and, the Research Institute for Innovation Management at Hosei University.

References

Chopra, S.: Supply Chain Management: Strategy, Planning, and Operation, Global Edition. Pearson (2018)

Christopher, M., Lee, H.: Mitigating supply chain risk through improved confidence. Int. J. Phys. Distrib. Logist. Manage. (2004)

Ghiani, G., Laporte, G., Musmanno, R.: Introduction to logistics systems planning and control. John Wiley & Sons (2004)

Golan, M.S., Jernegan, L.H., Linkov, I.: Trends and applications of resilience analytics in supply chain modeling: systematic literature review in the context of the COVID-19 pandemic. Environ. Syst. Decis. **40**(2), 222–243 (2020)

Ivanov, D., Dolgui, A.: A digital supply chain twin for managing the disruption risks and resilience in the era of Industry 4.0. Prod. Plann. Control **32**(9), 775–788 (2021)

JETRO: 2021 Fiscal Year "Survey on Overseas Business Expansion of Japanese Companies" (2022)

Krugman, P.: Increasing returns and economic geography. J. Polit. Econ. **99**(3), 483–499 (1991)

Li, R.: The impact of the transcontinental freight railway. China-Eur. Railw. Express Logist. Bus. Monthly 56–63 (2018)

Meersman, H., Nazemzadeh, M.: The contribution of transport infrastructure to economic activity: the case of Belgium. Case Stud. Transp. Policy **5**(2), 316–324 (2017)

Ramya, S., Devadas, V.: Integration of GIS, AHP and TOPSIS in evaluating suitable locations for industrial development: a case of Tehri Garhwal district, Uttarakhand, India. J. Cleaner Prod. **238**, 117872 (2019)

Saidi, S., Hammami, S.: The role of transport and logistics to attract foreign direct investment in the developing countries. In: 2011 4th International Conference on Logistics, pp. 484–489, IEEE (2011)

Vinokurov, E., Tsukarev, T.: The Belt and Road Initiative and the transit countries: an economic assessment of land transport corridors. Area Dev. Policy **3**(1), 93–113 (2018)

Wang, Z.: From crisis to nationalism? The conditioned effects of the COVID-19 crisis on neo-nationalism in Europe. Chin. Political Sci. Rev. **6**(1), 20-39 (2021)

Simulation Analysis for the Promotion of Organic Cotton with the Aim of Solving Social Issues in the Clothing Supply Chain

Rei Kinoshita, Tomomi Kito(✉), and Shunichi Ohmori

Graduate School of Creative Science and Engineering, Waseda University, 3-4-1 Okubo, Shinjuku-ku, Tokyo, Japan
tomomi.kito@gmail.com

Abstract. In the apparel supply chain, cotton farmers at the top of the chain bear the greatest burden. To address the issue of farmer suicides caused by debt from non-organic cotton cultivation and promote the adoption of organic farming, support for conversion to organic methods is being advocated. There are also efforts to increase consumer ethical awareness. However, the effectiveness and sustainability of these promotional strategies have not been thoroughly examined. This study develops a system dynamics model based on a fact-finding survey conducted in India. Simulation results using this model reveal that the problem cannot be sustainably resolved by solely continuing current initiatives. Moreover, the study demonstrates that a simultaneous, gradual approach to implementing support for transition and raising consumer awareness could offer a long-term and sustainable solution to this problem.

Keywords: System dynamics · simulation · apparel supply chain · sustainability

1 Introduction

The burdens faced by cotton farmers at the uppermost level of the apparel supply chain have become a serious social issue (Amaeshik et al. 2008). Farmers face immense pressure to produce large quantities of cotton, while enduring extremely harsh working conditions and severe economic challenges. In India, the world's largest producer of cotton, many farmers rely on expensive genetically modified seeds and extensive pesticide usage to achieve higher cotton yields (Kranthi and Stone 2020). The expenses incurred in purchasing these inputs often plunge farmers into heavy debt, exacerbating the distressing reality of farmer suicides before they can harvest and profit from their crops. The estimated number of such cases reaches a staggering 10,000 per year (Accidental Deaths and Suicides in India 2022). Farmers also confront severe health problems resulting from the abuse of chemical fertilizers and pesticides.

To solve this serious problem, a shift in cotton cultivation from non-organic to organic methods is being promoted (Chatna Organic). Organic farming eliminates the need for pesticides, chemical fertilizers, and expensive seeds, eliminating farmers' debt and health

M. Dragomir et al. (Eds.): ICPR 2023, LNPE, pp. 236–245, 2025.
https://doi.org/10.1007/978-3-031-92082-0_26

problems, while achieving cotton yields comparable to those of non-organic farming (Eyhorn et al. 2011). On the other hand, there are significant barriers for farmers to make the transition. Converting to organic farming requires substantial initial investment, and even after the investment, a few years of the in-transit period is required, during which farmers can only trade cotton at non-organic cotton prices (i.e., at a lower price). Additionally, organic farming demands advanced techniques and knowledge. To reduce these barriers, several support organizations, are providing transition assistance to farmers, including subsidizing their conversion costs and educating them on the necessary knowledge and skills. However, support activities have not expanded and the dire situation in India remains unimproved. It has not been fully examined whether the continuation of such support activities will really lead to a fundamental solution to this social problem and contribute to sustainable cotton cultivation in the first place.

Organic cotton production remains low, accounting for only 1.4% of cotton production in 2021 (Textile Exchange 2022). To encourage people to purchase organic cotton products even at higher prices, support organizations are working with apparel companies to develop/market organic cotton products and to educate consumers to raise their ethical awareness. Nowadays, many brands, including fast fashion brands, offer organic cotton products. However, in fact, organic cotton is of the same quality as non-organic one and has no direct benefits for consumers such as being gentle on the skin (although many consumers misunderstand this). Much literature indicates a positive impact of consumers' sustainability awareness on their willingness to pay (e.g., (Ellis et al. 2012)), but with the prerequisite that organic products are at a premium, which will disappear as they become widespread. The goal should be a 'healthy' market that is sustainable in the long run, even if consumers have economically rational preferences. However, the current market is far from that state.

The core research questions of this paper, therefore, are set as follows: Can the long-term diffusion of organically grown cotton and the reduction of farmer suicides be achieved through the current approaches? Namely, (a) by increasing the ethical awareness of consumers, or (b) by providing continued support to encourage farmers to switch to organic cultivation? Additionally, if the answers to (a) and (b) are negative, (c) how can this be accomplished?

This study addresses these questions using a system dynamics (SD) approach, which is a powerful tool for simulating and examining complex and dynamic systems (Sterman 2001). Firstly, based on real data on cotton cultivation in India and information obtained from interviews with organizations involved in farmer support, this paper constructs an SD model in the form of a causal loop diagram (CLD) that abstracts the complexity of the real situation. Next, the SD model is implemented as a flow-stock model, enabling us to simulate multiple scenarios. Our simulation demonstrates that sustainable diffusion of organic cotton while reducing farmer suicides cannot be achieved through the continuation of existing support measures alone (i.e., answering 'no' to the above research questions (a) and (b)). Furthermore, this paper pursues the research question (c) by proposing improvement measures and verifies their effectiveness through simulations.

The rest of the paper is organized as follows. The next section provides an overview of the current status of sustainable supply chain management (SSCM) research in the clothing and textile industry and the SD approach to SSCM, as background for this study. Section 3 introduces the SD model structure and settings of the simulations for the above research questions (a) – (c). Section 4 provides and discusses the simulation results. Finally, Sect. 5 summarizes the implications derived from the simulations and concludes the paper.

2 Literature Review

2.1 SSCM for the Apparel Industry

Behind its glamorous economic scale, the clothing and textile industry has a devastating negative impact on the environment, including enormous pollution due to the abuse of chemicals, massive consumption of natural resources in the production process, and extensive disposal of post-consumer products (Gwilt and Rissanen 2011). The industry also has a significant negative social impact. Excessive cost-cutting and resulting high profit margins have come at the expense of labor rights and health and safety (Caniato et al. 2012). Shocking scandals regarding labor exploitation and human rights violations by major companies hit the media from time to time, severely damaging their brand credibility (Croom et al. 2018).

SSCM is a holistic concept/approach that aims to enhance not only the financial value of stakeholder companies but also their environmental and social value (Seuring 2013). Although a number of SSCM practices are implemented in the industry, a systematic literature review by Cai and Choi (2020) revealed that current efforts in this industry contribute unevenly to the triple bottom line (i.e., financial, environmental, and social aspects) (Elkington 1994), with social aspects being particularly neglected. Companies are economically rational by nature and try to strategically link their contribution to sustainability to their own economic value. Thus, in the case of cotton, it is no wonder that many companies manufacture products containing very small amounts of organic cotton (as it is more expensive) and sell them at higher prices. However, a truly sustainable system is one in which farmers produce organic cotton without incurring debt and that is traded at a fair price, consumers pay for the quality they get (not the elusive 'premium'), and then organic cotton becomes more widespread. The realization of such a system, where economic, social and environmental values are linked and compatible, is still a distant goal (Elkington 2018).

2.2 System Dynamics Approach to SSCM

An SD model is a model that describes a dynamic system composed of a complex interplay of economic, social, and environmental elements as a set of (typically non-linear) relationships. It is suitable for supporting macroscopic and long-term decision making by testing and evaluating policies to achieve the desired system behavior (Sterman 2001). In recent years, there has been an emphasis on quantitative modeling for sustainability research, with the SD approach being shown to be particularly suitable (Nabavi 2017). However, although research in this area has increased recently, it is still insufficient relative to its importance. Additionally, although verification and validation are inevitable to ensure the scientific rigor of SD models, many studies present only CLDs (i.e., causal relationships among system components) without providing SSCM simulations using flow-stock models calibrated based on real data (Rebs et al. 2019). As highlighted by Rebs et al. (2019), SD simulations addressing SSCM should be based on validated models that encompass multiple stakeholders and explicitly include aspects of their social values.

3 System Dynamics Model

3.1 Causal Loop Diagram (CLD)

Figure 1 illustrates the CLD corresponding to the system captures the key elements of the problem structure concerning cotton farmers, as outlined below. Initially, farmers make investments in cotton cultivation and realize profits from their production after a few years. In the event that organic cotton becomes more lucrative, non-organic farmers are motivated to switch to organic production. However, such a conversion necessitates a substantial additional investment, alongside significant technical challenges associated with organic farming. The prices of organic and non-organic cotton are determined by the interplay of supply and demand in the market. Additionally, consumers' heightened awareness of sustainability contributes to an increased demand for organic cotton. Farmers often resort to borrowing funds to finance their investments, but if their debts accumulate to an overwhelming extent, they may be driven to suicide.

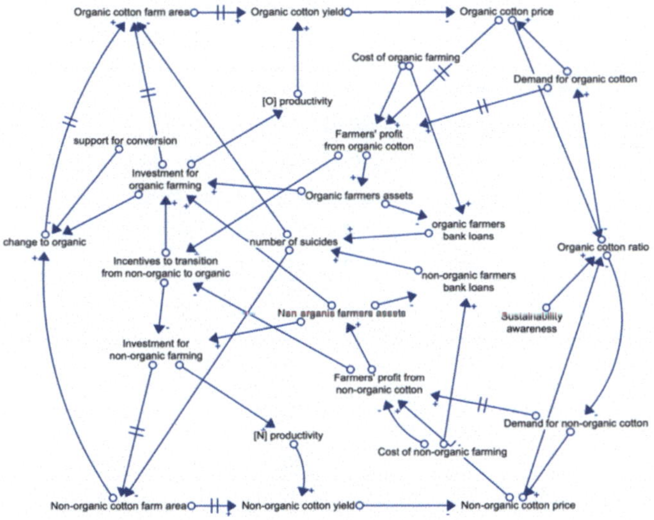

Fig. 1. Causal loop diagram of the model

3.2 Stock-Flow Diagram

The stock-flow diagram (SFD) of the model, which comprises various variables such as stock, flow, constants, and auxiliary variables, is presented in Fig. 2. Table 1 provides a summary of the variables and equations employed in this model. To evaluate the model's performance, we conducted a historical test using empirical data spanning the years 2014 to 2022, obtained from several reliable public data sources (Accidental Deaths and Suicides in India 2022) (Textile Exchange 2022). The data encompassed information on the yield of organic and non-organic cotton, farmland area, and the number of farmer suicides. The simulation output closely matched the actual data values, demonstrating a strong alignment with the time-series trends. Consequently, the model was deemed valid for predicting and analyzing future scenarios.

Table 1. Variables and equations of the SD model

Variable	Equation
Organic farm area	$S^o_{t-1} + \frac{1}{2}S^o_{t-3} \times \frac{1}{1+e^{-\left(\frac{INV^o_{t-3}}{100}-1\right)}} - \left(\frac{1}{4}S^o_{t-1} + 1000PRE^o_{t-1}\right) + CH_t$
Non-Organic farm area	$S^n_{t-1} + \frac{1}{2}S^n_{t-3} \times \frac{1}{1+e^{-\left(\frac{INV^n_{t-3}}{100}-1\right)}} - \left(\frac{1}{4}S^n_{t-1} + 1000PRE^n_{t-1}\right) - CH_t$
CH_t	$S^n_{t-3} \times SUP \times \frac{1}{1+e^{-\left(\frac{INV^{n\to o}_{t-3}}{100}-1\right)}}$
Organic farmers' investment to organic farming	$INV^{o\to o}_t = \frac{1}{10}A^o_t$
Non-organic farmers' investment to organic and non-organic farming	$INV^{n\to o}_t = \frac{1}{10}A^o_t \times INC_t$ $INV^{n\to n}_t = \frac{1}{10}A^o_t \times (1 - INC_t)$
Organic farmers' assets	$A^o_t = PF_{t-1} + L_{t-1} - INV^{o\to o}_{t-1} - EX_{t-1}$
Non-organic farmers' assets	$A^n_t = PF_{t-1} + L_{t-1} - INV^{n\to o}_{t-1} - EX_{t-1}$
Other expenses	$EX_t = CO + RP_t$
Bank loans	$BL_t = 3 \times \frac{EX^2_{t-1}}{A_{t-1}} - RP_t$
Pressure of loans	$PRE_t = \frac{\frac{1}{10}BL_{t-3}}{A_t}$
Profit	$PF_t = \frac{D_t \times PR_t}{S_t}$
Price	$PR_t = PR_{t-1} + \frac{3}{10}PR_{t-1} \times \frac{1}{1+e^{-5\left(\frac{D_{t-1}}{Y_{t-1}}-1\right)}} - \frac{3}{20}PR_{t-1}$
Demand	$D_{o,t} = AD \times OR_t$ $D_{n,t} = AD \times (1 - OR_t)$

(*continued*)

Table 1. (*continued*)

Variable	Equation	
Yields	$Y_t = S_t \times PR_t$	
Other variables and initial settings		
Conversion support: SUP_t Consumer awareness: AW_t		
$S_0^o = 276736$	$S_0^n = 10850000$	$A_0 = 1500$
$PR_0^o = 800$	$PR_0^n = 460$	$AD = 5907871$
PR_t^o : *random number with mean of* $0.362 \times (1 + \frac{0.1}{1 + e^{-INV_t^o}}$ *and variance of* 0.131		
PR_t^n : *random number with mean of* $0.467 \times (1 + \frac{0.1}{1 + e^{-INV_t^o}}$ *and variance of* 0.041		
$OR_t = \frac{1}{1+e^{-5(1-\frac{PR_t^0}{PR_t^n+10}}})$		
Scenario settings		
$SUP_t = 0.001(const.), AW_t = 10(const.)$ $SUP_t = 0.001(const.), AW_t = 50(const.)$ $SUP_t = 0.015(const.), AW_t = 10(const.)$ $SUP_t = SUP_{t-1} \times 1.1, SUP_0 = 0.001, AW_t = 10$ $SUP_t = SUP_{t-1} \times 1.1, SUP_0 = 0.001, AW_t = 50$		

3.3 Scenario to Be Simulated

The following five scenarios were simulated and analyzed to address the research questions (a) - (c):

(A) Current situation (i.e., simulating the model shown in Fig. 2 as it is).
(B) High level of consumers' sustainability awareness (corresponding to research question (a)).
(C) Reduction of barriers to conversion to organic cultivation through farmer support (corresponding to (b)).
(D) Gradual provision of support for conversion (first approach corresponding to (c)).
(E) Gradual provision of support for conversion while maintaining a high level of consumer's sustainability awareness (second approach corresponding to (c)).

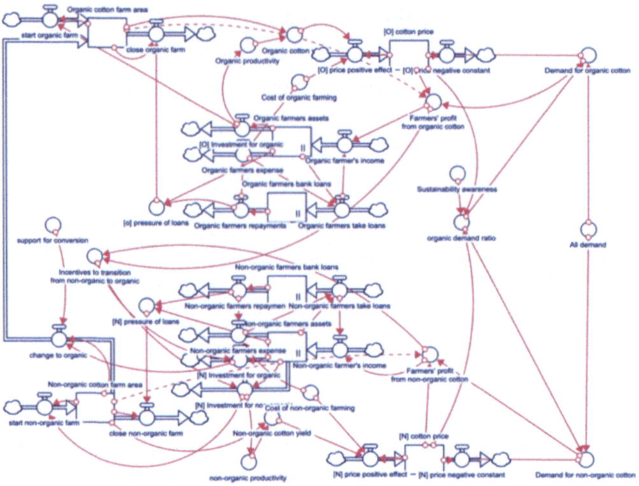

Fig. 2. Flow-stock diagram of the SD model

4 Simulation Results and Analysis

The simulation results for the next 40 years, based on scenarios (A–E), are summarized in Fig. 3. The figure's left panels depict the changes in the area of farmland under organic and non-organic cultivation. It should be noted that the area of organic farmland can be increased through further investments by organic farmers, or by the conversion of non-organic farmers to organic production. Therefore, the right panels of the figure show the changes in the area of farmland converted from non-organic to organic cultivation.

Additionally, Fig. 4 illustrates the change in the area of farmland driven out of business due to debt (represented as an indicator reflecting the number of suicides) compared to (A) – i.e., the continuation of the status quo.

In scenario (A), organic cotton is not projected to gain significant popularity in the future. Conversely, if consumer awareness increases (scenario (B)), the organic farmland area increases. However, the conversion from non-organic to organic cultivation does not progress, and existing organic farmers simply expand their current farmland. As a result, the number of suicides is expected to remain unchanged or potentially increase. Moving onto scenario (C), where farmers receive active support for conversion, the conversion initially occurs, leading to a significant increase in the area of organic farmland. However, it subsequently declines, indicating that this scenario is not sustainable. In other words, the answer to research questions (a) and (b) is negative, suggesting that the current initiatives alone will not lead to long-term improvement.

On the other hand, if the conversion support is gradually promoted (scenario (D)), the conversion progresses, and organic production continues to grow without exacerbating the situation. Furthermore, when consumer awareness is raised and gradual conversion support is provided (scenario (E)), the impact is even greater than in scenario (D), leading to sustainable conversion. The results of this study indicate that scenarios (D) and (E) could significantly contribute to reducing farmer suicides.

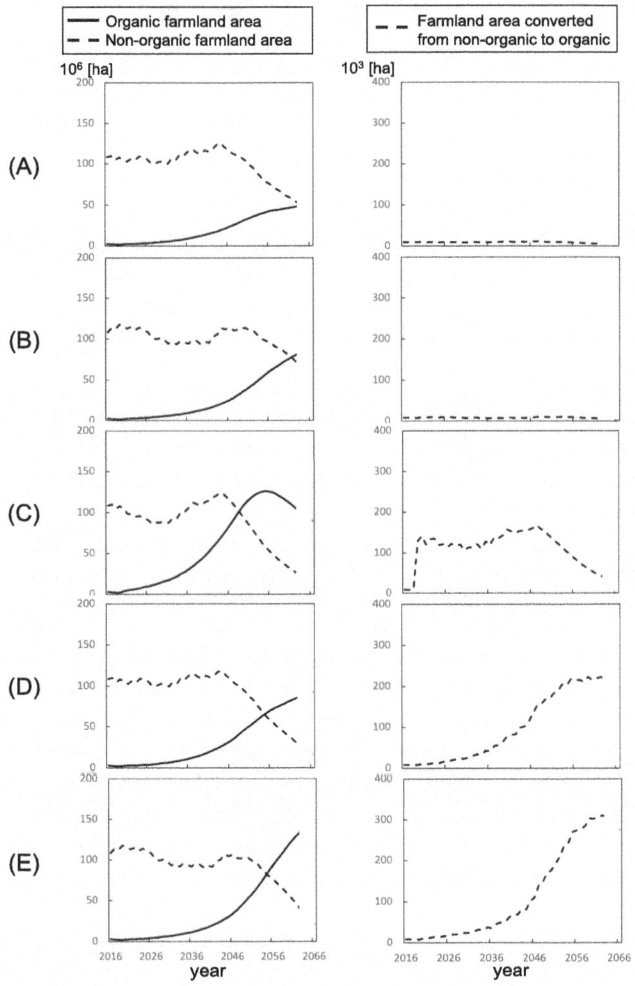

Fig. 3. Simulation results of the five scenarios (A-E)

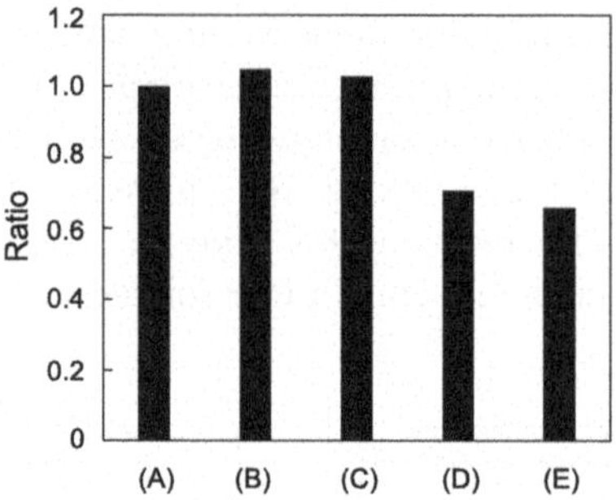

Fig. 4. Ratio of the farmland area out of business after 2050 when the result in (A) is set to 1

5 Summary and Conclusion

This study addressed the environmental and social challenge of converting cotton cultivation from inorganic to organic within the apparel supply chain. The goal was to find a solution that would benefit farmers while ensuring economic sustainability for the entire system. The following key findings were obtained from the system dynamics (SD) simulation constructed using real data:

- Merely improving consumers' ethical awareness is insufficient to achieve the sustainable diffusion of organically grown cotton.
- Hastily supporting the conversion to organic farming may yield temporary improvements but could have negative long-term effects.
- Gradual support for conversion holds the potential for sustainable diffusion of organic cotton and a reduction in farmer suicides. This process can be further enhanced by concurrently promoting consumer awareness.

Achieving sustainable supply chain management (SSCM), which entails balancing corporate economic profit with environmental and social values, is particularly challenging in the apparel industry due to its lengthy value chain and the vulnerable position of farmers upstream. The key findings of this study imply the limitation of the point-wise intervention with respect to time and stakeholders, and the need for the holistic approach toward the SSCM. The results of this study also demonstrate the effectiveness of the SD approach (with certain limitations) in addressing this challenge and offer feasible improvement measures that contribute to the advancement of SSCM research.

References

Amaeshik, M.K., Osuji, O.K., Nnodim, P.: Corporate social responsibility in supply chains of global brands: a boundaryless responsibility? Clarifications, exceptions and implications. J. Bus. Ethics **81**, 223–234 (2008)

Kranthi, K.R., Stone, G.D.: Long-term impacts of Bt cotton in India. Nature Plants **6**, 188–196 (2020)

Accidental Deaths and Suicides in India (2022). https://data.opencity.in/dataset/accidental-deaths-and-suicides-in-india-2022

Eyhorn, F., Ramarkrishman, M., Mäder, P.: The viability of cotton-based organic farming systems in India. Int. J. Agric. Sustain. **5**, 25–38 (2011)

Chetna Organic. https://www.chetnaorganic.org.in

Textile Exchange. Organic Cotton Market Report (2022)

Ellis, J.L., McCracken, V.A., Skuza, N.: Insights into willingness to pay for organic cotton apparel. J. Fash. Mark. Manag. **16**(3), 290–305 (2012)

Sterman, J.D.: System dynamics modeling: tools for learning in a complex world. Calif. Manage. Rev. **43**(4), 8–25 (2001)

Gwilt, A., Rissanen, T.: Shaping Sustainable Fashion: Changing the Way we Make and Use Clothes. Earthscan Publishing, London (2011)

Caniato, F., Caridi, M., Crippa, L., Moretto, A.: Environmental sustainability in fashion supply chains: an exploratory case based research. Int. J. Prod. Econ. **135**, 659–670 (2012)

Croom, S., Vidal, N., Septic, W., Marshall, D., McCarthy, L.: Impact of social sustainability orientation and supply chain operational performance. Int. J. Oper. Prod. Manag. **38**(12), 2344–2366 (2018)

Seuring, S.: A review of modelling approaches for sustainable supply chain management. Decis. Support. Syst. **54**(4), 1513–1520 (2013)

Cai, Y.-J., Choi, T.-M.: A united nations' sustainable development goals perspective for sustainable textile and apparel supply chain management. Transp. Res. Part E **141**, 102010 (2020)

Elkington, J.: 25 years ago I coined the phrase "triple bottom line." Here's why it's time to rethink it. Harv. Bus. Rev. (2018)

Nabavi, E., Daniell, K.A., Najafi, H.: Boundary matters: the potential of system dynamics to support sustainability? J. Clean. Prod. **140**, 312–323 (2017)

Rebs, T., Thiel, D., Brandenburg, M., Sehring, S.: Impacts of stakeholder influences and dynamic capabilities on the sustainability performance of supply chains: a system dynamics model. J. Bus. Econ. **89**, 893–926 (2019)

Modeling of Disassembly Parts Selection for Recycling, GHG Saving Rates and Cost Using Linear Physical Programming

Yuki Kinoshita[1](✉), Hiromasa Ijuin[2], and Tetsuo Yamada[3]

[1] Department of Informatics, Faculty of Engineering, Kindai University, 1 Takaya Umenobe, Higashi-Hiroshima, Hiroshima 739-2116, Japan
kinoshita@hiro.kindai.ac.jp

[2] Department of Industrial and Systems Engineering, Tokyo University of Science, 2641 Yamazaki, Noda, Chiba 278-8510, Japan

[3] Department of Informatics, The University of Electro-Communications, 1-5-1 Chofugaoka, Chofu, Tokyo 182-8585, Japan

Abstract. Material recycling contributes not only material circulation but also reduction of greenhouse gas (GHG) emissions since GHG emissions for a virgin material production can be saved by using recycled materials. Partial disassembly is required to maximize GHG saving rate defined as the saved volumes of GHG emissions by recycling, and recycling rate, and to minimize cost. This study proposes a multi-objective disassembly parts selection for cost, recycling and GHG saving rates using Linear Physical Programming (LPP). The LPP is one of the effective methods for solving multi-objective problems, and can obtain one satisfactory solution by expressing preferences of the decision maker (DM) for each objective. The proposed model was validated using a vacuum cleaner consisted of 23 parts, and found one satisfactory solution, which had over 70.00[%] and 80.36[%] of recycling and GHG saving rates and achieved over 50% reduction of total cost compared to complete disassembly.

Keywords: Low Carbon Emission · Reverse Supply Chain · End-of-life (EOL) Products · Disassembly · Life Cycle Inventory (LCI) Database · Multi Criteria Decision Making

1 Introduction

For a sustainable production, material recycling needs to be enhanced environmentally friendly and economically. Material recycling also contributes to reduction of Greenhouse Gas (GHG) emissions at a material production phase since GHG emissions for a virgin material production can be saved by using recycled materials [1–3]. Thus, both environmental indexes, namely, recycling and GHG saving rates should be considered for material recycling simultaneously. Disassembly of end-of-life (EOL) products is an essential and important phase, although each recovery option such as remanufacturing, reuse, material recycling has its own characteristics [4]. Disassembly can be defined as a

M. Dragomir et al. (Eds.): ICPR 2023, LNPE, pp. 246–254, 2025.
https://doi.org/10.1007/978-3-031-92082-0_27

systematic method to separate a product into parts, components, subassemblies or other grouping [5]. The disassembly task is usually conducted manually due to its complex tasks and a mix of different types and qualities of EOL products. Due to the labor cost of manual disassembly, the disassembly cost tends to be higher [6]. Hence, partial or selective disassembly to retrieve targeted parts is obvious suitable for recyclers since it will bring economic benefit compared to complete disassembly [7–9]. The partial disassembly and disassembly line balancing were addressed, simultaneously [7–10]. Chao et al. [7] proposed a two-sided disassembly line balancing method taking account of a disassembly tool for removing a part. The two-sided line indicated that the workstations are symmetrically positioned in both side of the conveyor, that is, the proposed model determined a disassembly line and a position of EOL products in the line simultaneously in order to assign each disassembly task for each workstation [7]. Mura et al. [8] presented a multi-objective line balancing method for maximizing productivity and a safety of the disassembly line. The safety objective function prioritized removing hazardous parts quickly so as to avoid eliminating the exposure of the greatest risks [8]. Yin et al. [9] proposed multimanned partial disassembly line balancing, which meant that there were multiple workers at one workstation. The proposed model assigned each disassembly task not only to a workstation but also a worker conducting its disassembly task [9]. Wang et al. [10] developed a multi-objective disassembly line balancing method using artificial bee colony algorithm for minimizing number of workstations, disassembly time and smoothness index with priority of removal of high-demand and hazardous parts. Güçdemir and Ilgin [4] focused on a disassembly sequencing problem, and proposed an integrated model of data groping using clustering and sequencing rules. The proposed model grouped the disassembly tasks using k-means clustering algorithm, and then, determined the disassembly sequencing for each grope with consideration of disassembly direction changes [4]. Kim et al. considered the variability of disassembly times in the partial disassembly problem removing the targeted parts decided in advance [11]. Even though those studies addressed the partial disassembly problems, their objective functions were economic aspects only such as maximizing profit and productivity or minimizing costs.

Kinoshita et al. [1, 12] and Hasegawa et al. [2] proposed bi-objective disassembly parts selection method for maximizing recycling or GHG saving rates and minimizing costs. These studies did not consider the recycling and GHG emissions simultaneously. Igarashi et al. [3] presented multi-objective disassembly system design to maximize these recycling and GHG saving rates and to minimize costs simultaneously, using ε constraint method. However, the ε constraint method needs many numerical trials for the decision maker (DM) to obtain the Pareto optimal solutions by changing value of ε. As one of effective methods of solving multi-objective problem, there is linear physical programming (LPP). The LPP can seeks one satisfactory solution based on the disability ranges of the DM for each criterion.

This study proposes a multi-objective disassembly parts selection for cost, recycling and GHG saving rates using LPP. The DM can obtain the most satisfactory solution for disassembly parts selection only by expressing his/her preferences regarding recycling, GHG saving rates and cost, respectively. Mathematical model is formulated by using LPP, and examined with numerical experiments.

2 Mathematical Model

Section 2 decribes overview of LPP, and formulates multi-objective disassembly parts selection using LPP. Subsection 2.1 introduces soft and hard classes to express preferences of the DM for each criterion, and LPP weight algorithem to determine mathematical weights based on the preferences of the DM. Subsection 2.2 formulates the proposed model using LPP to determine the combinantions of disassebled parts for minimizing costs, and maximizing recycling and GHG saving rates, simultenously.

2.1 Overview of LPP

LPP is developed by Messac, Gupta and Akbulut in 1996 [13] as a multi-objective optimization method. One of the features of LPP is that the DM only have to express his/her preferences for each criterion using a class function in LPP instead of determination of mathematical weight directly [13]. The class function is consisted of 4 different types of classes with two cases, hard and soft, which refer to the relationships between criterion value and the preference. The 4 hard classes are "Class-1H Must be smaller", "Class-2H Must be larger", "Class-3H Must be equal", and "Class-4H Must be in range". Then, there are only acceptable and unacceptable ranges in 4 hard classes.

With respect to soft classes, there are "Class-1S Smaller is better", "Class-2S Larger is better", "Class-3S Value is better", and "Class-4S Range is better". Each soft class has 6 different types of desirability ranges, namely, "Ideal", "Desirable", "Tolerable", "Undesirable", "Highly Undesirable", and "Unacceptable". The way of setting desirability ranges depends on each type of soft classes.

To solve multi-objective problem using LPP, the DM needs to select and determine desirability ranges based on the preferences of each criterion. In a case of criterion belonged to hard classes, it can be expressed as simple constraints since they have only acceptable and unacceptable ranges. In terms of soft classes, the mathematical weight is calculated based on LPP weight algorithm and set desirability ranges. The LPP weight algorithm is constructed based on one principle called One vs. Others (OVO) criteria rule, which prioritizes improvement of worth criterion firstly [14]. Owing to this LPP weight algorithm, the DM is relieved from mathematical weight allocation processes, and get correct mathematical weight to reflect the preferences by only setting desirability ranges.

2.2 Formulation of Multi-objective Disassembly Parts Selection Using LPP

A summary of notations in the proposed multi-disassembly parts selection using LPP is as follows:

J	:	Set of parts
P_j	:	Set of immediately preceding parts of part j
j	:	Index of parts
p	:	Index of objectives belonged to soft classes ($p = 1, 2, 3$)

(*continued*)

(*continued*)

s	:	Index of disability ranges in LPP ($s = 2, 3, 4, 5$)
c_j	:	Recycling cost at part j
r_j	:	Recycling rate at part j
e_j	:	GHG saving rate at part j
t_{ps}^{+}/t_{ps}^{-}	:	Positive/Negative limit to the sth range of objective p
$\tilde{w}_{ps}^{+}/\tilde{w}_{ps}^{-}$	:	Positive/Negative deviation weight of the sth range of objective p
x_j	:	Binary value: if part j is selected for disassembly, 0 otherwise
d_{ps}^{+}/d_{ps}^{-}	:	Positive/Negative deviation valuable from the sth range limit of objective p

The proposed multi-objective disassembly parts selection has 3 objective functions as shown (1), (2), and (3). First objective function is to minimize total cost (TC). Second and third objective functions is to maximize total recycling rate (R) and total GHG saving rates (E). The total recycling rate (R) indicates a ratio of the recycled weight against the whole product weight [12]. The total GHG saving rate means a ratio of saved amount of GHG emissions by material recycling compared to that emitted for a virgin material production. Only selected part j is removed from a EOL product manually for material recycling [1, 3]. Other non-selected parts are assumed to be destroyed for disposal without cost. Equation (4) ensures all preceding parts of part j have to be disassembled.

$$TC = \sum\nolimits_{j \epsilon J} c_j x_j \rightarrow Min \tag{1}$$

$$R = \sum\nolimits_{j \epsilon J} r_j x_j \rightarrow Max \tag{2}$$

$$E = \sum\nolimits_{j \epsilon J} e_j x_j \rightarrow Max \tag{3}$$

$$x_j - x_i \leq 0 j \in J, i \in P_j \tag{4}$$

To solve this multi-objective problem using LPP, objective functions (1), (2), and (3) are transported to constraints by introducing positive/negative limit to the sth range of objective p (t_{ps}^{+}/t_{ps}^{-}), and deviation valuable from the sth range limit of objective p (d_{ps}^{+}/d_{ps}^{-}) as shown in Eqs. (5)–(10). Since total cost (TC) belongs to "Class-1S Smaller is better", by minimizing deviation valuables (d_{1s}^{+}), the proposed model tries to seek a solution in a better range, that is, a solution with lower total cost as shown in Eqs. (5) and (6).

$$TC - d_{1s}^{+} \leq t_{1(s-1)}^{+} (s = 2,3,4,5) \tag{5}$$

$$TC \leq t_{15}^{+} \tag{6}$$

In contrast to total cost (TC), total recycling rate (R) and total GHG saving rate (E) are belonged to "Class-2S Larger is better". Thus, by minimizing deviation valuables (d_{2s}^- and d_{3s}^-), the total recycling rate (R) and GHG saving rate (E) are maximized to be in a better range.

$$R + d_{2s}^- \geq t_{2(s-1)}^- (s = 2,3,4,5) \tag{7}$$

$$R \geq t_{25}^- \tag{8}$$

$$E + d_{3s}^- \geq t_{3(s-1)}^- (s = 2,3,4,5) \tag{9}$$

$$E \geq t_{35}^- \tag{10}$$

Using LPP weight algorithm, the mathematical weights to reflect the preferences of the DM for each criterion are calculated. The aggregate objective function of LPP is formulated using the calculated mathematical weight and deviation valuable for each criterion as shown in Eq. (11). Equation (12) enforces that all deviation valuables are not negative.

$$\sum\nolimits_{p\in\{1,2,3\}} \sum\nolimits_{s\in\{2,3,4,5\}} \left(\tilde{w}_{ps}^+ d_{ps}^+ + \tilde{w}_{ps}^- d_{ps}^-\right) \rightarrow Min \tag{11}$$

$$d_{ps}^+, d_{ps}^- \geq 0 (p = 1, 2, 3s = 2,3,4,5) \tag{12}$$

3 Numerical Experiments

A case example of proposed model is illustrated in Sect. 3 using a vacuum cleaner as a product example. The vacuum cleaner is consisted of 23 parts. The estimated methods of cost, recycling, and GHG saving rates for each part j are described in Subsect. 3.1. Subsection 3.1 also set the disability ranges of each criterion, and calculates the mathematical weights. The results obtained using LPP is showed and discussed in Subsect. 3.2.

3.1 Product Example

Table 1 lists estimated cost, recycling and GHG saving rates of each part in the vacuum cleaner. The recycling cost and rate was estimated using Recyclability Evaluation Method, which was an estimation method and software for recycling cost and rate for each part [14, 15]. According to the REM, the recycling cost and rate can be estimated based on the material type, disassembly task, and weight of each part [14, 15].

Table 1. Cost, recycling and GHG saving rate of each part

No.	Part name	Material type	Recycling cost	Recycling rate [%]	GHG saving rate [%]
1	Wheel	PP	21.77	0.99	1.31
2	Wheel stopper	PP	20.06	0.24	0.32
3	Upper nozzle	PP	17.49	3.52	2.33
4	Lower nozzle	PP	17.49	2.89	1.91
5	Nozzle	PP	17.49	2.41	1.60
6	Right handle	PP	13.37	3.42	2.27
7	Switch	PVC	13.37	0.32	0.19
8	Left handle	PP	17.49	3.62	2.40
9	Left body	PP	36.51	13.10	8.68
10	Right body	PP	17.49	12.58	8.34
11	Dust case cover	PMMA	17.49	2.56	3.22
12	Mesh filter	cloth/Fiber	18.41	0.00	19.11
13	Connection pipe	Al/Al alloy	17.31	3.30	2.05
14	Dust case	PMMA	17.49	12.29	15.49
15	Exhaust tube	PVC	17.49	2.24	1.34
16	Upper filter	cloth/Fiber	18.37	0.00	18.38
17	Lower filter	PP	17.49	2.05	1.36
18	Protection cap	ABS	17.49	1.56	1.46
19	Motor	Motor	10.5	19.14	0.00
20	Rubber of outer frame of fan	Rubber	18.63	0.00	2.44
21	Outer frame of fan	Al/Al alloy	8.96	3.85	2.40
22	Lower fan	PP	17.49	1.06	0.70
23	Fan	Al/Al alloy	12.52	4.34	2.70
Total			402.17	95.48	100.00
Average			17.49	4.15	4.35
Standard deviation			4.98	4.97	5.58

The GHG saving rate was calculated based on the GHG emissions for a virgin material production, which was estimated using the LCI database with the Asian international I/O table listing the GHG emission intensity [16]. By inputting calculated material procurement cost, GHG emissions at a virgin material production was calculated [16]. The GHG saving rate was defined as a ration of saved GHG emissions by replacing recycled materials with virgin material [1]. Note that part #19 "motor" was excluded from calculation of GHG saving rate since it has too high GHG emissions (95% of the whole product) as well as [1, 3]. Moreover, the part #19 "motor" is assumed to be always selected for disassembly.

By setting 5 limits for each objective function, 6 different desirability ranges for each objective function were set as shown in Table 2. Using the set desirability ranges and LPP weight algorithm, the mathematical weight could be calculated as shown in Table 3. By using the listed data as shown in Tables 1 and 3, the numerical experiments were conducted. All numerical experiments were performed on the same laptop PC (Windows 10 with Intel(R) Core(TM) i5-7200U CPU@2.50 GHz) by optimization python library named PuLP [17].

Table 2. Disability ranges for each objective function

	Cost	Recycling rate[%]	GHG saving rate[%]
Ideal	<=50	=>90	=>90
Desirable	(50,100]	[65, 90)	[70, 90)
Tolerable	(100, 250]	[50,65)	[50, 70)
Undesirable	(250, 300]	[30, 50)	[40, 50)
Highly undesirable	(300, 400]	[15, 30)	[30, 40)
Unacceptable	>400	<15	<30

Table 3. Calculated mathematical weight using LPP weight algorithm

Objective	$\tilde{w}_{p2}$	$\tilde{w}_{p3}$	$\tilde{w}_{p4}$	$\tilde{w}_{p5}$
Total cost ($p = 1$)	4.000	0.027	32.45	18.61
Total recycling rate ($p = 2$)	8.000	32.267	50.937	276.044
Total GHG saving rate ($p = 3$)	10.000	20.200	152.201	368.464

3.2 Results

The 11 out of 23 parts were selected for disassembly. The selected parts were shown in Fig. 1. The non-selected parts in the vacuum clear were described with red slash in Fig. 1. By selecting 11 parts as shown in Fig. 1, the total cost (TC), total recycling rate (R), and total GHG saving rate (E) became 201.92, 70.00[%], 80.36[%], respectively. That is, the total cost (TC) was in "Tolerable" range as shown in Table 2. On the other hand, both total recycling rate (R), and total GHG saving rate (E) were in the "Desirable" range as shown in Table 2.

From Table 1, the total cost (TC) when all parts are disassembled is over 400. Compared to the solution obtained from LPP, it found out that LPP could seek one effective solution, which saved over 50% total costs from all parts disassembled but kept higher total recycling and GHG saving rates such as 70.00[%] and 80.36[%].

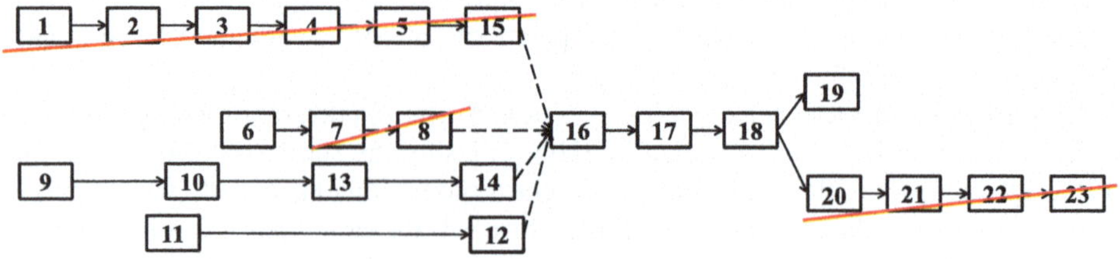

Fig. 1. Disassembly precedence relationships with results of parts selection by LPP

4 Counclusions

This study proposed a multi-objective disassembly parts selection for minimizing cost and maximizing recycling and GHG saving rates simultaneously using LPP. The DM can obtain one satisfactory solution by expressing his/her preferences for each objective function as desirability ranges. A vacuum cleaner consisted of 23 parts was used to illustrate design example. The obtained solution using LPP could save over 50% total cost but keep higher recycling and GHG saving rates such as 70% and 80% compared to a case of complete disassembly.

Future studies should apply the proposed method to other assembly products for validation, and investigate the effects of the disassembly parts selection on disassembly production schedule to meet demand of reused parts and recycled weights [18].

Acknowledgments. This study was partially supported by the Japan Society for the Promotion of Science (JSPS), KAKENHI, Grant-in-Aid for Scientific Research (B), JP23K26329, from 2023 to 2025, Grant-in-Aid for Scientific Research (B), JP23K22987, from 2022 to 2025, and Grant-in-Aid for JSPS Fellows, JP22KJ1363, from 2023 to 2024.

References

1. Kinoshita, Y., Yamada, T., Gupta, S.M., Ishigaki, A., Inoue, M.: Analysis of cost effectiveness by material type for CO_2 saving and recycling rates in disassembly parts selection using goal programming. J. Adv. Mech. Design Syst. Manuf. **12**(3), 1–18 (2018)
2. Hasegawa, S., Kinoshita, Y., Yamada, T., Bracke, S.: Life cycle option selection of disassembly parts for material-based CO_2 saving rate and recovery cost: analysis of different market value and labor cost for reused parts in German and Japanese cases. Int. J. Prod. Econ. **213**, 229–242 (2019)
3. Igarashi, K., Yamada, T., Gupta, S.M., Inoue, M., Itsubo, N.: Disassembly system modeling and design with parts selection for cost, recycling and CO_2 saving rates using multi criteria optimization. J. Manuf. Syst. **38**, 151–164 (2016)
4. Güçdemir, H., Ilgin, M.A.: A part grouping-based approach for disassembly sequencing. J. Eng. Res. **11**(1), 100026, 1–6 (2023)
5. Özceylan, E., Kalayci, C.B., Güngör, A., Gupta, S.M.: Disassembly line balancing problem: a review of the state of the art and future directions. Int. J. Prod. Res. **57**(15–16), 4805–4827 (2019)
6. Yamada, T.: Design issues for inverse manufacturing systems in Japanese cases. In: Northeast Decision Science Institute 2008 Annual Meeting, Brooklyn, NY, USA, pp. 551–556 (2008)
7. Chao, B., Liang, P., Zhang, C., Guo, H.: Multi-objective optimization for mixed-model two-sided disassembly line balancing problem considering partial destructive mode. Mathematics **11**(1299), 1–17 (2023)
8. Dalle Mura, M., Pistolesi, F., Dini, G., Lazzerini, B.: End-of-life product disassembly with priority-based extraction of dangerous parts. J. Intell. Manuf. **32**(3), 837–854 (2020). https://doi.org/10.1007/s10845-020-01592-z
9. Yin, T., Zhang, Z., Wu, T., Zeng, Y., Zhang, Y., Liu, J.: Multimanned partial disassembly line balancing optimization considering end-of-life states of products and skill differences of workers. J. Manuf. Syst. **66**, 107–126 (2023)

10. Wang, S., Guo, X., Liu, J.: An efficient hybrid artificial bee colony algorithm for disassembly line balancing problem with sequence-dependent part removal times. Eng. Optim. **51**(11), 1920–1937 (2019)
11. Kim, H.W., Park, C., Lee, D.H.: Selective disassembly sequencing with random operation times in parallel disassembly environment. Int. J. Prod. Res. **56**(24), 7243–7257 (2018)
12. Kinoshita, Y., Yamada, T., Gupta, S.M., Ishigaki, A., Inoue, M.: Disassembly parts selection and analysis for recycling rate and cost by goal programming. J. Adv. Mech. Design Syst. Manuf. **10**(3), 1–15 (2016)
13. Messac, A., Gupta, S.M., Akbulut, B.: Linear physical programming: a new approach to multiple objective optimization. Trans. Oper. Res. **8**, 39–59 (1996)
14. Hiroshige, Y., Nishi, T., Ohashi, T.: Recyclability evaluation method (REM) and its applications. In: 2002 ASME International Mechanical Engineering Congress & Exposition, New Orleans, LA, USA, pp. 315–320 (2002)
15. Hitachi, Ltd., EcoAssist. http://www.ecoassist.com/HTML_n/option/rem/rem_tr/ppframe.htm. Accessed 25 June 2023). (in Japanese)
16. Yoshizaki, Y., Yamada, T., Itsubo, N., Inoue, M.: Material based low-carbon and economic supplier selection with estimation of GHG emissions and affordable cost increment for parts production among multiple Asian countries. J.Jpn. Industr. Manage. Assoc. **66**, 435–442 (2016)
17. COIN-OR: Computational Infrastructure for Operations Research. https://www.coin-or.org/. Accessed 25 June 2023
18. Kinoshita, Y., Yamada, T., Gupta, S.: Design of disassembly-to-order system for reused components and recycled materials using linear physical programming. Int. J. Sustain. Manuf. **4**(2/3/4), 121–149 (2020)

Overvcoming SMEs' Barriers in Adopting Crowdworking with Quality Management Approaches

K. Lange[1](✉), R. Refflinghaus[1], A. Hupe[2], and U. Bretschneider[2]

[1] Department of Quality and Process Management, University of Kassel, Heinrich-Plett-Str. 40, 34132 Kassel, Germany
kirsten.lange@uni-kassel.de

[2] Department of Information Systems, University of Kassel, Pfannkuchstr. 1, 34121 Kassel, Germany

Abstract. This paper describes a concept to adapt the design of quality management in small and medium sized enterprises (SMEs) to better exploit the potential of crowdworking. It shows how quality management, especially for SMEs, can be organized to make efficient and targeted use of crowdworking and continue to ensure that customer requirements are met. The different forms of crowdworking will be shown. Barriers of SMEs in adopting crowdworking will be identified. A modular reference process model for the quality-oriented use of crowdworking in SMEs will be developed. In addition, concrete recommendations for action to adapt or supplement the design of quality management in SMEs for the efficient and effective use of crowd-based mechanisms are iterative developed.

Keywords: crowdworking · quality management · SME · reference model

1 Introduction

The importance of crowdworking has increased significantly in the last years [1]. Originally, crowd working was a niche area with a focus on software development. However, this area has grown, now including writing texts, for instance product descriptions, planning and designing tasks for most diverse aspects, and even product development. Within this context, platforms offering jobs for crowdworking have become well-established and jobs are offered from all kinds of companies, multinationals as well as start-ups. Nevertheless, it is essential for the companies offering crowdwork to assure the quality of the outsourced work tasks. Since the crowdworkers are not employees of the companies they are not involved in the established (quality) management processes and their performance is unknown to the company. The motivation of this work is to answer the question how quality management and corresponding methods can contribute to ensure the "quality" of crowdworking.

The research question that the paper is based on is "How does a process for introducing crowdwork in SMEs look like when ensuring quality standards of completed tasks?".

M. Dragomir et al. (Eds.): ICPR 2023, LNPE, pp. 255–263, 2025.
https://doi.org/10.1007/978-3-031-92082-0_28

2 State of the Art

2.1 Definition and Actors of the Crowdworking Process

Crowdworking is defined here as a form of digital labor where individuals receive financial compensation for their contributions and achievements, contribute significantly to their income, and act as self-employed agents with flexibility in their working arrangements and time [2].

The roles in crowdworking are the crowdsourcer, the crowdworking platform and the crowdsourcee. The crowdsourcer, who initiates the process, can be a company, an individual, an organization or a group [2]. The crowd to which the task is outsourced consists of a large number of crowdsourcees who process it. The processing of the task can be done by a group of Internet users or by an individual, the so-called crowd [3]. The task is processed via online platforms, which are usually provided by an intermediary, but can also be established by the crowdsourcer itself [4].

2.2 Crowdworking Types

Crowdworking can be categorized into internal, external, and hybrid types, all utilizing a flexible pool of workers to achieve fast, cost-effective, and optimal solutions [5, 6].

Internal crowdworking utilizes a company's internal workforce (crowd), who already have an employment contract [7]. Tasks for crowdworking are advertised through internal platforms or email, with the company having the choice to restrict access to specific groups or open it up to the entire workforce, including employees from other locations [3]. The main advantages of internal crowdworking include knowledge retention, expanded work horizons, potential discovery and development, improved communication, motivation, and identification with the company. However, there can be drawbacks such as competition and dissatisfaction among the employees due to selective pre-selection [8].

External crowdworking involves outsourcing tasks to individuals outside the company, providing access to a diverse pool of skilled labor [4]. Tasks are advertised through either the company's own platform or an external platform provider [5]. The impact of crowdworking on internal service provision varies depending on the type of task outsourced. Routine tasks can free up employees to focus on core competencies, while complex projects or innovation tasks benefit from external knowledge and creative input, reducing operational blindness [8, 9]. However, external crowdworking can negatively affect internal employee motivation and contribute to the "not invented here syndrome". It also presents challenges in communication, coordination, quality control, and data protection [6, 10].

Hybrid crowdworking combines internal and external crowds, combining the benefits of both approaches. This approach reduces the "not invented here syndrome" and employee dissatisfaction, as internal solutions are supplemented rather than replaced by external crowdworkers. Moreover, the likelihood of achieving a higher-quality solution increases with a larger and more diverse pool of crowdworkers [3, 11].

3 Barriers of SMEs in Adopting Crowdworking

In order to theoretically develop a reference process for how small and medium-sized enterprises (SMEs) can successfully include crowdwork in their business model, existing barriers were identified. Identifying those helps to precisely develop the process and to identify possible strategies and solutions to overcome existing barriers.

3.1 Method

A series of 20 qualitative Interviews was conducted to identify barriers in the context of online labor platforms.

The methodology for conducting the interviews was based on the guided interview approach. The guided interview is a semi-structured form of data collection in which a pre- formulated guide serves as orientation and structuring aid. [12].

The qualitative interviews were conducted in four phases, which determine the structure of the guide [12]:

In the information phase, the interviewees were informed about the study and its objectives as well as about the treatment of the data, and the interviewee's consent to use the data was obtained.

The warm-up or introductory phase (warm-up) aimed to facilitate the respondents' entry into the interview situation and the research topic. Here, the interviewer asked as open and broad a question as possible at the beginning, so that the interviewee got into telling the story and thus overcame the often observed initial shyness about the interview, the unfamiliar communication situation (and its recording).

In the main phase of the interview, the barriers of adopting crowdworking were discussed in communicative exchange with the interviewee. The questions could be either deductive, i.e., developed from prior knowledge, or inductive, so that they were derived from new information from the interview.

The interviews were concluded with a fade-out and closing phase in which the interviews were reflected upon and the participants were given the opportunity to cite additional information or relevant aspects that had not been addressed so far.

For this research paper, the goal is to understand the strategies employed by SMEs to overcome barriers in crowdworking. The authors chose conducted interviews as a form of expert consultation to gather insights. The study included eight experts who were managing directors, crowdworking experts, held higher management positions in SMEs, or were decision makers. Following, the data were evaluated and analyzed to present the results of the analysis.

3.2 Barriers of Adopting Crowdworking

Three main barriers for SMEs in adopting crowdworking could be derived.

The first barrier is related to quality issues, stemming from the risk of poor performance and errors by the crowd. This poses a significant challenge for SMEs, particularly those operating in niche markets with a limited customer base. Unlike larger companies, SMEs may struggle to recover from the loss of clients resulting from quality flaws. Conflicting interests between SMEs and crowdsourcees contribute to this risk, as SMEs seek high-quality solutions at low prices, while crowdsourcees prioritize quick completion with minimal effort. This misalignment often leads to lower-than-expected quality. The credibility of qualifications and experiences presented by crowdworkers on online platforms is also lower compared to traditional job applications, creating uncertainty that requires intensive internal evaluation. However, the crowd's involvement in multiple projects and organizations makes it difficult for them to familiarize themselves with different internal processes and quality expectations. Moreover, important quality and project management sessions, like kick-off meetings, cannot be effectively implemented when tasks are outsourced to the crowd.

The second barrier identified to implementing crowdworking in SMEs includes the reluctance of both management and employees. Management's traditional and conservative mindsets, coupled with a distrust of digital technologies, hinder their interest in innovation and their engagement with digital initiatives. This reluctance results in a resistance to adopting crowdworking, as managers prefer hiring permanent employees and face challenges in coordinating and assigning tasks to crowdworkers. Other contributing factors include a lack of awareness, fear of changing business models, trust and confidentiality concerns, and the absence of suitable platforms. Negative incidents and aversion to external ideas further impede the adoption of crowdworking. To address these barriers, strategies are needed to make on-demand workforces more attractive while addressing managers' concerns about losing SME identity.

Employees also exhibit reluctance toward crowdworking due to their contradiction with the close personal relationships and intra-organizational bonds that are integral to SMEs' company culture. The introduction of flexible working arrangements raises fears of job loss and a negative impact on the working atmosphere among permanent staff. Increased competition and pressure between permanent employees and crowdworkers can lead to conflicts and a sense of devaluation. In family-owned SMEs, where social relationships between owners and employees are crucial, there is an additional reluctance stemming from concerns about compromising core values and social ties when tasks are delegated to the crowd.

The third barrier identified to applying crowdworking in SMEs is the increased risk of losing sensitive information. SMEs often protect their secrets and sensitive information through patents, trade secrets, and copyrights. These intellectual property rights are crucial for the value creation of SMEs, especially when they rely on a single patented product that drives their success. SMEs fear that crowdworking may result in the theft or takeover of important information. Unlike larger organizations that have more diversification, SMEs face the risk of bankruptcy if a patent or product idea leaves their organization. Additionally, the weak data protection practices in the platform's country of origin raise concerns among interviewees regarding the transfer of data and the location of servers.

4 Developing a Reference Process Model

To enhance the appeal of crowdworking for SMEs, a theoretical model was established, which was further developed iteratively to a reference process model.

The validity of the model was ensured by conducting tests with three SMEs and making necessary adaptations. Following the completion of organizational testing, minor adjustments were made to strike a balance between generality and specificity.

4.1 Iterative Development of the Process Model

The purpose of the conducted interviews was to examine the relevant measures and framework conditions for crowdworking in companies/SMEs, ensuring the fulfillment of customer requirements.

The results from this formed the foundation for a basic theoretical Business Process Model (BPM). The objective was to identify the framework conditions in companies/SMEs that hold significance for crowdworking, the quality management measures, methods, and techniques presently employed for quality planning, assurance, and error management, as well as the existing shortcomings and areas of improvement concerning crowdworking from the perspective of companies/SMEs.

In another step, the basic BPM again was used to include three crowdworking tasks in industrial companies, two of which were in SMEs. Two tasks were processed as external crowdworking and one task as internal crowdworking. The extent to which communication between crowdworkers and SMEs took place or was required when processing the task, how the actual processing takes place, and what measures and processes (should) take place to assess the work performed were investigated. It was also examined how the adoption of the results or their use in the SMEs could be ensured. Patterns and typical procedures in the process flow of the various crowdworking tasks were identified and the reference processes adapted accordingly.

Finally the BPM was validated with the participating SMEs and depeloped into a modular reference process model. The changes identified in this way were adapted into the reference process model.

4.2 Modular Reference Process Model

In the following, the most important process steps of a crowdworking process are explained and excerpts from the process model are shown.

Change Management

As crowdworking is a new form of work, the introduction of crowdworking requires extensive training at staff and management level. In these trainings, the opportunities and risks as well as the implementation for each application (internal/external/hybrid) should be transparently presented and illustrated. If staff or management express fears, coaching or counselling can be provided. Another important aspect for the successful implementation of a project is the appointment of a crowdworking manager who coordinates, monitors, documents and optimizes the process. The requirements and areas of responsibility should be set out in a profile and communicated (Fig. 1).

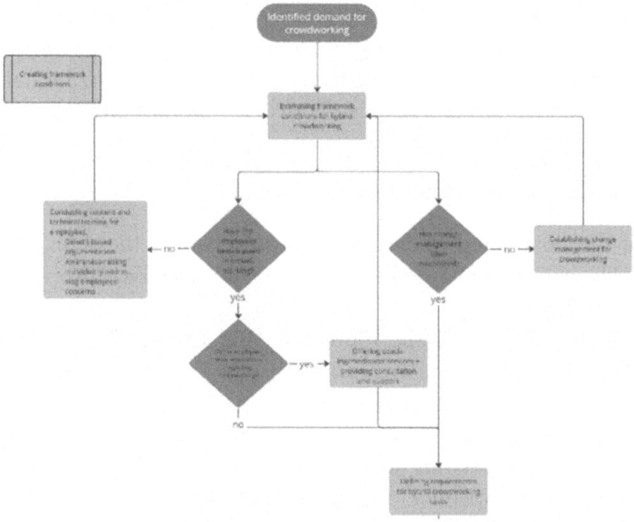

Fig. 1. Creating framework mechanisms

Defining Process Modalities

Defining process modalities should be coordinated by a task manager and constitutes quality planning within the framework of quality management. For external crowdworkers, it is first necessary to conclude a crowdworking contract. For all types of crowdworking, further details of the task should then be communicated to the crowdworker and the crowdworker's understanding should be checked. This step should be carefully repeated until the respective crowdworker signals their understanding. After that, interim results should be agreed with the respective crowdworker and regular appointments for consultation should be set (Fig. 2).

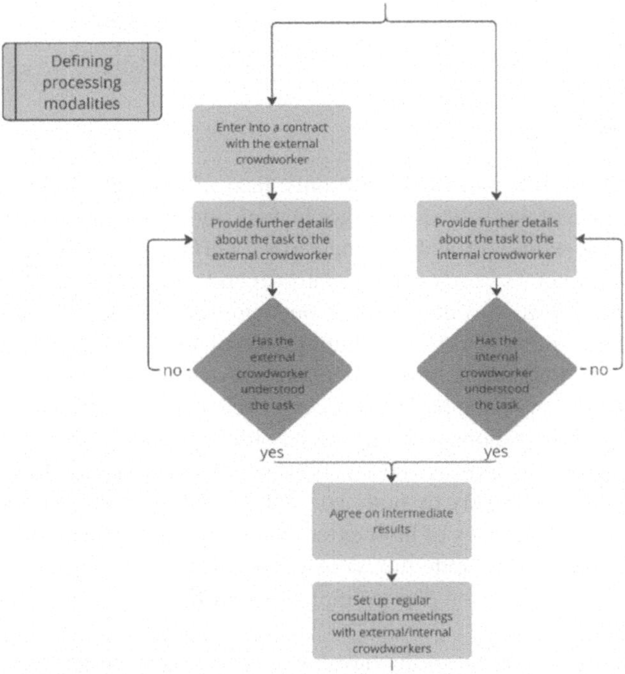

Fig. 2. Defining process modalities

5 Recommendations for Action

Based on the previous chapters, recommendations for action are summed up in the following to enable successful implementation of the crowdworking process. The recommendations are also intended to ensure qualitatively satisfactory project results for external, internal and hybrid crowdworking.

To successfully implement crowdworking, effective change management is crucial. It is important to involve employees in the introduction of crowdworking, particularly through open communication about its implementation and providing training on crowdworking introduction and project execution. Clearly defining responsibilities and conducting pilot projects or trial phases for crowdworking are also essential. Quality management findings from surveys suggest the need to identify requirements for potential crowdworkers, establish acceptance criteria and project milestones, conduct project risk analysis, and ensure a continuous improvement process (CIP). These measures can help overcome any reluctance from management and employees.

Since crowdworking represents a new approach to dividing and managing work, affecting process organization and personal work behavior, training becomes necessary. Training should cover the advantages and disadvantages of crowdworking processes and foster an understanding of the potential benefits. In cases of ongoing employee reluctance, coaching or counseling sessions can be offered.

When implementing crowdworking, it is important to ensure that the coordination effort is lower compared to in-house implementation [13]. This effort includes clearly defining the task, selecting a suitable crowdworking platform, uploading the task description, and choosing a candidate from the applicants [14].

6 Conclusions

In the context of digitization, SMEs not only face technological challenges but also encounter new forms of work due to changes in work and process structures. One of these new digital forms of work is crowdworking. Crowdworking refers to the paid processing of outsourced tasks via a crowdworking platform and can be utilized in three different use cases. Internal crowdworking involves outsourcing tasks to individual employees, departments, or the entire workforce within the company. This approach offers potential synergy effects, enhancing productivity, effectiveness, and employee satisfaction. External crowdworking, on the other hand, outsources tasks to a crowd outside the company, allowing for agile responses to employee bottlenecks or workloads. It also presents opportunities for reducing time and costs. Hybrid crowdworking involves processing the outsourced task simultaneously by both external and internal crowdworkers. This approach leverages the advantages of both external and internal crowdworking, increasing the likelihood of success. Despite the recognized potentials of crowdworking, companies often proceed with caution due to implementation, execution, and quality control challenges.

Through the dissemination of the developed model, data on the implementation of internal and hybrid projects could be collected and evaluated in the future. The question arises whether similar discrepancies between usage, awareness and suitability occur as with other crowdworking models. In addition, it would be interesting to investigate whether new potential for improvement can be discovered through the use of the hybrid and internal crowd model.

- Strengths of the study: first that provides a process model focusing on SMEs and their challenges, testing of the process after developing it
- Limitations: process tested only with three SMEs, but for future research we are planning to completing more tests in SMEs to further increase the validity of our process.

Acknowledgments. The CORNET promotion plan (No. 21758 N / 1) of the Research Community for Quality (FQS), August-Schanz- Str. 21A, 60433 Frankfurt/Main has been funded by the AiF within the programme for sponsorship by Industrial Joint Research (IGF) of the German Federal Ministry of Economic Affairs and Energy based on an enactment of the German Parliament. In the context of this paper, we would like to thank the project sponsor and the funding agency for their funding initiative.

References

1. World Bank Group, The global opportunity in online outsourcing (2015). http://documents.worldbank.org/curated/en/138371468000900555/pdf/ACS14228-ESW-white-cover-P149016-Box391478B-PUBLIC-World-Bank-Global-OO-Study-WB-Rpt-FinalS.pdf. Accessed 25 May 2023

2. Durward, D., Blohm, I., Leimeister, J.M.: Crowd Work. Bus. Inf. Syst. Eng. **58**(4), 281–286 (2016)
3. Warter, J.: Crowdwork, ÖGB Verlag, pp 34–43. Vienna (2016)
4. Feldmann, C., Hemsen, P., Giard, N.: Crowdworking: Einflüsse der Arbeitsbedingungen auf die Motivation der Crowd Worker, p. 5. Forschungsschwerpunkt Digitale Zukunft, Universität Bielefeld, Bielefeld. Universität Paderborn, Germany (2018)
5. Durward, D.: The future of digital labor: exploring crowd work as a new phenomenon in information systems. In: Research on IT, Service, Innovation, Collaboration, Kassel, Germany, vol. 18, pp.1,38,39 (2020)
6. Kleemann, F.: Consumers @work. Zum neuen Verhältnis von Unternehmen und Usern im Web 2.0. Frankfurt, Germany, p. 11, 84 (2012)
7. Däubler, W.: Digitalisierung und Arbeitsrecht. Internet, Arbeit 4.0 und Crowdwork, p.45. Frankfurt a.M., Germany (2018)
8. Leimeister, J.M., Durward, D., Simmert, B.: Die interne Crowd - Mitarbeiter/innen im Spannungsfeld neuer Arbeitsformen. Study der Hans-Böckler-Stiftung, Düsseldorf, Germany, vol. 436, pp. 39–42 (2020)
9. Gassmann, O.: Crowdsourcing. Innovations- management mit Schwarmintelligenz, 2nd edn., pp. 17–18 (2013)
10. Hauschild, J., et al.: Innovationsmanagement, 7th edn., pp. 38-39 (2023)
11. Varona, D., Capretz, L.F.: Using the DELPHI method for model for role assignment in the software industry. In: IECON 2021 – 47th Annual Conference of the IEEE Industrial Electronics Society, pp. 1–7 (2021)
12. Misoch, S.: Qualitative Interviews, 2nd edn., pp. 65-70 (2019)
13. Qin, S., van der Velde, D., Chatzakis, E., McStea, T., Smith, N.: Exploring barriers and opportunities in adopting crowdsourcing based new product development in manufacturing SMEs. Chin. J. Mech. Eng. **6**(29), 1052–1066 (2016)
14. Greineder, M., Blohm, I.: A process theory on transformation of work through internal crowdsourcing. In: Proceedings 2020 (1), p. 12694 (2020)

Ontology-Based Engineering – An Overview

J. Lentes(✉)

Digital Engineering, Fraunhofer IAO, Nobelstr. 12, Stuttgart, Baden-Württemberg, Germany
Joachim.Lentes@iao.fraunhofer.de

Abstract. The application of ontologies is a promising approach to advance knowledge sharing, interoperability and creational work in product creation. In this contribution, literature research is done to identify high and low addressed fields of the research area Ontology-Based Engineering. For this, two approximations to the state of the art are elaborated. At first, the search in Scopus and Web of Science. Second, to broaden the search results in Google Scholar. Identified main areas of work are the engineering of products and production systems, whereby other phases of the product lifecycle were addressed by work, two – excluding manufacturing and recycling. Currently trending areas, which could be addressed by ontology-related work, like Artificial Intelligence and Digital Twins, or even Cognitive Digital Twins are not targeted by the identified publications.

Keywords: Ontology-Based Engineering · Semantic Technologies · Product Lifecycle Phases

1 Introduction

Companies act in an ever competitive, dynamic and uncertain environment, resulting in high challenges on the development and manufacturing of products. One approach to cope with the rising challenges is to support development, planning and manufacturing with innovative Information Technologies like semantic technologies, i.e. ontologies. Typically, ontologies are used to support.

- communication,
- interoperability, and
- systems engineering (here, in a broader sense: content creation) [25].

In this contribution, communication support is extended to knowledge sharing and reuse, as this is a common objective of the application of ontologies, e.g. in [4, 10, 19, 23, 28].

Since the publication of the well-known definition of an ontology as „an explicit specification of a conceptualization" by Gruber [8] and his subsequent work with Olsen on an ontology for engineering mathematics [9], thirty years of research and technical development have passed. In the contribution, an overview about essential work done to support engineering by means of ontologies (i.e. Ontology-Based Engineering, OBE) is given based on an analysis of the literature.

M. Dragomir et al. (Eds.): ICPR 2023, LNPE, pp. 264–270, 2025.
https://doi.org/10.1007/978-3-031-92082-0_29

The literature research is done in two steps. In a first approximation, contributions are identified by searching for "Ontology-Based Engineering" in Web of Science [29] and Scopus [31] resulting in a rather short list of hits, which are streamlined by removing duplicates and entries not relevant to this work (as they are not concerned with the creation of goods but with software development and learning, respectively) and then analysed. In a second approximation, a full-text search for "Ontology-Based Engineering" is done in Google Scholar [30], resulting in a list of hits which are streamlined likewise and analysed. For The contribution ends with a summary and an outlook.

The databases were queried as of June 10th, 2023. Consequently, figures derived for the year 2023 are incomplete and the real values at the end of the year may substantially differ from the ones described in this contribution.

2 First Approximation

The search for "Ontology-Based Engineering" in Web of Science and Scopus resulted, after streamlining, in 12 contributions, the respective publication years are illustrated in Fig. 1, showing a peak in 2010 and a cluster of publications in the last years since 2019. For 2023, one has to be aware that a good part of publications is potentially still missing.

Bermejo-Alonso et al. developed a set of ontologies to support the development of autonomous systems by means of a dedicated software system for and subsumed the resulting approach as Ontology-Based Engineering [3]. A benefit of the application of an ontology was the sharing and reuse of knowledge between the developers.

In 2014 a group concerned with Ontology Based Engineering was founded [7]. The group addressed "the use of ontologies and semantic technologies in engineering applications" and had (or has) the objectives, to bring together related individuals and teams, to identify common views or differences and to develop a roadmap for harmonization and potential areas of research and development.

A large share of the recent publication cluster (six out of eight publications) was done around Mas and Arista, who are concerned with Models for Manufacturing (e.g. [20]) and the design of industrial systems, e.g. for assembly [2]. So, also this group of publications is based on a rather broad view on the term engineering which is more related with the development of artifacts along product creation than focused on product engineering itself.

Other publications mentioning the term Ontology-Based Engineering addresses the support for runtime failure detection [19], decision support for sustainable and smart manufacturing [1], and OBE for smart grid automation applications [26].

To summarize the result of the first approximation, i.e. the keyword search in Scopus and Web of Science, Ontology-Based engineering is seen as using ontologies and semantic technologies for the development of material and immaterial artefacts in relation to product creation in a broader sense. Thereby the focus is clearly on the engineering of products and production systems, including assembly systems.

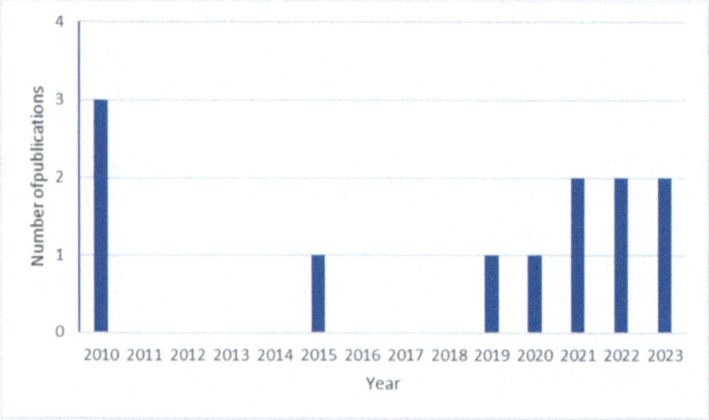

Fig. 1. Number of publications by year (1st approximation)

3 Second Approximation

The search for "Ontology-Based Engineering" in Google Scholar, followed by a streamlining as described above resulted in 43 hits with a timely distribution as shown in Fig. 2. The figure shows two main time clusters of publications, one around 2012 and especially one starting with 2019, even without having the final numbers for 2023.

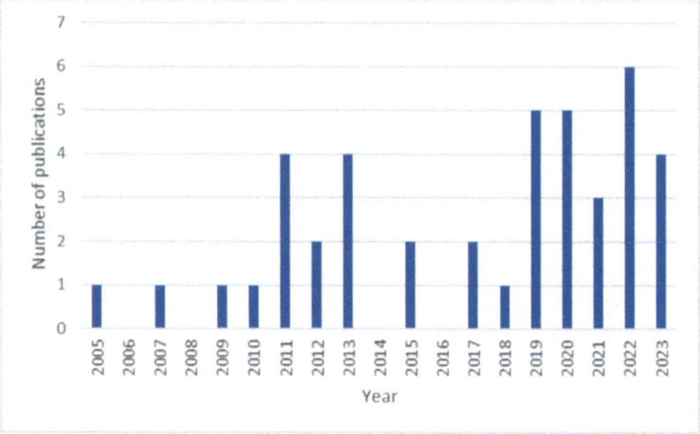

Fig. 2. Number of publications per year (2nd approximation)

When comparing the objectives of the application of an ontology in the identified publications as illustrated in Fig. 3, it is obvious that interoperability is seldom a reason, especially when considering that multiple assignments are possible. Only three publications were identified which solely address interoperability ([14, 21, 22]), where the others address at least knowledge sharing, too ([12, 13, 18]). The low number of work (publications) which address interoperability is in contrast to the known problem of having island-like software systems with silos for data and information, which results in fragmented information flows.

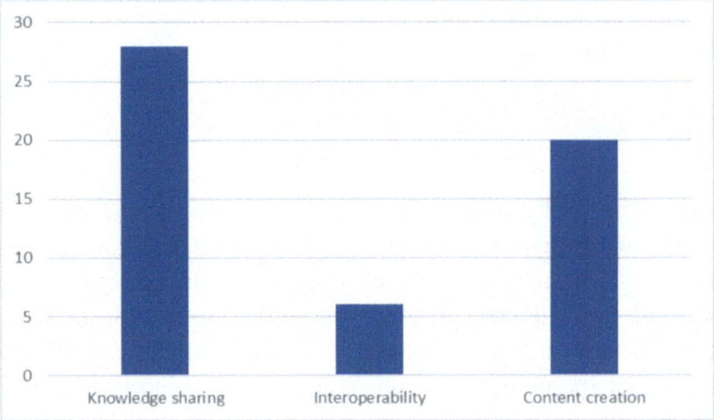

Fig. 3. Number of publications by objective of the ontology application

Considering the search phrase for the literature research, it is not surprising that most of the identified work (18 publications) addresses product development, see Fig. 4. Also for the product lifecycle phases Manufacturing Engineering and Assembly Planning as well as Operations a substantial amount of work (publications) were identified. Not in the focus but with contributions were the phases Research and (early) Development, Logistics, Supply Chain Management (SCM) and Service. No publications that address Manufacturing or Recycling were identified. Of the 43 identified publications, 3 are cross-phased and address both of Product Development as well as Manufacturing Engineering and Assembly Planning ([6, 16, 17]) and & are agnostic in what concerns the Product Lifecycle Phase ([5, 7, 11, 15, 18, 24]). On the other side, no publication addresses the phases Manufacturing and Recycling.

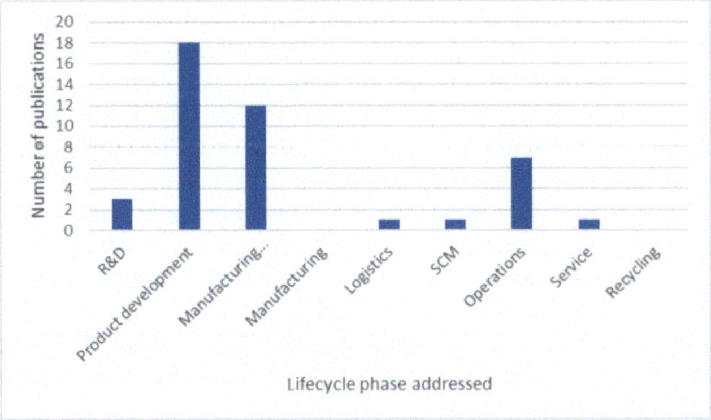

Fig. 4. Product Lifecycle phases addressed by the identified publications

For the combination of the perspective's objective for the use of ontologies and product lifecycle phases, no work could be identified which addresses manufacturing, SCM, and recycling for knowledge sharing; R&D, product development, manufacturing, logistics, service, and recycling for interoperability: and manufacturing, logistics, SCM, and recycling for content creation.

4 Conclusion

Concerning the identified publications, there seems to be a recent trend towards the theme Ontology-Based Engineering as illustrated in Figs. 1 and 2, especially when considering that a good part of publications of 2023 is still missing in current database states and is even not published, yet. Thereby, most of the publications address the engineering of products and production systems and operations, R&D, logistics, SCM and service are addressed, too, but not intensively. No work concerning manufacturing and recycling was identified. Consequently, there seems to be room for work which addresses these product lifecycle phases – potentially by connecting it with engineering-related lifecycle phases. Furthermore, no work could be identified that targets current trends like Artificial Intelligence, Digital Twins, or even the combination of the areas like Combined AI or Cognitive Digital Twins (e.g. as mentioned in [27]).

Concerning the literature research itself, an appropriate redefinition of the search terms could result in a broader view of the area of consideration. For example, omitting the quotation marks around Ontology-Based Engineering would result in 2483 hits in Scopus (17192 in Web of Science and round about 191,000hits in Google Scholar). Considering these large amounts of publications found with this alternative unprecise search phrase, a reasonable search phrase by using Boolean operators could be purposeful.

References

1. Aliabadi, M.M., Huang, Y.: A decision support framework for sustainable and smart manufacturing (2020)
2. Arista, R., Zheng, X., Lu, J., Mas, F.: An ontology-based engineering system to support aircraft manufacturing system design. J. Manuf. Syst. **68**, 270–288 (2023)
3. Bermejo-Alonso, J., Sanz, R., Rodriguez, M., Hernandez, C.: Ontology-based engineering of autonomous systems. In: Bauer, M., Mauri, J., Dini, O. (eds.) SIXTH International Conference on Autonomic and Autonomous Systems: ICAS 2010, PROCEEDINGS, pp. 47–51. IARIA, IEEE Comp Soc (2010)
4. Chakrabarty, S., Chougule, R., Lesperance, R.M.: Ontology-guided knowledge retrieval in an automobile assembly environment. Int. J. Adv. Manuf. Technol. **44**(11), 1237–1249 (2009)
5. Davis, M.: The business value of semantic technology. Semantic Web Applications for National Security (SWANS). , Crystal City (2005)
6. Demoly, F., Dutartre, O., Yan, X.-T., Eynard, B., Kiritsis, D., Gomes, S.: Product relationships management enabler for concurrent engineering and product lifecycle management. Comput. Ind. **64**(7), 833–848 (2013)
7. El Kadiri, S., Terkaj, W., Urwin, E.N., Palmer, C., Kiritsis, D., Young, R.: Ontology in engineering applications. In: Cuel, R. and Young, R. (eds.) Formal Ontologies Meet Industry, pp. 126–137. Springer, Cham (2015)
8. Gruber, T.R.: A translation approach to portable ontology specifications. Knowl. Acquis. **5**(2), 199–220 (1993)
9. Gruber, T.R., Olsen, G.R.: An Ontology for Engineering Mathematics (1994). http://ksl-web.stanford.edu/knowledge-sharing/papers/engmath.html

10. Imran, M.: Towards an Assembly Reference Ontology for Assembly Knowledge Sharing. Loughborough University (2013)
11. Kitamura, Y., Mizoguchi, R.: Ontological characterization of functions: perspectives for capturing functions and modeling guidelines. AI EDAM **27**(3), 259–269 (2013)
12. Kovalenko, O., Moser, T.: Using explicit and machine-understandable engineering knowledge for defect detection in automation systems engineering. Presented at the Proceedings of International Doctoral Symposium on Software Engineering and Advanced Applications (IDoSEAA 2011), (2011)
13. Lentes, J.: Towards an ontology for a lightweight support system for production system rough planning. IFAC PapersOnline **55**(10), 3010–3015 (2022)
14. Lu, Y., Panetto, H., Ni, Y., Gu, X.: Ontology alignment for networked enterprise information system interoperability in supply chain environment. Int. J. Comput. Integr. Manuf. **26**(1–2), 140–151 (2011)
15. Manzoor, S., et al.: Ontology-based knowledge representation in robotic systems: a survey oriented toward applications. Appl. Sci. **11**(10), 4324 (2021)
16. Markaj, A., Fay, A., Schoch, N., Stark, K., Hoernicke, M.: Intention-based engineering for the early design phases and the automation of modular process plants. In: 2022 IEEE 27th International Conference on Emerging Technologies and Factory Automation (ETFA), pp. 1–8. IEEE, Stuttgart, Germany (2022)
17. Markusheska, N., et al.: Implementing a system architecture model for automated aircraft cabin assembly processes. CEAS Aeronaut. J. **13**(3), 689–703 (2022)
18. Mas, F., Arista, R., Oliva, M., Hiebert, B., Gilkerson, I.: An updated review of PLM impact on US and EU aerospace industry. In: 2021 IEEE International Conference on Engineering, Technology and Innovation (ICE/ITMC), pp. 1–5 (2021)
19. Melik-Merkumians, M., Moser, T., Schatten, A., Zoitl, A., Biffl, S.: Knowledge-based runtime failure detection for industrial automation systems. In: CEUR Workshop Proceedings, pp. 108–119 (2010)
20. Morales-Palm, D., Oliva, M., Arista, R., Vallellano, C., Mas, F.: Enhanced metamodels approach supporting models for manufacturing (MfM) methodology. In: CEUR Workshop Proceedings (2022)
21. Pignaton de Freitas, E., ET AL.: Ontological concepts for information sharing in cloud robotics. J. Ambient Intell. Human. Comput. **14**(5), 4921–4932 (2023)
22. Ryabinin, K., Chuprina, S., Kolesnik, M.: Calibration and monitoring of IoT devices by means of embedded scientific visualization tools. In: Shi, Y., Fu, H., Tian, Y., Krzhizhanovskaya, V.V., Lees, M.H., Dongarra, J., Sloot, P.M.A. (eds.) Computational Science – ICCS 2018, pp. 655–668. Springer, Cham (2018)
23. Sanfilippo, E.M., Belkadi, F., Bernard, A.: Ontology-based knowledge representation for additive manufacturing. Comput. Ind. (2019)
24. Strasser, T.I., Pröstl Andrén, F.: Engineering and validating cyber-physical energy systems: needs, status quo, and research trends. In: Mařík, V., et al. (eds.) Industrial Applications of Holonic and Multi-Agent Systems, pp. 13–26. Springer International Publishing, Cham (2019)
25. Uschold, M., Gruninger, M.: Ontologies: principles, methods and applications. Knowl. Eng. Rev. **11**(2), 93–136 (1996)
26. Zanabria, C., Andren, F.P., Strasser, T.I., Kastner, W.: A model-driven and ontology-based engineering approach for smart grid automation applications. In: 45TH Annual Conference of the IEEE Industrial Electronics Society (IECON 2019), pp. 6635–6641. IEEE; Univ Nova Lisboa; IEEE Ind Elect Soc (2019)
27. Zheng, X., et al.: A semantic-driven tradespace framework to accelerate aircraft manufacturing system design. In: Review (2022)

28. Zheng, X., et al.: Development of an application ontology for knowledge management to support aircraft assembly system design. In: Sanfilippo, E.M. (ed.) Joint Ontology Workshops, JOWO 2021. Proceedings. Online resource : Episode VII: The Bolzano Summer of Knowledge; co-located with the 12th International Conference on Formal Ontology in Information Systems (FOIS 2021), and the 12th International Conference on Biomedical Ontologies (ICBO 2021), Virtual & Bozen-Bolzano, Italy, 11–18 September 2021 Online im WWW, 2021 (CEUR workshop proceedings 2969). (2021)
29. Document search - Web of Science Core Collection. https://www.webofscience.com/wos/woscc/basic-search. Accessed 10 June 2023
30. Google Scholar. https://scholar.google.de/. Accessed 10 June 2023
31. Scopus - Document search. https://www.scopus.com/search/form.uri?display=basic#basic. Accessed 10 June 2023

A Preliminary Study for Ontology-Based Controller in Factory Automation

Yu-Ju Lin, Chi-Yuan Tseng, Chiao-Yueh Liu, and Chin-Yin Huang(✉)

Department of Industrial Engineering and Enterprise Information, Tunghai University, Taichung, Taiwan

huangcy@thu.edu.tw

Abstract. Since the 1970s, Programmable Logic Controller (PLC) as a typical example of sequence control system has gradually become the mainstream technology in factory automation. However, the inflexibility and low maintainability of PLC bring a difficulty to adopt a PLC in modern factory automation with agility to handle unexpected internal and external events. Recent research shows that ontology can provide better knowledge representation and reasoning, in comparison with expert system/knowledge-based system. Ontology provides high flexibility and maintainability when it is facilitated as a controller for factory automation. Examples could be to add a new machine to replace an old machine, or to add an urgent order with customized processes to the shop floor. This research proposes a preliminary ontology controller integrating with Internet of Things (IoTs), cloud computing, and sematic model and reasoners. The controller is implemented in a laboratory pick-and-place station. The results indicate the ontology-based controller can handle unexpected events without a need of interruption to adjust/recode the program. This research intends to show a direction for the controller of factory automation. The success of the research paves a new way to develop a further large and complicated controller for factory automation.

Keywords: Smart Manufacturing · Manufacturing Control System · Ontology · Owlready2

1 Introduction

In the past, the manufacturing industry primarily mass produces goods. However, the current trend is toward smart manufacturing with extreme flexibility [1]. As a result, smart manufacturing [2] systems have become increasingly popular and are expected to be a significant trend in the future.

Smart manufacturing is a complex topic that is being studied by many scholars from different perspectives [3]. As manufacturing systems today have an ever-growing range of resources, it becomes crucial to integrate equipment and imbue them with intelligent capabilities through technologies such as the Internet of Things (IoT) and Artificial Intelligence (AI). Doing so can significantly enhance overall efficiency.

M. Dragomir et al. (Eds.): ICPR 2023, LNPE, pp. 271–280, 2025.
https://doi.org/10.1007/978-3-031-92082-0_30

Despite the rise of Industry 4.0, most research still heavily relies on the ISA-95 [4]architecture, which uses signal control technology from Industrial 3.0 based on Programmable Logic Controllers (PLC) [5]. For SCADA (Supervisory Control and Data Acquisition) [6] in ISA-95 is a concept borrowed from continuous process control used in chemical plants, but it only provides monitoring capabilities. Consequently, most smart manufacturing research focuses on analysing quality and machine information through big data analysis, while expert systems related to artificial intelligence are often overlooked. Nonetheless, deep learning is widely used in the field of smart manufacturing for diagnosing and preventing issues related to machine failure and product quality.

PLC in ISA-95 plays a crucial role in controlling machines, supervising production processes, and collecting real-time equipment status information on the shop floor. Meanwhile, to meet changing production demands, Manufacturing Execution System (MES) [7] in ISA-95 must be adaptable and responsive, enabling timely responses to customer needs. However, there are several challenges that need to be addressed for factory automation in Industry 4.0, such as difficulty in dynamically dispatching resources, lack of information sharing among various equipment, and difficulty in effectively utilizing production line data collected through sensors.

To address these challenges, this study proposes an intelligent monitoring knowledge ontology inference model using Python programming language. The model can make real-time decisions based on the status of resources on the production site and IoT devices, monitoring production equipment and providing proactive feedback to the production site. Additionally, the study aims to integrate with the Synology private cloud database for real-time signal upload and storage, enabling information sharing and flexible production processes.

Using smart manufacturing technology as a verification field for the blueprint of a smart factory, this study develops inference rules for the classification operation process of self-made electromechanical integrated devices and analyses the application of ontological manufacturing to processes and line changeovers. The study also develops a manufacturing control system based on the ontology and verifies the feasibility of the ontology model application through the on-site classification system.

2 System Architecture

The system architecture of this research in Fig. 1 consists of three main layers: the physical layer, the data layer, and the knowledge inference layer. The resources in the physical layer located on the shop floor include Automated Guided Vehicles (AGVs), Automated Storage and Retrieval Systems (AS/RS), conveyors, sensors, machines, and robots. Related data for those resources are stored in a database and mapped to the ontology model in the knowledge inference layer. The data layer is linked to the equipment data in the physical layer and collects sensor data for the resources in the physical layer. The Owlready2 API is used to update the ontology model and perform knowledge inference.

To construct the ontology inference model, the owlready2 module is used with the Python programming language to build the ontology model for the resources in the physical layer and construct association rules for inference. Sensor data is uploaded to

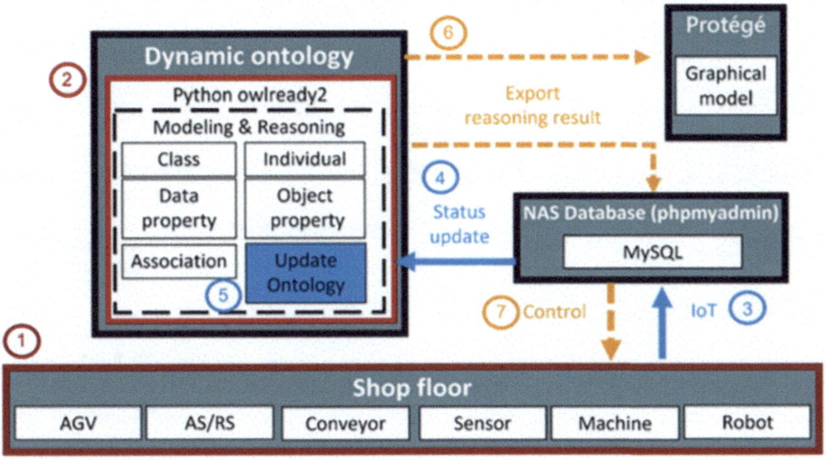

Fig. 1. System Architecture

the Synology® NAS private cloud database through the Linkit™ 7697 MCU chip as the sensor endpoint interface. The resource status is dynamically updated in real-time based on the signals monitored by the sensors, and the inference model is updated accordingly through the database to align with the resource situation [8].

The inference results are converted into control signals using the MySQLdb module in Python and returned to the database. The results are also visualized in Protégé to provide a clear structure for users. Finally, real-time control on the equipment is achieved by updating the database. Then, data status of the database drives the Linkit™ 7697 MCU to control the resources in the physical layer.

3 Ontology Model

In this study, the ontology class, data property, and individuals were created using the Python package Owlready2. The main resources in the classification of the operating system are the motor for horizontal screw movement (Motor), conveyor belt (Conveyor), sensor (Sensor), and gripper (Gripper).

The inferred classes (Class) are mainly used to store the results of inference, including the high placement area (Place_High), low placement area (Place_Low), possible pick-up action (Possible_to_PickUp), possible placement in high area (Possible_to_PlaceHigh), possible placement in low area (Possible_to_PlaceLow), gripper up (Gripper_Up), and gripper down (Gripper_Down).

The main classes are further classified into screw positioning sensors (Location_Sensor), gripper up/down positioning sensors (Gripper_UpDown_Sensor), and object detection sensors for the conveyor belt (Object_Detection_Sensor). The models in Protégé are shown in Fig. 2 and 3.

Data property are established to describe numerical data, such as whether the motor is on or whether the conveyor belt is on. Corresponding properties are created in the data

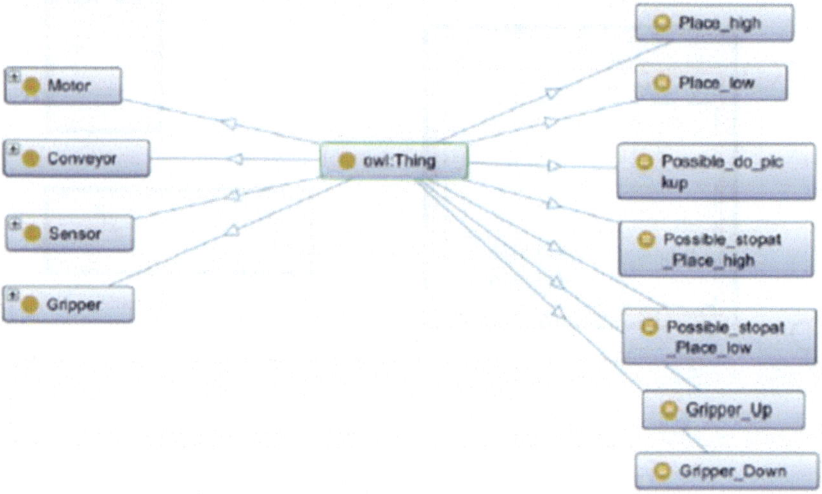

Fig. 2. Ontology classes in Protégé

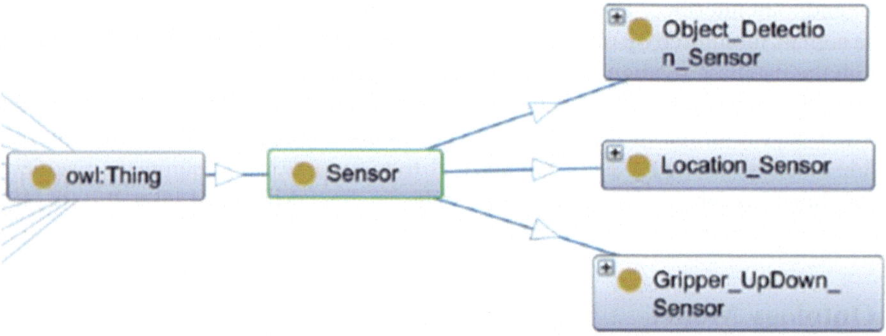

Fig. 3. Subclasses of Sensor in Protégé

properties and the range of property values is set. Figure 4 shows data properties in the ontology model.

Individuals represent the physical devices on the shop floor. The main resources in the case study (described later) are motors for horizontal screw movement, conveyor belts, grippers, and three types of sensors: positioning, gripper up/down, and detection sensors. The corresponding individuals are created. The visual model in Protégé is shown in Fig. 5.

This study aimed to develop a control for classifying and transporting workpieces in a mechatronic system. Ontology inference logic is used instead of PLC for the control. The signals from the sensors are uploaded to a cloud database via IoT modules. Finally, the Python program can infer all possible decision actions using a reasoner and store the updated system status in the cloud database.

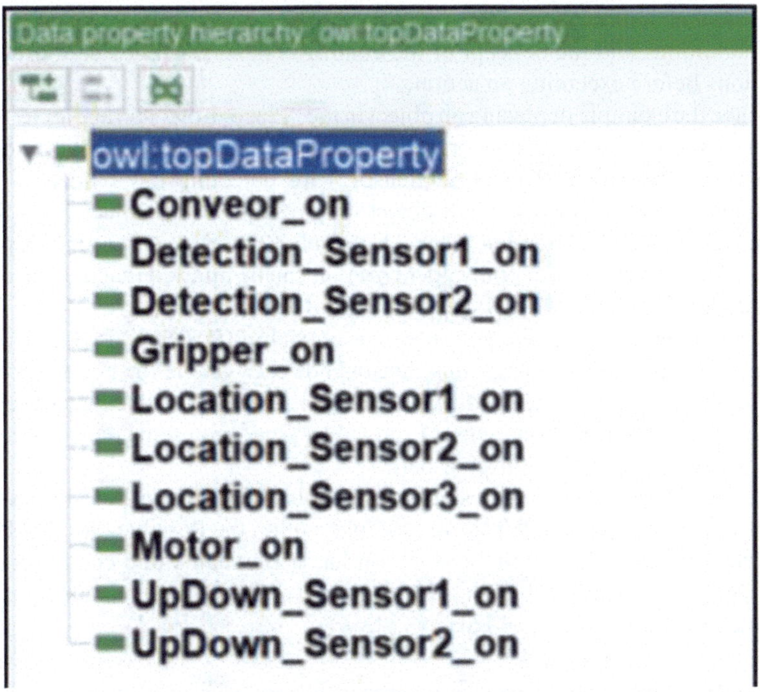

Fig. 4. Data Property in Protégé

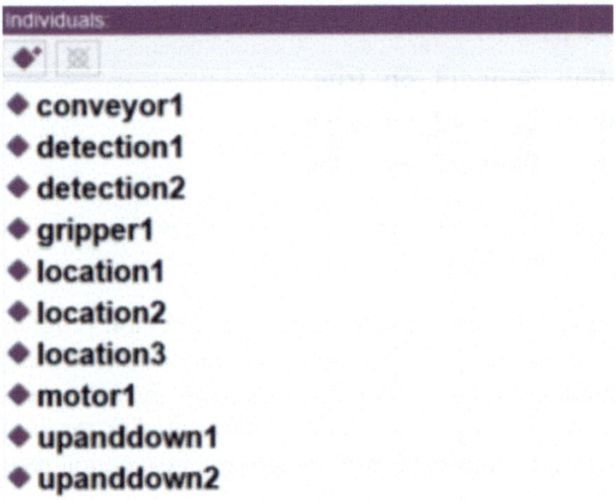

Fig. 5. Individuals in Protégé

Association rules required for inference can also be constructed. By utilizing the built-in "equivalent_to" feature in the Owlready2 package module, association rules can be established to describe the relationship between individual instances of numerical properties and specific classes. The rules established in the case study are based on the field conditions and the concept of the control system. Each device must meet the preconditions before executing an action.

Let's take the example of placing an object in the "Place_Low" storage area. There are two material sensors on the conveyor belt: Sensor 1 (Detection_Sensor1) for detecting high objects and Sensor 2 (Detection_Sensor2) for detecting low objects. When the material sensors on the conveyor belt detect the presence of an object, with Sensor 1 displaying a False value (Detection_Sensor1_on.value(False)) and Sensor 2 displaying a True value (Detection_Sensor2_on.value(True)), it can be inferred that the object should be placed in the "Place_Low" storage area.

Next, let's continue with the possible pick-up action (Possible_do_PickUp). Since there is a positioning sensor (Location_Sensor1) directly above the conveyor belt, this positioning sensor determines whether the gripper is currently above the conveyor belt. In the rule conditions, Location_Sensor1_on.value(True) represents the detection of the positioning sensor above the conveyor belt.

Figure 6 and 7 show the output results connected to Protégé from the Python side. Taking the class Possible_do_PickUp as an example: within the Possible_do_PickUp class, the numerical property Location_Sensor1_on has a Boolean value condition of True, Location_Sensor2_on has a Boolean value condition of False, and Location_Sensor3_on has a Boolean value condition of False.

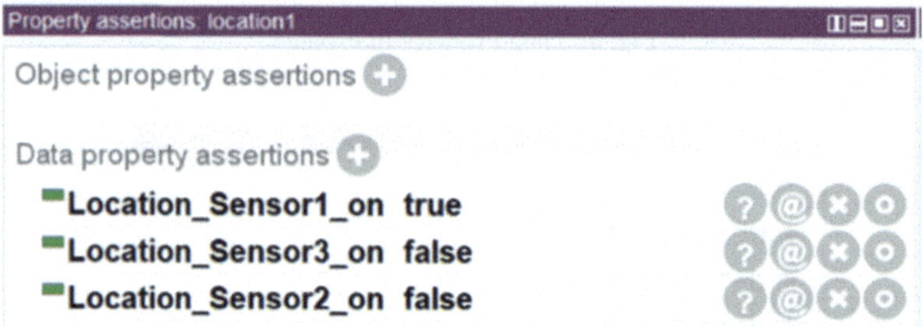

Fig. 6. Protégé displays attached data property in individual

In the dynamically inferred model created in the Python program, the output displayed in Protégé, as shown in Fig. 6, represents the individual (Individual) location1 with accompanying numerical properties. Figure 7 displays the inference rules and results within the Possible_do_PickUp class. As location1 satisfies the inference rules, it is displayed in the Instances, consistent with the program execution results.

4 Emulation on a Lab Case Study

Firstly, the detection of incoming materials is performed. In Fig. 8 (A), because the gripper is positioned above the conveyor and detects the presence of incoming materials, it infers that the gripper should go down. Figure 8 (B) shows a successful control of the gripper.

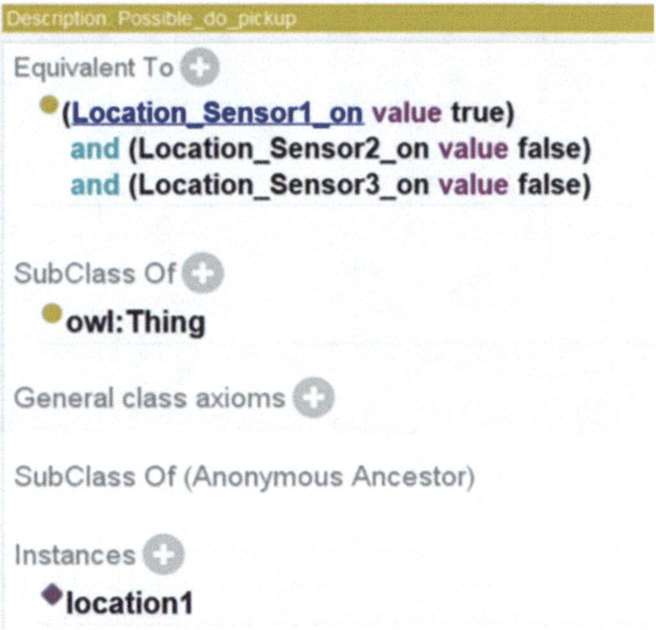

Fig. 7. Protégé displays inference constraints and results in class

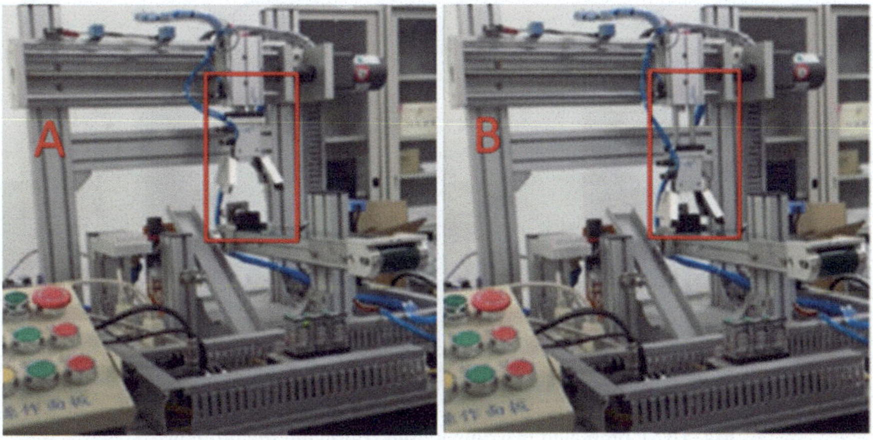

Fig. 8. Detection of incoming materials to determine gripper's lowering position

When the conditions for accessing the storage area are met, the system infers which type of storage area to go to. For example, in Fig. 9 (A), if the gripper has executed a command to hold an object, the system infers that it should go to the High storage area (Place_High). Figure 9 (B) shows the gripper execute accordingly based on the inferred result to move to Place_High.

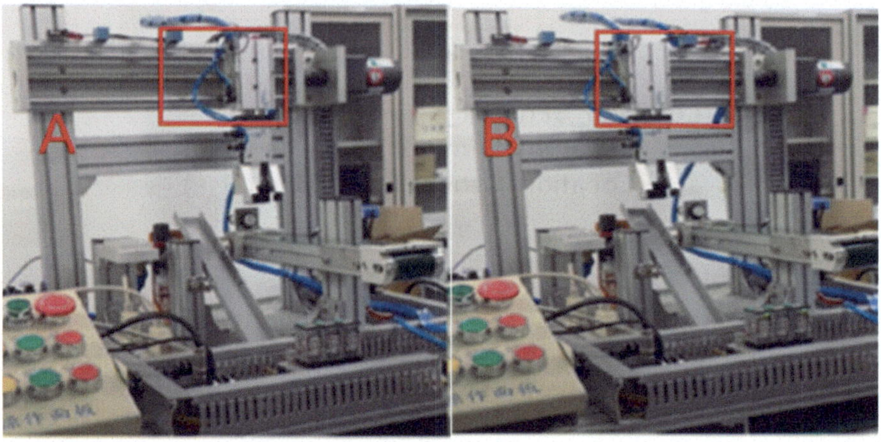

Fig. 9. Moving the gripper according to the inferred command

Finally, Fig. 10 (A) shows the gripper arrives at the High storage area (Place_High) with an object held, and the system infers that the object should be placed. Figure 10 (B) shows that by following the inference, a successful placement of the material is finished in the storage area.

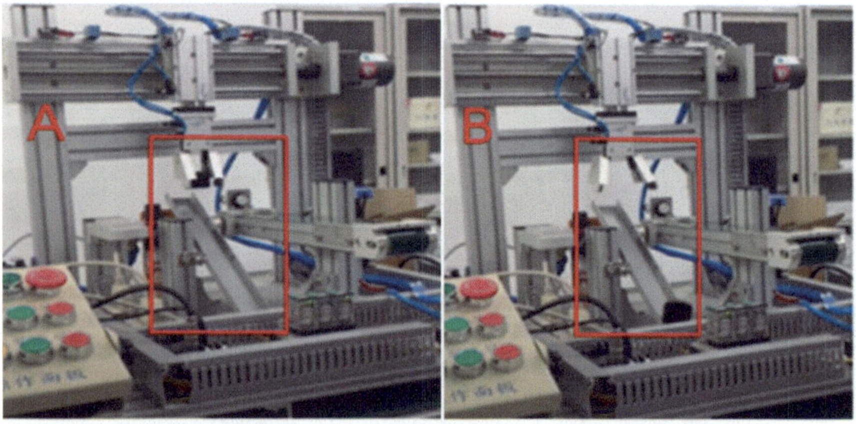

Fig. 10. Completed placement after following the inference

5 Conclusions

In the era of Industry 4.0, smart factories have become a norm for the manufacturing industry. As the manufacturing process becoming increasingly complex, machines with logic thinking are critical. It requires more inference rules to enhance the system's level of intelligence.

This study utilizes the Owlready2 Python API to link ontologies. When sensors transmit status information to the database in the cloud through wired or wireless IoT, the Python program retrieves the manufacturing status from the database. The individuals in the ontology, such as AGV, Machine, Robot, are with data properties, such as on/off/idle, pressure, temperature, vibration, queue length. Then, by enabling the reasoner, all possible decision actions can be inferred and stored in the database in the cloud, and then trigger to control the resources in the physical layer.

In summary, this research has completed the following results:

1. Providing a method of control using ontology for factory automation, demonstrating the applicability of knowledge ontology in factory control and providing a direction for the use of ontology.
2. By integrating the Owlready2 API [9] with Python and linking it to a cloud database, this study achieves real-time monitoring of resources through the sharing of production signals via the updating data in the cloud database.
3. The inference rules used for machine control are universal and not limited to specific operating environments, making them applicable to different types of industries.

Compared to using PLC for control, this research using ontology for machine control has the following advantages:

1. Building ontology models, inference rules, and control systems using Python provides users with a good development environment. In the future, the user interface can also effectively integrate related applications, such as MCU or databases, to make it flexible and coordinated.
2. The signals for machine control in this study are the results of ontology inference. Those signals are transmitted in and out through IoT and cloud databases. The control is independent of machine control specifications and the operating systems.
3. By using the rich semantics of ontology rules, this research has eliminated the limitation of PLC. Ontology-based control can provide detailed descriptions of things, properties, and relationships used in the process.

This study has provided preliminary validation for the use of ontology and knowledge inference in machine control. However, the real-time capabilities of ontology are limited by computational resources and device performance. In the future, further research will be conducted to apply this control system to different types of production operations, use more powerful IoT and cloud processors to solve the issue of poor real-time inference, and explore how advancements in technology can overcome hardware constraints and enable knowledge inference to be a more flexible, understandable, constructible, and humanized control system for factory automation.

Acknowledgments. The research of this paper is supported by the funding projects 111-2221-E-029-013-MY3, 110-2221-E-029-025, and 109-2221-E-029-015 from National Science and Technology Council, Taiwan.

References

1. Coyle, J.J., Novack, R.A., Gibson, B.J., Langley, C.J.: Supply chain management: a logistics perspective. Cengage Learning (2021)
2. Tao, F., Qi, Q., Liu, A., Kusiak, A.: Data-driven smart manufacturing. J. Manuf. Syst. **48**, 157–169 (2018)
3. Gölzer, P., Fritzsche, A.: Data-driven operations management: organisational implications of the digital transformation in industrial practice. Prod. Plan. Control **28**(16), 1332–1343 (2017)
4. Daneels, A., Salter, W.: What is SCADA? Paper presented at the International Conference on Accelerator and Large Experimental Physical Control Systems, Trieste, Italy (1999)
5. Lin, Y.-J., Tan, C.-F., Huang, C.-Y.: Integration of logic controller with IoT to form a manufacturing edge computing environment: a premise. Procedia Manuf. **39**, 398–405 (2019)
6. Sverko, M., Grbac, T.G., Mikuc, M.: Scada systems with focus on continuous manufacturing and steel industry: a survey on architectures, standards, challenges and industry 5.0. IEEE Access **10**, 109395–109430 (2022)
7. Almada-Lobo, F.: The industry 4.0 revolution and the future of manufacturing execution systems (MES). J. Innov. Manag. **3**(4), 16–21 (2015)
8. Lin, Y.J., Lan, C.-B., Huang, C.-Y.: A realization of cyber-physical manufacturing control system through industrial internet of things. Procedia Manuf. **39**, 287–293 (2019)
9. Lamy, J.-B.: Owlready: ontology-oriented programming in Python with automatic classification and high level constructs for biomedical ontologies. Artif. Intell. Med.Intell. Med. **80**, 11–28 (2017)

A Proposed Ontological OPC-UA in Semiconductor Cleanroom Monitoring System

Yu-Ju Lin, Ting-Yi Weng, Ting-Kai Liang, and Chin-Yin Huang(✉)

Department of Industrial Engineering and Enterprise Information, Tunghai University, Taichung, Taiwan

huangcy@thu.edu.tw

Abstract. With the development of Industry 4.0 and the rapid changes in the internal/external environment, it is necessary to have common platform and language for communication and co-work. OPC-UA (Open Platform Communications Unified Architecture) is a platform-independent protocol that allows communication between different hardware/software systems. OPC-UA uses a client-server architecture, where a client can communicate with one or more servers to access data or perform operations. However, the application of OPC-UA is domain-related. To construct an OPC-UA protocol platform for industrial communications requires domain and OPC-UA experts to apply the domain Companion Specifications of OPC-UA. The integration of various domain companion specifications may become a problem when the project includes various domains. By applying ontology to specify the elements of OPC-UA, this research standardizes OPC-UA. The standardized ontological OPC-UA is more readable, adaptable, and transferable. To demonstrate the advantages of ontological OPC-UA, this research applied it to a semiconductor cleanroom monitoring system. The results show the quick adaptability and expandability of the ontological OPC-UA.

Keywords: companion specifications · Internet of Thing · ontology · OPC-UA · owlready2

1 Introduction

The trend of Industry 4.0 development requires production lines to be flexible and adaptable to cope with the challenges of the manufacturing environment [1]. However, the communication between the machines from different vendors has become a major obstacle for the manufacturers to achieve smart manufacturing. In the German Industry 4.0 project, the use of the OPC-UA communication protocol is initiated to integrate equipment and application systems in the factory, and to implement the communication of reconfigurable manufacturing systems. OPC-UA has shortcomings such as lack of logic for semantic description in data models and lack of browsing functionality [2]. Therefore, this study integrates the complementary specifications released by the OPC Foundation to establish a standardized complementary specification. Then, this research utilizes the

M. Dragomir et al. (Eds.): ICPR 2023, LNPE, pp. 281–292, 2025.
https://doi.org/10.1007/978-3-031-92082-0_31

advantages of ontology to establish a knowledge model to ensure that data can be successfully delivered. Finally, taking the example of cleanroom environment monitoring in the semiconductor industry, Python is used as the integration platform to collect and integrate sensing/control data.

2 Literature Review

2.1 Ontology-Based OPC-UA Applications

The OPC-UA communication protocol provides an extensible information model that helps address interoperability issues at the automation level. As a NoSQL database, OPC-UA can store large amounts of data and achieve high throughput [3]. Due to the unstructured nature of its data, which is difficult to access directly, and the lack of fully defined semantics, automatic reasoning and knowledge inference on the OPC-UA address space are not possible. Therefore, some researchers have proposed using ontology to enhance semantic representation and apply it to specific domains of OPC-UA models. In this study, we integrate these advantages to investigate how to establish a standardized data model for widespread use in the manufacturing industry. Table 1 summarizes the application scenarios of ontology-based OPC-UA in recent years.

Table 1. Application Scenarios of Ontology-Based OPC-UA

Research	Year	Application context
Bunte, Niagemann, & Stein [4]	2018	Production System
Steindl, Frühwirth, & Kastner [5]	2019	Boilers in Industrial Environments
Bakakeu et al. [6]	2020	Factory Equipment
Steindl & Kastner [2]	2021	Packing material regenerator

2.2 Semiconductor Industry Clean Room

The semiconductor industry needs to prevent external contamination in order to improve the performance of components and product yield, and therefore requires manufacturing processes to be carried out in a clean environment. As a result, cleanroom environmental monitoring is very important. International Organization for Standardization (ISO) 14644-1, Federal Standard (FS) 209E in the United States, and Japanese Industrial Standards (JIS) 9920 are standards for cleanrooms. Environmental factors in cleanrooms include particulate matter, pressure differentials, airflow, temperature and humidity, noise, illumination, and vibrations. Different ISO classes are distinguished based on particle counts as a measure of cleanliness level. In this study, environmental monitoring of cleanrooms in semiconductor factories is taken as an example, and a standardized model is constructed using the OPC-UA communication protocol to investigate the feasibility of using this specification.

3 Method

In order to establish a standardized data model specification that can be widely applied in various fields, this study develop three layers in the system architecture: Communication Layer, Mapping Layer, and Application Layer.

3.1 System Architecture

The architecture shown in Fig. 1 is based on a Python-based OPC-UA system, which collects data from local sensors/machines and communicates with the server through communication and mapping layers. The mapping layer uses the OPC-UA data model proposed by Pfrommer in 2017 to establish relationships among the entities in the ontology based on the data collected by the communication layer [3]. In the application layer, Python language is used for environment integration, to receive data information collected by environment sensors and make real-time adjustments based on the on-site conditions. Finally, the results are presented to the users through a user interface.

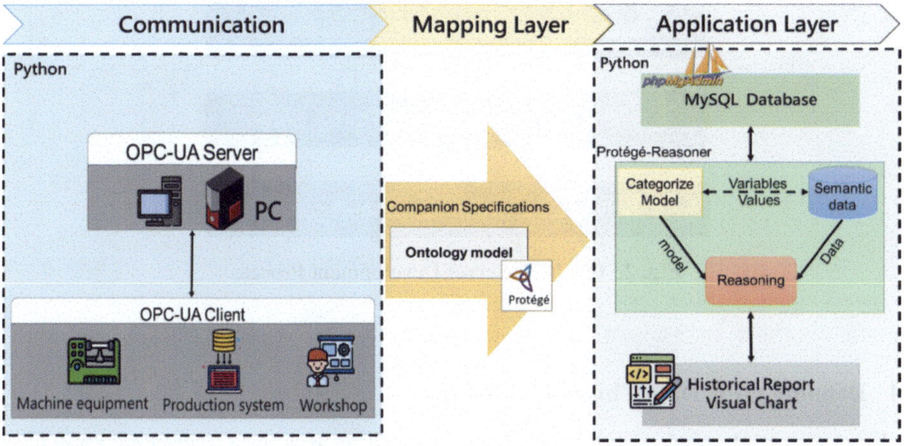

Fig. 1. System Architecture

3.2 Establishing OPC-UA Communication Protocol with Python

This study utilizes the OPC-UA communication protocol and the free open-source package FreeOpcUa to create a communication architecture between a server and client based on Python, for acquiring data from factory equipment, and edited using PyCharm Community Edition. The next step will be to implement the monitoring of a cleanroom environment and provide instructions on how to set up the Python OPC-UA server and establish connection with Python OPC-UA client.

3.3 OPC-UA Sever and Client

The access methods of Server and Client in OPC-UA are designed to minimize redundant data type determination and target reading. The OPC-UA establishment process shown in the following diagrams mainly involves configuring the connection method between the Server (Fig. 2) and Client (Fig. 3), as well as establishing security mechanisms.

When connecting to an OPC-UA Server from a Client, it is necessary to configure security policies, input the IP of the Server as the URL for connection, and use subscription methods to read and write values of Tags. If the connection to the Server is not needed, it can be terminated. The diagram below illustrates the process of an OPC-UA Client connecting to a Server.

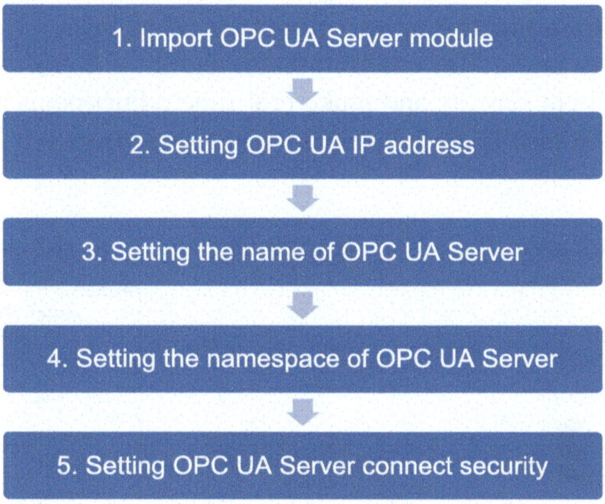

Fig. 2. OPC-UA Server Development Process

3.4 Defining OPC Data Through Ontology

In this research, Protégé, a widely used ontology editor, was used to create a knowledge ontology. Ontologies using classes, individuals, and properties to describe knowledge in various domains. A Python ontology libraries Owlready2 was adopted for developing the integrated system in this research [7]. The OPC-UA Core ontologies model by A. Perzylo was used as a reference to compile commonly used parameters in the manufacturing environment, and to implement ontology-based OPC-UA data access [8]. The projects and contents are summarized in Table 2.

4 Scenario Simulation

4.1 Scenario Settings

In this study, the experimental environment was set up in a cleanroom with a cleanliness standard of Class 1,000 and a particle size of 0.5 µm. The information ontology model constructed in this study was applied, and values returned based on the actual situation

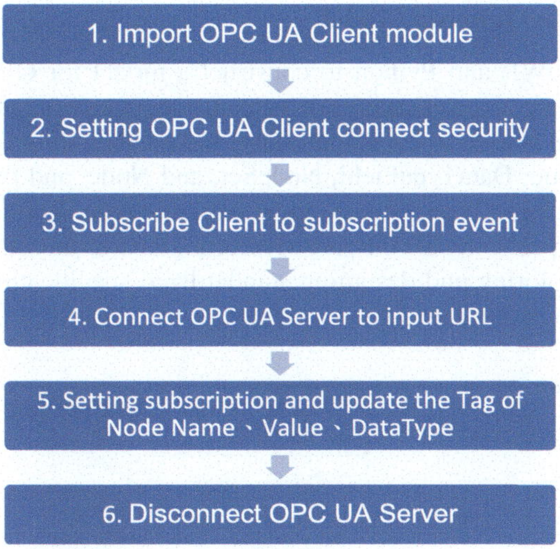

Fig. 3. OPC-UA Client Connecting Process

Table 2. Project and Contents

Number	Project	Content
1	OPC-UA Core	• UA Data Type Definition • UA Role Permission • UA Value • UA Data Type Field • UA Node Set • UA Node
2	Node	• UA Object Type • UA Object • UA View • UA Variable Type • UA Reference Type • UA Method • UA Variable • UA Data Type
3	Object	• Create common object classes: measurement, operation, system • Create common objects: UI, other systems, sensors
4	Variable	• Create common parameters: Angle, distance, feed rate, humidity, coordinate position, pressure, speed, temperature, transaction time
5	Reference	• Built according to the specifications of the OPC-UA Foundation

were used for decision making. If any factor in the cleanroom exceeds or falls below this standard, real-time notifications are sent to the operating machines and personnel [9].

4.2 Constructing an Ontology Model

This study utilizes OWL and Python to construct a model for OPC-UA objects, and standardizes the design of specifications of a cleanroom in a semiconductor factory. The Owlready2 in Python is used to create ontology classes, including DataTypeDefinition, RolePermission, Value, DataTypeField, NodeSet, and Node, and further subclasses of Node class such as ObjectType, Object, View, VariableType, ReferenceType, Method, Variable, and DataType. Nodes are the core of information transmission in OPC-UA. This study focuses on exploring and designing standardized specifications, and the overall ontology model is shown in Fig. 4.

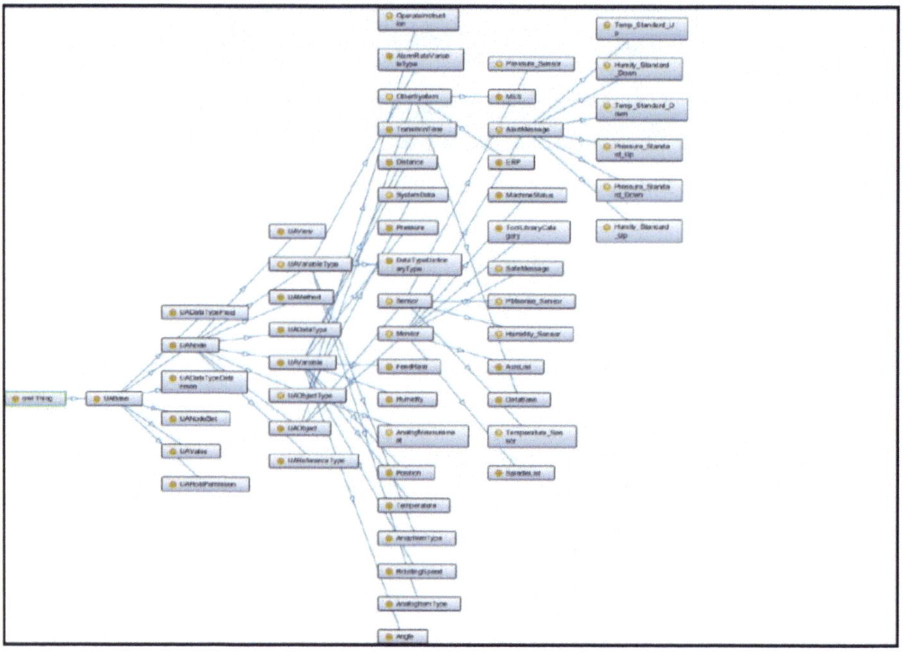

Fig. 4. The Overall Ontology Model

4.3 Establishing Inference Rules

In addition to refining the associations between OPC-UA data models, OPC-UA was used for cleanroom environment monitoring and provided an alarm system to notify production line personnel of the location of errors for timely resolution. OPC-UA utilizes nodes for basic information exchange, forming a node set to establish associations, such as the relationships between variables and objects.

There are four types of sensors used in cleanroom environment monitoring: temperature, pressure, humidity, and number of dust particles (Fig. 5).

This study established rules for the alarm system, utilizing the explicitness feature of ontology to provide users with detailed information about error messages for better

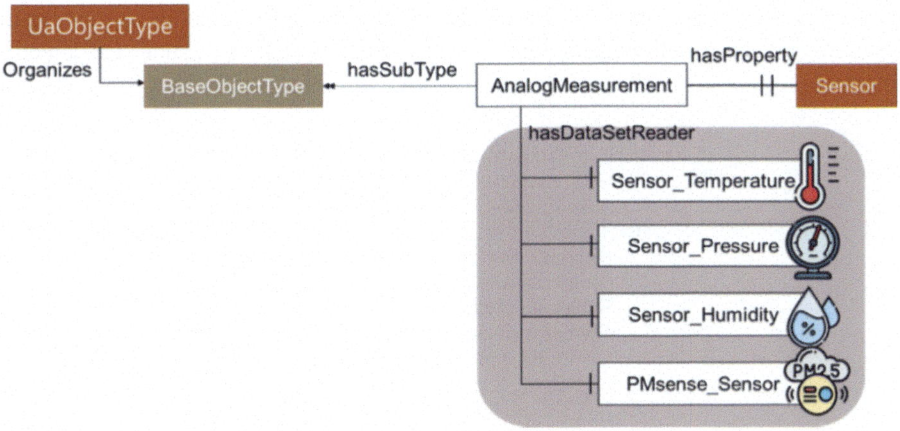

Fig. 5. Object association diagram

understanding of the error causes. The system categorizes messages into safety messages and alarm messages, where messages with the DataProperty userExecutable attribute set to True are considered safety messages, and those with the attribute set to False are considered alarm messages. In the alarm messages, error codes are classified based on the Description attribute, and the corresponding codes are pushed into their respective categories. Table 3 lists the code-to-classification mapping.

Table 3. Table of Classification Code

Error code	Description	Classification
1	Exceed the standard temperature	Temp_Standard_Up
2	Below the standard temperature	Temp_Standard_Down
3	Exceed the standard pressure	Pressure_Standard_Up
4	Below the standard pressure	Pressure_Standard_Down
5	Exceed the standard humidity	Humidity_Standard_Up
6	Below the standard humidity	Humidity_Standard_Down
7	Exceed the standard PM	PM_Standard_Up

This study has established a custom-developed OPC-UA data transmission system that uploads data to a MySQL database in association with the ontology model. The data table in the database includes fields for recording time, temperature, humidity, pressure, number of dust particles, and current time, which are used to accurately infer whether the cleanroom environment comply with the standard.

Using the Python programming language, the environmental factors of the cleanroom are determined to be either in a safe state or an alarm state through the rules of database correlation. The results are pushed to the corresponding Sensor class, and any issues related to the alarm state are displayed, enabling the technicians to solve the environmental abnormalities quickly. Figure 6 shows an example of codes for detecting anomalies.

```
#判斷溫度是否在正常範圍內
if 22 <= TemperatureValue <= 24 :
    TemperatureMessage.userExecutable=[True]
else:
    TemperatureMessage.userExecutable=[False]
    if TemperatureValue > 24:
        Temp1.description =[1]
    elif TemperatureValue < 22:
        Temp1.description =[2]

#判斷濕度是否在正常範圍內
if 45 <= HumitidyValue <= 55 :
    HumitidyMessage.userExecutable=[True]
else:
    HumitidyMessage.userExecutable=[False]
    if HumitidyValue > 55:
        Hum1.description =[3]
    elif HumitidyValue < 45:
        Hum1.description =[4]
#判斷壓力是否在正常範圍內
if 10 <= PressureValue <= 25 :
    PressureMessage.userExecutable=[True]
else:
    PressureMessage.userExecutable=[False]
    if PressureValue > 25:
        Press1.description =[5]
    elif PressureValue < 10:
        Press1.description =[6]
#判斷落塵數量是否在正常範圍內
if PMValue < 1000 :
    PMMessage.userExecutable=[True]
else:
    PMMessage.userExecutable=[False]
    if PMValue > 1000 :
        PM1.description =[7]
```

Fig. 6. Codes for Detecting abnormality

4.4 Inference Verification

This study chose to use Python as an integration platform to examine the feasibility of the OPC-UA model. In the OPC-UA address space, node classes are defined using node types, which establish relationships between objects through references. Hierarchical structures can be established between node types by using “Organizes”, and the actual objects and object types are bound together by using “HasTypeDefinition” references. The “Sensor” object class has corresponding sensor node identifier attributes, including objects, data, and methods related to the received values and actual object values. Finally, the inference results from Protégé correctly push the sensor classes into the object classes, and the values of the actual objects are also pushed into the corresponding classes, as shown in Fig. 7.

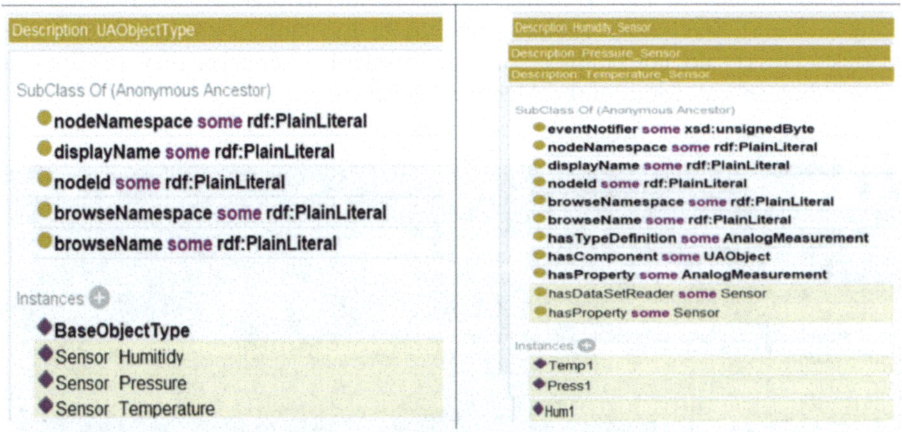

Fig. 7. Ontology Inference Results (Part 1)

According to the boolean value of the userExecutable attribute, it indicates whether the value needs to trigger an alert message (Fig. 8). If the value exceeds the range, it will be pushed an AlertMessage, as in the example displaying humidity and temperature information; otherwise, if the value is within the range, it will be pushed as an example a SafeMessage.

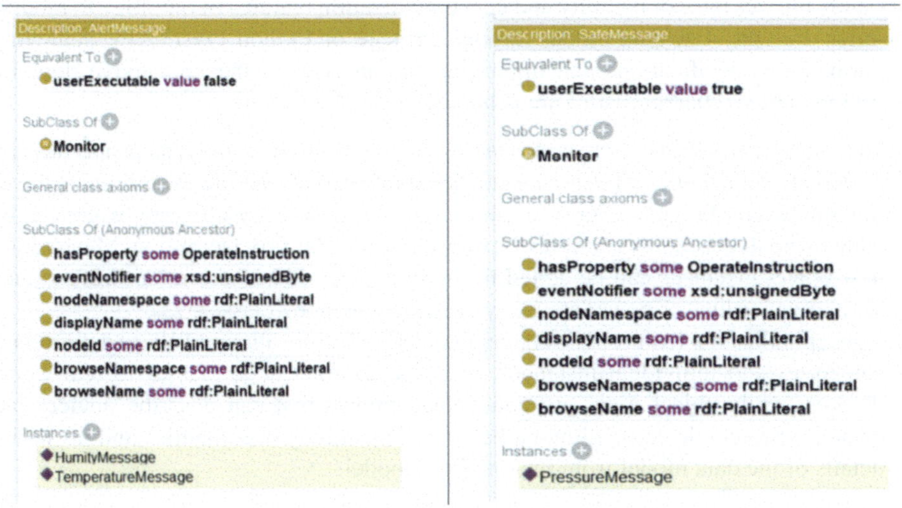

Fig. 8. Ontology Inference Results (Part 2)

5 Case Illustration

5.1 A Simulation Case

The display of data in Fig. 9 from the cleanroom environment obtained by connecting to the MySQL database shows that the temperature (test4.Temp1) and pressure (test4.Press1) do not comply with the standard. Therefore, they fall into the corresponding subcategories of the AlertMessage class.

Fig. 9. Inference Result

5.2 Discussion

This study utilizes the extensibility and conceptualization features of ontology to modify and infer the OPC-UA data model through Protégé or Python OWL API, achieving standardization specifications and improving the analysis of the knowledge domain. The following two characteristics are achieved in this framework:

(1) Ontology extensibility: New node classes, object properties, and data properties can be added, and a relational structure can be established to validate the inference information model of objects. For example, a new object class for ultrapure water can be added with new variables, and can be referred using HasTypeDefinition. Meanwhile, new object instances can be added to monitor other cleanroom monitoring values such as gas values or power status by creating new node classes.

(2) Conceptualization feature: The conceptualization feature allows OWL to define class categories and individual instances, mapping abstract node classes to real-world devices, while object types are conceptual models that can describe devices but cannot represent devices, allowing better understanding of semantics and browsing details of the data model from the ontology model.

6 Conclusions

In the era of Industry 4.0 and the new norm after the COVID-19 pandemic, the manufacturing system needs to address the issues of connectivity and real-time information exchange between machines, manufacturing system, and environment. While the OPC-UA (Open Platform Communications Unified Architecture) protocol can solve these problems, the current specifications are domain-specific, and unmatched issues may arise when integrating multiple systems and different types of machines. Therefore, this study proposes a standardized ontology specification, which uses programs to create an ontology model and employs knowledge inference to validate its correctness. Case example verification is conducted in a cleanroom context to realize an ontology-based control system. In summary, this study presents:

(1) Using the Owlready2 Python module to create a knowledge model and systematically categorize and configure contextual data, enabling real-time updates of contextual information.
(2) Establishing generic rules for ontology models and leveraging the high extensibility and flexibility provided by the programming module to adapt to different industries and environments, and achieve knowledge-driven reasoning through data sharing.
(3) Verifying the feasibility of the ontology specification through case scenarios and solving the problem of version compliance.

This study has verified the ontology-based OPC-UA data model specification, which can interpret the semantic meaning of received data through knowledge-based inference. In the future, it can be applied to handle emergency/monitoring issues in the remote yet connected operation systems.

Acknowledgments. The research of this paper is supported by the funding projects 111-2221-E-029-013-MY3, 110-2221-E-029-025, and 109-2221-E-029-015 from National Science and Technology Council, Taiwan.

References

1. Bakakeu, J., et al.: Automated reasoning and knowledge inference on OPC UA information models. In: 2019 IEEE International Conference on Industrial Cyber Physical Systems (ICPS), pp. 53–60. IEEE (2019)
2. Steindl, G., Kastner, W.: Transforming OPC UA information models into domain-specific ontologies. In: 2021 4th IEEE International Conference on Industrial Cyber-Physical Systems (ICPS), pp. 191–196. IEEE (2021)
3. Pfrommer, J.: Semantic interoperability at big-data scale with the open62541 OPC UA implementation. In: Interoperability and Open-Source Solutions for the Internet of Things: Second International Workshop, InterOSS-IoT 2016, Held in Conjunction with IoT 2016, Stuttgart, Germany, November 7, 2016, Invited Papers 2, pp. 173–185. Springer (2017)
4. Bunte, A., Niagemann, O., Stein, B.: Integrating owl ontologies for smart services into automationml and opc ua. In: 2018 IEEE 23rd International Conference on Emerging Technologies and Factory Automation (ETFA), pp. 1383–1390. IEEE (2018)

5. Steindl, G., Frühwirth, T., Kastner, W.: Ontology-based opc ua data access via custom property functions. In: 2019 24th IEEE International Conference on Emerging Technologies and Factory Automation (ETFA), pp. 95–101. IEEE (2019)
6. Bakakeu, J., Schäfer, F., Franke, J., Baer, S., Klos, H.-H., Peschke, J.: Reasoning over opc ua information models using graph embedding and reinforcement learning. In: 2020 Third International Conference on Artificial Intelligence for Industries (AI4I), pp. 40–47. IEEE (2020)
7. Jean-Baptiste, L.: Ontologies with Python: Programming OWL 2. 0 Ontologies with Python and Owlready2. Apress, Pune (2021)
8. Perzylo, A., Profanter, S., Rickert, M., Knoll, A.: Opc ua nodeset ontologies as a pillar of representing semantic digital twins of manufacturing resources. In: 2019 24th IEEE International Conference on Emerging Technologies and Factory Automation (ETFA), pp. 1085–1092. IEEE (2019)
9. Toyo_Automation: TOYO 2018 Clean Room Type (2018). https://accu-techusa.com/wp-content/uploads/2020/10/Clean-Room-Actuators-20180302.pdf

Exploring Models' Interoperability in Digital Thread and Twin: A Proposal for an Ontology-Driven Approach Based on Intra-And Cross-Organizational Enterprise Integration

Zhengyu Liu[1(✉)], Arkopaul Sarkar[1], Sina Namaki Araghi[1], Bernard Archimede[1], Rebecca Arista[2], and Mohamed Hedi Karray[1]

[1] Université de Technologie Tarbes Occitanie Pyrénées, 65000 Tarbes, France
zhengyu.liu@enit.fr

[2] Airbus SAS, 31700 Blagnac, Midi-Pyrénées, France

Abstract. The increasing adoption and investment in Model-Based Systems Engineering (MBSE) and Digital Thread and Digital Twin (DT&T) demand interoperability among models across the organization. Model interoperability is a critical challenge in the development of DT&T due to the diversity of models and standards used in different phases of product or system design life cycle (PLC/SDLC). Moreover, this challenge is further complicated by the complexity of the differences among various organizations and enterprises. This paper proposes an ontology-driven approach to address the challenge of model interoperability and emphasizes the significance of intra-and cross-organizational Enterprise Integration (EI) and trustworthy decision-making in the development of DT&T. The proposed approach investigates models' interoperability, transformation, and portability and introduces potential research avenues to add explainability throughout the digital model's lifecycle. Furthermore, the paper emphasizes the need for standard evaluation criteria or measurements to evaluate the maturity of EI and to compare different model translation and portability methodologies.

Keywords: Ontology · MBSE · Digital twin · Digital thread · Interoperability

1 Introduction

Enterprise Integration (EI) has long been a crucial aspect of business strategy for large-scale manufacturing industries, enabling improvements in productivity, customer experience, and cost reduction. With the emergence of Industry 4.0 and AI, the industry's digital landscape has undergone significant changes in how data from the product life cycle is managed and computed, necessitating a renewed vision of EI and upgrades to existing practices [1].

M. Dragomir et al. (Eds.): ICPR 2023, LNPE, pp. 293–300, 2025.
https://doi.org/10.1007/978-3-031-92082-0_32

On the one hand, the growing application of Model-Based Systems Engineering (MBSE) in the field of production engineering and operation has led to a significant transformation in the practices of the product or system development life cycle (PLC/SDLC). Within this context, achieving EI through MBSE hinges crucially on the interoperability of models. However, as Digital Thread and Digital Twin (DT&T) become popular, realizing interoperability has also emerged as a significant challenge that requires consideration. One of the main obstacles is the lack of a standardized modeling language and data exchange format that blocks the integration of different models and simulation tools. Another would be the diversity of modeling approaches and techniques employed across various industries and applications, which complicates the attempts to establish common ground [2].

Throughout the phases of SDLC, the lack of interoperability lead to further obstacles such as: (i) the dependency on specific commercial software applications could possibly lead to vendor lock-in; (ii) the difficulty of achieving global-scale integration due to customizations such as SysML stereotypes results in isolated model silos; (iii) the complexity, heterogeneity and dynamic updates of the underlying models constantly prevents decision-makers from having complete visibility and understanding of the model-driven workflows and DT&T. These issues in integration cause decision making to suffer from lack of visibility, lack of continuity, and therefore less trustworthy and make difficult for industries, especially for small medium enterprises to justify the investment in DT&T.

The necessity to address the interoperability problem leads to this ontology-driven approach proposed by this study, which focuses on intra- and cross-organizational EI as well as trustworthy decision-making. Establishing standard evaluation criteria or metrics for assessing EI maturity and comparing various model translation and portability methodologies is also a crucial aspect for future advancement. The proposed approach aims to assist organizations in creating interoperable, portable digital twins and threads that can be seamlessly transformed across different systems. By addressing the challenges of interoperability and standardization, this approach seeks to enhance the efficiency and effectiveness of EI, ultimately promoting better collaboration, reduced complexity, and streamlined development processes across diverse domains.

Therefore, the research goal defined in the scale of this study is to develop a meta-model considering domain semantics for model integration (MIS) based on rigorous ontological analysis to uniformly represent different types of models (system architecture, simulation, and operation) generated by different applications.

2 State of the ART

2.1 Model Interoperability in DT&T

Model interoperability is crucial in DT&T through MBSE. In its report on 'Digital Twin System Interoperability Framework' [3] the digital twin consortium recommends various types of models as candidates for interoperability, e.g., physics-based, information, simulation, connection, and meta-models. This report also makes a crucial recommendation of treating "everything as system" which is incorporated in the underlying assumptions of this project by weaving MBSE practice in its framework.

The incompatibility of tools creates risks of industrial silos, hindering the consistent application of digital twin definitions across industries. Current MBSE methods for system integration in DT&T (e.g., DSL-based [4], file and API-based [5]) are diverse and case-specific, limiting their reusability. Such diversity comes from the difference in the scopes and the purposes of the domain and application [6]. Consequently, these case-specific solutions cannot be easily reused.

This paper proposes to unify past efforts of diverse models and tools by applying a generalized semantic model as the reference language for representing the underlying models into DT&T and a flexible reference architecture to be deployed in a vast number of organizational scenarios.

2.2 Model Interoperability in MBSE

Louis Fussell's vision for MBSE includes the integration of beyond-system models and human behavioral models with domain-specific languages, extensive model reuse, and standard SE services [7]. The modeling languages and MBSE toolchains are the most important in terms of EI. Ma et al. claimed that SysML and UML dominate the MBSE market [8] as SysML and its foundational language UML together have 60% coverage. Despite being able to provide instant interoperability as a common meta-modelling language, SysML lacks scalability to integrate and reuse domain-specific system models and is hard to be integrated with simulation models.

Consequently, this paper materializes Fussell's vision by employing semantically driven model interoperability to mitigate tool incompatibility and language heterogeneity across the SDLC based on common semantics instead of a common modeling standard.

2.3 Model Translation and Transformation for EI

Model translation and transformation are used to solve tool incompatibility in SDLC by generating lower-level models from higher-level abstractions or reusing/integrating models across different contexts. Model Translation is an ideal solution for solving the incompatibility among tools used in SDLC, for which earlier solutions used graph grammar, XSLT [9], and UML. Currently, most of the system modelling applications support XMI as a common format for exchanging models. XMI is part of OMG's MetaObject Facility (MOF) [10], which integrates it into industry-standard Model-driven Architecture (MDA) [11]. Despite the availability of UML, SysML, and Unified Profile for DoDAF and MODAF (UPDM) in XMI format, it is not a complete solution, as it has limited expression ability which may not fully capture the semantics in models [12]; also, it focuses on the model transmission and may not have equal capacity for model transformation or portability across different phases in SDLC [13].

Model transformation is applied in both generating lower-level models from higher-level abstractions. Although the MDA framework has received widespread attention and is still actively developing, it faces several criticisms, including its reliance on numerous incomplete standards, being overly idealistic and rigid [14], as well as uncertainties in value generation. The OMG proposed QVT (Query/View/Transformation) [15] as a standard set of languages for model transformation, with several implementations found, including Eclipse M2M and MagicDraw as plugins. These transformation languages are generic, necessitating intensive development for each specific case.

As to conclude from the state of the art, the existing approaches to enterprise integration, which include DT&T and MBSE, have limitations in interoperability and standardization across domains, thereby impeding the seamless integration of systems and sharing of data. This work aims at developing a holistic approach to integrating systems by developing interoperable models and ontologies that enable cross-domain interoperability, as well as providing generic routines for model translation and transformation, to enable cross-domain interoperability and support forward-looking decision-making initiatives.

3 Methodology

At the core of this methodology lies Semantics for Model Interoperability (MIS) which provides a set of common vocabulary to annotate models and mapping rules for translation and transformation. The MIS is also depicted in Fig. 1 which describes the outline of the method. Based on it, an explainable decision-making strategy is also going to be developed, and be tested, and evaluated together as a complete methodology.

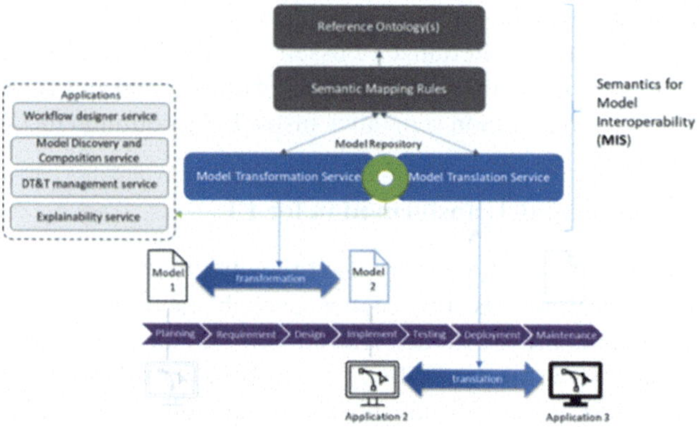

Fig. 1. General Architecture of the methodology

After the literature review of the reference model and their usages, ontologies containing the upper-level semantics will be developed based on MBSE and other industrial standards and modeling languages (e.g., UML, SysML, simulation, and other physics-based models). The objective is to establish meta-model semantics to provide reference to the domain-level semantics. And the core idea is to capture platform-independent semantics in Manufacturing Core Ontology adopting standard modeling language and integrating reference ontologies for System engineering.

In the next stage of the process, presented in Fig. 2, the domain and application-specific ontologies will be developed or integrated to support the translation and transformation of system models across SDLC phases. The term "system models" concerns a wide range of models, such as requirement, architecture, simulation, and quality models, which represent data that are input and output of the models, i.e., product and asset information, simulation results, multi-disciplinary analysis, and optimization trials, process log and sensor-generated data. Some of the existing ontologies [16] are going to be reused as they provide some generic concepts for various domains, however, application-specific concepts still need to be generated from actual data and models.

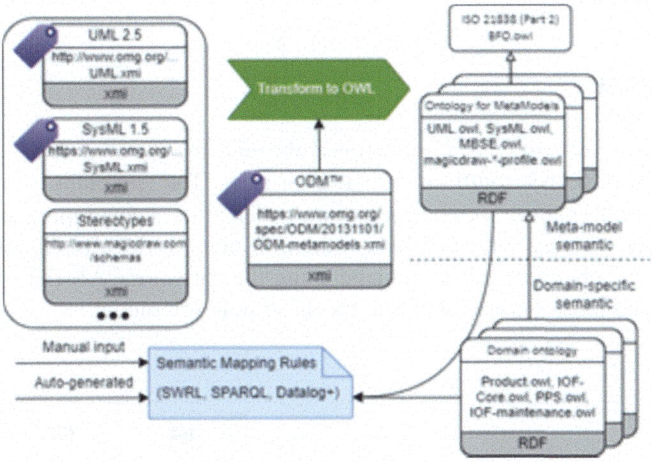

Fig. 2. Domain and application-specific semantic model, using existing ontologies

As illustrated in Fig. 3, the final step in MIS is to devise the semantic rules for mapping concepts belonging to various domains and application-specific ontology. Mapping rules will dictate the translation and transformation of one model to another and will be specific to different translation or transformation scenarios and organized in different sets with close association with the translation and transformation services. The mapping rules will be generated either manually (for generic rules related to meta-models and rules requiring domain-specific knowledge) or automatically (for model-specific concepts) during the translation or transformation process.

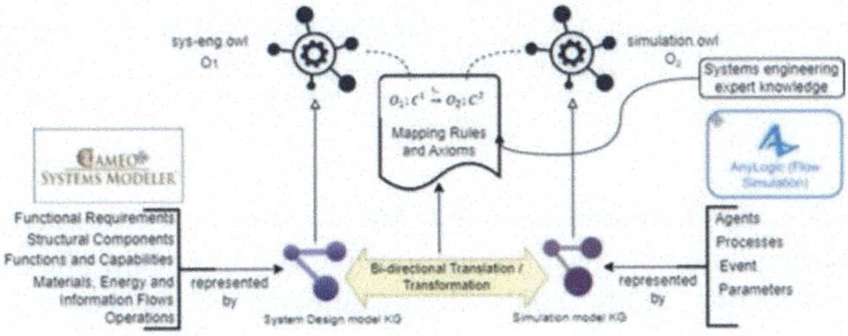

Fig. 3. Details in MIS, Model translation/transformation with semantic mapping rules

Within MIS, the developed multi-layered reference ontology(s) aims to capture highly abstract meta-models to domain-specific model semantics with rich axiomatic definitions and taxonomy. The objective of these ontologies is to guarantee a uniform interpretation of a wide range of models annotated by them. Moreover, these ontologies will establish a common set of terms to write mapping rules that are either domain-dependent or domain-independent, in line with the semantic rule standard. These rules can then be assessed for their consistency against the axiomatic definitions' constraints embedded in the ontologies.

4 Experiment

In this section, a case study is demonstrated through the development of a test plant to present a model translation/transformation process. The test plant is part of project CHAIKMAT in Laboratoire Genie Production, which aims at developing an explainable decision-making strategy for a flexible production line whose general activity flow is shown in Fig. 4. This process is visualized using the software CAMEO for the system design model and the software Anylogic for the simulation model.

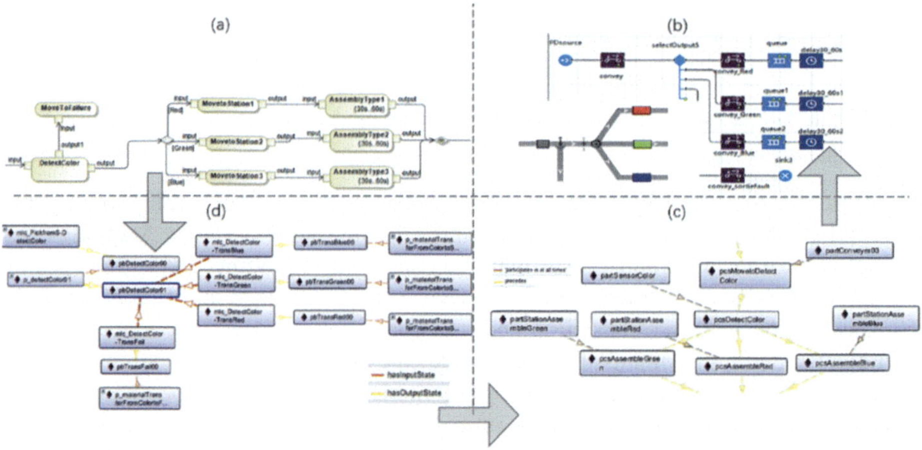

Fig. 4. Development process of a part related to color sorting and assembly of the test production line, following the arrow order: (a) CAMEO design activity diagram; (d) Ontology diagram of design model using Protégé; (c) Ontology diagram for simulation model; (b) Simulation model by Anylogic

The system design is first developed in CAMEO, starting from specifying the function to fulfill the requirements of a production line of three main phases: fabrication, assembly, and packaging. Additionally, the assembly is composed of a color sorting process which allows different assemblies based on colors. The packaging process has also a shape sorting process to square and round products in different packages. The following step is to draw the technical diagram and decomposition diagram to design the composition of such production line which leads to the activity diagram of detailed material flow shown below.

Four upper-level classes are extracted from reference ontology for system design: process, process boundaries, material location change (MLC), and function. The instances in these classes are annotated with the following relations: a process realizes a function and has two corresponding process boundaries as input and output states. MLC describes the material flows from one process to the next. It also has two process boundaries as input and output state, where its input state is the output state of the previous process, and its output state is also the input state of the following process.

The knowledge graph of the corresponding simulation model is automatically generated from the system design thanks to a set of upper-level as wel l as domain-specific semantic mapping rules that align the concepts between system design and simulation ontology. The simulation model assigns required resources and deduces process sequences as well as transfer configurations e.g., process duration from the system design model. Although the resulting knowledge graph will follow the same structure as described, the extraction process needs an application-specific parser. However, we developed a parser for XMI-encoded SysML, that enjoys universal support from a wide variety of modern system design editors, including Cameo.

Based on the knowledge graph generated for simulation, an application-specific model is generated. As Anylogic is chosen for this experiment, the corresponding translation routine encodes a discrete event model generated from the knowledge graph. It is to be noted that these translations are application specific, e.g., another set of translation routine specific to Simio [17] can generate a Simio-compatible model from the same knowledge graph. The Simio parameters will be mapped and appended to the Anylogic model to configure the simulation. However, the runtime configuration of the simulation model will be separate from the translation module and thus provide flexibility in running the same model multiple times with different parameters for experimentation purposes.

5 Conclusion and Future Work

The paper introduces an ontology-driven approach in the context of a research project to tackle the challenge of model interoperability and emphasizes the importance of intra-and cross-organizational EI and trustworthy decision-making in the development of DT&T. The proposed approach investigates models' interoperability, transformation, and portability and introduces explainability techniques throughout the digital model's lifecycle. The paper highlights the need for standard evaluation criteria or measurements to evaluate the maturity of EI and to compare different model translation and portability methodologies. The proposed approach can assist organizations to develop interoperable DT&T that can be integrated across different systems and make trustworthy decisions based on digital models. This paper stays at a conceptual level, and it illustrates the method at its primary phases while the team is aiming at improving the results in the upcoming months.

In the future, the team plans to establish an explainable decision-making strategy to support enterprise implementation. Support services for making business-critical decisions by leveraging explainable AI aiming to broaden the application of XAI will be developed by using semantic reasoning on the model and data knowledge graph to derive various types of explanations. These explanations will be presented to users through reporting services in multiple formats. Additionally, the team will explore process diagnosis, improvement, and optimization, as well as conformance checking and quality assurance, using evidence-based and interview-based approaches.

References

1. Archimède, B., Ducq, Y., Young, B., Karray, H. (eds.) Enterprise Interoperability IX: Interoperability in the Era of Artificial Intelligence, vol. 10. In: Proceedings of the I-ESA Conferences, vol. 10. Springer, Cham (2023). https://doi.org/10.1007/978-3-030-90387-9
2. Frechette, S.P.: Model based enterprise for manufacturing. In: 44th CIRP International Conference on Manufacturing Systems, p. 6. Omnipress Madison, USA, 2011
3. Budiardjo, A., Migliori, D.: Digital-Twin-System-Interoperability-Framework-12072021.pdf, Digital Twin Consortium. https://www.digitaltwinconsortium.org/pdf/Digital-Twin-System-Interoperability-Framework-12072021.pdf. Accessed 15 June 2023
4. Chaudhary, H.A.A., Margaria, T.: DSL-based Interoperability and Integration in the Smart Manufacturing Digital Thread, vol. 81, 2022
5. Platenius-Mohr, M., Malakuti, S., Grüner, S., Schmitt, J., Goldschmidt, T.: File- and API-based interoperability of digital twins by model transformation: an IIoT case study using asset administration shell. Futur. Gener. Comput. Syst. **113**, 94–105 (2020). https://doi.org/10.1016/j.future.2020.07.004
6. Glatt, M., Sinnwell, C., Yi, L., Donohoe, S., Ravani, B., Aurich, J.C.: Modeling and implementation of a digital twin of material flows based on physics simulation. J. Manuf. Syst. **58**, 231–245 (2021). https://doi.org/10.1016/j.jmsy.2020.04.015
7. Fussell, L.: Integration of Model-Based Systems Engineering and Programmatic Analysis Tools, presented at the INCOSE IS 2007, San Diego, CA USA, 2007. https://www.nasa.gov/sites/default/files/atoms/files/18_mbse_and_pa.pdf. Accessed 23 Apr 2023
8. Ma, J., Wang, G., Lu, J., Vangheluwe, H., Kiritsis, D., Yan, Y.: Systematic literature review of MBSE tool-chains. Appl. Sci. **12**(7), 3431 (2022). https://doi.org/10.3390/app12073431
9. Peltier, M., Ziserman, F., Bézivin, J.: On levels of model transformation. In : XML Europe, pp. 1–17 (2000)
10. "MetaObject Facility | Object Management Group," OMG. https://www.omg.org/mof/
11. "Model Driven Architecture (MDA) | Object Management Group." https://www.omg.org/mda/. Accessed 23 Apr 2023
12. Jouault, F., Kurtev, I.: Transforming models with ATL. In: Bruel, J.M. (eds.) Satellite Events at the MoDELS 2005 Conference. MODELS 2005. LNCS, vol. 3844. Springer, Berlin, Heidelberg (2006). https://doi.org/10.1007/11663430_14
13. Czarnecki, K., Helsen, S.: Feature-based survey of model transformation approaches. IBM Syst. J. **45**(3), 621–645 (2006). https://doi.org/10.1147/sj.453.0621
14. "UML - Unified or Universal Modeling Language? https://www.jot.fm/issues/issue_2003_01/column1/. Accessed 24 Apr 2023
15. "About the MOF Query/View/Transformation Specification Version 1.3." https://www.omg.org/spec/QVT/. Accessed 24 Apr 2023
16. Karray, M., et al.: The industrial ontologies foundry (IOF) perspectives. In: Proceedings: Industrial Ontology Foundry (IOF) Achieving Data Interoperability Workshop, International Conference on Interoperability for Enterprise Systems and Applications, Tarbes, France, 17–24 March 2020 (2021)
17. Simulation, Production Planning and Scheduling Software | Simio. https://www.simio.com/. Accessed 24 Apr 2023

Implementation of Value Stream Mapping and Mathematical Model for Patient Scheduling in Radiotherapy Treatment

Wapee Manopiniwes[1](✉), Takashi Irohara[2], Wannapha Nobnop[3], and Pooriwat Muangwong[3]

[1] Department of Industrial Engineering, Faculty of Engineering, Chiang Mai University, Chiang Mai, Thailand
wapee.m@cmu.ac.th
[2] Faculty of Science and Technology, Sophia University, Tokyo, Japan
[3] Department of Radiology, Faculty of Medicine, Chiang Mai University, Chiang Mai, Thailand

Abstract. This paper is based on the improvement of patient scheduling in the radiotherapy department. We started with the application of Value Stream Mapping (VSM) to the entire patient flow in curative cases. The scheduling plan in radiotherapy treatment is necessary for curative cases, whereas in the palliative cases, it may not be required to the same extent. The VSM analysis of the patient scheduling process revealed that there are several operations in the process that are causing delays. The patients are waiting too long to get into several treatments called fractions. In this context, VSM can be used to map out the operations involved in the patient scheduling process as well as identify any bottlenecks or inefficiencies. As a solution, we optimized the scheduling plan through linear programming, thereby synchronizing simulation, contouring, dose calculation, and plan approve operations of radiotherapy care. The goal of our formulation was to find the optimal patient schedule that minimizes the maximum completion time across all patients, while also factoring in a penalty score.

Keywords: Patient Scheduling · Value Stream mapping · Radiotherapy Treatment · Healthcare Engineering

1 Introduction

Radiotherapy, which involves the use of high-energy radiation to destroy cancer cells, is a critical treatment modality for cancer patients. The success of radiotherapy treatment depends not only on the accuracy of radiation delivery but also on the appropriate scheduling and coordination of patient appointments [1]. Efficient patient scheduling minimizes idle time for radiotherapy machines and medical personnel while ensuring that patients receive treatment in a timely manner. However, the scheduling of radiotherapy patients is a complex task due to various factors, including the availability of treatment machines, the specific requirements of each patient, and the unpredictable nature of treatment sessions [2]. Traditional scheduling methods often rely on manual

M. Dragomir et al. (Eds.): ICPR 2023, LNPE, pp. 301–309, 2025.
https://doi.org/10.1007/978-3-031-92082-0_33

decision-making, which can be time-consuming and prone to errors. Consequently, there is a need for innovative approaches that optimize the scheduling process in radiotherapy.

Value Stream Mapping is a lean management tool widely used in manufacturing and service industries to analyze and improve process efficiency. It provides a visual representation of the entire process flow, highlighting areas of waste and inefficiency. By identifying and eliminating non-value-added activities, VSM allows organizations to streamline processes, reduce lead times, and improve overall performance. Adapting the principles of VSM to the healthcare context offers significant potential for optimizing patient scheduling in radiotherapy treatment [3, 4]. This information serves as a foundation for the development of an optimized scheduling model. By formulating the patient scheduling problem as a mathematical model, it becomes possible to consider multiple constraints and objectives simultaneously, such as the availability of treatment machines, patient-specific requirements, and staff workload [5]. Mathematical models enable the exploration of various scheduling scenarios, allowing decision-makers to make informed choices based on objective criteria [6].

The primary objective of this research is to propose and implement an innovative approach to patient scheduling in radiotherapy treatment using the combined methodologies of Value Stream Mapping and mathematical modeling. The study aims to:

- Analyze the existing patient scheduling process in radiotherapy treatment, identifying bottlenecks, inefficiencies, and areas for improvement.
- Develop a mathematical model that integrates multiple constraints and objectives to optimize patient scheduling.
- Implement the proposed scheduling model in a real-world radiotherapy treatment setting and evaluate its effectiveness in terms of patient waiting times, resource utilization, and overall process efficiency.
- Provide insights and recommendations for healthcare organizations seeking to improve patient scheduling processes in radiotherapy treatment.

By integrating the principles of Value Stream Mapping and mathematical modeling, this research offers a novel approach to patient scheduling in radiotherapy treatment. The outcomes of this study have the potential to significantly enhance patient care, improve resource utilization, and contribute to the ongoing optimization of healthcare processes.

2 Literature Review

Efficient patient scheduling plays a crucial role in the delivery of high-quality healthcare services, particularly in radiotherapy treatment, where timely treatment sessions are critical for cancer patients. This section presents a review of relevant literature on patient scheduling in healthcare, the application of Value Stream Mapping in healthcare process improvement, and the utilization of mathematical modelling for patient scheduling optimization.

2.1 Patient Scheduling in Healthcare

Patient scheduling has been extensively studied across various healthcare domains, including hospitals, clinics, and specialized treatment centers. Researchers have highlighted the importance of efficient scheduling in reducing patient waiting times, enhancing resource utilization, and improving overall patient satisfaction. In the context of radiotherapy treatment, several studies have investigated patient scheduling challenges and proposed different approaches. For example, [7] conducted a study on patient scheduling in radiation therapy centers and identified the variability of treatment session durations as a significant factor affecting scheduling efficiency. They emphasized the need for adaptive scheduling strategies to accommodate unpredictable treatment times. The article [8] introduced an innovative framework for patient scheduling in healthcare. By integrating Integer Programming, Mixed Integer Programming, and Constraint Programming, the study optimized radiotherapy schedules to reduce patient waiting times, improve appointment consistency, and better allocate healthcare resources. This research highlighted the critical role of advanced scheduling strategies in addressing challenges in healthcare systems.

2.2 VSM in Healthcare Process Improvement

Lean production principles have been increasingly applied in healthcare to enhance efficiency and reduce waste. Value Stream Mapping, originally developed within the context of lean manufacturing, has gained attention as a valuable tool for process improvement in healthcare settings. VSM enables the identification and elimination of non-value-added activities, streamlining processes, and reducing waste. In the healthcare sector, VSM has been applied to various processes, including patient flow, medication administration, and surgical procedures. For instance, [4] utilized VSM to analyze the patient discharge process in a hospital, identifying bottlenecks and areas for improvement. Their study demonstrated that VSM helped streamline the discharge process, reduce delays, and enhance patient satisfaction. While the application of VSM in radiotherapy treatment scheduling is relatively limited, some studies have explored its potential benefits. [9] utilized VSM to analyze the scheduling process in a radiotherapy center and identified several areas of waste and inefficiency. By implementing VSM-based improvements, they achieved a significant reduction in patient waiting times and improved resource utilization. [9] demonstrated the use of simulation-based value stream mapping to improve patient flow in the physical therapy and rehabilitation department of a public hospital, highlighting significant reductions in process inefficiencies.

2.3 Mathematical Modeling for Patient Scheduling

Mathematical modeling has been widely employed to optimize patient scheduling in healthcare settings. By formulating the scheduling problem as a mathematical model, various constraints and objectives can be considered simultaneously, leading to optimal scheduling solutions. In radiotherapy treatment, researchers have developed mathematical models to address different aspects of patient scheduling. For example, [10] proposed a mixed-integer linear programming model to optimize the allocation of radiotherapy machines to patient groups. Their model aimed to maximize machine utilization while satisfying patient-specific constraints. Furthermore, [11] proposed an innovative approach that focuses on improving the treatment planning process. By simulating different scenarios and evaluating various strategies, the model aims to identify bottlenecks and inefficiencies in the treatment workflow. Recently, a multi-criteria decision model for radiotherapy treatment planning was proposed by [12]. The authors aim to find a high-quality Pareto-optimal treatment plan from the mathematical model and present these alternatives to the radiotherapy oncologist, who will choose a single one.

To the best of our knowledge, there are no study has explored the integration of VSM and mathematical modeling specifically for patient scheduling in radiotherapy treatment. This research aims to bridge this gap by proposing a novel approach that leverages the strengths of both methodologies to enhance patient care, improve resource utilization, and streamline the scheduling process.

3 VSM of Radiotherapy Treatment

While VSM is commonly associated with manufacturing, it can also be applied to service industries, including healthcare such as radiotherapy treatment. In this study, our focus was specifically on curative care to address the need for improvement in patient scheduling as palliative care does not require scheduling in the same manner. Curative radiotherapy aims to eradicate the tumor or achieve long-term disease control, while palliative radiotherapy only focuses on alleviating symptoms and improving the patient's quality of life. We used VSM to analyze processes of curative radiotherapy by visualizing processes by visualizing the flow of materials, information, and activities as shown in Fig. 1. Before the delivery of treatment, there are four steps in planning: simulation, contouring, dose calculation, and plan approve which are also considered in our mathematical formulation in this study.

(1) Simulation: In this step, imaging techniques such as computed tomography (CT) or magnetic resonance imaging (MRI) are used to visualize the tumor and surrounding anatomical structures. These images help in accurately defining the target volume and critical organs at risk. Radiation oncologists, with the assistance of medical physicists and radiologists, contour the tumor and relevant structures on the images.
(2) Contouring: During contouring, radiation oncologists or medical dosimetrists manually identify and draw the contours of the tumor and nearby organs or structures that need to be spared from radiation. This involves identifying the boundaries and shapes of the target tumor volume and critical structures, such as organs at risk (OARs), including the spinal cord, lungs, heart, and other surrounding healthy tissues.
(3) Dose Calculation: After the treatment plan is designed, the next step involves calculating the actual dose distribution throughout the patient's anatomy. Complex algorithms and computer systems are employed to accurately calculate the radiation dose delivered to the target volume and surrounding tissues. The dose calculation considers the patient's anatomy, treatment beam characteristics, and tissue response models.

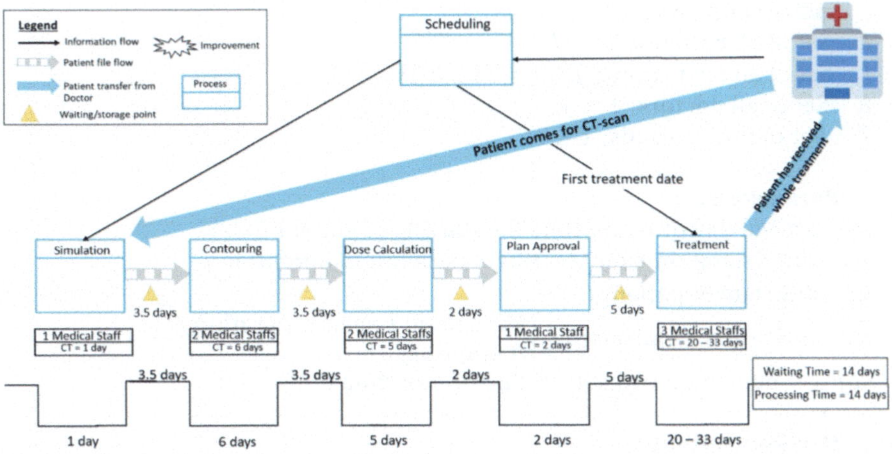

Fig. 1. Value Stream Mapping for Curative Radiotherapy Treatment Process

(4) Plan Approval: Once the dose calculation is complete, the treatment plan undergoes a thorough evaluation and quality assurance process. The plan is reviewed to ensure that the dose distribution meets the treatment goals and constraints set by the radiation oncologist. Any necessary adjustments or refinements are made to ensure the plan meets the required standards before treatment delivery.

From VSM of the current planning, the lead time of 28 days refers to the duration between the initial referral or consultation of a patient and the start of their radiotherapy treatment. The 28-day lead time allows healthcare providers to carefully plan and individualize treatment for each patient based on their specific needs. It also provides sufficient time for coordination between different departments and professionals involved in the radiotherapy process. However, reducing the lead time is often a goal in radiotherapy departments to ensure prompt treatment initiation and improve overall patient satisfaction. In certain types of cancer, particularly aggressive ones, may require immediate intervention. Initiating treatment earlier can help prevent further tumour growth and improve patient outcomes. Shortening lead time also allows for better utilization of equipment, staff, and facilities, as treatment slots can be utilized more effectively. This can lead to increased patient throughput and reduced waiting times for other patients.

4 Mathematical Model

This section provides an overview of the mathematical formulation related to patient scheduling in the context of radiotherapy.

Indices and Sets:

I Set of all patients; $i \in I$
J Set of all operations; $j \in J = \{1,2,3,4\}$
K Set of all doctors; $k \in K$
T Set of time periods; $t \in T$

Parameters:

a_i arrival day or ready time for starting of patient i
p_{ij} processing time for treatment patient i in operation j
d_i due date of patient i
r_{ij} operation requirement $\begin{cases} 1 \text{ if patient } i \text{ needs operation } j \\ 0 \text{ otherwise} \end{cases}$
cap_{kt} maximum capacity of doctor k in time t

Decision variables:

Z_{ijkt} $\begin{cases} 1 \text{ if patient } i \text{ is assigned to operation } j \text{ by doctor } k \text{ in time period } t \\ 0 \text{ otherwise} \end{cases}$
X_{ij} treatment completion time of patient i in operation j
Y_{ijk} treatment completion time of patient i in operation j on doctor k
C^{max} maximum completion time of all patients (makespan)

Model Formulation:

$$\text{Minimize} \quad C^{max} \tag{1}$$

Subject to

$$C^{max} \geq X_{i4} \quad ; \forall i \tag{2}$$

$$\sum_{i \in I} \sum_{j \in J} Z_{ijkt} \leq cap_{kt} \quad ; \forall k, t \tag{3}$$

$$\sum_{t \in T} \sum_{k \in K} Z_{ijkt} = r_{ij} \quad ; \forall i, j \tag{4}$$

$$Y_{ijk} - Y_{i,j-1,k} \geq \sum_{k \in K} p_{ij} Z_{ijkt} \quad ; \forall i, j, k, t; j \neq 1 \tag{5}$$

$$Y_{ijk} \geq a_i + p_{ij} \quad ; \forall i, j, k \tag{6}$$

$$Y_{i4k} \leq d_i \quad ; \forall i, k \tag{7}$$

$$X_{ij} \geq Y_{ijk} \quad ; \forall i, j, k \tag{8}$$

$$r_{ij} X_{ij} = \sum_{t \in T} \sum_{k \in K} p_{ij} Z_{ijkt} \quad ; \forall i, j \tag{9}$$

$$Z_{ijkt} \in (0, 1) \quad ; \forall i, j, k, t \tag{10}$$

$$X_{ij} \; and \; Y_{ijk} \in \text{Integer} \quad ; \forall i, j, k \tag{11}$$

Objective function (1) is to minimize the makespan of patient flow. Constraint (2) specifies the makespan (C^{max}) which is a maximum completion time of all patients. Constraint (3) constraint of doctor that can treat operation in a period less than or equal to capacity. Constraint (4) ensures that a patient must be treated for each operation by only one doctor for one period. Constraint (5) is a precedence constraint to ensure that the completion time of any operation must be greater than or equal to its ready time plus its processing time. Constraint (6) states that the completion time of a patient who treats which operation at which doctor must be equal to or larger than the patient's ready time plus processing time of that operation. Constraint (7) specifies the last operation cannot exceed the due date for each patient. Constraint (8) explains the completion time of a patient undergoing a specific operation and being treated by a particular doctor. Constraint (9) considers the combination of time period and doctor to determine the total processing time. Constraint (10) specifies that decision variables are binary. Constraint (11) specifies that decision variables are integers.

5 Results Discussion and Conclusion

In this study, a numerical example is given to illustrate the solution methods of the proposed radiotherapy planning and scheduling model. The nature of this problem can be likened to a flow shop scheduling problem, where each patient must follow a predetermined sequence, and the doctors remain fixed at each operation.

To make the problem practical, computational experiments are executed using simulated data according to the real planning process of curative care patients. Three main parameters the number of patients, list of operations, and processing time are generated along with database. Other parameters, such as doctor capacity and operation requirement are supposed with fix number. In the actual problem, each schedule is on a weekly basis over a planning horizon but in the experiment, it was about a month of preparation.

Table 1. Experimental Results

Case	Description	Patient Flow (days)	Waiting Time (days)	Tardiness (days)	C^{max} (days)
Reality	6D (OT allowed)	28.00	14.00	5	45
1	6D	22.83	9.42	10	44
2	6D - shared	21.25	7.83	5	40
3	7D	19.83	6.42	0	38
4	7D - shared	19.17	5.75	0	37

To clarify how modifying the size of an instance can influence, researchers have generated an example of a radiotherapy planning and scheduling problem with twelve patients. The patient arrival pattern in the system is characterized by one patient per day. Table 1 presents the experimental results of four cases, comparing them with the reality case that includes overtime (OT) allowance of doctors. Six doctors are used for model analysis in which each doctor's capacity is set to one patient. The addition of one more doctor in operation of contouring is also proposed in case 3 and 4. We also explore the potential benefits of resource sharing in case 2 and 4.

In addition, unspecified parameters are both random and fixed. Example parameters are shown in Table 1. The patient scheduling mathematical model is solved by the LINGO program version 14.0 by exact method. Patient flow refers to the average total lead time that patients must wait until the fraction treatment starts. It shows the same tendency with patient's waiting time. In the context of this specific problem, tardiness is considered when there is a delay in meeting schedule appointments beyond the designated 28-day timeframe.

The result shows better patient's lead time, waiting time, and makespan when comparing with the reality. It appears that some patients are unable to commerce their treatment within 28 days with six medical staffs (6D) while all patients can meet their appointment and start their treatment accordingly when an addition of medical staff (7D) is permitted. However, the department aims to save costs by implementing resource sharing instead of relying on OT. Comparing case 3 and the reality with the same number of medical staffs, the resource sharing strategy shows a better performance in terms of patient flow, waiting time, and makespan. This approach can also help prevent burnout and fatigue among healthcare professionals, leading to better job satisfaction and improved quality of care.

In conclusion, the implementation of Value Stream Mapping (VSM) and a mathematical model for patient scheduling in radiotherapy treatment offers significant benefits. VSM enables the identification and elimination of process inefficiencies, while the mathematical model optimizes patient scheduling for improved resource utilization and reduced waiting times. This research provides valuable insights and tools for enhancing the efficiency and quality of radiotherapy treatment through effective patient scheduling. Future research in the field of patient scheduling in radiotherapy treatment can explore several promising directions. Conducting research on multi-objective optimization techniques can help strike a balance between minimizing patient waiting times, optimizing resource utilization, and improving treatment plan quality. This can involve exploring trade-offs and developing algorithms that consider multiple objectives simultaneously. Furthermore, evaluating the impact of patient urgencies and personalized treatment plans on scheduling algorithms can lead to more patient-centric scheduling approaches. Finally, examining the influence of organizational factors, such as staffing levels and resource allocation, on patient scheduling can provide insights into optimizing operational efficiency in radiotherapy departments.

Acknowledgment. Sophia Lecturing-Research Grant by the Science and Technology Exchange Committee (STEC), Sophia University, Tokyo, Japan.

References

1. Frimodig, S., Enqvist, P., Carlsson, M., et al.: Comparing optimization methods for radiation therapy patient scheduling using different objectives. Oper. Res. Forum **4**, 83 (2023). https://doi.org/10.1007/s43069-023-00251-2
2. Vieira, B., Demirtas, D., van de Kamer, J.B., et al.: Radiotherapy treatment scheduling considering time window preferences. Health Care Manag. Sci. **23**, 520–534 (2020). https://doi.org/10.1007/s10729-020-09510-8
3. Marin-Garcia, J.A., Vidal-Carreras, P.I., Garcia-Sabater, J.J.: The role of value stream mapping in healthcare services: a scoping review. Int. J. Environ. Res. Public Health **18**(3), 951 (2021). https://doi.org/10.3390/ijerph18030951
4. Vidal-Carreras, P.I., Garcia-Sabater, J.J., Marin-Garcia, J.A.: Applying value stream mapping to improve the delivery of patient care in the oncology day hospital. Int. J. Environ. Res. Public Health **19**(7), 4265 (2022). https://doi.org/10.3390/ijerph19074265
5. Saure, A., Patrick, J., Tyldesley, S., Puterman, M.L.: Dynamic multi-appointment patient scheduling for radiation therapy. Eur. J. Oper. Res. **223**(2), 573–584 (2012). https://doi.org/10.1016/j.ejor.2012.06.046
6. Conforti, D., Guerriero, F., Guido, R.: Optimization models for radiotherapy patient scheduling. 4OR **6**, 263–278 (2008). https://doi.org/10.1007/s10288-007-0050-8
7. Gupta, T., Kannan, S., Ghosh-Laskar, S., Agarwal, J.P.: Systematic review and meta-analyses of intensity-modulated radiation therapy versus conventional two-dimensional and/or three-dimensional radiotherapy in curative-intent management of head and neck squamous cell carcinoma. PLoS ONE **13**(7), e0200137 (2018). https://doi.org/10.1371/journal.pone.0200137
8. Pham, T.S., Rousseau, L.M., De Causmaecker, P.: A two-phase approach for the Radiotherapy Scheduling Problem. Health Care Manag. Sci. **25**, 191–207 (2022). https://doi.org/10.1007/s10729-021-09579-9
9. Doğan, N.Ö., Unutulmaz, O.: Lean production in healthcare: a simulation-based value stream mapping in the physical therapy and rehabilitation department of a public hospital. Total Qual. Manag. Bus. Excell. **27**(1–2), 64–80 (2016). https://doi.org/10.1080/14783363.2014.945312
10. Vieira, B., Demirtas, D., van de Kamer, J.B., Hans, E.W., Jongste, W., van Harten, W.: Radiotherapy treatment scheduling: implementing operations research into clinical practice. PLoS ONE **16**(2), e0247428 (2021). https://doi.org/10.1371/journal.pone.0247428
11. Babashov, V., et al.: Reducing patient waiting times for radiation therapy and improving the treatment planning process: a discrete-event simulation model. Clin. Oncol. **29**(6), 385–391 (2017). https://doi.org/10.1016/j.clon.2017.02.001
12. Dias, L.C., et al.: Learning target-based preferences through additive models: an application in radiotherapy treatment planning. Eur. J. Oper. Res. **302**(1), 270–279 (2022). https://doi.org/10.1016/j.ejor.2021.12.019

Decision-Making Model for Packing Substandard Agricultural Products in D2C

Kotomichi Matsuno(✉), Takahiro Ohno, and Yoshikuni Edagawa

Department of Business Design and Management, Waseda University, 3-4-1 Okubo, Shinjuku-ku, Tokyo, Japan
matsuno@ris.ac.jp

Abstract. In recent years, Japanese farmers are gradually increasing their use of direct sales channels to deliver agricultural products directly to consumers to increase their profits. In addition, with the widespread use of the Internet, e-commerce is being put into practice in agriculture. However, the share of direct sales in the total volume of agricultural products shipped is still very small. At present, most of the farmers send their harvested agricultural products to Japan Agricultural Cooperative in order to focus on cultivation activities and ensure stable income. On the other hand, many substandard agricultural products are returned to the farmers due to the strict purchasing standards set by Japan Agricultural Cooperative, especially for fruits such as peaches, the color, sweetness, shape, etc. are strictly controlled by sorting machines and visual inspection. As a result, problems such as decreased income for agricultural producers and increased food loss due to waste disposal are on the rise. Selling substandard agricultural products directly to consumers through D2C platforms is considered a good approach to solve such problems. However, the variation in the condition of substandard products is quite large. Packing based on individual experience results in a significant difference in product value per package, which can lead to customer dissatisfaction. In this study, we focus on the direct sale of substandard peaches, assuming that attributes such as color and fruit spots are quantified by new sorting machines. To improve customer satisfaction, a decision-making model for packing peaches that maximizes the value of each package while minimizing the difference in product value between packages is proposed. The effectiveness of the proposed algorithm is confirmed under different distribution of quantity and quality.

Keywords: Substandard Agricultural Products · Direct-to-Consumer · Optimal Decision-Making

1 Introduction

Agricultural producers in Japan face challenges such as labour shortages due to an ageing population, and lacking replacements. These problems make them difficult to secure stable income [1]. In addition, the producers are facing low purchase prices and strict criteria for the purchase of agricultural products. As a result, many substandard items are returned to the producers. In particular, expensive fruit such as peaches are

M. Dragomir et al. (Eds.): ICPR 2023, LNPE, pp. 310–318, 2025.
https://doi.org/10.1007/978-3-031-92082-0_34

rigorously inspected for colour, sugar content, shape and other factors using automated sorting machines as well as visual inspection. Therefore, a large number of peaches that do not meet the criteria and cannot be purchased by agricultural cooperatives are identified and returned. Some of the returned peaches are delivered to food processing factories who producing cut fruits or juice. However, the purchase prices offered by these factories are significantly lower than that for fresh consumption. One possible solution to address this problem is to sell the substandard peaches directly to consumers through direct sales channels such as farmers' markets and e-commerce platforms [2]. However, there are no clear rules for the packing process, so it is common for individuals to rely on their own experience when packing. Since substandard products can vary widely in condition, the value of each package can vary widely. This often leads to customer dissatisfaction even if the price is about half that of standard products.

Multi-period packing operation in direct sales of substandard peaches is discussed in this study. Group classification by considering customer evaluations of peach appearance, such as color and fruit spot is performed, and a decision model for packing and allocation based on the objective function of maximizing the overall improvement in product value and minimizing the differences in product value between packages is proposed.

2 Literature Review

Ang et al. discussed the sea cargo mix problem in international ocean container shipping industry and designed characteristics of the cargo mix problem for the carrier in a multi-period planning horizon and formulated it as a multi-dimensional multiple knapsack problem (MDMKP) [3]. Lau et al. modelled Availabe-to-Promise problem in logistics scheduling as the multi-period multi-dimensional knapsack problem. Based on insights obtained from properties for a special case of a single-dimensional problem, a two-phase heuristic for solving the multi-dimensional problem was proposed [4]. Since customers' evaluation of the appearance of agricultural products varies widely, there are only a few works that address the multi-period multiple problem in the agricultural products. Kurano et al. discussed a food packing problem in an automated food packing system. The problem is to select some peppers whose weights have been measured and to bring the total weight as close as possible to the target weight per bag, but not less than the target weight per bag. This problem was modelled as a knapsack problem and solved using dynamic programming [5].

In this study, a multi-period, multi-objective, multiple knapsack problem of selling substandard peaches that considers not only the weight but also the consumers' evaluation of the appearance attributes is defined. A greedy algorithm that is close to the current working is also proposed to solve and analyze the problem.

3 Classification of Substandard Peaches Based on Appearance Attributes

Based on the results of the interview with Company T, which grows and sells peaches, the following attributes of peach appearance considered in this study are listed in Table 1.

For standard grade peaches, there is no significant variation in appearance as they meet the purchasing criteria of the agricultural cooperative. On the other hand, for non-standard grade peaches, there is a greater difference in appearance, resulting in a greater variation in value. In this study, based on the four attributes shown in Table 1, each attribute is divided into three levels. The L9 orthogonal array is used to perform grouping based on the combinations of attributes. It is possible to divide all substandard peaches into 9 groups with different 9 value levels. The decision making of packing is based on the 9 value levels.

Table 1. Appearance attributes of peach

Attributes	Description
Degree of Coloring	Percentage of red color from top to the total surface
Fruit Spot	White spots on the fruit surface (More white spots, higher sugar content)
Stem Color	Green or yellow green means unripe. Cream-colored means ripe
Shape Balance	Symmetrical, nicely rounded one has less flavor variation

Although sorting machines that can determine factors such as color ratio and shape are already used in practical applications, this study assumes the introduction of a sorting machine that can accurately classify and sort peaches into 9 groups based on above 4 attributes.

4 Decision-Making Model for Packing Substandard Peaches

In this study, we define the packing and selling of substandard peaches as a multi-period, multi-objective, multiple knapsack allocation problem. We propose a decision-making model of packing using a greedy algorithm that closely resembles the working logic based on the operator's intuition.

4.1 Objective Function and Constraints of the Decision Model

The first step is to define the variables and related formulations for designing the algorithm. The following Eq. (1) shows the value of peach j belonging to group k. The value of nap sack i allocated to each customer in period t is shown by Eq. (2), and Eq. (3) indicates whether or not a peach from appearance group k is in knapsack i in period t.

$$v_{kj} = w_j \cdot s_k \tag{1}$$

$$V_{ti} = \sum_{k,j \in N_t} v_{kj} x_{tij}, x_{tij} \in \{0, 1\}, \forall t, i, j \tag{2}$$

$$y_{tik} = \begin{cases} 0(g_{tik} = 0) \\ 1(g_{tik} \geq 1) \end{cases} y_{tik} \in \{0, 1\}, \forall t, i, k \tag{3}$$

s_k: The reference value of peaches belonging to group k and the average reference value of all groups are set as half of the market price for standard-grade peaches.

w_j: The size (weight) of peach j in group k determines its value, the larger peaches having higher value.

g_{tik}: The number of peaches of appearance group k contained in the knapsack i of period t.

The objective of the model is to maximize the overall value of all packs as well as minimize the difference in value among all packs. The constraints of this model include capacity restrictions for each pack and limitations on the number of varieties. The following Eqs. (4)–(7) show the objective function and the constraints:

Objective Function:

$$\text{Maximize} \sum_{t=1}^{T} \sum_{c=1}^{C} \sum_{k,j \in N_t} v_{kj} x_{tij} \tag{4}$$

$$\text{Minimize } \max(f_{1t}f_{2t}, \ldots, f_{Ct}) - \min(f_{1t}, f_{2t}, \ldots, f_{Ct}) \tag{5}$$

Subject to:

$$\alpha \leq \sum_{k \in N_t} y_{tik} \forall t, i \tag{6}$$

$$ca_{min} \leq \sum_{j \in N_t} w_j x_{tij} \leq ca_{max} \forall t, i \tag{7}$$

N_t: The population of substandard peaches in period t.

α: The lower limit on the number of groups that can be included in a pack.

C: Total number of customers in period t.

ca_{min}: Minimum pack capacity.

ca_{max}: Maximum pack capacity.

4.2 Design of Algorithms Based on the Greedy Method

The knapsack and allocation problem addressed in this study can be decomposed into two processes: the pack filling process and the pack allocation process to customers. Figure 1 shows the pack filling algorithm and Fig. 2 shows the pack allocation. Both are based on the greedy method.

Number of packs for period t is determined based on Eq. (8), Eq. (9) shows the number of packs for period t if several customers have already received T packs.

$$np = \begin{cases} C & (n_t \geq ca_{max} C) \\ \frac{n_t}{c_{max}} & (n_t < ca_{max} C) \end{cases} \tag{8}$$

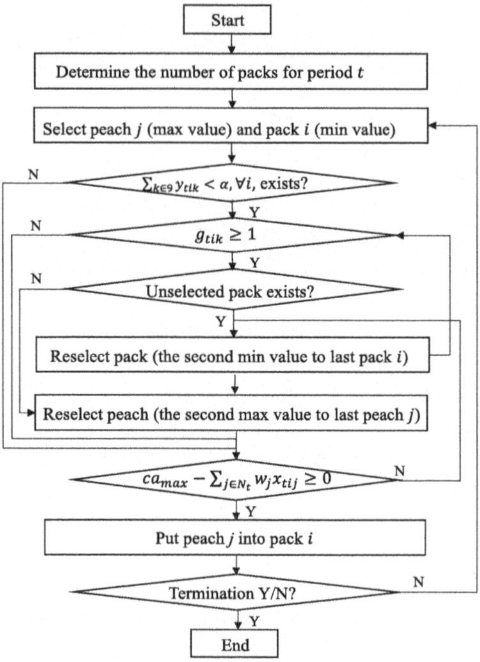

Fig. 1. Algorithm for pack filling operations

$$np = \min\left(\frac{n_t}{c_{min}}, q\right)(9) \tag{9}$$

np: Number of packs

n_t: Total number of substandard peaches for period t

q: Number of customers who can receive packs

Rules for determining whether to continue the packing operation are set as below, if either is satisfied, the packing operation will be completed.

1. All n peaches have already put into packs.
2. Capacity limits of all packs have been reached. $ca_{max} - \sum_{j \in N_t} w_j x_{tij} \le w_j, \forall i, j$

The detail of calculation of Step 2 and Step 3 in the pack allocation process is shown as below,

Step 2: Calculation of average value of packs allocated by period $t' - 1$ ($r \notin C$)

The average value of the total value of knapsacks allocated to customers not included in set C up to period $t' - 1$ can be calculated by Eq. (10).

$$ave.value = \sum_{r \notin C} \sum_{t \in t' - 1} V_{ti}/(C - l) \tag{10}$$

C: Total number of customers in period.t

l: Total number of customers who belonging to set C in period t

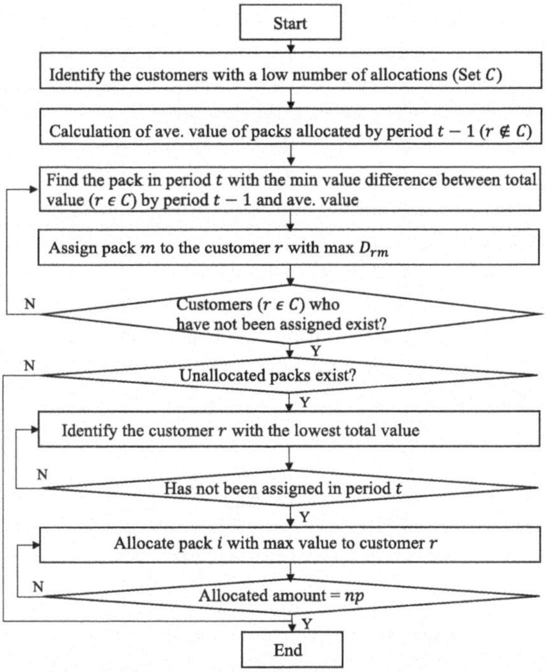

Fig. 2. Algorithm for pack allocation

Step 3: Find the pack in period t with the min value difference between total value ($r \in C$) by period $t-1$ and average value.

Equation (10) shows the difference between the total value of customer $r(r \in C)$ who has been assigned by pack i and the average value calculated by Eq. (9). Equation (11) shows the minimum value of d_{ri} among all customers belonging to set C.

$$d_{ri} = \sqrt{\left(\sum_{r \notin C} \sum_{t \in t'-1} V_{ti}/(C-l) - \sum_{t \in t'} V_{tir}\right)^2} \tag{11}$$

$$D_{rm} = min(d_{r1}, d_{r2}, \ldots, d_{rI}) \tag{12}$$

$t\prime$: The current period

5 Numerical Experiment and Results

5.1 Experimental Conditions and Summary

In this section, the effectiveness of proposed algorithm for pack filling operations and pack allocation is to be confirmed in different situations such as different distribution of substandard peach quantity and different distribution of appearance attributes. The experimental conditions and criteria parameters are set as below,

(1) Probability of fewer, normal, and more substandard products: 20%, 60%, 20%
(2) The average number of peaches in each of the nine groups, divided by appearance attribute, is multiplied by a factor of 0.4 to 0.8 if there are few substandard peaches, or 1.2 to 1.6 if there are many substandard peaches.
(3) Average and standard deviation of the log-normal distribution which show the quantity per group are set as same value, which are 2.5 and 0.8.
(4) Minimum and Maximum diameter range of substandard peach: 4, 8
(5) Average and standard deviation of diameter of all substandard peaches which is following a log-normal distribution: 1.85, 0.8
(6) Value of peaches: sale price (diameter is from 4 to 8), 0.8 $\times$ sale price (diameter is lower than 4), 1.2 $\times$ sale price (diameter is higher than 8)
(7) Sale price for each group: 100 (A), 105 (B), 110 (C), 115 (D), 120 (E), 125 (F), 130 (G), 135 (H), 140 (I)
(8) $ca_{min} = 4$, $ca_{max} = 10$
(9) $\alpha = 4$
(10) Total periods:10
(11) Number of total customers: 10
(12) Number of program execution for each case: 50

Numerical experiment program for decision-making algorithm designed in this research is created by Visual Studio Code (Python extension for VS Code). The experiment is conducted in the following experimental environment:
CPU: 3.1 GHz 6-Core Intel Core i5
Memory: 8 GB 2667 MHz DDR4
OS: MacOS Ventura 13.4
Programming Software: Visual Studio Code ver. 1.78.2 (Universal).

5.2 Results

Total value of all packs for all customers in all periods, average value of the difference between all value and the average value and program processing time are considered the valuation indicators.

Figure 3 shows the difference among three different situations which are listed in Table 2. The probability of more substandard peaches is higher, the total value of all packs for all customers is higher, and the average of the value difference is greatly reduced at the same time.

Table 2. Different distribution of substandard peach quantity

	Probability of fewer	Probability of normal	Probability of more
Fewer	0.6	0.4	0
Normal	0.2	0.6	0.2
More	0	0.4	0.6

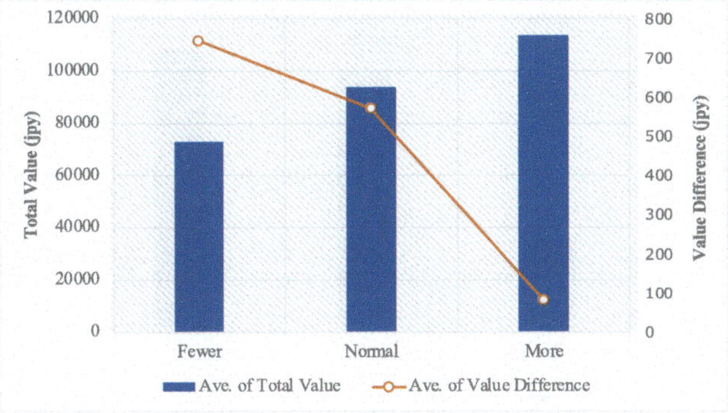

Fig. 3. Total value and value difference comparison based on different distribution of quantity

The program processing time of the 3 situation is 1.94, 5.15, 8.69 respectively. The processing time of Normal is 2.65 times that of Fewer, and the processing time of More is 1.69 times that of Normal. It is confirmed that the workers cost more time on the manual work of filling and allocation packs under the situation of more substandard peaches.

Table 3. Proportional difference of quantity of peaches in different groups (Mean and variance of lognormal distribution)

	Gr. A, B, C	Gr. D, E, F	Gr. G, H, I
Low-quality	3 (+20%)	2.5	2 (-20%)
Normal	2.5	2.5	2.5
High-quality	2 (-20%)	2.5	3 (+20%)
Variance of lognormal distribution of all situations: 0.8			

Figure 4 shows the difference between three different situations which are listed in Table 3. The average value of peaches in group A is the lowest and the average value of peaches in group I is the highest.

From Fig. 4, as the proportion of higher quality substandard peaches is higher, the total value of all packs for all customers becomes higher. In addition, the average of the value difference increases from the situation of higher proportion of low-quality substandard peaches to normal and becomes lower from normal to the situation of higher proportion of high-quality substandard peaches. According to the packing rule in this study, the closer the number of peaches in different groups, the larger the value difference between the packs. The reason is considered that the greater the difference in quantity among all groups, the greater the number of possible combinations and the smaller the difference in value among the packages under this experimental condition.

In addition, there is no obvious difference among the program processing time of the 3 situation which are 5.46, 5.15, 5.89 respectively, in the case of near quantity of all groups, work execution is confirmed to be relatively fast.

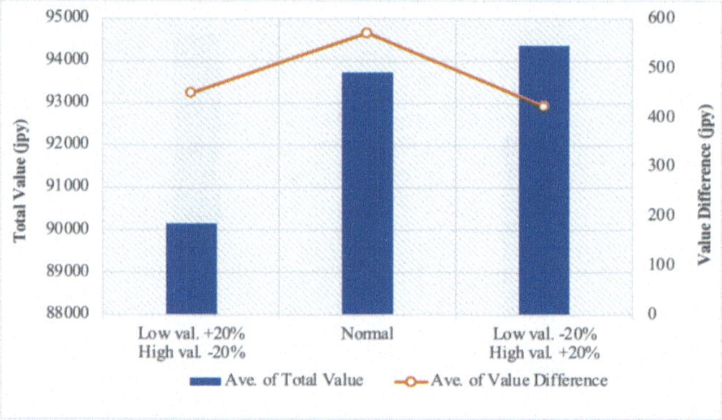

Fig. 4. Total value and value difference comparison based on different proportion of quality

6 Conclusions

In order to improve customer satisfaction in purchasing packs of substandard peaches, this study proposes a decision-making model based on a greedy algorithm for packing substandard peaches that maximizes the value of each package while minimizing the difference in product value between packs. The effectiveness of the proposed algorithm is confirmed under different distribution of quantity and quality. This study has developed a prototype decision-making model for small scale customers. Future work will focus on larger scale customers and develop an allocation function that considers the customer's evaluation of the peaches according to appearance attributes.

Acknowledgments. This study was supported by JSPS KAKENHI Grant Number JP22K04608 and JP23H01641.

References

1. Ministry of Agriculture, Forestry and Fisheries (Japan), Japan's strategy for sustainable food systems (Strategy MIDORI), and climate-smart agriculture (2023)
2. Yamamoto, Y., Yuzawa, Y.: A study on actual situation and effect of farmer's market in roadside station from viewpoint of regional promotion function – a case study in Kanto region. J. City Plann. Inst. Japan **47**(3), 985–990 (2012)
3. Ang, J.S., Cao, C., Ye, H.Q.: Model and algorithms for multi-period sea cargo mix problem. Eur. J. Oper. Res. **180**(3), 1381–1393 (2007)
4. Lau, H.C., Lim, M.K.: Multi-period multi-dimensional knapsack problem and its application to available-to-promise. In: Proceedings of International Symposium on Scheduling (ISS), pp.94–99 (2004)
5. Kurano, Y., Saito, R.: Heuristic algorithms with rounded weights for a combinatorial food packing problem. J. Adv. Mech. Des. Syst. Manuf. **11**(1) (2017). https://doi.org/10.1299/jamdsm.2017jamdsm0003

Internal Process Optimization for E-commerce D2C

G. Pagnoncelli[1], P. Carpejani[1(✉)], E. Pinheiro de Lima[2], U. Tortato[1,3], S. E. Gouvea da Costa[2], and F. Deschamps[1,4]

[1] Industrial and Systems Engineering, Pontifical Catholic University of Parana, 1155 Imaculada Conceicao, Prado Velho, Curitiba, Parana, Brazil
p.carpejani@pucpr.br

[2] Industrial and Systems Engineering, Federal University of Technology, Sete de Setembro, Reboucas, Curitiba, Parana 3165, Brazil

[3] Business School, Pontifical Catholic University of Parana, Imaculada Conceicao, Prado Velho, Curitiba, Parana, Brazil

[4] Mechanical Engineering, Federal University of Parana, 230 Coronel Francisco Dos Santos, Jardim das Americas, Curitiba, Parana, Brazil

Abstract. Corporations increasingly need to improve their processes to face challenges. Therefore, companies that do not manage their strategic planning and operations constantly end up wasting opportunities. In this circumstance, it is necessary to use performance measures that manage the conservation of processes related to the admitted strategies. Based on this, a study was carried out applied in the area of D2C (Direct-to-Consumer), in the sectors of Logistics and PPC (Production Planning and Control) in the home appliance company in order to reduce the time in some daily processes and use the Power BI tool to manage indicators that contain information on consumer interactions in relation to Logistics service and the availability of products for the consumer in the virtual store. For this, the DMAIC methodology was used, which can assist in the structured resolution of problems, seeking root causes, and providing actions that contribute to the elimination and reduction of problems. To achieve the objective, some quality tools such as A3, 5W2H, 5 Whys, Brainstorming, Ishikawa Diagram, Stratification, Benchmarking, and Indicators were used. With this, the lack of connection between all the data that is collected made it possible to identify that it is an impacting factor in the problem. In this way, it was proposed to use Power BI to create a dashboard with indicators that contain all the information requested daily. The solution developed also included a reduction in the performance of activities, being approximately 75% in both sectors. The application was validated by stakeholders and provided a more strategic vision for the company in decision-making and, consequently, adding value to end consumers.

Keywords: D2C · E-commerce · Business Intelligence

M. Dragomir et al. (Eds.): ICPR 2023, LNPE, pp. 319–327, 2025.
https://doi.org/10.1007/978-3-031-92082-0_35

1 Introduction

Electronic commerce or e-commerce, according to Assunção et al. (2019), is about the progress of trade itself, which over the years in search of new consumers and suppliers, contributed to the conquest of new methods. E-commerce proposes to use electronic means for commercialization, but the highlight of these commercial operations occurred as from the internet, where in a more facilitating way, it united buyers and sellers from anywhere through virtual stores. In this context, e-commerce operations need to evolve according to market needs, and from strategic actions, seek innovations in the internal processes of companies (Assunção et al. 2019).

Among the four most common e-commerce modalities there is the Direct to Consumer (D2C), whence the sale of products is made directly to the final consumer without intermediaries, strengthening the relationship between company and customer. Therefore, having good planning to meet consumer demand and keep them satisfied with the service provided is one of the greatest care to be taken by companies (Varela 2022).

In order to improve strategic actions for the e-commerce area of a company, we intend to improve the internal processes related to D2C.

One of the problems identified in the sector of the company studied was the lack of unified management of feedbacks coming from consumers related to the provision of D2C (Direct-to-Consumer) logistics services. Furthermore, the lack of clarified information in the data concerning the accuracy of products available for the company's e-commerce is also a problem.

Therefore, based on the need to obtain relevant information in a more effective and agile way, the project sought to optimize some current procedures in the D2C e-commerce sector in a company that manufactures home appliances, thinking mainly in contributing to strategic actions and improvements that as a final consequence are beneficial to consumers.

The objectives were defined in a way to be mainly efficient for the internal and external stakeholders, improving the e-commerce internal processes, reducing time, and increasing agility for strategic and operational decision-making.

2 Research Design

The techniques that will be described correspond to the Quality Management, Strategic Operations Management and Information System tools that were used to achieve the objective.

To analyze the problem and develop the proposal, some steps were necessary, guided by the DMAIC tool, instrumented from the A3 report. The purpose of this tool is to contribute to the development of the project, considering planning, execution, and control phases. In addition, it provides the use of other auxiliary tools in each stage of the problem analysis. The A3 report aims to identify and propose a solution to the problem. It enables the identification of the root cause, its nature, and the mapping of appropriate countermeasures to solve the problem described during the root cause analysis. It contains a step where it is necessary to put the actions into practice and observe whether the problem has been solved. The report consists of the following steps: initial considerations, current state, analysis, proposed improvement, action plan, and follow-up (Durward 2009).

As described by Lobo (2010), Brainstorming is a technique for generating cooperative ideas by a group of people. Thoughts. Its concept assumes that generating ideas collectively is more effective in solving problems because in a society there are several opinions and perceptions of the same problem. Contains three stages for its development, the first is the development of the problem, the second is the creative phase, and the last is the critical phase.

The Ishikawa Diagram is one of the Quality tools formed by the problem and its possible causes. This tool is used mainly to seek, organize, categorize, highlight, and present the causes in a graphic and grouped manner. Having as groups of causes, which are classified by the 6Ms: method, raw material, labor, machine, measurement and environment (Forlogic, 2016).

According to author Paladini (2019), benchmarking is innovative When practices are defined, the organization studies ways to apply them in its processes to improve its performance.

The method boils down to questioning five times the why of a problem. Terner (2008) states that the primary questioning should be structured using the described problem itself, which automatically answers why the problem is happening. In the second why, the answer to the first why should be used as a consequence of the first why. And so on and so forth until the root cause is reached.

The 5W2H is a quality tool created in the Japanese automobile industry and is applicable in several areas within an organization, including strategic planning to execute actions. Its use assisted in clarifying problems, and decision making, and acted as a checklist containing seven specific questions (Napoleão 2018).

The Standard Operating Procedure is referred to as a detailed file with instructions for routine activities. The SOP, being a tool focused on eliminating waste and acting on best standardization practices, helps in several applications. But each document, whether checklist or flowchart, is specific to a particular task, that is, a SOP is needed for each application (Coutinho 2020).

The term indicator means to discover, to demonstrate, to indicate. Indicators inform the progression toward an established goal. It is also understood as a resource that makes a trend or fact more evident, which is not immediately perceptible (Van Bellen 2005).

Indicators are informative, which show a system's properties, simplifying its facts (Van Bellen 2004).

BI (Business Intelligence) main goal is to provide interactive access to and manipulation of data leading to appropriate analyses for better decision making. Its process consists of turning data into information; information into data, and finally data into actions (Sharda et al. 2019).

Stratification is a very useful feature during data inspection. It is based on the division of a group into several subgroups with distinctive properties. From the stratification of data, it is idealized to understand how the variation of factors (equipment, inputs, people, methods, measures, and environmental conditions) influence the problem to be investigated (CarpinettI 2016).

According to Barbier (2002), it is a continually changing process in spirals of reflection and action, where each spiral includes: Diagnosing a practical problem to improve or solve, formulating and developing strategies based on effectiveness, understanding the new situation, and proceeding to the same steps for the new practical situation.

3 Implementing

This part of the article will describe the business scenario and introduce the organization and preparation of implementation data. The application of the study focuses on two sectors within the same company.

3.1 Research Area: The Company

The company selected for the development of the project is a Swedish multinational manufacturer of household appliances, the second largest in the world, with sales of 60 million products per year. Founded in 1919, it is currently present in 150 countries, with seven production plants in Latin America: Argentina, Brazil and Chile, employing 11 thousand employees.

The company's purpose is to promote a sustainable and healthy life for the whole world. Therefore it has three pillars as its objectives: to act in a sustainable way by making available efficient and intelligent solutions, to create better experiences from the products developed, and to always improve, seeking innovations and opportunities.

The project was developed in the Supply Chain sector, in the D2C area situated in the factory which is located in Brazil, in the State of Paraná.

Both sectors value the service provided to the consumer, considering their satisfaction and excellence in service. While sector 1 of Logistics is responsible for ensuring the transportation of products to the consumer, respecting the defined lead time, and delivering a product in perfect condition, sector 2 of PPC (Production Planning and Control), is responsible for ensuring that the product will be available to the customer when he wants to buy or provide the stock forecast for that specific model that is in fulfillment. The contact between the areas is daily, especially in the resolution of cases in which the customer service of a product was not met in the desired manner.

3.2 Preparation

In the definition stage, the objective was to clarify the problem. From a Brainstorming, it was defined what the situation was, taking into account some questions such as place of occurrence, when and how often the problem occurs, people and sectors affected, and finally what the current standard was.

Then, in the measurement stage, a goal that needs to be achieved is defined, as well as a survey of the possible causes of the problem. The Ishikawa Diagram tool was used to measure the current system and survey the causes.

After the survey of the causes, it was possible to proceed to the analysis stage, in which the 5 whys were used to find and analyze the root cause of the problem. The use of this method was effective and extremely important to assist in the development of the mapping of countermeasure actions.

In the next step, improvement, the goal is to outline improvement countermeasures that should be executed to eliminate the causes identified in the analysis step. To facilitate and bring objectivity to the execution of the plans, the 5W2H was used in a lean way, to make the development clearer and more effective (Hans et al. 2004).

Finally, in the control stage, the performance of the process returns was verified to reach the goal set in the measurement stage, and the results achieved were evaluated. Furthermore, it was necessary to create a standardization of the procedures and processes developed, to ensure the effectiveness and continuity of the improvement. Figure 1 shows an example of instrumentation.

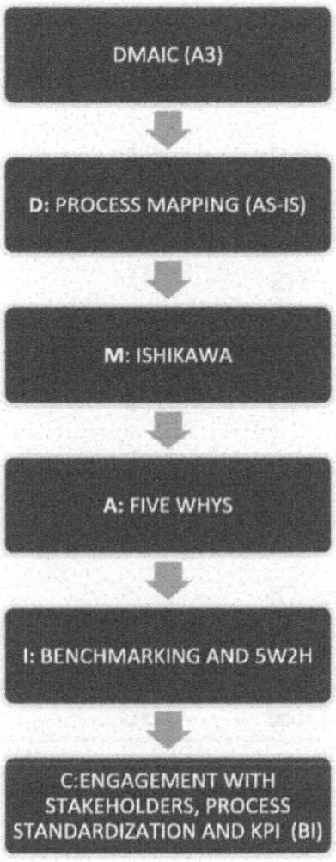

Fig. 1. Application example

The example shown in Fig. 1 was repeated for Sector 1 and also for Sector 2. Depending on the sector applied, different quality tools were used. However, as a major guideline, the DMAIC process was completed via A3.

4 Discussions and Results

The relationship between the Logistics sector, sector 1, and the Customer Service Management (PPC) sector 2 in D2C e-commerce is essential to ensure customer satisfaction. The Logistics sector takes care of product transportation and delivery, while the PPC Service Management sector ensures product availability and provides stock forecasts. The interaction between the areas is daily, especially in solving customer service problems.

Based on the studies and actions taken, it was possible to analyze the results in the D2C area of the company. The application of the project brought benefits such as the reduction of information compilation time and an improved strategic vision for the company. Both sectors identified the need for an indicator management for D2C, leading to improvements and standardization of activities.

The study started with sector 1, Logistics, whence actions were required due to the lack of an existing process and non-standardized reporting systems. The changes made resulted in a reduction in activity time greater than the established goal of 76%.

Previously, an activity of 3 h and 10 min was reduced to only 45 min, as can be seen in Table 1:

Table 1. Activity performance time

Activities	Before time (min)	Present time (min)	Reduction time (min)
Extract Zendesk Report	10	2	-8
Extract Trustvox Report	10	2	-8
Extract Delivery Report	20	2	-18
Zendesk Report Analysis	35	5	-30
Trustvox Report Analysis	50	5	-45
Delivery Report Analysis	20	20	0
Unify Reports	15	5	-10
Generate Visual Indicators	30	NA	-30
Update Dashboard	NA	2	2
Update Monthly Number of Orders	NA	2	2
Total	190	45	-145

The automation in receiving the reports via e-mail and the standardization facilitated the extraction of the data, reducing the time required to 45 min per day. This represents a significant reduction from 836 h per year to 198 h, equivalent to almost 80 days of work for one employee.

In addition, a Power BI dashboard was created with relevant indicators, such as the number of complaints, reasons, evaluations, problem resolution rate, consumer recommendations, location with the highest evaluation rate, dissatisfaction recurrence, and evaluation rate per carrier, among others. These indicators help in strategic decision-making and in reducing the number of complaints (Pacheco 2019).

However, due to the absence of a previous process of data analysis and management, it is still difficult to have a comparative view and measure the results in relation to the company's reputation among consumers. It is believed that after a year of implementation of the improvements, it will be possible to obtain a more accurate assessment of the impacts generated.

Aiming to improve service management in D2C e-commerce, a study was also conducted in sector 2, D2C PPC. The need for a strategic and agile view of product availability information was identified. Interactive indicators were created using Business Intelligence to provide this strategic and operational vision. This resulted in a 75% reduction in the number of clicks required for each analysis, providing greater efficiency. The indicators allowed for autonomy in the data search, with the dashboard link being sent in no more than 2 min. Tables 2 and 3 below show this click reduction. It was decided to measure the number of clicks due to a lack of effectiveness in reporting the time it took before for each analysis requested because each piece of information presented a different complexity. Now that the search for this data is done autonomously, the only time that will depend on the analyst is to send the dashboard link when necessary. This will take a maximum of 2 min to be available.

Table 2. Previous sector 2 analysis time

Action	Clicks
Open folder	7
Open unfold	1
Search for data in the file	6
Analyze data	5
Check information	5
Check SAP	8
Total clicks per analysis	32

Table 3 shows the process improvement based on the number of clicks to perform such activity. That is, the reduction of clicks will provide process improvement through time management.

Table 3. Analysis time after improvement Sector 2

Action	Clicks
Click on the power BI link	1
Send the link to the solicitor	3
Possible interactions with graphics	4
Total clicks per analysis	8

The Tables above represent the steps needed to obtain the requested information. The goal was to eliminate the need for clicks, allowing everyone to access the Dashboard. The indicators in Power BI streamlined the supply of data to the Logistics sector and allowed managers and directors to monitor the performance of e-commerce. The reduction of the time to update the information will be addressed in the future.

The indicator in Power BI, besides helping to streamline the provision of data for the Logistics sector and other sectors, also allowed managers and directors to observe how the company's performance is about e-commerce and to observe issues such as sales, meeting the quota, quantity of product available in a given period, the reason for the fulfillment, average daily sales, over-quota rates, fulfillment projections, days of stock available, etc.

The results of both sectors were measured by feedback, and stakeholders and employees reported positive aspects, such as agility in decision-making, ease of interaction with the dashboard, and improvement in customer satisfaction. The Business Intelligence tool made up for the lack of management indicators, allowing dynamic analysis to be available at any time. Access to this strategic information will impact the management and decisions of senior management, adding greater value to the company.

The improvements in the Logistics and PPC Service Management sectors provided benefits such as reduced information compilation time and an improved strategic vision. The collaboration between the areas was fundamental to solving customer service problems.

5 Conclusions

The paper aimed to develop improvements in the D2C area, in customer service management sectors in e-commerce, which required better visibility of information.

The objectives were achieved through the use of quality tools and techniques, such as brainstorming, Ishikawa Diagram, and analysis of the 5 whys. The Business Intelligence tool, Power Bi, played a key role in the development of the project, allowing data analysis and visual presentation of information, enabling process improvements, time reduction, and elimination of manual activities.

The application of the improvements resulted in qualitative benefits, such as a reduction in repetitive and recurring work, in addition to ensuring a better consumer experience, because with the compilation of the data, it was possible to speed up the response time and identify the biggest offenders and outline actions to minimize the impacts. Making the annual comparison, we noticed a reduction in the evaluations that had the highest negative index, such as carriers that had a 50% to 80% reduction in complaints due to the actions taken, which were made possible by the data indicators to ensure a better service. The general index of positive evaluations had an increase of 20%, besides a reduction in the resolution time of the most critical cases. There were also financial benefits, such as a reduction in work time and a reduction in economic losses (Hans et al. 2005).

In addition, it was possible to make information available more technologically and visually, using operational and strategic indicators, resulting in a significant reduction in the time required for the compilation of data related to customer experience, logistics service, and e-store inventory (Ramesh et al. 2019). This translated into hours saved over the year, such as 80 days of reduced work in sector 1 and a 75% reduction in time in both sectors.

In summary, this applied project was successful in improving the e-commerce customer service management area through the use of quality tools and the implementation of a Business Intelligence dashboard. The benefits were both qualitative and financial, considering that the employee's labor could be spent on other activities and because the strategic vision is very useful in measuring risks and avoiding waste. For future research, it is recommended that digital transformation tools be explored, as they will contribute to the operation's resilience (Carpejani et al. 2020). The project contributed to ergonomics by reducing the workload of employees performing a repetitive activity and providing a better consumer experience with the brand.

References

Assunção, W., Fagundes, P., Révillion, A.: Comércio Eletrônico, Sagah, Porto Alegre, Rio Grande do Sul (2019)

Lobo, R.: Gestão da Qualidade. 2 Edição. Editora Saraiva (2020)

Coutinho, T.: Procedimento Operacional Padrão traz para sua empresa. Voitto (2020)

Van Bellen, H.M.: Desenvolvimento sustentável: uma descrição das principais ferramentas de avaliação. Ambiente Sociedade **7**(1), 67–87 (2004)

Van Bellen, H.M.: Indicadores de sustentabilidade: uma análise comparativa, Fundação Getúlio Vargas, Rio de Janeiro, Rio de Janeiro (2005)

Paladini, E.P.: Gestão da qualidade - Teoria e Prática. 4. ed. Barueri: Grupo Gen (2019)

Carpeneti, L.C.R.: Gestão da qualidade - Conceitos e Técnicas. 3 Edição, Grupo Gen, Curitiba, Paraná (2016)

Sharda, R., Delen, D., Turban, E.: Business Intelligence e Análise de Dados para Gestão do Negócio. 4 Edição, Grupo A (2019)

Napoleão., Ferramentas da Qualidade. Diagrama de Ishikawa (2018)

Forlogic., Ferramentas da Qualidade. 5W2H (2016)

Barbier., A pesquisa ação, Brasília, Distrito Federal (2002)

Carpejani, P., et al.: The use of digital transformation as a sustainable mechanism: an automotive industry case. In: World Sustainability Series (2020). https://doi.org/10.1007/978-3-030-26759-9_6

Intelligent Additive-Subtractive Manufacturing for Resilient Production

M. Palalić(✉), F. Bieg, C. Maucher, K. Güzel, and H.-C. Möhring

Institute for Machine Tools (IfW), University of Stuttgart, Holzgartenstraße 17, 70174 Stuttgart, Germany

marijana.palalic@ifw.uni-stuttgart.de

Abstract. Additive manufacturing is a key enabler for fast-reacting and therefore resilient production of complex components. A vast variety of materials can be processed and function-integrated parts with dedicated geometrical and physical properties can be realized. In order to fulfill functional requirements, post-treatment terms of machining are necessary in most cases, leading to an additive-subtractive process chain. With the aim to achieve a first-part-right production, intelligent strategies, comprising process monitoring and data analytics, simulation and process control are required and have to be implemented throughout the entire process chain. This contribution introduces related intelligent approaches with an emphasis on the targeted application of artificial intelligence methods.

Keywords: additive-subtractive manufacturing · direct energy deposition · process monitoring · image processing

1 Introduction

Additive manufacturing technology (AM) enables the realization of component structures that are technically very difficult, ecologically or economically inefficient, or physically impossible to manufacture with other manufacturing processes [1, 2]. Additive manufacturing technology thus opens up new degrees of freedom in the design of products and systems. In addition, AM enables the responsive, flexible, and thus resilient production of small series and individual one-offs [3]. With component structures that are optimized in terms of their topology but also their density, lightweight construction potential can be tapped, and thus, for example, mobility systems with lower energy consumption can be realized [4, 5]. Due to the ability to also create internal structures in components, AM contributes to making diverse possibilities for functional integration feasible and usable [6]. Functional integration can also consist of using AM to set specific stiffness and damping properties of components that are particularly advantageous for the respective application [7, 8]. Finally, AM enables structures to be built up additively on existing or prefabricated substrates [9] and thus, among other things, the repair and regeneration of complex, cost-intensive functional components. The main areas of application for AM are therefore in the automotive industry, aerospace technology, medical technology, the consumer goods and electronics industry, production technology, and

M. Dragomir et al. (Eds.): ICPR 2023, LNPE, pp. 328–335, 2025.
https://doi.org/10.1007/978-3-031-92082-0_36

general mechanical engineering, as well as in many other branches of industry. Using AM, three-dimensional objects are created directly based on CAD (Computer Aided Design) data by building up materials in layers [10, 11]. A wide variety of materials can be used [12, 13]. Laser-based Directed Energy Deposition (DED-LB) is a promising additive manufacturing technique that has significant potential for the fabrication of large-scale, complex components with improved material properties.

2 Metal Powder Characterization

However, the success of the DED-LB process depends to a large extent on the selection and monitoring of the powdered materials, since these play a crucial role in determining the final quality and mechanical properties of the parts manufactured. The characteristics of the powder directly impact the quality of the printed parts. Monitoring and controlling the powder properties, such as particle size distribution, and purity, ensure consistent results and reduce the risk of defects during the process.

The size distribution and morphology of the powder significantly affect the packing density, porosity, and fluidity of the powder. Both directly affect the behavior of the weld pool during the process. The typical particle size for the DED-LB process is between 30–150 μm [14]. Furthermore, these properties can affect the surface finish of the additively manufactured parts [15, 16]. The shape and size of the powder particles can influence the roughness of the printed surface, impacting the need for post-processing steps. Additionally, the microstructure and mechanical properties of the final printed part are also influenced. Post-processing steps, such as heat treatment, may be required to optimize the properties of the material, and the initial powder characteristics play a significant role in defining these parameters.

In case of defects in the printed part, understanding the powder properties can help diagnose the root cause. For instance, inconsistent flow due to powder agglomeration or poor powder quality could lead to voids or incomplete fusion. Therefore, the utilized powder is of importance not only for the DED-LB process itself but also for subsequent process steps.

3 Experimental Setup and Process Monitoring

Although additive manufacturing technology in combination with traditional machining operations removes multiple obstacles when it comes to the manufacturing of complex parts and specific part properties in an economical and ecological manner, there are still many effects within the processes, and in particular interactions that have not yet been fully understood and currently elude comprehensive observation, control, and process regulation solutions. There is particular potential for the implementation of innovative metrological solutions in combination with different methods of artificial intelligence and machine learning techniques.

3.1 Melt Pool Image Monitoring

Identification of the melt pool characteristics such as shape and geometry is an essential part of additive process control and optimization [17]. The goal of the optimal process is to maintain a constant bead geometry for maximum geometric accuracy of the part [18] (Fig. 1).

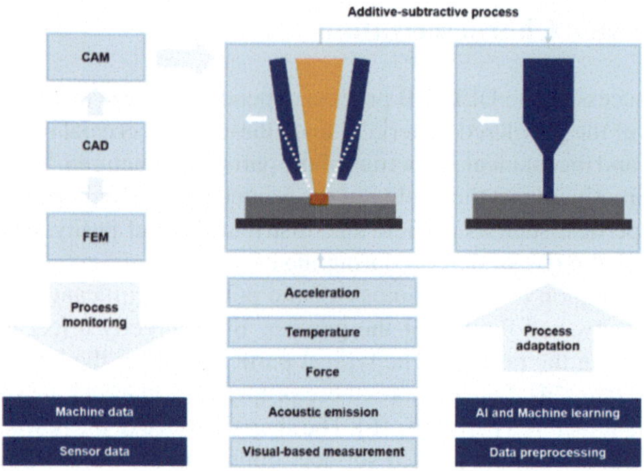

Fig. 1. A concept of monitoring and adaption of a hybrid process

The experiments are performed on a hybrid machine from DMG MORI (Lasertec 65 DED). The machine combines the additive process of laser metal deposition (LMD) and 5-axis milling. The images of the melt pool were taken by the commercially available DMG MORI hybrid machine-integrated coaxial system with the CCD camera (Manta G-033, Allied Vision). The melt pool radiation is reflected from a dichroic mirror to the sensors of the CCD camera which is transferring the captured images to the laser PC via Ethernet [19]. The acquisition rate used for image acquisition was 30 frames per second.

Images were saved locally and analyzed offline after the process terminated. The 8-bit melt pool images were pre-processed so that all the reflections coming from the base plate are filtered and only a certain region of interest (ROI) is selected for further investigation. The captured images have dimensions of 246x328 pixels.

In this experiment, the investigated sample was a thin-walled triangle structure represented in Fig. 2. The triangular structure was built using three different scanning patterns to achieve different build qualities of the samples. In Fig. 2, the first sample shows most clearly how the inappropriate strategy leads to significant geometric distortions. The process parameters used to manufacture the samples were laser power of 1600 W, powder feed rate of 20 g/min, and scanning speed of 1000 mm/min. Laser power was adjusted in the first layers from the maximum to the final laser power in order to assure optimal printing conditions at the beginning of the process.

Different melt pool information such as width, size, and mean intensity are extracted from the images using different image segmentation techniques. Image segmentation is a method of partitioning an image into different regions often based on the properties of the pixels in the image. The calculation of the melt pool width is based on the Circular Hough Transform (CHT) algorithm. CHT algorithm is used for the detection of circles as well as circular arcs in digitized images and the method is inherently robust even when the boundary is faint or partially occluded [20].

In Fig. 3 the extraction of the circle calculated based on the arc of the melt pool geometry using the CHT algorithm is represented.

A circular shape, as shown in Fig. 3(b), is not able to accurately represent the entire area of the melt pool and its shape. In order to better describe the aforementioned features of the melt pool, the k-means clustering algorithm was used for image segmentation. The k-means clustering algorithm is an unsupervised algorithm that is often employed for image segmentation. The algorithm divides a set of data into clusters through two different phases.

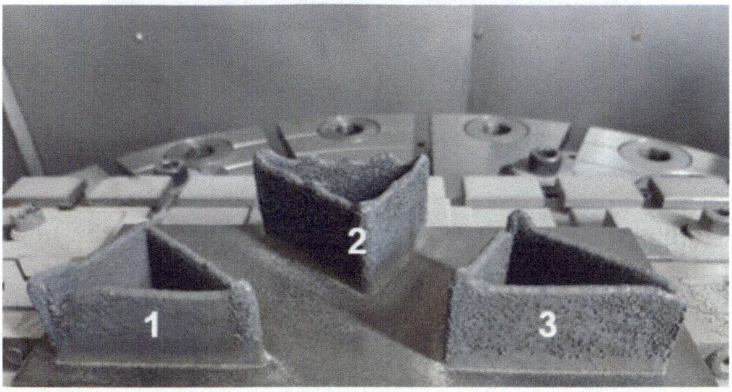

Fig. 2. Thin-wall triangle structures manufactured with different scan musters 1) same starting point in the layers, 2) start point offset, 3) each side with its own starting point

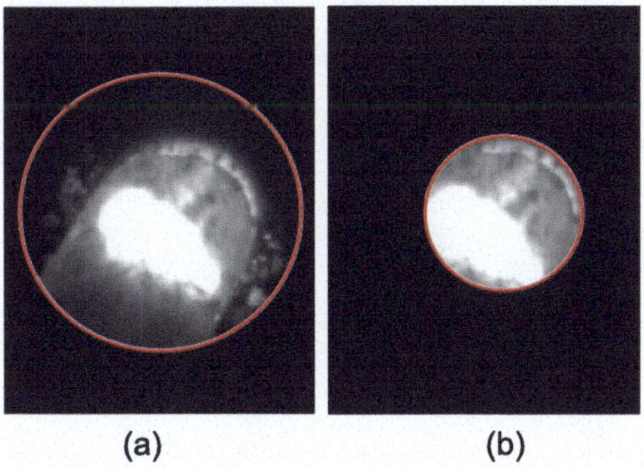

Fig. 3. Image processing a) ROI, b) CHT image segmentation for the calculation of the melt pool width

In the first phase, the k centroid is calculated and in the second phase, each point is appointed to the cluster, which has the nearest centroid from the respective data point

[21]. Using this method, the melt pool area in pixels is separated from the background by dividing the image into two clusters, Fig. 4.

The dynamical conditions in the melt pool are different at the beginning of the process than at a later stage, which can also be seen in Fig. 4.

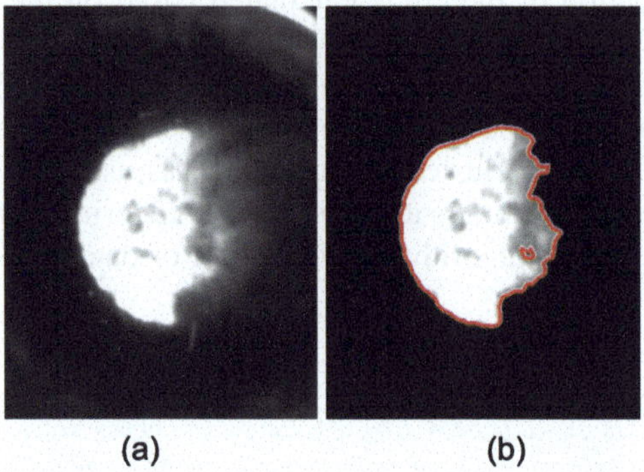

Fig. 4. Image segmentation using the k-means clustering algorithm a) the raw unfiltered image of the melt pool, b) the segmented image of the pool

Also, as mentioned earlier, the laser power is adjusted in the first layers of the sample, which affects the size of the melt pool in this transition phase. After the first 10 layers in structure buildup, the melt pool becomes larger as the process progresses, as represented in Fig. 5. In this paper, the perspective of combining different image segmentation methods to monitor the melt pool geometry is presented. The two methods, which extract two different features of the molten pool, provide valuable information on the process dynamics.

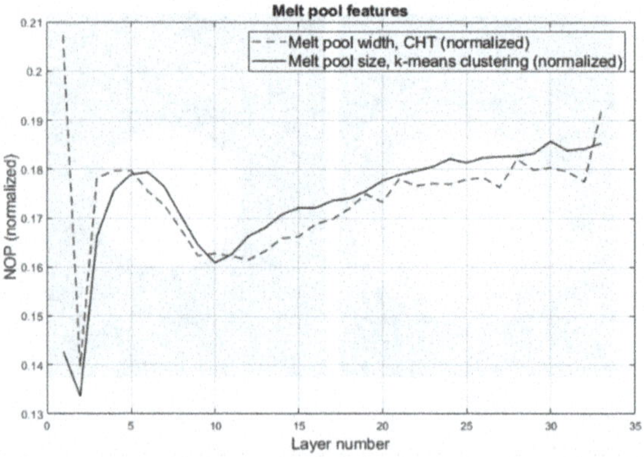

Fig. 5. Melt pool features extracted using CHT and K-means clustering techniques for image segmentation

4 DED-LB Process Outside of the Process Window

In additive manufacturing processes, the production of defect-free parts depends on the used process parameters. Furthermore, the scanning strategy used, especially the degree of overlap between the individual melt tracks, has a significant influence on the density and the geometrical accuracy of the created components.

Another important parameter is the working distance, respectively the focus distance. The working distance is the distance between the nozzle tip and the substrate surface. The focal distance represents the length between the nozzle and the focal plane, which is located just below the substrate surface so that part of the substrate is melted and becomes part of the melt pool. The process parameters influence one another and have a small operating window in which the production of defect-free and dimensionally accurate structures is possible.

In this article, the effect of performing the DED-LB process outside the preset process window was examined, for the use of an insufficient laser power and path distance factor.

Besides an incorrect parameter set, insufficient laser power or heat input during the DED-LB process can result in the collapse of the previously deposited structure. The collapse of the thin-wall structure results from a too-high heat input. The melt pool is deeper and longer, which collapses the structure. After the collapse of the structure the working distance between the nozzle and the substrate surface increases, and the focal plane is raised above the substrate surface. The focal point on the substrate surface thus becomes larger and the heat input decreases and the melt pool is smaller. Less powder will be molten at these positions of the structure. Therefore, the layer thickness is decreasing, and the collapse of the structure is increasing even more until there is no real structure built up.

For the generation of the test samples in the chapters before, a laser power of around 1800–2500 W is used to produce the thin wall structures. These parameters are recommended by the machine manufacturer to produce parts with maximum density and minimum defects. To show the influence of insufficient laser power, a laser power of 500–900 W was used, and cubic structures with an edge length of 25 mm were produced. For this investigation, a path distance factor of 0.6 x width of the deposited melt track was used.

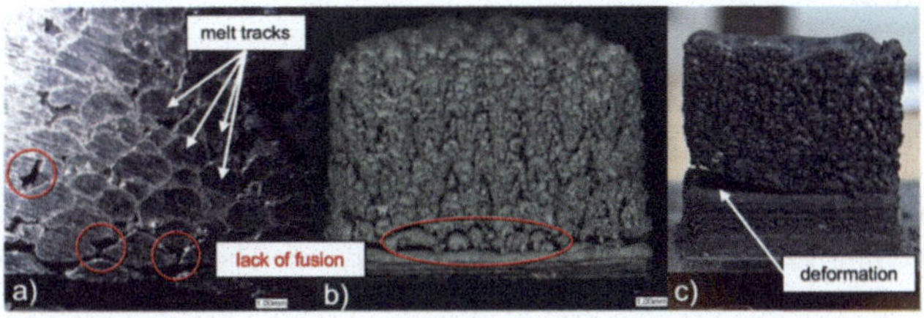

Fig. 6. Lack of fusion in (a) cross-section, (b) surface, (c) connection

The use of insufficient laser power results in the creation of interlayer porosity. The heat input into the substrate and the blow in powder material is too low to create a bonding between the individual melt tracks among each other, which can be seen in

Fig. 6 a) and b) and between the substrate and the deposited structure, shown in Fig. 6 c).

Because of the lack of fusion between the substrate and the deposited structure, the bottom of the cubic parts deforms and raise from the substrate. This deformation results from the thermal stresses generated in the deposited structure due to the high cooling rates during the solidification. When using optimal laser powers, these deformations do not occur because the bottom layers of the structure are firmly fused to the substrate, Fig. 6 c). The lack of fusion is amplified by an increasing path distance factor. The melt tracks don't overlap and lie next to each other. Because of the low laser power and the lack of fusion, unmolten powder particles remain in the structure as pores. A too-low laser power leads to a reduced density. At a path distance of 2 x the width of the deposited track, there is no connection between the tracks. The structure is held together by the melt tracks of the overlying layer.

With the help of the lack of fusion between the melt tracks and the layers among each other, interlayer porosity can be selectively created in DED-LB fabricated structures with the help of too-low laser power. With the help of these graded porous structures, superior device properties can be achieved by fabricating them with locally optimized dynamic and thermal properties, such as a thermo-symmetric structure.

5 Discussion and Conclusions

Even though impressive research work has already been carried out with respect to the detailed analysis of the various process technologies separately, intensive research is still required to achieve a comprehensive understanding of the complex process interrelationships. The ultimate goal is to enable targeted implementation of the process chains in the single-part production of parts with predefined functional properties. One of the most challenging tasks is the implementation of in-situ monitoring systems that enable controlled regulation of the individual process steps. Special potential lies here in the implementation of innovative metrological solutions in combination with various artificial intelligence methods. In this paper, the authors have given insight into some steps of process chain monitoring and observation, such as powder characterization, melt pool monitoring, and additive-subtractive process investigation by forcing the process outside the process window.

Acknowledgments. The presented research results were elaborated within the research projects in AISA (Artificial Intelligence Software Academy) with support from the Ministry of Science, Research and Arts Baden-Württemberg. Furthermore the authors thank the Ministry of Science, Research and Arts of the Federal State of Baden-Württemberg for the financial support of the projects within the *InnovationsCampus Mobilität der Zukunft*. The authors are sincerely grateful for the funding of these research initiatives.

References

1. Thompson, M.K., et al.: Design for additive manufacturing: trends, opportunities, considerations, and constraints. CIRP Ann. **65**(2), 737–760 (2016)

2. Schmidt, M.T., Wegener, K., Overmeyer, L., Vollertsen, F., Levy, G.N., et al.: Laser based additive manufacturing in industry and academia. CIRP Ann. **66**(2), 561–583 (2017)
3. Jankovics, D., Barari, A.: Customization of automotive structural components using additive manufacturing and topology optimization. IFAC PapersOnLine **52–10**, 212–217 (2019)
4. Blakey-Milner, B., et al.: Metal additive manufacturing in aerospace – a review. Mater. Des. **209** (2021)
5. Ullah, A.M.M.S., et al.: A system for designing and 3D printing of porous structures. CIRP Ann. **69**, 113–116 (2020)
6. Moehring, H.C., et al.: The additive-subtractive process chain - a review. J. Mach. Eng. (2023). https://doi.org/10.36897/jme/162041
7. Chen, Y., Liou, F.: Additive manufacturing of metal functionally graded materials: a review. In: Proceedings of the 29th Annual International Solid Freeform Fabrication Symposium, pp. 1215–1231, 2018
8. Li, Y., et al.: A review on functionally graded materials and structures via additive manufacturing: from multi-scale design to versatile functional properties. Adv. Mater. Technol. **5**, 1900981 (2020)
9. Soshi, M., et al.: Innovative grid molding and cooling using an additive and subtractive hybrid CNC machine tool. CIRP Ann.-Manuf. Techn. **66**, 401–404 (2017)
10. Lakner, T., et al.: Additively manufactured milling tool with focused cutting fluid supply. Proc. CIRP **81**, 464–469 (2019)
11. Syam, W.P., et al.: Design and analysis of strut-based lattice structures for vibration isolation. Prec. Eng. **52**, 494–506 (2018)
12. Maliaris, G., et al.: Novel additively manufactured bio-inspired 3D structures for impact energy damping. CIRP Ann. **70**, 199–202 (2021)
13. Lehmhus, D., et al.: Customized smartness: a survey on links between additive manufacturing and sensor integration. Proc. Techn. **26**, 284–301 (2016)
14. Pinkerton, A.J.: Laser direct metal deposition: theory and applications in manufacturing and maintenance. In: Advances in Laser Materials Processing: Technology, Research and Application|Adv. in Laser Mater. Process.: Technol., Res. and Appl.: Technology, research and applications, pp. 461–491 (2010). https://doi.org/10.1533/9781845699819.6.461
15. Tan, J.H., et al.: An overview of powder granulometry on feedstock and part performance in the selective laser melting process. Addit. Manuf. **18**, 228–255 (2017). https://doi.org/10.1016/j.addma.2017.10.011
16. Brika, S.E., et al.: Influence of particle morphology and size distribution on the powder flowability and laser powder bed fusion manufacturability of Ti-6Al-4V alloy. Addit. Manuf. **31**, 100929 (2020). https://doi.org/10.1016/j.addma.2019.100929
17. Akbar, P., et al.: MeltpoolNet: melt pool characteristic prediction in Metal Additive Manufacturing using machine learning. Addit. Manuf. **55**, 102817 (2022). https://doi.org/10.1016/j.addma.2022.102817
18. Tyralla, D., et al.: A multi-parameter control of track geometry and melt pool size for laser metal deposition. Procedia CIRP **94**, 430–435 (2020). https://doi.org/10.1016/j.procir.2020.09.159
19. Kledwig, C., et al.: Analysis of melt pool characteristics and process parameters using a coaxial monitoring system during directed energy deposition in additive manufacturing. Materials (Basel, Switzerland) **12**(2) (2019). https://doi.org/10.3390/ma12020308
20. Luo, D., et al.: Circular hough transform for roundless measurement of objects. Pattern Recogn. Elsevier Sci. **28**(11), 1745–1749 (1995)
21. Dhanachandra, N., et al.: Image Segmentation Using K-means Clustering Algorithm and Subtractive Clustering Algorithm. Procedia Comput. Sci. **54**, 764–771 (2015). https://doi.org/10.1016/j.procs.2015.06.090

Simulation-Based Evaluation of Optimized Reconfigurable Cellular Manufacturing Deploying a Diversified Workforce

Julian Perwitz[1](✉), Thomas Sobottka[1,2], Ádám Szaller[3,4], and Fazel Ansari[1,2]

[1] Factory Planning and Production Management, Fraunhofer Austria Research GmbH, Theresianumgasse 7, 1040 Vienna, Austria
julian.perwitz@fraunhofer.at

[2] Research Group of Production and Maintenance Management, TU Wien, Theresianumgasse 27, 1040 Vienna, Austria

[3] EPIC InnoLabs Nonprofit Ltd., Kende Str. 13-17, 1111 Budapest, Hungary

[4] Institute for Computer Science and Control, Kende Str. 13-17, 1111 Budapest, Hungary

Abstract. This paper compares the performance of reconfigurable cellular manufacturing systems (RCMS) and linear systems when employing a diversified workforce with different work speeds.

Due to demographic and labor market trends manufacturing companies increasingly struggle to find skilled personnel for efficient manufacturing operations usually organized in tact-driven production lines. Concepts for flexible manufacturing can foster the integration of a more diversified workforce. Yet, this potential is largely untapped in practice, as the benefits as well as the circumstances under which they can be realized are unknown.

To address this gap, this paper proposes a simulation-based approach to compare the performance of reconfigurable cellular manufacturing systems (RCMS) and linear systems under different scenarios of workforce diversity. The approach is demonstrated within a dynamic multi-variant engine assembly use case, using discrete event simulation. The results confirm the feasibility of the method and point to the advantages of RCMS with different work speeds.

Keywords: flexible manufacturing · planning and control · discrete event simulation · human-centric production

1 Introduction and Motivation

Manufacturing companies face considerable economic and technological challenges, such as volatile markets, unstable economics, increasing competition as well as growing product and process complexity [1]. Societal challenges also include increasing personnel risks due to demographic developments and growing competition in the labor market, as well as changing expectations and fluctuations in the available workforce [2]. Hence, the ability to integrate a diversified workforce with different skills, speeds, strengths and weaknesses is a considerable advantage for manufacturing companies in the future.

M. Dragomir et al. (Eds.): ICPR 2023, LNPE, pp. 336–344, 2025.
https://doi.org/10.1007/978-3-031-92082-0_37

Conventional tact and linear manufacturing systems struggle in this aspect, as different work speeds or worker performance can result in blockages and waiting times and thus lower system Productivity [3]. Flexible manufacturing systems seem to be suited for this, however, specific investigations that quantify such advantage are lacking [4]. Therefore, this paper examines (1) the extent to which workforce diversity affects the productivity of manufacturing systems and (2) the potential advantages of flexible manufacturing systems over traditional production lines in this regard.

The rest of this paper is structured as follows: related work on flexible manufacturing systems, diversity aspects and other relevant workforce-related matters are reviewed in Sect. 2. This is followed by a presentation of the developed simulation approach in the method Sect. 3, followed by a simulation-based evaluation of diversity scenarios in a use-case in Sect. 4. A discussion of the results and an outlook conclude this paper.

2 Related Work

This section highlights related work for two relevant aspects of the concepts behind the scenarios analyzed herein: (1) The characteristics and differences of linear and flexible manufacturing systems and (2) how workforce planning accounts for diversified workforce capabilities, as both are major modeling background for the method and evaluation used in Sects. 3 and 4.

2.1 From Line to Matrix Production

The evolution of manufacturing and assembly systems is driven by partly contradicting requirements such as efficiency and cost pressure, flexibility and changeability as well as fluctuating production volume [5]. Therefore, various manufacturing concepts have been developed, to provide more flexibility, compared to dedicated manufacturing systems, inter alia, single or mixed-multimodel assembly flow lines (AFL). The evolution of flexible systems contains cellular manufacturing systems (CMS), reconfigurable manufacturing systems (RMS) as well as reconfigurable cellular manufacturing systems (RCMS). RCMS is also known as matrix-structured manufacturing systems (MMS) or simply matrix production [6]. A RCMS is characterized by flexibly linked reconfigurable modules, which result in a highly dynamic system that controls and conditionally balances temporary fluctuations in product variant and volume [7, 8]. Its characteristic degrees of freedom of routing and operation flexibilities allow for increased productivity, depending on the flexibility requirements and the actual manufacturing system setup [8, 9]. The concept of RCMS is not only seen as beneficial by decision-makers to tackle current and future challenges [1] but already has been realized in various forms with considerable technological maturity [10].

2.2 Workforce Diversity, Workforce Planning and Workforce Reconfiguration Strategies

In manufacturing, workforce diversity is described by human variability factors, such as age, gender, physical measures, qualification and skills [4]. Factors such as these are important in determining workforce productivity, i.e., the output of goods generated in a given time frame by a specific worker or group of workers [11].

Workforce planning determines on a system level the available workforce capacities as well as the assignment of workers to processes based on human variability factors, particularly skill [12]. Therefore, it includes the problem of process-capacity allocations.

Workforce reconfiguration strategies contemplate the deployment of utility workers, temporary workers, walking workers, bucket brigades or cross-trained workers. They issue optimization measures by addition, removal or interchange of process capabilities and workforce capacities [13, 14].

The aforementioned aspects, i.e. workforce diversity, planning and reconfiguration, have an impact on manufacturing system performance as well as workforce productivity. They can cause significant challenges for resilient system design as well as efficient production planning and control, depending on the labor market situation [2, 14]. Notably, there is little research on how manufacturing systems react to heterogeneous work speeds and workforce productivity by a diversified workforce. This is particularly true for flexible manufacturing systems, such as RCMS, where optimized workforce planning and workforce reconfiguration measures become even more complex, due to individual constraints and additional degrees of freedom [14].

The paper at hand addresses such questions by contributing a simulation-based method to comparatively evaluate the performance of AFL and RCMS concerning workforce diversity. It further demonstrates the synergetic aspects of RCMS with workforce reconfiguration measures (e.g. cross-trained work), to integrate and heuristically optimize a manufacturing system with a diversified workforce. The goal is to improve understanding of the application potential and decision-making on RCMS.

3 Method

This section describes how the linear and flexible manufacturing systems are set up and modeled in a simulation, which will be used in Sect. 4 to compare the system performance when faced with varying work speeds. The comparison is based on a mixed-multimodel engine assembly use case from an industrial partner. The following approach is used to model, compare and evaluate an AFL and an equivalent RCMS in a basic and a diversity scenario. It allows to systematically solve the various decision problems, such as segmentation, allocation or setup, in a sequential manner.

1. Product, process and capacity segmentations
2. Process-capacity and capacity-station allocations
3. Physical setup of stations
4. Heuristic optimization of process-capacity and capacity-station allocations

First, four reference products (V01-V04) are defined to sufficiently cover the existing product spectrum of engines. Methods-time Measurement is used to determine the processing times of all assembly steps. Subsequently, these are aggregated differently to derive the AFL and RCMS with 10 (AFL P01-P10) and 14 (RCMS P01-P14) processes respectively (c.f. Fig. 1). These process segmentations are conducted differently along either organizational (in the case of AFL) or functional, technical and logistical (in the case of RCMS) criteria, as they need to consider matters of tact orientation, line balancing, assembly restrictions, logistical material disposition, technology deployment and tool availability [15]. The aim is to achieve a fair setup for comparison where both settings are designed and optimized accordingly to the respective manufacturing system. The processing times are modeled with stochastic deviations via a triangular distribution of $\pm 10\%$ to represent common human and manufacturing system dynamics. Capacities are modeled for a workforce of ten workers (C01-C10) in both systems. In the use case, the work speed is modeled for every workforce capacity utilizing a linear process-independent factor. Hence, each combination of a process, product and capacity – an action – transforms a specific product based on the individual work speeds of the capacities [2]. The production program consists of a predefined mix of products with a random sequence that resembles actual production sequences in a real-life system. New jobs are released according to the principle of continuous work-in-progress (CONWIP), ensuring an even workload in the system and preventing both overflowing and starving; this is particularly important in the RCMS as routing flexibility can lead to overloading the production system. CONWIP permanently issues eight units of V01 and V02 as well as six units of V03 and V04. Setup processes and capacity or station failure are neglected in the use case, as they would make it harder to identify the effect of the diversity scenario. Evaluating the effects of machine failures on AFL and RCMS has been investigated by the authors in a previous publication [9].

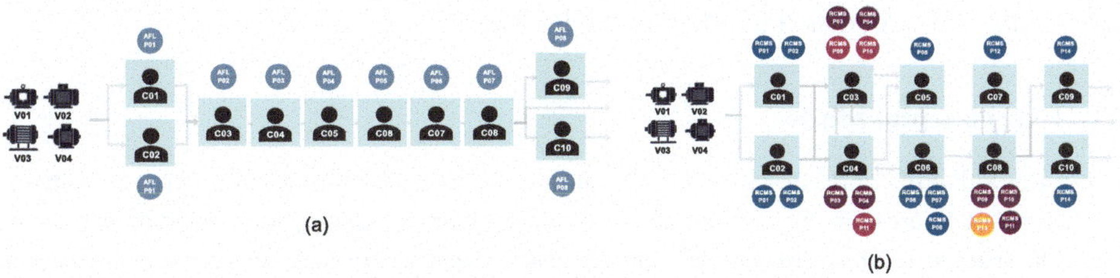

Fig. 1. Setup of the AFL (a) and RCMS (b)

Second, workforce planning assigns initial process-capacity and capacity-station allocations, i.e., how many workers are assigned to certain processes. Like the process segmentations, these allocations are conducted along different criteria. However, there are far more degrees of freedom for the process-capacity allocations in the RCMS, as operation sequences do not constitute a constraint.

Third, the production setup is determined, based on layout restrictions. Both the AFL and the RCMS are given the same space for workstations and work-in-progress.

Fourth, the process-capacity allocations and capacity-station allocations are heuristically optimized in the simulation model. This is an iterative process, with evaluations in the simulation followed by adaptions in the capacity allocations. It is especially important for RCMS where dynamic effects make it less obvious, how and where the operators should be assigned to tasks. Figure 1 outlines the setup of the AFL (a) and RCMS (b).

The AFL, characterized by a sequential arrangement of processes, capacities and stations, only allows for low flexibility (e.g., skipping a workstation if the processes offered by it are not required for a given product type). The RCMS, with a finer process segmentation and refined allocation, can utilize potentials from operation sequence and routing flexibilities. However, due to various entanglements of the assembled product components, none of the products in this case study allow for any operation sequence flexibility; this could potentially increase the flexibility of the RCMS setup if the manufacturing technology enables this. Therefore, in this use case the additional routing flexibility of the RCMS is solely achieved through redundant process-capacity allocations, created by the workforce reconfiguration strategy of cross-training and offering certain processing capabilities by multiple operators and workstations (e.g., Fig. 1, (b) shows how the processing steps RCMS P11 and RCMS P11 are offered by operators C03 and C08, due to cross-training).

The simulation model is implemented in Simo® version 14.22. An easily adaptable RCMS simulation framework was set up for modeling both manufacturing systems in a previous publication by the authors [9]. Additional features, such as a job pool with a queue, which sequences the jobs for release, and linear factors to alter differences in work speed on a process level are added for the experimental study.

4 Evaluation

This section presents the comparative performance analysis for varying work speeds, focusing first on the experimental study with the simulation, followed by an analysis of how the flexibility comes into effect in the RCMS.

4.1 Experimental Study

The modeled AFL and RCMS are validated and evaluated based on reference output, workforce utilization and given constraints with a sample size of ten experiments each. Two scenarios are set up for the AFL and RCMS respectively to analyze the influence of a diversified workforce. The basic scenario features a workforce of ten workers, all at full work speed (100%). The diversity scenario contemplates ten workers, with two workers (C03 and C05) performing considerably slower (70% work speed), while one (C09) is outperforming everyone else (110% work speed). Both the process-capacity and the capacity-station allocations are heuristically optimized for each setup in the simulation model. Table 1 shows the results of the experimental study.

Most notably, the loss of output of the RCMS in the diversity scenario (-5%) equals the total decrease in workforce productivity of the diversified workforce. There are no additional losses in workforce productivity, due to the manufacturing system flexibility. Contrary to that, the major output loss of the AFL in the diversity scenario (-23%) is caused by tact-bleeding, despite heuristic optimization measures on the process-capacity allocations to adjust the system to the altered staff capability. Frequently adjusting a tact-oriented process segmentation or allocation, based on fluctuating workforce diversity levels seems to be unfeasible, due to high planning and implementation expenses. In the use case, the AFL is not able to balance the workload for the diversified workforce and is therefore hampered by productivity losses caused by production system dynamics and (temporary) bottlenecks.

Table 1. Design of experiments and results

	Simulated AFL	Simulated RCMS
Basic Scenario		
Output [Units]	104	105
Workforce Utilization	86%	87%
Diversity Scenario		
Output [Units]	80 (–23%)	100 (–5%)
Workforce Utilization	77% (–9%)	87% (–0%)

Further investigations of the sensitivity on varying work speeds more or less, proved the robustness of the RCMS' performance, while the productivity losses of the AFL remained significant.

4.2 Flexibility Mechanism

The better relative performance of the RCMS in the diversity scenario is founded on refined targeted process-capacity allocations, which affect multiple processes using purposely created process entanglements of first and second grade. Figure 2 outlines these entanglements along capacities and processes as three cross-trained allocations are added. Entanglements of first grade are directly connected to cross-trained allocations, while entanglements of second grade are only indirectly affected. The combination of routing flexibility and such process entanglements between capacities allows for compensation of fluctuating workloads from the production program, mitigates potential efficiency losses of a diversified workforce in a manufacturing environment and overall ensures high capacity utilization.

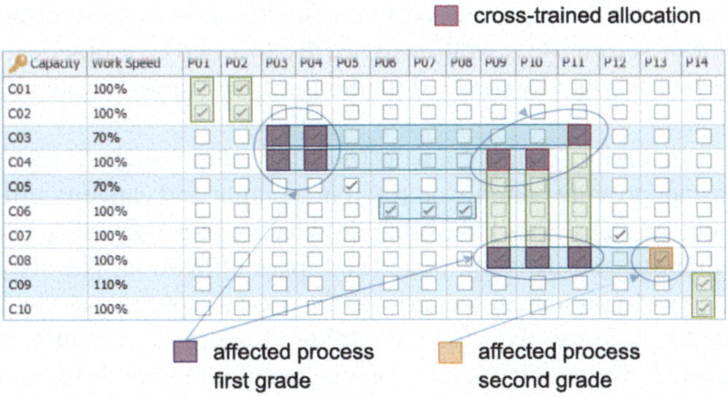

Fig. 2. Process-capacity allocations and entanglements through cross-trained allocations in the RCMS

5 Discussion and Outlook

The results confirm the feasibility of the method. They exemplify how workforce diversity can be modeled and (1) outline the impact of a diversified workforce with varying work speeds on different manufacturing systems' performances in a use case. Furthermore, (2) the presented study showcases that compared to conventional manufacturing, e.g. AFLs, flexible systems, e.g. RCMS', together with combined measures of workforce planning and reconfiguration, e.g. cross-training, allow for optimized integration of a diversified workforce. Consequently, manufacturing system resilience is strengthened and the impact of personnel risks is mitigated, i.e. RCMSs allow for better integration of a wider range of employees without major system productivity loss.

However, there are various limitations to be considered. RMCSs are most likely to outperform conventional AFLs when various constraints regarding market, product and production are existent at the same time. These include short-term market fluctuations in product demand (i.e. volatility and volume) or frequent changes in product design and production process need. Diversity aspects in production, such as persistent work speed differences by a diversified workforce, temporary in-line training measures or the need to integrate performance-diverse technologies or assistance systems, further favor the deployment of a RCMS. For the presented use case, the RCMS most probably does not provide an economic cost advantage, due to increased expenses on material disposition, transport, required personnel coordination efforts as well as hardware-software-integration costs for production planning and system control. Only when high flexibility requirements, as in the diversity scenario apply, the RCMS advantages justify the effort required to make these flexible systems work efficiently in practice. The additional effort includes a digital optimal planning aide (this would in most cases require some form of rolling horizon planning method, e.g. a digital twin with an automatically optimized planning function), cross-training of operators on multiple processes, a flexible intralogistics system that can transport products between stations in a RCMS, some form of visualization and worker-guidance system to help operators navigate their tasks in a system with changing processing requirements. Most of these can be tackled with digitalization, but it reinforces the notion that benefits analysis such as this can only be the starting point of setting up a well-designed RCMS before a practical implementation – this must be followed by detailed planning of the support systems, as described above.

For the AFLs diversity scenario, a workforce reconfiguration strategy, for instance, bucket brigades, can provide benefits on reference output and capacity utilization, however, this particular strategy does have other drawbacks on process responsibility, traceability, tool availability as well as equipment and space utilization. The diversity scenario currently represents a hypothetical scenario, as the use-case partner, as well as most companies, are more likely to use workforce planning measures to recruit workers that can comply with defined productivity levels, rather than to integrate different work speeds. However, the increasingly challenging task of finding and replacing suitably skilled and trainable personnel, also with demographic change, is likely to increase the relevance to integrate a diversified workforce.

This paper uses heuristic rule-based optimization for the aforementioned decision problems on segmentation, allocation and setup. Further technical work can be conducted on advanced planning and control as well as system design, particularly to ensure fast reactivity as well as efficient multi-criteria and energy optimization of RCMS through more sophisticated heuristic condition-based re-routing logics, metaheuristics or reinforcement learning [16].

Aside from the technical aspects, the flexibility considered in this paper also has a pronounced social and human dimension that is not the focus of the current work but should be considered in future research in this field. Two major aspects in this field are considerations of fairness and privacy. The aspect of fairness could bring up questions of payment according to workload vs. payment according to individual ability and its compatibility with the need for economic competitiveness. The privacy concerns include the value of transparent data (and collecting data) on individual skills, abilities and performance to be able to optimally plan with a diverse workforce vs. the need for employees to keep their strengths and weaknesses private to be safe from potential exploitation by the employers. These are just two of the major human factor considerations that should be investigated when looking at implementing measures to include a diversified workforce.

Acknowledgments. The authors would like to acknowledge the financial support of the European Commission for funding the H2020 research project EPIC under grant No. 739592.

References

1. Fries, C., Fechter, M., Nick, G., Szaller, Á., et al.: First results of a survey on manufacturing of the. Future **180**, 142 (2021)
2. Ihlenfeldt, S., Wunderlich, T., Süße, M., Hellmich, A., et al.: Increasing Resilience of Production Systems by Integrated Design 11, p. 8457, 2021
3. Fletcher, S.R., Baines, T.S., Harrison, D.K.: An investigation of production workers' performance variations and the potential impact of attitudes, vol. 35, p. 1113, 2008
4. Katiraee, N., Battini, D., Battaia, O., Calzavara, M.: Human diversity factors in production system modelling and design: state of the art and future researches, vol. 52, p. 2544, 2019
5. Koren, Y.: The Global Manufacturing Revolution: Product-Process-Business Integration and Reconfigurable Systems. Wiley, Hoboken, New Jersey (2010)
6. Schönemann, M., Herrmann, C., Greschke, P., Thiede, S. : Simulation of matrix-structured manufacturing systems, vol. 37, p. 104, 2015
7. Greschke, P., Schönemann, M., Thiede, S., Herrmann, C.: Matrix Structures for High Volumes and Flexibility in Production Systems, vol. 17, p. 160, 2014
8. Hofmann, C., Brakemeier, N., Krahe, C., Stricker, N., et al.: The Impact of Routing and Operation Flexibility on the Performance of Matrix Production Compared to a Production Line. In : Advances in Production Research, p. 155. Springer International Publishing, Cham (2019)
9. Perwitz, J., Sobottka, T., Beicher, J.-N., Gaal, A.: Simulation-based evaluation of performance benefits from flexibility in assembly systems and matrix production, vol. 107, p. 693, 2022
10. Forschungsbeirat Industrie 4.0/acatech – Deutsche Akademie der Technikwissenschaften (Hrsg.), 2022. Umsetzung von cyber-physischen Matrixproduktionssystemen
11. Khobreh, M., Ansari, F., Seidenberg, U.: 2020. A Knowledge-Based Approach for Linking Workforce Experience and Labor Productivity in Smart Factory Industry 4.0, in p. 17 (2020)

12. Bruecker, P.D., van den Bergh, J., Beliën, J., Demeulemeester, E.: Workforce planning incorporating skills: state of the art, vol. 243, p. 1, 2015
13. Mohammed, O., Nadia, B., Gerard, J.G.: Integrating workers' differences into workforce planning, vol. 63, p. 1096, 2012
14. Hashemi-Petroodi, S.E., Dolgui, A., Kovalev, S., Kovalyov, M.Y., et al.: Workforce reconfiguration strategies in manufacturing systems: a state of the art, vol. 59, p. 6721, 2021
15. Foith-Förster, P., Bauernhansl, T.: Axiomatic design of matrix production systems, vol. 1174, p. 1174, 2021
16. Czarnetzki, L., Laflamme, C., Halbwidl, C., Gunther, L.C., et al.: Optimisation of Matrix Production System: Reconfiguration with Reinforcement Learning.(submitted to KI 2023 – 46th German Conference on Artificial Intelligence), 2023

Tailored ALNS to Optimize Real-World Logistic Services for Dependent Patients

Francesco Pilati[1](✉), Riccardo Tronconi[1], and Karl Franz Doerner[2]

[1] Department of Industrial Engineering, University of Trento, Via Sommarive 9, Trento, Italy
francesco.pilati@unitn.it

[2] Department of Business Decisions & Analytics, University of Vienna, Oskar-Morgenstern Platz 1, Vienna, Austria

Abstract. In recent years, the number of disabled people have been growing drastically. This trend makes the existing healthcare logistic services inadequate to meet patients demand and, thus, a model to improve the efficiency of this service is necessary. To fulfil this need, the goal of this research is the development of a metaheuristic algorithm to solve a static Dial-a-Ride problem (DARP) with homogeneous requests and a single depot. Indeed, this paper presents the implementation of an Adaptive Large Neighbourhood Search (ALNS) algorithm with some destroys and repair operators tailored for the targeted problem to reach a high-quality solution in an acceptable time consistent with the need of a real healthcare department. This metaheuristic algorithm is compared with the current benchmark in literature and, also, it reports different Key Performance Indicators about the logistic service such as the average time worked and the total waiting time.

Keywords: Dependent patient logistics · healthcare service · adaptive large neighbourhood search · social sustainability

1 Introduction

Today, people with disability represent around 15% of the world population and this number is expected to grow in the next few years due to the increasing of both people aging and noncommunicable diseases [1]. Current healthcare logistic systems are struggling in efficiently manage this growing demand from disabled people. The main reason is that logistic has been always ignored in healthcare despite its relevance and impact in the efficiency of this public service [2]. To improve healthcare systems, such as logistics, mathematical methods have been developed through the years to model specific aspects and simulate different scenarios [3]. Among the healthcare logistic services, patients transportation is one of the most common problems to address which in research is identified as Dial-a-Ride problem (DARP). DARP is a combinatorial problem which differs from the Vehicle Routing Problem since the former optimizes also a social aspect, besides the economic one [4]. To define this social aspect, DARP usually emphasizes the minimization of patients inconvenience, expressed in terms of patients waiting time and ride time, thus neglecting the drivers working conditions which are rarely inserted in the problem objective function [5].

M. Dragomir et al. (Eds.): ICPR 2023, LNPE, pp. 345–353, 2025.
https://doi.org/10.1007/978-3-031-92082-0_38

The goal of this paper is the development of a mathematical model and a tailored metaheuristic algorithm to improve the efficiency of a healthcare logistic service. In particular, this logistic service consists in the dependent patients transportation problem, which well represents a realistic case study based on Austrian Red Cross (ARC) situation. The presented work aims at minimizing the total travel time, but it also analyses the effect of different routing solutions on the patients waiting time and drivers working conditions. The targeted DARP is firstly defined with a mathematical model and, then, solved through a tailored metaheuristic algorithm, whose performance is compared with the literature benchmark.

The paper is organized as it follows. In Sect. 2 a literature review on DARP is presented. Section 3 describes the characteristics of the targeted problem, as well as its objective function while Sect. 4 defines the proposed ALNS algorithm and its operators. Section 5 presents the case study applied to validate and test the developed metaheuristic algorithm and the results obtained. Finally, Sect. 6 draws the conclusions and proposes further research both in the algorithm and in the characteristics of the problem.

2 Literature Review

In literature, the problem of non-autonomous patients transportation falls into the DARP study, defined as the door-to-door transportation of passengers with personalized needs [6]. One of the most common DARP categories is the static version, which does not consider the time in which a request is performed and, thus, it supposes all the requests known in advance [7, 8]. Different types of metaheuristic algorithms have been developed to solve static DARPs. One of the most adopted metaheuristic algorithm is the tabu search, which consists in shaking a reference solution to find a new one, and keep in memory the latest solutions to avoid falling in a local minimum [9, 10]. Other contributions, instead, adopt a deterministic annealing algorithm in which a deteriorating solution can be accepted if the worsening in the objective function is lower than a deterministic threshold dependent on the algorithm advancement [11, 12]. These two types of metaheuristic algorithms are static, i.e. they cannot change during their run according to the past performance. To deal with this issue, novel researches on DARP adopt the Adaptive Large Neighbourhood Search (ALNS) algorithm which includes different destroy-repair operators and, at each iteration, it chooses them with a probability based on their past performance [13]. In [14], ALNS is implemented to solve a DARP and compared with other two metaheuristics to analyse the different performances. ALNS is also applied to solve DARP for a Ride Hailing scenario, i.e. passengers request a vehicle for a single trip from a pickup to a drop-off location, in an American case study [15]. The main drawback of many contributions which adopt ALNS is the implementation of standard destroy-repair operators which does not consider the characteristics of the tackled problem.

To fill these gaps in the literature, the presented work develop and implement an ALNS algorithm with tailored destroy-repair operators to solve a single-depot DARP with homogeneous requests and vehicles which represents a possible real-world case study. In detail, the developed metaheuristic includes some destroy and repair operators which consider the peculiar features of the targeted problem, such as the time windows or the ride time of the patients. In addition, this paper reports some peculiar Key Performance Indicators (KPIs) which measure the service quality for patients and the drivers conditions.

3 Problem Description

The tackled problem regards the dependent-patients transportation from home to an healthcare facility for a medical treatment, or the reverse transportation from the facility to their home. As stated in the first section, this type of problem is called DARP and, in particular, the targeted problem is a static DARP, i.e. all the transportation requests are known in advance. This problem is distinguished by a single depot from and to which each healthcare vehicle starts and ends its journey. In addition, requests are homogeneous since all the patients request the same transportation mode. Consequently, the fleet of vehicles is homogeneous too, all including the same pieces of equipment needed to fulfil the placed requests and all distinguished by the same capacity. Each healthcare vehicle is assigned to a single driver who is distinguished by a fixed working shift which cannot be violated. Each transportation request includes one patient and it is characterized by a time window within which the vehicle should arrive at the assigned location. In detail, these time windows cannot be violated on the right-hand side, i.e. vehicles cannot arrive after the latest arrival time. However, the early arrival is allowed and, in this circumstance, the vehicle would wait at the assigned location until the earliest arrival time. Thus, if the vehicle is carrying some patients, they would also wait outside the destination until the earliest time of its time window is reached.

Regarding the objective function, it aims at minimizing the total travel time spent cs in the solution s of this transportation service. Since the metaheuristic algorithm could provide infeasible solutions during the search, an evaluation function fs is computed to compare the different solutions (Eq. 1). This function includes, besides the objective function, the total violation of load (qs), ride time (rs), shift (ds) and time windows (ts). All these violations are weighted through dynamic parameters (respectively α, β, γ and ε) which are initialized with value 1 and increase or decrease of a value δ at each algorithm iteration. In detail, if solution s violates a specific constraint, the corresponding value is multiplied by δ, otherwise it is divided by it.

$$fs = cs + \alpha * qs + \beta * rs + \gamma * ds + \varepsilon * ts \tag{1}$$

It is important to clarify that a solution can be selected as the final best one only if it is feasible, i.e. $f(s) = c(s)$.

4 Metaheuristic Algorithm

Due to its several features and constraints, the targeted DARP is a NP-hard problem and, thus, it cannot be solved optimally for large instances. For this reason, an ALNS algorithm has been developed and implemented with tailored destroy and repair operators to solve this problem and find an efficient global best solution (s_best). ALNS is a metaheuristic algorithm based on different destroy and repair operators which are alternatively chosen at each iteration to provide a new solution. The peculiarity of this method is that the operators selection is adapted to their past performances.

In detail, after the generation of an initial reference solution s^* through a construction heuristic (described in Sect. 4.1), a destroy-repair operators pair is selected to modify s^* and find a new solution s'. If the new solution is promising, i.e. cs' at most 2% worse than cs^*, an intra-route local search is applied to further improve its quality. Then, an acceptance criteria (described in Sect. 4.3) is checked to measure the performance of the destroy and repair operators selected. The selection of the these operators is defined by a roulette wheel based on the weight w of each one (Eq. 2).

$$wi / \sum wk \tag{2}$$

where i represents the specific operator and k iterates over all the operators. In detail, at each iteration, according to the performance of the new generated solution, the score π of the adopted operators is updated with one of the following values:

- $\sigma 1 \rightarrow$ if the operators pair provides a new global best solution
- $\sigma 2 \rightarrow$ if the operators pair provides a new reference solution, but not best.
- $\sigma 3 \rightarrow$ in all the other circumstances

Every constant time interval *t_seg*, dependent on the dimension of the studied instance, the operators weights are updated through the following equation (Eq. 3):

$$wi, \mathrm{j} + 1 = wi, \mathrm{j} \, * \, (1 - \rho) + \rho * \, (\pi \mathrm{i} \, / \theta \mathrm{i}) \tag{3}$$

where i is the specific operator, j represents the time intervals in which the algorithm is divided, ρ is the reaction factor and θ is the number of times the specific operator i is selected in the current time interval j.

4.1 Initial Solution

The construction heuristic starts by sorting the transportation requests according to their time windows, putting first the earliest requests. Then, following this order, one request per route is inserted. The remaining ones are sequentially inserted in the route which minimizes one of the following four distances, randomly selected for each request individually:

- Distance between the pickup of the last request and the pickup node of the new one
- Distance between the pickup node of the last request and the delivery node of the new one
- Distance between the delivery node of the last request and the pickup node of the new one
- Distance between the delivery node of the last request and the delivery node of the new one

From the description of this construction heuristic, it is deducted that the generated initial solution can be infeasible with respect to time windows or shift violations.

4.2 Destroy and Repair Operators

The body of the developed ALNS is the implementation of different destroy and repair operators, including some general-purpose and some tailored for this specific DARP.

The destroy operators remove a specific number of requests which is randomly selected between a range defined by the contribution of [16]:

- Random removal: it removes random requests from random routes.
- Worst removal: it removes the worst requests in terms of evaluation function. In particular, it sequentially removes the request that provides the largest decrement in the evaluation function after its removal.
- Related removal: this operator displaces the first request randomly. Then, it removes the requests most related with the already-removed ones, where the relatedness R of two requests is defined by the travel time between them and the similarity of their time windows.
- Load violation removal: this operator removes the requests distinguished by the largest load violation. If it is null, the requests which generates the largest reduction of the objective function are selected for removal.
- Time-based removal: this operator removes the requests which cause the largest shift violation. If there is no shift violation, the requests distinguished by the largest violation of ride-time are displaced. If ride-time violation is null too, the requests which generates the largest reduction of the objective function are selected for removal.

The repair operators insert each request removed by the destroy ones:

- Best insertion: this operator inserts the requests in the order with which they are removed by the destroy operator. It finds the three best positions among all the routes for the current request and it selects one of them through a roulette wheel selection. The positions are evaluated in terms of evaluation function increase and, thus, the best position is the one that increase this function the least.
- Random insertion: same as "Best insertion" but here the requests are selected randomly from the list of removed requests.
- Regret-2 insertion: for each removed request, this operator finds the two best positions and, then, it computes the regret value as the difference in the evaluation function between these two positions. Finally, this operator selects the request which maximize this value.
- Zero-load insertion: it creates a list of positions in which the vehicle load is null. Then, this operator chooses randomly one of these positions and inserts a random request there. After each insertion, the new zero-load positions are updated.

4.3 Acceptance and Stopping Criteria

As described in the first paragraph of Sect. 4, at each iteration the ALNS generates a new solution *s'* which is compared to the reference one *s** and to the best one *s_best*. In detail, if the new solution is feasible and better than the best, i.e. fs' $< f$s_best, the former becomes the new global best solution. In addition, if *s'* is better than the *s**, it also becomes the new reference solution. On the contrary, if no improving solutions are found for a time *t_last*, which depends on the dimension of the considered instance, *s'* becomes the new reference solution in any case to shake the solution and move from the local optimum.

The developed ALNS stops when the run time reaches the time limit *t_tot*, which is based on the number of nodes of the tackled problem, and it provides the global best solution found up there.

5 Case Study and Results

The proposed DARP is a representation of a real-world scenario which can occur in the healthcare transportation service of ARC. However, to validate and test the presented work, the developed metaheuristic algorithm is implemented with the instances presented by [17]. In addition, Table 1 lists the values of the constant problem parameters adopted in the aforementioned case study.

Table 1. ALNS parameters

Parameter	Value
σ1	10
σ2	2
σ3	0
ρ	0.1
δ	UNIF(0.05–0.1)

The results obtained with the developed ALNS are compared to the ones of [17] which represents the current benchmark for this type of DARP. As shown in Table 2, the performance of the proposed metaheuristic algorithm is similar to the benchmark one since it is at most 5% worse in 18 out of 20 tested instances. Indeed, in two of the biggest instances (marked in bold type in Table 2) the developed ALNS is still distinguished by inefficient results which need to be improved.

Besides the comparison with the current benchmark, this paper analyses two performance indicators related to service quality and drivers condition. The first indicator is the average waiting time for each patient which provides a quantitative measure on patients satisfaction (Fig. 1). In this problem, the average waiting time varies drastically among the different instances, moving from almost 1 min/patient to more than 13 min/patient.

The second evaluated aspect is the time worked by drivers to quantitatively measure their working conditions. In detail, for each instance the minimum, average and maximum time worked by drivers are computed to study if the workload are balanced among them (Fig. 2). Results show that, apart from the first instance, the other ones are quite balanced since the range box are not widely spread, bounded between 200 min/day and 480 min/day.

Table 2. Comparison between proposed ALNS and current benchmark in all the instances of [17]

Instance	N° of requests	fs_best benchmark	fs_best ALNS	% worse
R1a	24	190.02	190.02	0.0
R2a	48	301.34	302.49	0.38
R3a	72	533.01	543.31	1.93
R4a	96	573.92	589.43	2.70
R5a	120	636.77	655.43	2.93
R6a	144	802.12	836.14	4.24
R7a	36	291.71	291.79	0.03
R8a	72	490.58	513.70	4.71
R9a	108	666.96	682.80	2.37
R10a	144	866.97	903.53	4.22
R1b	24	164.46	164.46	0.0
R2b	48	296.65	303.44	2.29
R3b	72	490.16	510.56	4.16
R4b	96	533.15	558.31	4.72
R5b	120	583.12	610.76	4.74
R6b	**144**	**742.12**	**796.39**	**7.29**
R7b	36	248.21	248.47	0.10
R8b	72	462.50	470.97	1.83
R9b	108	600.18	627.93	4.62
R10b	**144**	**796.90**	**876.98**	**10.05**

Fig. 1. Average waiting time in each tested instance

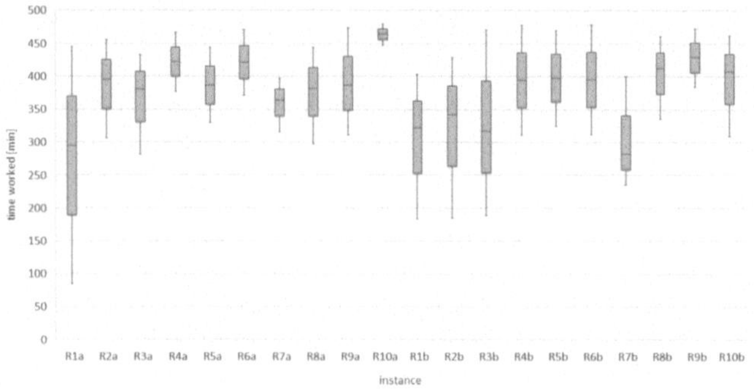

Fig. 2. Range box of the time worked by drivers in each tested instance

6 Conclusions

In the presented paper, a realistic transportation problem of non-autonomous patients is tackled. In detail, this problem is solved as a single-depot DARP with homogeneous requests and vehicles, with the objective of minimizing the total travel time. The peculiarity of this work is, firstly, the consideration of additional aspects in the solution analysis, such as the patients waiting time and the drivers conditions in terms of time worked. Another novel characteristic of this research is the implementation of an ALNS metaheuristic algorithm based on tailored destroy and repair operators, which is adopted to solve large instances of the targeted problem. The developed metaheuristic presents performances similar to the ones of the literature benchmark (at most 5% worse) for 18 out of 20 tested instances. Furthermore, results show that the patients average waiting time ranges between 1 and 13 min/patients among the different instances while the time worked by drivers is quite balanced between the least and the most employed workers.

Further research can be carried out about this type of transportation problem. First of all, the developed ALNS should be improved to be efficient also in the biggest instances of the literature benchmark. Furthermore, additional features can be applied to the targeted problem, such as heterogenous requests or multiple depots, to make it closer to the reality. Finally, the developed ALNS can be tested and validated with a real dataset provided by specific healthcare department, such as the ARC, to analyse its performance in a real-world scenario.

References

1. United Nations (UN)., Factsheet on persons with disabilities. https://www.un.org/development/desa/disabilities/resources/factsheet-on-persons-with-disabilities.html. Accessed 24 May 2023
2. Beaudry, A., Laporte, G., Melo, T., Nickel, S.: Dynamic transportation of patients in hospitals. OR Spectrum **32**, 77–107 (2008)
3. Vanagas, G., Krilavicius, T., Man, K.L.: Mathematical modeling and models for optimal decision-making in health care. Comput. Math. Methods Med. (2019)
4. Berbeglia, G., Cordeau, G.F., Laporte, G.: Dynamic pickup and delivery problems. Eur. J. Oper. Res. **202**, 8–15 (2010)

5. Guerriero, F., Pezzella, F., Pisacane, O., Trollini, L.: Multi-objective optimization in dial-a-ride public transportation. Transp. Res. Procedia **3**, 299–308 (2014)
6. Toth, P., Vigo, D.: Vehicle routing: problems, methods, and applications, 2th edition, Society for industrial and applied mathematics eds., (2014)
7. Cullen, F.H., Jarvis, J.J., Ratliff, H.D.: Set partitioning based heuristics for interactive routing. Networks **11**, 125–143 (1981)
8. Melis, L., Sorensen, K.: The static on-demand bus routing problem: large neighborhood search for a dial-a-ride problem with bus station assignment. Int. Trans. Op. Res. **29**, 1417–1453 (2022)
9. Aldaihani, M., Dessouky, M.M.: Hybrid scheduling methods for paratransit operations. Comput. Ind. Eng. **45**, 75–96 (2003)
10. Kirchler, D., Calvo, R.W.: A granular tabu search algorithm for the dial-a-ride problem. Transp. Res. Part B **56**, 120–135 (2013)
11. Braekers, K., Caris, A., Janssens, G.K.: Exact and meta-heuristic approach for a general heterogeneous dial-a-ride problem with multiple depots. Transp. Res. Part B **67**, 166–186 (2014)
12. Su, Y., Dupin, N., Puchinger, J.: A deterministic annealing local search for the electric autonomous dial-a-ride problem. Eur. J. Oper. Res. **309**, 1091–1111 (2023)
13. Mara, S.T.W., Norcahyo, R., Jodiawan, P., Lusiantoro, L., Rifai, A.P.: A survey of adaptive large neighborhood search algorithms and applications. Comput. Oper. Res. **146**, 105903 (2022)
14. Masmoudi, M.A., Hosny, M., Braekers, K., Dammak, A.: Three effective metaheuristics to solve the multi-depot multi-trip heterogeneous dial-a-ride problem. Transp. Res. Part E **96**, 60–80 (2016)
15. Syed, A.A., Kaltenhaeuser, B., Gaponova, I., Bogenberger, K.: Asynchronous adaptive large neighborhood search algorithm for dynamic matching problem in ride hailing services. In: IEEE Intelligent Transportation Systems Conference (ITSC), pp. 3006–3012, Auckland, New Zealand (2019)
16. Ropke, S., Pisinger, D.: An adaptive large neighborhood search heuristic for the pickup and delivery problem with time windows. Transp. Sci. **40**(3), 455–472 (2006)
17. Parragh, S.N., Doerner, K.F., Hartl, R.F.: Variable neighborhood search for the dial-a-ride problem. Comput. Oper. Res.Oper. Res. **37**, 1129–1138 (2010)

Production Research in Romania: Trends, Needs and Agenda

D. Popescu, M. Dragomir(✉), S. Popescu, C. Neamțu, and D. Dragomir

Technical University of Cluj-Napoca, Bd. Muncii 103-105, Cluj-Napoca, Romania
mihai.dragomir@muri.utcluj.ro

Abstract. An examination is performed on the future trends that are developing in the field of production research, that are pertinent to the academic and industrial needs in Romania. In the proposed approach, we include, for an analysis based on experience, the most relevant directions and investigations for research demarches, and we apply the strategic forecasting principles of Hoshin Kanri. The outcome of this process can form the basis of a future research agenda in the field of production, including topics related to manufacturing systems, digital transformation, innovative technologies, quality management and sustainability aspects.

Keywords: production research · production engineering · production management

1 Introduction

In the past 30 years, Romania has undergone many economic, social, and cultural changes, as it moved away from the previous political and economic system to become a full member of the European Union. One of the most challenging domains during this transition has been without a doubt the manufacturing sector. Once the backbone of the planned communist economy, albeit under non-market oriented and non-environmentally friendly conditions, the production industry has shrunk considerably in terms of volume of goods produced, functional factories and employed workforce. At the same time, the companies that managed to survive in this domain, as well as the ones that appeared in the new capitalist economic system, were forced to become efficient, innovative, and connected to the needs of their customers, in order to remain competitive.

The present paper draws upon the experience of these decades within research projects in the field of production engineering and management and proposes a perspective for the future based on the work performed at the Technical University of Cluj-Napoca, while at the same time integrating the trends present in the country as a whole. The final result takes the form of a scientifically grounded projection over the next 10 to 15 years in this field, with the goal to enable future researchers to come up with new creative solutions without repeating the mistakes of the past.

M. Dragomir et al. (Eds.): ICPR 2023, LNPE, pp. 354–359, 2025.
https://doi.org/10.1007/978-3-031-92082-0_39

2 Methodology

The research methodology employed is rooted on personal research and consultancy experience distillation, combined with actual project management experience in the private sector, and strategic thinking in terms of future options for companies and universities active in production research (Fig. 1).

We estimate that the experience processing that has been used as input in our research includes over 50 companies in various domains of the manufacturing sector that our research team has supported in the past 20 years. This includes metal fabrication, automotive, machinery, furniture, textile, plastics, electronics and industrial goods manufacturers from across Transylvania. On the other hand, the hands-on experience we have had in various companies and project management roles, played an important role in introducing the needs of the market into our strategic foresight exercise based on the Hoshin Kanri methodology.

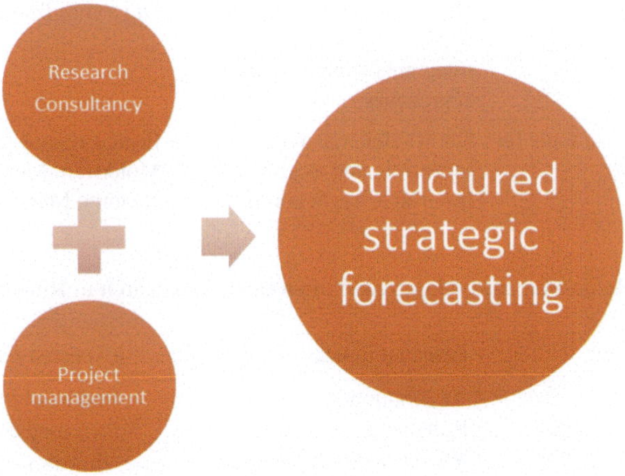

Fig. 1. Summary of the research methodology implemented

3 Results

The analysis has been divided into two steps:

- a survey of published research results from Romanian researchers superimposed on the detected main current research themes and topics (Table 1) and
- proposed framework of tentative research themes and topics in the future, together with an estimation of the most important results (Table 2)

In the first step, we can see 5 main directions that have proved interesting for research teams and for their industrial and/or academic partners. The first two columns detail an aggregation of the most common topics at the level of research themes and subjects addressed, while the last column exemplifies these structural elements with some relevant examples published by Romanian authors in the past years. A thorough literature review should identify more publications, but we believe the research axes would remain the same.

Table 1. Current landscape in production research in Romania

Current research themes	Investigation topics	Representative examples
Industry 4.0 readiness and technologies	Shifting technology trends Managing and reducing costs Limitations of technology	Țîțu, Doicin et al., 2023 [1] Doicin, Ulmeanu et al., 2021 [3] Haba and Oancea, 2023 [10] Alexa, Pîslaru et al., 2022 [14]
Manufacturing processes deployment	Technical problem solving Production system management Enabling cost-based competition	Stan, Turcanu et al., 2020 [4] Popan, Bâlc et al., 2022 [6] Braga, Udroiu et al, 2022 [8] Mitru, Semenescu et al. [13]
Environmental and social impact of production	Accelerating pace of change Increasing requirements Limited available know-how	Kifor&Grigore, 2023 [2] Drăghici, Dursun et al., 2022 [5] Cioca, Kot et al., 2023 [7]
Diffusion and transformation of products and production	New challenges and risks Development of skills and talents Integrated solutions and approaches	Crăiuț, Bungău et al., 2022 [9] Dragomir, Blagu et al., 2022 [15]
Production engineering tools for medical and cultural applications	Knowledge transfer Incentives and motivations Innovative applications	Pîslă, Crișan et al., 2023 [11] Moldovan, Moldovan et al. [12] Neamțu, Măruțoiu et al. [16]

Table 2. Forecasted landscape in production research in Romania

Future research themes	Potential topics	Expected results
Industry 4.0 and 5.0 implementation and results	AI adoption Robotics evolution VR/AR maturation	Improved processes Increased economic efficiency Improved productivity
Production sustainability and resilience	Low carbon production Circular economy	ESG performance Limiting climate change Resource saving and reuse
Production of physical goods in a VUCA World	Digital transformation Production systems engineering Smart products and processes	Flexibility, reconfigurability Scarce inputs adaptability Producers as change agents
Integration of production within society	Human factors in production Friendly production facilities Globalized supply chains	Future-work environments Alignment to societal goals Pozitivation of localized
Competence sharing across industries	Production to services Production to agriculture Production to public sector	Economic system transformation Production renaissance Future proofing of industries

The second step was generated through successive iterations using the specific Hoshin Kanri deployment stages of the X matrix tool, going from long term planning to intermediary and then to operational planning, but adapted to the needs of a more impalpable topic.

The principal factors to be considered in the brainstorming sessions were related to industrial development policies, regional aspects, international cooperation, as well as technical trends and market transformations that are expected to take place.

4 Discussion

It is apparent from our analysis that production research, including engineering and management topics, is in the process of being aligned with international trends. There is still a considerable component dedicated to manufacturing technologies and materials processing, but interdisciplinary aspects related to transforming European production into a more competitive one are becoming more important. There is a detectable shift from the compartmentalized treatment of classic or even modern subjects, such as environmental impact, to a more integrated approach that connects the challenges of product development, process improvement, markets and communities, into a holistic way of thinking. This can enable better and more impactful results in the long run, with improved visibility and more chances to make relevant contributions on an international scale.

One component which is not well represented deals with scheduling, optimization and supply chain performance, which are all connected to advanced mathematical modeling. As more and more practical solutions become available at low costs and will be implemented by companies, it is also expected that this field, should it become more important, will mostly have a pragmatic approach to it, rather than an abstract one. The impact of digital and virtualized technologies will continue to be of great importance, but the human aspects of production, as well as the transformational effect this domain has upon the economy and society as a whole, will probably be at the forefront of preoccupations. We also expect knowledge sharing to intensify across industries, as production has already learned important lessons from its challenging transformations that other fields will also face as part of their drive for more resilience in the face of local and global threats. Of course, we should also mention that other important issues have been dealt with or might become important, that are not covered here, as the space for this paper is limited.

References

1. Țîțu, A.M., cin, C.V., Ionescu, N., Gusan, V.: Application of the PFMEA methodology in a production flow with collaborative robots. In: Karabegovic, I., Kovačević, A., Mandzuka, S. (eds.) New Technologies, Development and Application VI. NT 2023. LNCS, vol. 687, pp. 27–36. Springer, Cham (2023). https://doi.org/10.1007/978-3-031-31066-9_3
2. Kifor, C.V., Grigore, N.A.: Circular economy approaches for electrical and conventional vehicles. Sustainability **15**(7), 6140 (2023). https://doi.org/10.3390/su15076140
3. Doicin, C.V., Ulmeanu, M.E., Rennie, A.E., Lupeanu, E.: Mass optimisation of 3D-printed specimens using multivariable regression analysis. Int. J. Rapid Manuf. **10**(1), 1–22 (2021)
4. Stan, F., Turcanu, A.M., Fetecau, C.: Analysis of viscoelastic behavior of polypropylene/carbon nanotube nanocomposites by instrumented indentation. Polymers **12**(11), 2535 (2020). https://doi.org/10.3390/polym12112535
5. Draghici, A., Dursun, S., Başol, O., Boatca, M.E., Gaureanu, A.: The mediating role of safety climate in the relationship between transformational safety leadership and safe behavior—the case of two companies in turkey and Romania. Sustainability **14**(14), 8464 (2022). https://doi.org/10.3390/su14148464
6. Popan, I.A., Balc, N., Popan, A.I., Cerciu, A.: Experimental studies on turning process, using the Prime turning processing strategy, proposed by Sandvik Coromant. In: IOP Conference Series: Materials Science and Engineering, vol. 1268, no. 1, 012018. IOP Publishing (2022)
7. Cioca, L.I., Kot, S., Bratu, M.L.: Optimizing the management system and ergonomics in organizations, appropriate to the personality of employees. Front. Psychol. **13**, 1105192 (2023)
8. Braga, I.C., Udroiu, R., Nedelcu, A.: Novel method for failure modes detection in UV-cured clear coated polymer for automotive interior mechatronic devices. Polymers **14**(18), 3811 (2022). https://doi.org/10.3390/polym14183811
9. Craiut, L., Bungau, C., Bungau, T., Grava, C., Otrisal, P., Radu, A.F.: Technology transfer, sustainability, and development, worldwide and in Romania. Sustainability **14**(23), 15728 (2022). https://doi.org/10.3390/su142315728
10. Haba, S.A., Oancea, G.: Study on design and manufacturing of an engine block using digital tools. In: Kyratsis, P., Manavis, A., Davim, J.P. (eds.) Computational Design and Digital Manufacturing. Management and Industrial Engineering, pp.167–203. Springer, Cham (2023). https://doi.org/10.1007/978-3-031-21167-6_9
11. Pisla, D., et al.: Structural study of a robotic system for sils surgery. In: Tarnita, D., Dumitru, N., Pisla, D., Carbone, G., Geonea, I. (eds.) New Trends in Medical and Service Robotics. MESROB 2023. Mechanisms and Machine Science, vol. 133, 20–31. Springer, Cham (2023). https://doi.org/10.1007/978-3-031-32446-8_3
12. Moldovan, F., Moldovan, L., Bataga, T.: Assessment of labor practices in healthcare using an innovatory framework for sustainability. Medicina **59**, 796 (2023). https://doi.org/10.3390/medicina59040796
13. Mitru, A., Semenescu, A., Simion, G., Scutelnicu, E., Voiculescu, I.: Study on the weldability of copper— 304L stainless steel dissimilar joint performed by robotic gas tungsten arc welding. Materials **15**, 5535 (2022). https://doi.org/10.3390/ma15165535
14. Alexa, L., Pîslaru, M., Avasilcăi, S.: From industry 4.0 to industry 5.0—an overview of european union enterprises. In: Draghici, A., Ivascu, L. (eds.) Sustainability and Innovation in Manufacturing Enterprises. Advances in Sustainability Science and Technology, pp. 221–231. Springer, Singapore (2022). https://doi.org/10.1007/978-981-16-7365-8_8

15. Dragomir, M., Blagu, D.A., Popescu, S., Fulea, M., Neamțu, C.: How well are manufacturing companies in Transylvania, Romania adapting to the low- carbon economy in order to become sustainable? Int. J. Environ. Res. Public Health **19**, 2118 (2022). https://doi.org/10.3390/ijerph19042118
16. Neamtu, C., et al.: Multidisciplinary investigation of the imperial gates of the 17th century wooden church in Sălișca, Cluj county Romania. Sustainability **10**, 1503 (2018). https://doi.org/10.3390/su10051503

Increasing Supply Chain Performance of a SME by Increasing Delivery Function

C. I. Ratiu[1] and C. I. Ucenic[2(✉)]

[1] Department, Technical University Cluj Napoca, Muncii Boulevard 103, Cluj Napoca, Romania
[2] Department MIE, Technical University Cluj Napoca, Muncii Boulevard 103, Cluj Napoca, Romania
camelia.ucenic@mis.utcluj.ro

Abstract. A higher supply chain performance is related with a better logistics. The prospective significant problems can be divided in smaller issues. Each issue can be analysed and improved in order to obtain a better performance level. SMEs have limited resources and this approach allow them to become more competitive. This study analyses the delivery function for a small Romanian firm which intends to implement the QDCM model. Suitable planning and execution is mandatory for delivery. The delivery function has four parameters: on time delivery, variance of supplier lead time, backlog efficiency and transportation efficiency. Data was collected for six months from 2022. The average value of delivery function was compared with the thresholds specified by literature review.

Keywords: Delivery function · modelling · performance · supply chain

1 Introduction

A well-managed and performant supply chain will provide goods and services to the customers according with their order specifications. The companies invest much time and efforts in order to develop strategies for improving supply chain performance. Every supply chain is characterised by three flows: material, cash-flow and information. Plan, source, make and deliver are the central activities from a supply chain.

The supply chain performance has an important impact on the goals of the company, in short-term as well as in long-term objectives. The optimisation of supply chain generates a decrease with about 15% for the costs of delivery [16].

Unexpected disturbances can occur in the supply chain. The last years putted a lot of pressure at local and global level for supply chains. The pandemic and the war generated shortage of raw material, delays in delivery, sanitary limitations. Only the best managed supply chains were able to adapt and face these problems. A large extend of theoretical and practical studies were done in order to assure a higher performance of supply chains and a better firms' competitiveness. The firms intend to develop agile and sustainable supply chains.

M. Dragomir et al. (Eds.): ICPR 2023, LNPE, pp. 360–368, 2025.
https://doi.org/10.1007/978-3-031-92082-0_40

2 Aspects Regarding the Delivery Performance

The delivery performance is one of the main aspects for a competitive supply chain. It has to be increased at company level as well as at whole supply chain level. The delivery performance is related with the material flow.

The delivery performance can be approached by different sights: the entire amount of accomplished deliveries, on time delivery, average time of a delivery, expected delays of delivery and others. Oracle made a study and one of its findings was that thirteen percent of the clients will not buy anymore from a company if the demand was not delivered at the planned moment [17].

Delivery performance management assumes to gather, to process, to investigate and to communicate the aspects related with the delivery processes. Proper measurement methods and reporting structures have to be developed and implemented.

The companies work in a dynamic environment. The lack of unlimited resources and capabilities obliged the firms to create and to use innovation tools and methods in the field of delivery performance. They use brainstorming, benchmarking, best practices and other innovative approaches to increase delivery performance [9].

A global survey disclosed that more than forty percent of companies intend to rise their expenditure in technology and about one third of the firms have implemented some types of artificial intelligence instruments in their daily supply chain and delivery activities [16].

The general delivery statistics for 2020 reveal that a standard delivery takes about three to five working days and the average time per delivery is twelve minutes. On another hand, if a customer has two or more bad delivery experiences with a company, the majority of them will change the firm and make contracts with a competitor of this firm [15].

A lot of research was done in the field of supply chain and delivery performance. The models which appraise delivery performance are grouped in two types: index based [1] and cost based models [2, 10].

Many sub-metrics were developed for the main indices. The managers from the companies have to decide which are the most suitable for their firm. Some of the sub-metrics of delivery performance are on time delivery [6], delivery speed [7], suppliers' delivery performance [8], and order fulfilment lead time [12].

Other authors proposed methods centred of the approach of supply chain delivery process as a linkage of activities which are required to be improved. This approach treats the delivery process as complete, without dividing it in the components activities and not considering the relationships between its activities [13].

A high level of delivery performance has a lot of positive influences as increased customer transaction volume and a better unit price [14].

3 Methodology

The study was done with the main goal of collecting reliable information for implementing in the future the QCDM (Quality, Cost, Delivery, Motivation) model. This managerial approach was implemented for the first time in the British automotive companies. The main advantage of it is that allows one to divide the main problem into several small problems. Dealing with smaller problems permit to the managers to solve faster the weak points and generate better solutions. QCDM is a cost effective method. The fact that it is easy to be practiced is another great advantage.

The quality is related with the customers' desires to receive a proper good at the ordered features and specifications. The second element of the model–the cost, is associated with the manufacturing and distributing the goods. The delivery has to provide the accurate volume of ordered products, at the specified moment and in the correct time. The motivation aspects cover the participation of workforces at the working place.

This work investigates only one aspect of the model, the delivery function. It is considered as a performance measure for supply chain. The delivery function was calculated for six months for a Romanian SME. Due to confidentiality reasons its name is not revealed. Four elements were considered for describing the delivery function: on time delivery OTD, percentage difference between expected time delay of delivery and real time delivery APD, backlog efficiency BE and transport efficiency TE. It was calculated the average value for each of them and compared with the theoretical thresholds in order to categorised from the point of view of performance level.

4 Case Study: Evaluation of Delivery Function

The starting point of this analysis is the model proposed by Fechete and Nedelcu [4, 5]. The model was adapted from the point of view of APD element.

The delivery function is described by the following equation:

$$DF = a \times OTD + b \times APD + c \times TE + d \times BE \quad (1)$$

DF–delivery function.
OTD–on time delivery.
APD–absolute percentage difference between expected delay time of delivery and real time delay.
TE–transport efficiency.
BE–backlog efficiency.
a, b, c,,d–the weight coefficients associated with each element

$$a + b + c + d = 1$$

The first step was to calculate the weight coefficients. It was done using the method of specialists. The selected specialists were the middle managers from commercial and marketing area. Each specialist gave a grade from one to ten for each element. The coefficient was calculated as the fraction between the score achieved by an element and the total score (Table 1).

$$a = \frac{47}{177} = 0.265$$

$$b = \frac{40}{177} = 0.225$$

$$c = \frac{46}{177} = 0.259$$

$$d = \frac{44}{177} = 0.248$$

Table 1. Score achieved by each element

	S1	S2	S3	S4	S5
OTD	9	10	9	10	9
APD	8	9	8	8	7
TE	9	9	8	10	10
BE	8	10	9	8	9

The final form of delivery function became:

$$DF = 0.265 \times OTD + 0.225 \times APD + 0.259 \times TE + 0.248 \times BE$$

4.1 On Time Delivery OTD

The on time delivery OTD is calculated as the percentage of orders delivered at the moment imposed by the contract out of total number of deliveries (Table 2).

$$OTD = \frac{\text{NODT}}{\text{TND}} \times 100 \quad (2)$$

OTD–on time delivery.
NODT–number of orders delivered in time
TND–total number of deliveries

Table 2. On time delivery

	NODT	TND	OTD
March	114	128	89.06
April	98	100	98
June	79	86	91.86
July	87	105	82.85
August	118	140	84.28
September	148	153	96.73

The lowest values of OTD are in July and August. A possible explanation can be the lack of personnel which prepare the orders, due to the holiday. The highest value is in April. It is sustained by the bonuses received for Easter, the motivation of the workers was increased. A better organisation of delivery for the holidays is strongly required. It can be done either by adding additional shifts for the workers or by bringing new persons with part-time jobs for the summer season.

4.2 Absolute Percentage Difference Between Expected Time Delay of Delivery and Real Time Delivery APD

Delivery process is affected by different internal and external factors. The time spent in the borders increased due to the war and because of the fact that Romania is not included in Schengen space. The delays in delivery are generated by the differences between the real time delivery and the planed time for delivery. A high quality of delivery time forecasting allows a better planning of all process. This aspect will offer a higher level of customers' satisfaction (Table 3).

$$APD = \frac{ED - RTD}{RTD} \times 100 \tag{3}$$

$$DD = \frac{(ED - RDT) \times TNO}{RDT} \tag{4}$$

APD–percentage difference between expected time delay of delivery real time delivery.
DD–delays in delivery.
ED–expected delay (forecasted delay).
RDT–real delivery time.
TNO–total number of orders.

Table 3. Calculation of DD and APD

	RDT (days)	ED (days)	DD	APD %
March	4.2	6.9	82.28	64.28
April	3.5	6	71.42	71.42
June	2.4	4.4	71.66	83.33
July	4.6	8.6	91.30	86.95
August	3.8	6	81.05	57.89
September	4	6	76.5	50

4.3 Transport Efficiency TE

The costs of transportation recorded unexpected variations. The increase of fuel price affected the transportation market. Another weak point was the lack of drivers and an increase of their wages. High inflation rates were registered in all countries (Table 4).

Transport efficiency TE

$$TE = \frac{\text{RTC} - \text{PTC}}{\text{PTC}} \times 100 \tag{5}$$

RTC–real transport cost.
PTC–planed transport cost.

Table 4. Transport efficiency TE

	RTC (€)	PTC (€)	TE (%)
March	2800	2055	36.25
April	3200	2055	55.71
June	3800	2055	84.91
July	3120	2055	51.82
August	2500	2055	21.65
September	2700	2055	31.38

Transport efficiency varies in a large range, from 21.65% to 84.91%. This fact suggest the necessity of finding cheaper transportation firms or renegotiate the existing contracts.

4.4 Backlog Efficiency BE

The backlog represents a feature which is necessary to reach a chosen level and to have desired capabilities. Backlog is a matter of change and has its particular metrics. The calculation of the backlog helps the managers to prioritize tasks and better organise the activity [3].

The backlog has to be regularly revised. The effort to reach a task is compared with the value obtained after accomplishing that task. The effort is quantified from the point of view of resources necessary to reach the goal. Backlog can have different signification for different teams / managers from the company. Its estimation can be done from three point of view: the difficulty of the assignment, the amount of work which is required and the uncertainty given by the analysed feature [11].

Backlog efficiency BE is the percentage of total backlog delivered this month divided with the total backlog in this month (Table 5).

$$BE = \frac{\text{BDM}}{\text{TBM}} \times 100 \tag{6}$$

BDM–backlog delivered this month.
TBM–total backlog of this month.
The highest backlog efficiency was in April and the worst in June.

Table 5. Backlog efficiency BE

	BDM	TBM	BE
March	12	15	80
April	19	21	90.47
June	14	26	53.84
July	28	32	87.5
August	9	14	64.28
September	16	22	72.72

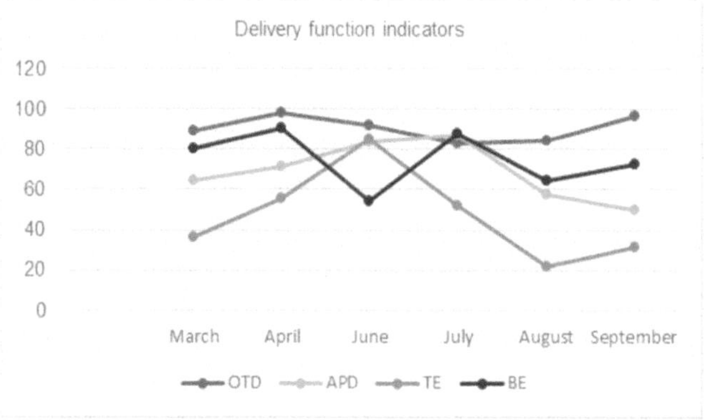

Fig. 1. Evolution of delivery indices

4.5 Delivery Function

The evolution of the delivery indices is illustrated in Fig. 1.

It can be observed that all indices have upward trends in April. The lowest values are registered in the summer, the months July and August have a downward trend. Three out of four indices have increased values in September.

The literature review mentions three thresholds for evaluation of performance. Non-performance is defined by values in the interval [1, 5]. An average performance has values in [5, 8], while high performance values are in the interval [8, 10].

The biggest fluctuation was observed at the backlog efficiency BE which varies in opposite direction every month. This fact suggests that the company was not able to sustain a constant effort for decreasing BE. The average value of BE is 74.8% which proves an average value of performance, but closed with the limit of high performance at 80%.

The greatest interval of variation is related with the transport efficiency. TE varies from 21,6% to 84.9%. Its average value is 46.95%. This value illustrates a non-performance situation. It is recommended to find alternative transportation solutions, especially for the summer time.

Good measures were implemented for APD. It had an upward trend in the first four months but the trend shifted to a downward trend in the last two months. The average value of APD is 68.97% which defines an average performance level.

Another element which registered a desired trend in the last two consecutive months is on time delivery OTD. OTD takes an average value of 92.46%, so it has a high performance. The last months present a continuous improvement for two (OTD and TE) out of four elements. More efforts must be done for improving BE and APD.

The last step of evaluation consists of calculation of delivery function. The computation is done using formula 1. The indices are weighted with the above calculated weights. The results are presented in Fig. 2.

The delivery function shows a downward trend from April to August. The company has to make a deeper analysis for these time interval in order to find the negative elements which decreased delivery function. Some efforts were done in September, fact which is illustrated by a strong increase of the function.

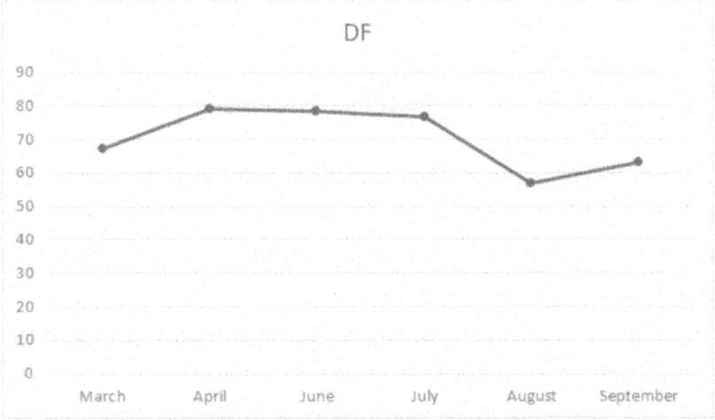

Fig. 2. Delivery function

The average value of delivery function is 70.2%. The analysed company is one with average performance from the point of view of delivery function.

5 Conclusions and Recommendations

The foremost types of supply chain metrics are time, quality and costs. This study had as starting point the QDMC model, with emphasis on delivery function. The evaluation of delivery function was done for a Romanian SME firm. The analysis covered six months from March to September 2022. The usage of a larger time interval was not possible because of the lack of coherent data.

Four elements were considered for describing the delivery function: on time delivery OTD, percentage difference between expected time delay of delivery and real time delivery APD, backlog efficiency BE and transport efficiency TE. The average values of each element and of delivery function were compared with the thresholds specified by the literature review.

OTD has an average value of 92.46%, meaning a high performance. The average value of BE is 74.8% which demonstrates an average value of performance. The average value of APD is 68.97% which defines an average performance level. TE average value is 46.95% and proves a non-performance situation.

Only one element is in the best performance category, two are in the medium performance and one is in non-performance case. The average value of delivery function is 70.2%.The firm has a medium performance delivery function but its value is closed to the threshold toward high performance. These values help the managers to prioritise the delivery function elements and the measures imposed for increasing them. A higher value of delivery function generates a higher performance of company's supply chain.

Further research can be done for a larger time interval when much data is available. Another improvement of the study is to add other elements in delivery function.

References

1. Bushuev, M.A.: Delivery performance improvement in two-stage supply chain. Int. J. Prod. Econ. **195**, 66–73 (2019)
2. Bushuev, M.A.: Optimal position of supply chaindelivery window: concepts and general conditions. Int. J. Prod. Econ. (2012)
3. Cebi, B.: How to manage backlog efficiently?. https://blog.producter.co/how-to-manage-backlog-efficiently/
4. Fechete, F.,Nedelcu, A.: Model for determining and optimizing delivery performance in industrial systems. In: MATEC Web of Conferences, vol. 94, p. 060006 (2017)
5. Fechete, F., Nedelcu, A.: Performance management assessment model for sustainable development. Sustainability **11**(10), 2779 (2019)
6. Katayama, H., Bennett, D.: Agility, adaptibility & leanness: a comparison of concepts and a study of practice. Int. J. Prod. Econ. **60–61**, 43–51 (2004)
7. Mason, R., Lalwani, C.: Integrating transportation into the supply chain to improve supply chain performance. In: Proceedings of 9th Annual Logistics Research Network Conference, pp. 370 – 378 (2004)
8. Morgan, C., Dewhurst, C.: Multiple retailer supplier performance: an exploratory investigation into using SPC techniques. Int. J. Prod. Econ. **111**, 13–26 (2008)
9. Rao, M., Rao, K., Muniswamy, V.V.: Delivery performance measurement in an integrated supply chain management: case study in batteries manufacturing firm. Serb. J. Manage. **6**(2), 205–220 (2011)
10. Rezapour, S., Allen, J., Mistree, F.: Uncertainty propagation in a supply chainor supply network. Transp. Res. **73**, 185–206 (2015)
11. Sowa, K., How to estimate product backlog efficiently (2019). https://medium.com/@concisesoftware/how-to-estimate-product-backlog-effectively-aefa4b9051aa
12. Tannock, J., Cao, B., Farr, R., Byrne, M.: Data-driven simulation of the Supply-chain – Insights from the aerospace sector. Int. J. Prod. Econ. **110**, 70–84 (2007)
13. Tao, L., Liang, A., Bushuev, M.: Supply chain delivery performance: a white boxperspective. Int. J. Prod. Res. (2023)
14. Xiaosong, D., Lu, G.: Exploring the impact of delivery performance on customer transaction volume and unit price: evidence from an assembly manufacturing supply chain. Prod. Oper. Manage. **26**, 880–902 (2017)
15. 101 Delivery Statistics for 2020, Internet resource. Accessed May 2023
16. How to Measure Delivery Performance (2021). www.elogii.com. Accessed Mar 2023
17. Oracle Global Survey, Setting the Bar Research (2019). Accessed June 2023

Systematic Scientific Literature Review and Bibliometric Analysis of Researches for Resilience in Production and Industrial Systems

Emre Bilgin Sari[1(✉)] and Dušan Šormaz[2]

[1] Dokuz Eylul University, Izmir, Turkey
emre.bilgin@deu.edu.tr
[2] Ohio University, Athens, OH, USA

Abstract. The concept of "resilience" is generally used in various disciplines to designate the meanings such as endurance, flexibility, adaptability, and recovery from hardship. For production and industry, "resilience" is used to mean that systems can stand up to crises or unexpected adversities and gain new capabilities by adapting to changes. In this study, systematic scientific literature review and the bibliometric analysis of the literature on industrial resilience have been conducted by determining the focus of scientific research in previous years regarding the concept of "resilience" for production and industrial systems. For analysis, production and industrial resilience articles from the Scopus database have been examined within the framework of selected criteria and subjected to bibliometric analysis and visualized with the VOSviewer scientific mapping technique. From the analyses will be made, as an interesting topic, publication trend of studies will be revealed. The most frequently used keyword has been discovered. In addition to this, the countries, institutions, publication services and researchers of published documents about industrial resilience have been determined.

Keywords: Scientific Literature Review · Bibliometric Analysis · Science Mapping · Resilience in Production · Industrial Resilient

1 Introduction

Recent COVID pandemic has exposed the vulnerability of companies in situations where significant disruptions happen in their operations, customer base or supply chains. Due to COVID pandemic three sets of disruption could be identified: a) reduction in workforce numbers due to illness, or limited mobility of workforce, b) reduction of customer base due to lack of customer needs due to reduced mobility of customer, and c) large variability of supply chain dynamics due to previous two factors. All of these factors have caused reduced level of economic activity, which resulted in reduced income, and lower bottom-line, but also on the other side has caused the need to understand effects of disruption, and methods to handle them into a concept called resiliency. It is defined as

M. Dragomir et al. (Eds.): ICPR 2023, LNPE, pp. 369–377, 2025.
https://doi.org/10.1007/978-3-031-92082-0_41

the ability to withstand adversity and bounce back from difficult life events. This paper provides a scientific literature review (SLR) of the literature of resilience in production and industrial systems. Section 2 provides a brief overview of the resilience concept in the context of production. Section 3 explains the methodology applied. Section 4 presents the analysis and results of the study. The paper ends with conclusions and references used.

2 Resilience in Production and Industrial Systems

The need for production systems to adjust to the changes in their operating environment has been recognized for many years. The concept of flexibility was introduced in early 1980s in [1] as as the ability of a manufacturing system to adapt to changes in environmental conditions and in the process requirements. This definition takes into account both the exogenous and the endogenous nature of flexibility: the former as consequence of the market's demand, the latter as exploitation of the opportunities offered by technological innovations. The concept of flexibility has been further classified into machine flexibility, process flexibility, product flexibility, and routing flexibility in [2]. These early works were followed by many other efforts to understand and model manufacturing adaptability, rapid manufacturing, with overall objective to develop method to face unexpected disasters. Resilience has started by work in overall economic recovery in face of disasters [3]. Another work proposes guiding principles for design and management of resilient manufacturing systems as: redundancy, total function, learning and training, and ontology modelling [4]. Since then, the body of research in production resilience has increased and this paper provide a systematic review and classification if that work in context of production and supply chains.

3 Methodological Approach

Systematic scientific and bibliometric analysis are major approaches in evidence-based practice. These methods are applied in management research to close the "research-practice gap" [5]. The methodology is used in these approaches are identifying, selecting, and analyzing secondary data [6, 7].

In SLR, there are a set of five steps by [7] that guides the performed review process. Question formulation, study location, study selection and evaluation, sample analysis and synthesis, result reporting. In addition to SLR, while bibliometric methods or analyses examine science various aspects, they also provide the opportunity to rank universities and institutions worldwide according to their scientific efficiency in manner subjects [8]. Bibliometric analysis creates an indirect approach in determining academic performance and quality by making use of the quantity of scientific publications, so that determinations can be made about the past status of the relevant research subject and the path they will follow in the future [9]. Therefore, bibliometric analysis evaluates scientific research both quantitatively and qualitatively.

Systematic scientific and bibliometric analysis reveal the scientific communication structure of the relevant subject, while examining and analysing of scientific publications taken from many online databases (WoS, Scopus, Google Scholar, Science Direct, etc.) containing bibliometric data [9, 10].

Bibliometric mapping is used to visualize the analysis, as a two- or three-dimensional representation of a field of science [11]. Science mapping analysis is carried out to determine the structure, evolution, and main actors of a scientific field. This technique is a spatial representation of disciplines, fields, areas of expertise, as well as how documents or authors relate to each other [12, 13]. In this study, bibliometric networks are used for mapping, based on data obtained from appropriate databases.

4 Analysis and Results

In this study, the procedure is starting from SLR, and as [7] explained SLR has five steps as question formulation, study location, study selection and evaluation, sample analysis and synthesis, result reporting. Figure 1 shows these steps.

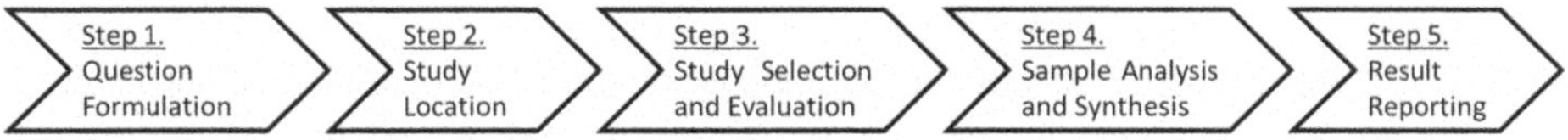

Fig. 1. Methodological Steps of Scientific Literature Review

- *Question Formulation*

Overall, this paper aims to provide a framework as an aspect of production and industrial research enhancing understanding of the resilience concept. Hence, the research question is about general definition of the concept resilience in production and industrial systems.

R.Q.: What are the key terms in the concept of "resilience in production and industrial systems"?

- *Study Location*

The purpose of locating study is searching relevant journal articles is to create a comprehensive list of core contributions in relation to the review questions [7]. It is expected that from the database selected to collect bibliometric data will have a high quality, reliable and comprehensive content. The Scopus database has been used for gathering publication data. The Scopus as operated by Elsevier includes all journals with Scientific Journal Rankings, and Journal Citation Reports, and their impact factors. With over 25,100 titles from more than 5,000 international publishers, the Scopus is an indexed database of scientific production [14].

- *Study Selection and Evaluation*

The research direction and history of Resilience in Production and Industrial Resilient has its roots during 1990s, there are studies on the topic from almost 30 years. The studies conducted between the years "1994–2023" formed the basis of the dataset. The research is conducted in 31st of May 2023.

The terms are categorized in 3 class as "resilience in", "resilient" and "resilience". Searching terms for the study are decided as "resilience in industry", "resilience in production", "resilience in manufacturing", "production resilient", "manufacturing resilient", "industrial resilience" and "industry resilience", and these terms are searched within all fields in all time range. Figure 2 exhibits the categorization of research terms.

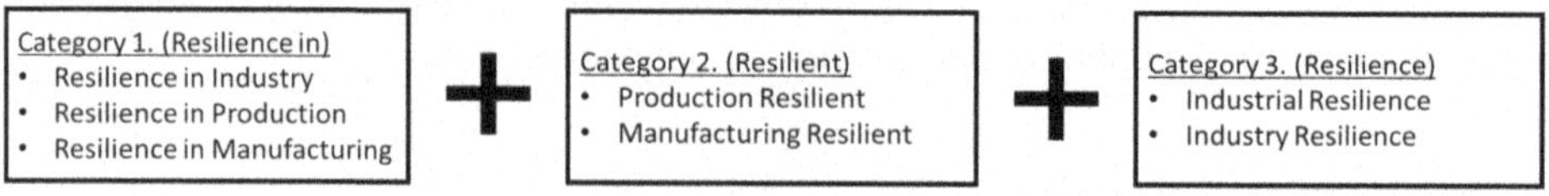

Fig. 2. The Categorization of Research Terms

After searching these terms in Scopus databases, the results obtained as totally 759 documents. While 585 of them are journal articles (77%), there are 78 conference papers (10%), 38 book chapters (5%) and 36 reviews (4%) and the remains are the other type of documents. Figure 1 exhibits the shares of document types, in this study articles are investigated. Figure 3 shows the annual publication of "resilience in production and industrial systems" articles over 30 years. It can be predicted that the number of publications showing an increasing trend will continue to increase in the future.

Fig. 3. Annually Publication Numbers over Years

While the subject area in which the articles are published is searched, the highest rate is found to be Business, Management and Accounting with 19%. It is followed by Social Science with 16%, Engineering with 14%, Environmental Science with 11%, Computer Science with 8% and Decision Sciences with 6%. Figure 4 is providing these ratios in subject areas. These data exhibit that, the social and technical documents about Resilience in Industry have major parts for research fields. As a socio-economic system, it is becoming the research interest of sciences during years from commencing.

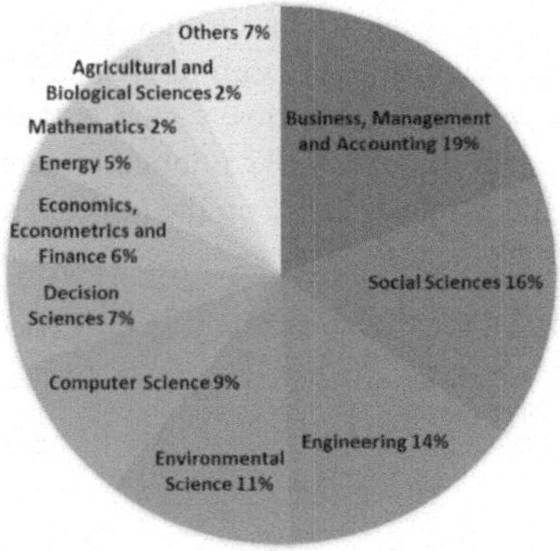

Fig. 4. Subject Areas of Publications about The Concept

In the totally investigated data, it is seen that the subject was published by 160 different sources. In all document types, "Sustainability" with 34 publications is the leading publisher organization. It is followed by "International Journal of Production Research" with 16 publications and "Regional Studies" with the 15 publications. It can be seen in Table 1, the lists of the sources that have published more than 5 times.

Table 1. The Sources of Search on the Concept

Source Title	Number of Document
Sustainability Switzerland	34
International Journal of Production Research	16
Regional Studies	15
European Planning Studies	11
Current Issues in Tourism	8
International Journal of Logistics Management	8
International Journal of Production Economics	8
Environmental Science and Pollution Research	7
Business Strategy and The Environment	6
Computers and Industrial Engineering	6
International Journal of Emerging Technologies in Learning	6
Annals of Operations Research	5
Benchmarking	5
Cambridge Journal of Regions Economy and Society	5
International Journal of Disaster Risk Reduction	5
International Journal of Emerging Markets	5
International Journal of Environmental Research and Public Health	5
Journal of Cleaner Production	5
Production Planning and Control	5
Sustainable Production and Consumption	5

In the data obtained from Scopus, it is seen that articles published on "resilience in production and industrial systems" have been published from 82 different countries around the worldwide. Among these countries, China is the leader with 132 publications, while United Kingdom follows with 87 publications and United States with 84 publications. Table 2 lists the countries study on the concept at least 10 publications.

Table 2. The Countries Study on the Concept

Country	Number of Document	Country	Number of Document
China	106	Iran	14
India	72	South Korea	14
United Kingdom	71	Taiwan	14
United States	57	Greece	13
Italy	46	Turkey	13
Australia	35	Finland	12
Germany	27	Canada	11
Spain	17	New Zealand	11
Netherlands	16	Norway	11
Hong Kong	15	Poland	11
Indonesia	15	Portugal	10
Sweden	15	Romania	10

Review the keywords of the studies published on the concept is important in terms of drawing the lines of the research field of the subject. In addition to this, during the analysis of 585 articles' keywords, 570 of them are published in English language, 7 Chinese, 3 Spanish, 2 Persian, 2 Russian. English written articles are focused on this research. The most frequently repeated keywords are listed in the examined studies as; there is "Resilience" in the first place with 129 times, and then "COVID-19" is used 103 times, "Supply Chains" is used 51 times, "Sustainability" is used 50 times, "Decision Making" is used 40 times. The list of top used 30 keywords is in Table 3.

Table 3. The list of Top Used 30 Keywords

Keyword	Number of Use	Keyword	Number of Use
Resilience	129	Supply Chain	20
COVID-19	103	Tourism	20
Supply Chains	51	Pandemic	19
Sustainability	50	Crisis Management	18
Decision Making	40	Manufacturing	18
Supply Chain Management	34	Industry 4.0	17
Sustainable Development	33	Business	15

(continued)

Table 3. (*continued*)

Keyword	Number of Use	Keyword	Number of Use
China	31	Economic Development	15
Supply Chain Resilience	31	Crisis	14
Regional Economy	25	Employment	14
Human	21	Industrial Research	14
Article	20	Supply Chain Resiliences	14
Manufacture	20	Conceptual Framework	13
Organizational Resilience	20	Humans	13
Risk Management	20	Innovation	13

- *Sample Analysis and Synthesis*

As sample analysis and synthesis, Sustainability and International Journal of Production Research which are two journals that have published many articles on the subject, are addressed. Figure 5 exhibits the choosing process of sample.

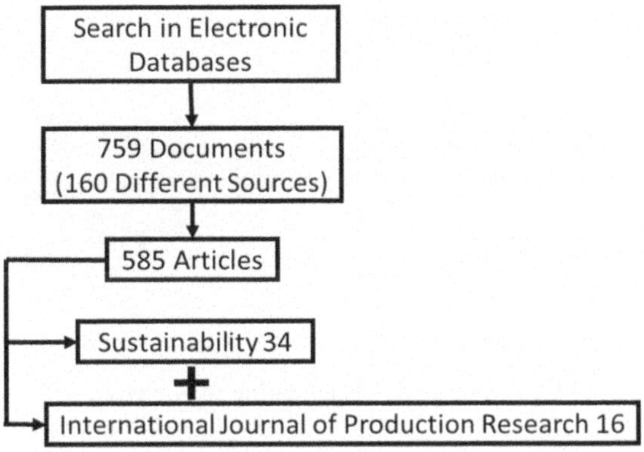

Fig. 5. Sampling Process

Totally 50 articles are investigated from these two sample journals, published between 2011 - 2023. In these articles subject area is spread in these topics in descending, energy, environmental science, social sciences, engineering, computer science, business, management and accounting, decision sciences. These subject areas are corresponding with overall data, but not completely identical ratios. Keywords for these 50 articles are similar to 570 English written articles in other journals.

- *Result Reporting*

As a result of SLR, scientific mapping has been conducted. The density of the most frequently used keywords are visualized in Fig. 6 and the networks between keywords is shown in Fig. 7. Bibliometric mapping analysis is prepared with Vosviewer.

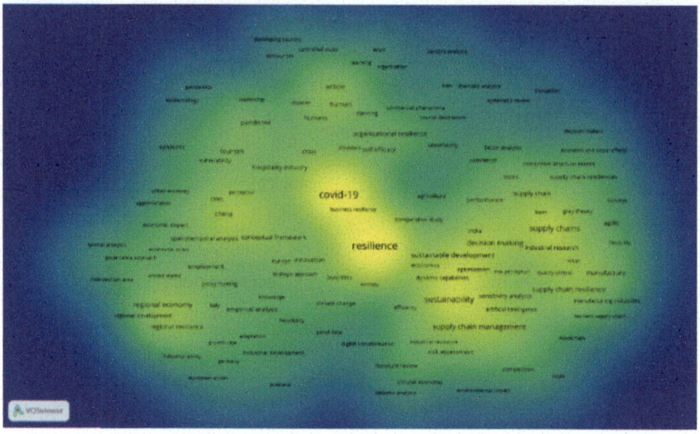

Fig. 6. Keyword Density Presentation

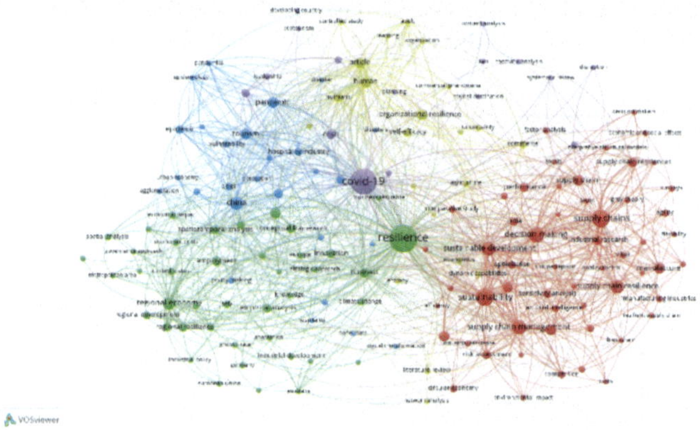

Fig. 7. Network Visualization of Keywords

5 Conclusion

Resilience in production and industrial system is a requirement when the systems are under the crises and unexpected situations. In this study, to draw the attention to importance of the issue, systematic scientific literature review and bibliometric analysis are made. The time period in which the subject was studied was found and attention was drawn to the increasing number of publications related to the subject. It has been determined that the subject has been published in the fields of social sciences and technical studies. The importance of the subject for social and technical research fields has been revealed. For the purpose of bibliometric analysis, the countries that published on the subject are analysed and the global orientation is examined. Keywords based on studies published on resilience in production and industrial system have been identified and density and network analyses have been made for these keywords.

Resilience in production and industrial system is a concept that has been studied in the literature for almost 30 years and the number of research is increasing day by day. In this study, a framework has been created for the studies on the subject in order to guide future researchers. While it comes to the fore that the Covid 19 pandemic studies are concentrated in the studies on the concept, the effect of the concept of resilience in unexpected situations that may occur in the future is presented.

References

1. Zelenović, D.M.: Flexibility—a condition for effective production systems. Int. J. Prod. Res. **20**(3), 319–337 (1982)
2. Browne, J., Dubois, D., Rathmill, K., Sethi, S.P., Stecke, K.E.: Classification of flexible manufacturing systems. FMS Mag. **2**(2), 114–117 (1984)
3. Rose, A.: Defining and measuring economic resilience to disasters. Disaster Prev. Manage. Int. J. **13**(4), 307–314 (2004)
4. Zhang, W.J., Van Luttervelt, C.A.: Toward a resilient manufacturing system. CIRP Ann. **60**(1), 469–472 (2011)
5. Rousseau, V., Aubé, C., Savoie, A.: Teamwork behaviors: a review and an integration of frameworks. Small Group Res. **37**(5), 540–570 (2006)
6. Tranfield, D., Denyer, D., Smart, P.: Towards a methodology for developing evidenceinformed management knowledge by means of systematic review. Br. J. Manag. **14**(3), 207–222 (2003)
7. Denyer, D., Tranfield, D.: Producing a systematic review. In: Buchanan, D., Bryman, A. (eds.) The Sage Handbook of Organizational Research Methods, pp. 671–689. Sage Publications Ltd., London (2009)
8. Ellegaard, O., Wallin, J.A.: The bibliometric analysis of scholarly production: how great is the impact? Scientometrics **105**, 1809–1831 (2015)
9. Cobo, M.J., López-Herrera, A.G., Herrera-Viedma, E., Herrera, F.: Science mapping software tools: review, analysis, and cooperative study among tools. J. Am. Soc. Inform. Sci. Technol. **62**(7), 1382–1402 (2011)
10. Glanzel, W.: Bibliometrics as a research field a course on theory and application of bibliometric indicators (2003)
11. Noyons, E.: Bibliometric mapping of science in a policy context. Scientometrics **50**, 83–98 (2001)
12. Noyons, E.C., Moed, H.F., Luwel, M.: Combining mapping and citation analysis for evaluative bibliometric purposes: a bibliometric study. J. Am. Soc. Inf. Sci. **50**(2), 115–131 (1999)
13. Small, H.: Visualizing science by citation mapping. J. Am. Soc. Inf. Sci. **50**(9), 799–813 (1999)
14. Elsevier. "Scopus Content Coverage Guide" (2020). https://www.elsevier.com/__data/assets/pdf_file/0007/69451/Scopus_ContentCoverage_Guide_WEB.pdf. Accessed July 2022

Seru Production System: An Optimization Model for Loading Problem in Multi-skilled Worker Environment

Emre Bilgin Sari[1](✉), Dušan Šormaz[2], and Omar Alhawari[2]

[1] Dokuz Eylul University, Izmir, Turkey
emre.bilgin@deu.edu.tr
[2] Ohio University, Athens, OH, USA

Abstract. The Seru Production System (SPS) is a Japanese cellular production system, that originated with the need for improvements in assembly processes of the electronic industry. Distinct from conventional production systems, SPS has lots of autonomous production units called seru. In these serus, there is the opportunity for producing different kinds of products, which helps to increase the flexibility of the production system. The workers who are working in these seru units, are cross-trained, so they are capable to complete the manufacturing or assembly process by themselves. In this study, we present a seru loading model in multi-skilled worker environment. The objective of the model is to minimize cycle time, while decisions are assignments of workers and product types to different serus with workers having different skill levels for different tasks and including product changeover setup times. The total number of seru, the number of workers and the number of product types and their demand are predetermined. The proposed math model has been implemented in open-source solver and results were reported in the paper.

Keywords: Seru Production System · Seru Loading Problem · Multi-Skilled Workers · Mixed Integer Non-Linear Mathematical Model

1 Introduction

Recently in global market conditions, the change in the demand of customers for a certain product with the developing technologies causes the development and trans-formation of modern production systems day by day. Fluctuations in demand are tried to be met with flexible production systems. Seru - assembly cells - structures emerged in Japan in the 90s with the search for mass production systems in the face of demand fluctuations. The Seru Production System, which was first used by Sony in 1992, was later applied by large electronics companies such as Canon, Panasonic, Fujitsu, Sharp, Yamaha, Hitachi [1]. As a production system based on the philosophy of just-in-time production, the production systems formed by the seru structures have been a system widely used in sectors where demand fluctuations in the product are high and product life cycles are shortened [2]. Principally, with the help of multi skilled workers, this system is giving the response to the fluctuating demand effect [3] so ensures the efficiency of assembly line and the flexibility of workshop production together [4].

M. Dragomir et al. (Eds.): ICPR 2023, LNPE, pp. 378–385, 2025.
https://doi.org/10.1007/978-3-031-92082-0_42

2 Seru Production System and Multi-Skilled Workers

Seru Production System (SPS) is defined as a human-oriented production system based on skilled workers [5]. "Seru" means cell in Japanese, and SPS is defined as a production system based on cellular and flexible manufacturing logic [6]. In SPS all tasks in the production process of a product are performed by one or more highly skilled workers by dividing traditional assembly lines into divisional seru [5, 6, 7]. Divisional seru, rotating seru, and yatai are three common types of SPS [1]. In the divisional seru, the tasks required to produce a product are shared by more than one worker in the seru. Workers can perform more than one operation according to their skill level. In the rotating seru, each worker can perform all the operations required to produce a product. There is more than one worker in rotating seru, and while the product moves on the U-type workbenches, the workers follow the product along the benches and perform the operations. A worker performs the relevant operation after the worker in front of him finishes the job. In this type of seru, workers are alternately and cross-trained, gaining the ability to perform all operations of a product. Yatai is a special case of the rotating seru type, and there is only one worker in the seru. This worker is in a fixed place and the products are transported around the worker, while duties are fulfilled [2, 5]. Seru types are shown in Fig. 1.

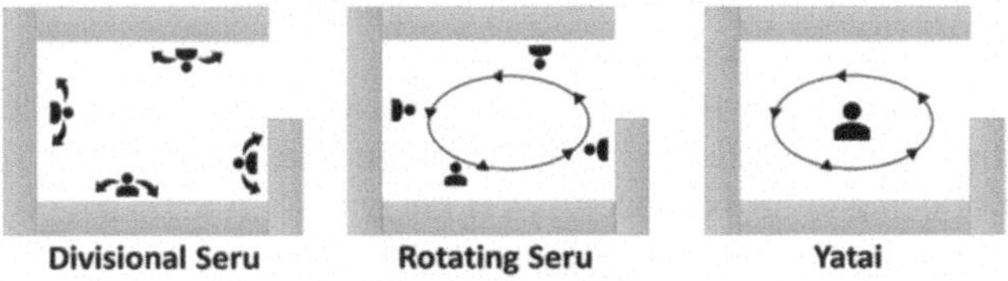

Fig. 1. Seru Types

In the Seru Production System, it is accepted that the workers who basically do a part of the work in the divisional seru. And then the workers are cross-trained and have reached the level of learning to do most of the tasks of the jobs, so the rotating seru application is started [8]. During cross-training, it is the case that workers have different skill levels and can perform the same task with different processing times according to these skill differences [9]. In this case, the assignment of multi skilled workers with different duration of work to the seru generates the main decision problem [4].

In Seru Production System literature, there are the implementations of the problem for worker assignment. [4], researched a worker seru assignment and worker-product assignment, in which a certain product type is produced in each batch, will balance the working time of the workers in each batch and minimize the total labor training time. In the study it is aimed to obtain an assignment chart and a mathematical model is proposed. [10] considered a production system consisting of seru located in geographically different places. From this model, a mathematical model that makes the decision of work-worker assignment for seru in different places is proposed. [11], are researched another seru loading problem, in that study, a production system consisting of completely divisional seru is considered. In the problem, it is aimed to determine the assignment of the jobs to the seru and the sequence of the assigned jobs in the seru. In the study conducted by [12]., job and worker assignments to the existing seru were considered together in the problem of seru loading. In the model, only a certain type of product will be produced in each seru and it is assumed that the product and seru numbers are the same. [13] studied on the problem of assigning workers and jobs to seru and jobs to workers was considered, and it is aimed to balance the workload distribution between seru and within seru.

3 Proposed Model

The model proposed in this study is based on the assignment of the jobs done in the seru and the workers who will do the jobs. Different fom the current studies in the literature in the proposed model, the number of seru, the number of workers and the number of product types vary independently from each other. In this model seru type is studied as rotating seru, that means in each rotating seru there can be at least two workers, otherwise it is called yatai. For worker assignment, it is defined multi skilled workers for each type of products, so their skill coefficient is varying and is helping the define processing time of each product type is produced which skilled worker. Some other assumptions of the proposed model as following:

- The number of seru existing in the system is determined.
- The number of workers is determined.
- The maximum number of workers that can work in a seru is determined.
- The types of products targeted to be produced are specific.
- The amount of demand for product types is known.
- Each product can be produced in only one batch. (Batches are not divided between seru)
- Production capacity is determined for each seru. (Maximum amount of product to be produced)
- Some amount of same product can be divided among workers assigned to the same seru.
- Processing time is known for each product type.
- The skill coefficient varies and known for each product type of each worker.
- Set up time is specific for each product type.

A mixed integer nonlinear mathematical model is created for the mathematical representation of the problem. The notation for the indices, parameters and decision variables used in this developed model is given below, and then the objective function and constraints in the mathematical model are defined.

Indices.

i operations
j workers
n stations

Parameters

D_n demand for product type n

S_n set up time for product type n
T_n processing time for product type n
L_{in} skill coefficient of worker i for product type n
C production capacity for seru
W maximum number of workers in seru
M minimum number of workers in seru

Decision Variables

X_{ij} worker i is assigned to seru j 1; in otherwise is 0.
Y_{nj} product type n is assigned to seru j 1; in otherwise is 0.
R_{ijn} number of product n produced in seru j by worker i
F_{nj} flow time of product type n in seru j
E_j end time of seru j
C_{max} maximum of end time for all seru

Objective Function

$$\text{MinimizeZ} = C_{max} \tag{1}$$

Constraints

$$\sum_{j=1}^{J} X_{ij} = 1 \;\; i = 1, 2, \ldots, I \tag{2}$$

$$\sum_{i=1}^{I} X_{ij} \le W \;\; j = 1, 2, \ldots, J \tag{3}$$

$$\sum_{i=1}^{I} X_{ij} \ge M \;\; j = 1, 2, \ldots, J \tag{4}$$

$$\sum_{j=1}^{J} Y_{nj} = 1 \;\; n = 1, 2, \ldots, N \tag{5}$$

$$F_{nj} \ge T_n * L_{in} * R_{ijn} \;\; i = 1, 2, \ldots, I;\; j = 1, 2, \ldots, J;\; n = 1, 2, \ldots, N \tag{6}$$

$$E_j \ge \sum_{n=1}^{N} F_{nj} + \sum_{n=1}^{N} S_n * Y_{nj} \;\; j = 1, 2, \ldots, J; \tag{7}$$

$$C_{max} \ge E_j \;\; j = 1, 2, \ldots, J; \tag{8}$$

$$\sum_{i=1}^{I} \sum_{j=1}^{J} R_{ijn} \ge D_n \;\; n = 1, 2, \ldots, N \tag{9}$$

$$\sum_{i=1}^{I} \sum_{n=1}^{N} R_{ijn} \le C \;\; j = 1, 2, \ldots, J \tag{10}$$

$$R_{ijn} \le C * X_{ij} * Y_{nj} \;\; i = 1, 2, \ldots, I;\; j = 1, 2, \ldots, J;\; n = 1, 2, \ldots, N \tag{11}$$

In the model, Eq. (1) is the objective function and aims to minimize the maximum completion time for the seru loading problem. Equation (2) shows that all workers are assigned to a seru. Equation (3) ensures that the total number of workers assigned a seru does not exceed maximum number of workers in seru. Equation (4) ensures that the total number of workers assigned a seru does not least the minimum number of workers in seru. Equation (5) shows that all product types are assigned to only one seru. Equation (6) explains that the flow time in each seru, flow time is vary-ing according to skill coefficient of the worker assigned to the product type and number of products are produced. Equation (7) shows that the end time for each seru with the summation of flow time and set up times in each seru. Equations (8) the con-straint gives the maximum completion time. Equation (9) ensures that the number of products produced in the system satisfy the demand of the product type. Equation (10) ensures that the total number of products produced in seru cannot exceed pro-duction capacity of seru. Equation (11) ensures that if there is no workers and product types in seru, the number of products are not capable.

4 Computational Experiment

In this study, to test the proposed model, it is made a computational experiment. In the founded example, there are 3 seru, 8 workers and 5 product types, as shown in Fig. 2. Also, these workers have different skill levels for producing these different product types. And the research question of the study is finding the which worker will assign to which seru, and in these seru these workers will produce which product.

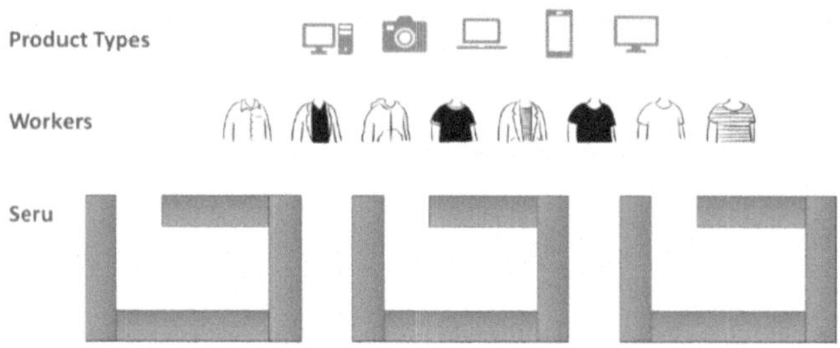

Fig. 2. Experimental Model

To manage this problem, there are some data parameters about experiment. Number of workers in seru is a key indicator for seru literature. In the proposed model, it is modeled to create rotating seru type, that means in each seru there will be at least 2 workers. That is why, the minimum number of workers is defined as 2. Additionally, it is advised to the maximum number of workers in a seru is limited about 15 [7, 14]. All parameters for seru in the computational experiment are given as below in Table 1.

In the problem, there are 5 different types of products. These products are assumed to known demand in the proposed model. And all these different product types have set up times, to make preparation in seru to produce that product type. Additionally, processing time is given according to first class worker data. Parameters for product types are given in Table 2.

In seru literature, workers are divided into four class, Liu et al., (2010), mentioned the implementation in Canon company about these workers classification, in their case, there

Table 1. Parameters for Seru

Parameters for Seru	Data
Minimum number of workers in Seru	2
Maximum number of workers in Seru	15
Production Capacity of Seru	1000

Table 2. Parameters for Product Types

Parameters for Product Types	n1	n2	n3	n4	n5
Demand for Product Type n	400	200	400	600	500
Set up Time for Product Type n (min.)	180	150	150	120	120
Processing Time of (1st Class Worker) for Product Type n (min.)	240	160	180	120	160

are 3rd class – 2nd class - 1st class – super class workers. In that ranking system workers in seru improve their performance for the super class worker level. In this implementation, 8 workers have different skill levels for different product type. In Table 3 it is shown that which worker has what kind of skill classification for each product type.

Table 3. Parameters for Workers Class

Worker\Product Type	n1	n2	n3	n4	n5
Worker1	Super	1st	1st	Super	Super
Worker2	2nd	1st	2nd	2nd	2nd
Worker3	3rd	Super	2nd	2nd	2nd
Worker4	3rd	3rd	3rd	2nd	3rd
Worker5	1st	3rd	Super	1st	1st
Worker6	2nd	3rd	2nd	2nd	1st
Worker7	1st	2nd	1st	Super	Super
Worker8	3rd	1st	2nd	2nd	2nd

5 Results

The results of the computational experiment with the help of the proposed model founded by using Gurobi solver with 217691 iterations. The solution of the program shows that, product type 1 will be produced in seru1 and worker1 and worker2 will be assigned to seru1 to produce product type 1. For seru2; worker3, worker5 and worker7 will be assigned and they will produce product type 3 and product type 4 in that seru2. The results for seru3 is like that; worker4, worker6, worker8 will be assigned to the seru 3 to produce product type 2 and product type 5. The result for experimental model is shown as in Fig. 3.

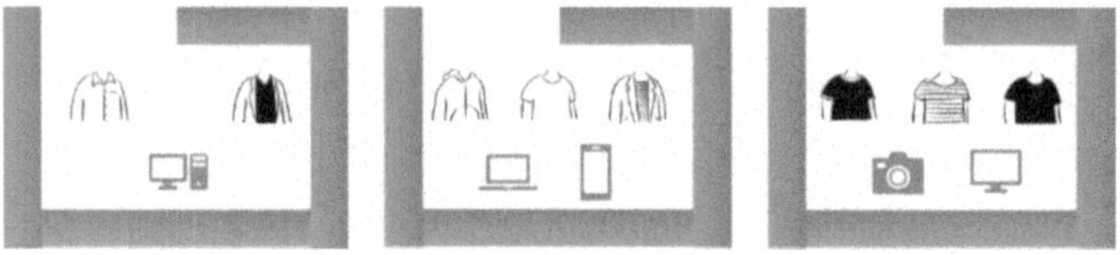

Fig. 3. Results for Experimental Model

Flow time, set up time and completion time for each seru is calculated as in Table 4. Objective function of the model is defined as minimizing the maximum completion time of all seru. This is a MinMax problem and assignments are made with this requirement.

Table 4. Objective of the model

Seru	Worker	Product Type	Flow Time	Set Up Time	Completion Time
Seru 1	Worker 1 Worker 2	n1	46080	180	46260
Seru 2	Worker 3 Worker 5 Worker 7	n3	23400	150	47070
		n4	23400	120	
Seru 3	Worker 4 Worker 6 Worker 8	n2	13216	150	47070
		n5	33584	120	

6 Conclusion

The Seru Production System (SPS) gains its advantages from cross-trained multi-skilled workers. While assigning multiskilled workers to the autonomous productions units(seru), the flexibility of the production system can be increased. So, in each separated seru there can be opportunity to produced different types of products. In this study, there is a model proposal for loading seru in multiskilled workers environment. Different from the literature, in this study, the number of seru, the number of workers and the number of product types are taken independently from each other. Additionally, model is founded for loading to rotating seru type, that means, each serus need to be assigned at least two workers and these workers will have different skill levels to complete each product type with their skill class. After model proposal, the experimental computation is made to model validation and to produce 5 types of products, 8 workers assigned to the 3 seru.

In this study, model proposal and model validation is targeted. That's why small number of example case is shown. After this study, the research can be improved to adding different variables and constraints and the problem size can be increased.

References

1. Liu, C., Stecke, K.E., Lian, J., Yin, Y.: An implementation framework for seru production. Int. Trans. Oper. Res. **21**(1), 1–19 (2014)
2. Stecke, K.E., Yin, Y., Kaku, I.: Seru production: an extension of just-in-time approach for volatile business environments. In: Analytical Approaches to Strategic Decision-Making: Interdisciplinary Considerations, pp. 45–58. IGI Global (2014)
3. Yu, Y., Wang, J., Ma, K., Sun, W.: Seru system balancing: definition, formulation, and exact solution. Comput. Ind. Eng. **122**, 318–325 (2018)
4. Liu, C., Yang, N., Li, W., Lian, J., Evans, S., Yin, Y.: Training and assignment of multi-skilled workers for implementing seru production systems. Int. J. Adv. Manuf. Technol. **69**(5), 937–959 (2013)
5. Stecke, K.E., Yin, Y., Kaku, I., Murase, Y.: Seru: the organizational extension of JIT for a super-talent factory. Int. J. Strateg. Decis. Sci. (IJSDS) **3**(1), 106–119 (2012)
6. Villa, A., Taurino, T.: From JIT to Seru, for a production as lean as possible. Procedia Eng. **63**, 956–965 (2013)
7. Liu, C., Lian, J., Yin, Y., Li, W.: Seru Seisan-an innovation of the production management mode in Japan. Asian J. Technol. Innov. **18**(2), 89–113 (2010)
8. Abdullah, M., Suer, G.A.: Consideration of skills in assembly lines and seru production systems. Asian J. Manag. Sci. Appl. **4**(2), 99–123 (2019)
9. Suzuki, Y.: Structure of the Japanese production system: elusiveness and reality. Asian Bus. Manag. **3**(2), 201–219 (2004)
10. Manupati, V.K., Deepthi, T.V., Ramakotaiah, K., Rao, S.S.: Reconfiguration of networked Seru production systems in an Indian perspective. In: Proceedings of the 2015 International Conference on Industrial Engineering and Operations Management, Dubai, B.A.E., 3–5 Mart (2015)
11. Luo, L., Zhang, Z., Yin, Y.: Seru loading with worker- operation assignment in single period. In: 2016 IEEE International Conference on Industrial Engineering and Engineering Management (IEEM), Bali, Indonesia, pp. 1055–1058 (2016)
12. Ying, K.C., Tsai, Y.J.: Minimising total cost for training and assigning multiskilled workers in seru production systems. Int. J. Prod. Res. **55**(10), 2978–2989 (2017)
13. Luo, L., Zhang, Z.: Modelling, and numerical analysis of seru loading problem under uncertainty. Eur. J. Ind. Eng. **11**(2), 185–204 (2017)
14. Miyake, D.I.: The shift from belt conveyor line to work-cell based assembly systems to cope with increasing demand variation in Japanese industries. Int. J. Automot. Technol. Manage. **6**(4), 419–439 (2006)

Ontology Modeling of Plan and Their Conformance to Manufacturing Execution

Arkopaul Sarkar[1,4](✉), Milos Drobnjakovic[2], Sina Namaki Araghi[1,4], Hedi Karray[1,4], and Dusan Sormaz[3]

[1] Laboratoire Genie de Production, Ecole Nationale d'Ingenieur de Tarbes, Tarbes, France
arkopaul.sarkar@enit.fr

[2] Faculty of Science, University of Bern, Bern, Switzerland

[3] ISE Department, Ohio University, Athens, USA

[4] Production Engineering Laboratory of Tarbes University of Technology (LGP-UTTOP), Tarbes, France

Abstract. Semantic data interoperability has attracted a lot of attention in recent years, with ontologies being one of the major tools to accomplish the goal of seamless data and application integration in manufacturing, industry and supply chains. Among the most challenging issues are semantic relations between manufacturing plans and their realization in actual production. This paper provides a model based on the IOF approach to the subject. The initial discussion is devoted to the semantic modelling of intentions, desires, plans and actions, from which we propose the specification/process approach to modelling plans and their realizations. The developed ontological model of plans includes relations between specifications and future occurrences; models of sequences, hierarchy, and alternatives; and conformance of executed processes to their planned counterparts as given by specifications. The model is illustrated by several small examples implemented in OWL language.

Keywords: Plan · Manufacturing · Process

1 Introduction

Industries adopted a number of manufacturing standards (e.g., PDDL, ISO18828-3, ISO15926-4), usually published by standardization bodies and manufacturing consortia, as a means to represent and exchange process plans as part of the product lifecycle in distributed manufacturing. In this paper, we addressed the problems in representing process models as part of plans. The crux of the problem is that the plan is generated in the planning phase as a directive to perform some processes in the future. However, many of the existing data standards often represent both a process as it should occur and a process as it occurred equivocally. This makes it difficult to interpret planning data separately from the process data in post-production data analysis, for example, answering the questions of whether a production adhered to a plan, when or to what extent the process steps followed the plan, and if there any unexpected output, then at which step the production process deviated from the plan.

M. Dragomir et al. (Eds.): ICPR 2023, LNPE, pp. 386–396, 2025.
https://doi.org/10.1007/978-3-031-92082-0_43

Ontology engineering is a superior method to model data due to its capability of expressing detailed truthmakers (necessary and sufficient conditions) in defining the concepts based on formal logic (e.g., OWL 2.0[1]). In the past, only a few ontologies were proposed by researchers to represent manufacturing plans, e.g., MIT activity and plan ontology[2], an ontology for assembly planning knowledge by Huang et al. [1], ontology-based production planning by Denkena et al. [2], process specification language (PSL) by Gruninger [3], Semantically-integrated process planning model (SIMPM) by Sormaz and Sarkar [4] and P-Plan ontology[3]. Among them, PSL and P-Plan both admit the planned activity and executed activity is two separate concepts, e.g., *Activity* vs *ActivityOccurrence* in PSL and *Activity* vs *Step* in P-Plan. However, both ontologies classify these concepts as primitives and their ontological commitments are not clear. SIMPM classify *Process Plan* as information and being motivated by Basic Formal Ontology (ISO/IEC 21838-2, a.k.a. BFO), as an *Information Content Entity* [5]. This proposal found reciprocation in the core ontology (IOF-Core[4]) of Industrial Ontology Foundry (IOF) [6], published this year. Although, IOF allowed users to model the basic structure of a plan by including the concepts, such as *Plan Specification*, *Action Specification*, *Objective Specification* in the core ontology, distinguishing them from the foundational concepts, e.g., *process* in BFO, the model still do not address how to relate actual production processes to plan, deduce their conformance to the plan, and most importantly, how to model complex details of plan, such as hierarchy of process, alternative plans and temporal ordering among processing steps.

In this work, we propose a model for representing process plan heavily reusing the concepts of IOF-Core to address the ontological grounding planning entities with some extensions to suitably capture information on hierarchy, temporal order, and alternatives, following a model for integrated process planning model, proposed by Sormaz as 3D planning model [7, 8].

2 Ontology Model of Plan

IOF-Core includes the concept *iof:PlanSpecification* (p-spec) to capture the information of plans. According to its definition, p-spec contains both *iof:ActionSpecification* (a-spec) and *iof:ObjectiveSpecification* (o-spec) as its part with only sufficient conditions (e.g., a-spec *iof:prescribes* some process) to relate to executed processes. Therefore, it is not necessary and sometimes not possible to assert a process instance if the p-spec is being asserted before any of the processes it prescribes has not occurred yet. In this situation, some planning entities can only be classified explicitly (e.g., asserting rivet-making-plan is a p-spec) lacking justification for such assertion.

In the proposed ontology (given in Fig. 1), we introduce the concept *plan* characterizing it as a psychological state of mind of the planner, specifically 'intention'. Belief, desire, and intention are types of intentionality, which are borne by the mind of some

[1] https://www.w3.org/TR/owl2-overview/.

[2] http://groups.csail.mit.edu/medg/projects/ontology/drafts/current-draft/plans.html.

[3] http://purl.org/net/p-plan#.

[4] https://spec.industrialontologies.org/iof/ontology/core/core/.

agent and govern their behavior. Among them, intention and desire could only be fulfilled by some process or states occurring in the future as both of these attitudes have world-to-mind 'direction of fit' [9], that they require the world to be in such a state that conforms to such attitude. With these considerations, *Intention* and *Desire* are classified as *bfo: 'realizable entity'*, based on the connotating that such realization is similar to the conformance of the state of mind to the world state. Intention and Desire can also be considered as propositional attitude, which is the mental state of an agent directed or about something.

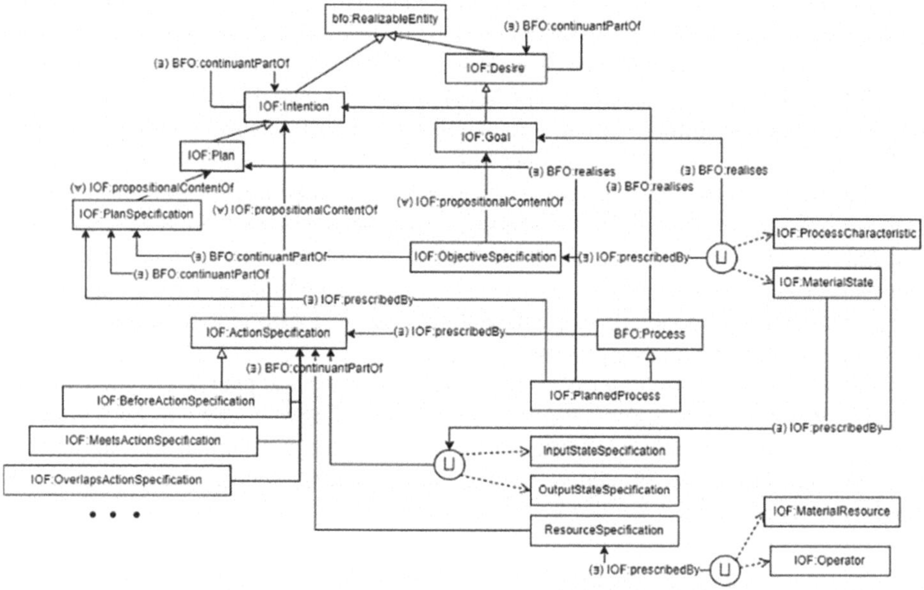

Fig. 1. Ontology model for representing plan

For example, in an illocutionary act *F(x)*, as proposed by Searle, '*F*' stands for the illocutionary force of utterance and '*x*' stands for propositional content. In earlier research on ontologizing Speech Act Theory, specific types of illocutionary acts were related to intentional moments [10]. In practice, the propositional content can be captured by some sentence tokens or information (as a reference) regarding some entities in reality. As a-spec captures the information of some process, they are modelled as 'propositional content of' (we introduce it as a new relationship as a type of 'aboutness' – sub-property of *iof:isAbout*) an Intention. Similarly, o-spec is information about some objective that is the 'propositional content of' a Desire. A *Plan* is classified as an Intention but is more directed than a simple intention, the propositional content of a plan contains information about some objective along with specifications on steps as a-specs that are required to be performed to achieve such objective, represented by o-spec.

$$A1: Intention(x)\rightarrow RealizableEntity(x)\wedge\exists y((Agent(y)\vee GroupOfAgent(y))\wedge bears(y,x))$$

$$A2: Desire(x)\rightarrow RealizableEntity(x)\wedge\exists y((Agent(y)\vee GroupOfAgent(y))\wedge bears(y,x))$$

$$A3: Intention(x)\wedge\exists y\,(Process(y)\wedge realizes(y,x))\rightarrow Plan(x)$$

$$A4: Desire(x)\wedge\exists y\,((ProcessCharacteristic(y)\vee MaterialState(y))\wedge realises(y,x))\rightarrow Goal(x)$$

$$A5: ObjectiveSpecification(x)\leftrightarrow InformationContentEntity(x)\wedge\forall y\, propositionalContentOf(x,y)\rightarrow Goal(y)$$

$$A6: ActionSpecification(x)\leftrightarrow InformationContentEntity(x)\wedge\forall y\, propositionalContentOf(x,y)\rightarrow Intention(y)$$

$$A7: PlanSpecification(x)\leftrightarrow InformationContentEntity(x)\wedge\forall y\,(propositionalContentOf(x,y)\rightarrow Plan(y))\wedge\exists a,o\,(ActionSpecification(a)\wedge continuantPartOfAtAllTime(a,x)\wedge ObjectiveSpecification(o)\wedge continuantPartOfAtAllTime(o,x))$$

We show a basic assertion of the plan below with the concepts discussed above. For example, Pixar plans to build a toy car "Lightning McQueen" which requires painting the car in red which will be done by General Motors, among many other steps.

2.1 Relationship with Future Occurrences

Intention being a realizable entity, they are realised when a process occurs as prescribed by the a-spec, that is the propositional content of an Intention. Similarly, some *iof:PlannedProcess* (p-process) occurrence that is prescribed by the p-spec, realizes the Plan that it is the propositional content of. It is to be noted that relating a process to its specification by *iof:prescribedBy* assumes that the process has occurred as per the specification, which may contain a number of details including the resources and operators to be used, the configurations of the machines, raw materials required for the process, and its input and output states. It depends on the application to ensure the satisfaction of these conditions. With some amendment to the latest version IOF, we declare that only *iof:MaterialState* (m-state) or *iof:ProcessCharacteristic* (p-char) can be prescribed by the o-spec, being more inclusive of the cases that o-spec may contain information about. Such m-stat or p-char *realizes* the Goal has the o-spec as the proposition content (Fig. 2).

$$A8: InformationContentEntity(x)\wedge\exists y\, Process(y)\wedge prescribes(x,y)\rightarrow ActionSpecification(x)$$

$$A9: InformationContentEntity(x)\wedge\exists\, PlannedProcess(y)\wedge prescribes(x,y)\rightarrow PlanSpecification(x)$$

$$A10: InformationContentEntity(x)\wedge\exists y\,((MaterialState(y)\vee ProcessCharacteristic(y))\wedge prescribes(x,y))\rightarrow ObjectiveSpecification(x)$$

2.2 Modelling Sequence, Hierarchy, and Alternatives in the Plan

In the section, we show that the proposed ontology can capture sequence, hierarchy, and alternatives in plans as described in [7].

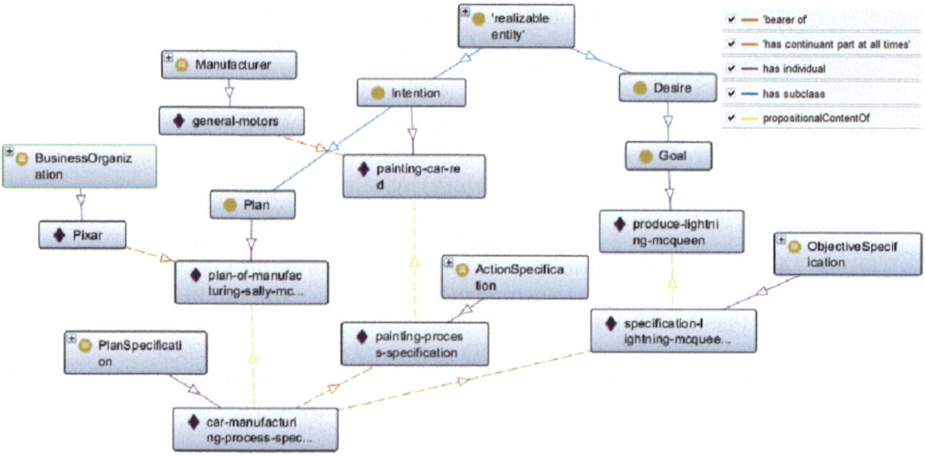

Fig. 2. Model of plan of production of a toy car

Temporal Order in the Steps of a Plan.
In order to make a-spec able to capture sequences among the steps of a plan, a-spec is subtyped into subclasses which will contain the information on the type of temporal relationship along with general process-specific information. For example, *BeforeActionSpecification* (ba-spec) allows the user to add another a-spec as part, denoting that a process prescribed by the instance of ba-spec needs to occur before the process that is prescribed by the a-spec that it is part of.

$$\begin{aligned} \text{A11: } & \text{Precedes}(x,y) \wedge \text{prescribedBy}(x,x') \wedge \text{prescribedBy}(y,y') \\ & \rightarrow \text{BeforeActionSpecification}(x') \wedge \text{ActionSpecification}(y') \\ & \wedge \text{hasContinuantPartAtAllTImes}(x',y') \end{aligned}$$

Similar to *ba-spec,* many other temporal relations can be added, e.g., 13 relationships from Allen's algebra. We provide an example to illustrate the modelling pattern for a p-spec with some ordering among the steps. For example, the p-spec to make coffee contains three steps: grinding coffee, boiling water, and brewing coffee. Among these, both grinding and boiling should be completed before brewing coffee.

Hierarchy in a Plan
A plan may be described in cascading details. For example, the plan for producing a car could be decomposed in sub-plans of producing the engine, the chassis, the body etc. Although it will be tempting to assert sub-plans as part of p-spec for the plan, the parthood relationship among two p-specs is not semantically equivalent to the parthood of the processes that the p-specs prescribes (Fig. 3).

Instead, we captured the semantic of the decomposition of a plan based on the parthood among the processes it is realized by. The following axiom says that if Intention *i'* is a part of Intention *i* then *i'* is one of other Intentions i_1 to i_n that are also part of *i* and if their exists some processes e_1 to e_n, each of which realizes plans from i_1 to i_n then there exists a process *e* which realizes plan *i* and every process from e_1 to e_n are

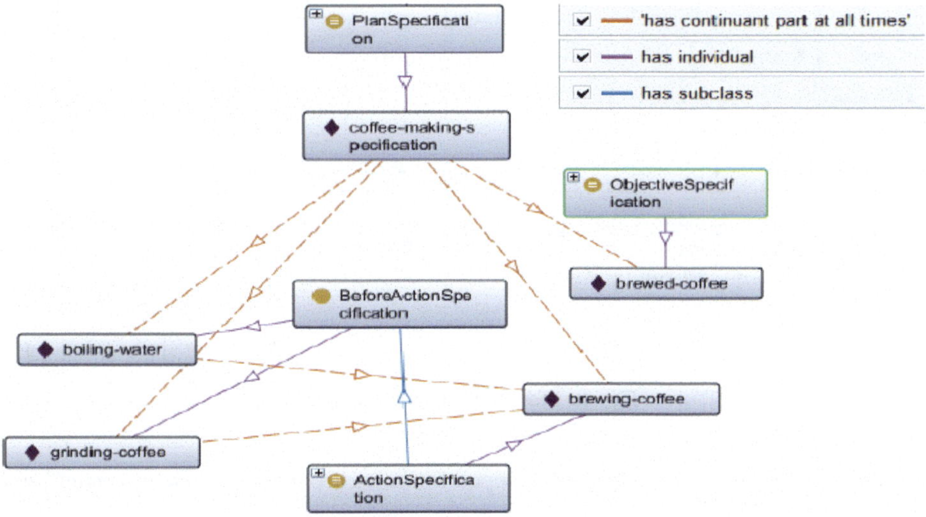

Fig. 3. Modeling ordered steps under plan specification

occurrent part of *e*. We can note from the axiom that the veracity of the parthood among plans cannot be ensured until the availability of the processes that realizes them. Still, this is not a hindrance in capturing the hierarchy of plans. We can also observe that the plan hierarchy can be developed at both a-spec and p-spec levels as both Intention and Plan can be decomposed. However, modeling plan hierarchy at the p-spec level requires corresponding sub-goals to be associated with the sub-plans. The sub-goal is a difficult concept to define because of the presence of a variety of cases involving p-char and m-state. We keep it out of the scope of this paper.

$$\begin{aligned} \text{A12: } & \forall i{:}\text{Intention}, i'{:}\text{Intention continuantPartOf}(i', i) \\ & \rightarrow \exists i_1{:}\text{Intention}, i_2{:}\text{Intention}, i_3{:}\text{Intention} \ldots i_n{:}\text{Intention} \\ & \wedge_{j=1}^{n} \text{continuantPartOf}(i_j, i) \wedge i' \in \text{list}(i_1, i_2, i_3, \ldots i_n) \wedge \\ & \exists e_1{:}\text{Process}, e_2{:}\text{Process}, e_3{:}\text{Process} \ldots e_n{:}\text{Process} \\ & \wedge_{j=1}^{n} \text{realizes}(e_j, i_j) \rightarrow \exists e \text{ realizes}(e, i) \\ & \wedge_{j=1}^{n} \text{occurrentPartOf}(e_j, e) \end{aligned}$$

In the following example we show decomposition at both p-spec and at the a-spec level. For example, a plan for manufacturing a car is composed of four sub-plans: manufacture body, manufacture chassis, assemble car, and paint car. Moreover, paint car sub-plan can have two steps: apply paint and dry paint (Fig. 4).

Alternative Plans

Sometimes, alternative plans may be developed for the same goal. This is trivial given the ontology models Goal and Plan are modelled independently while being linked only at the specification level. Therefore, many instances of plan may be present each of which having the same o-spec (propositional content of the same Goal) as part of their corresponding plan specification. The following example illustrates the scenario. For

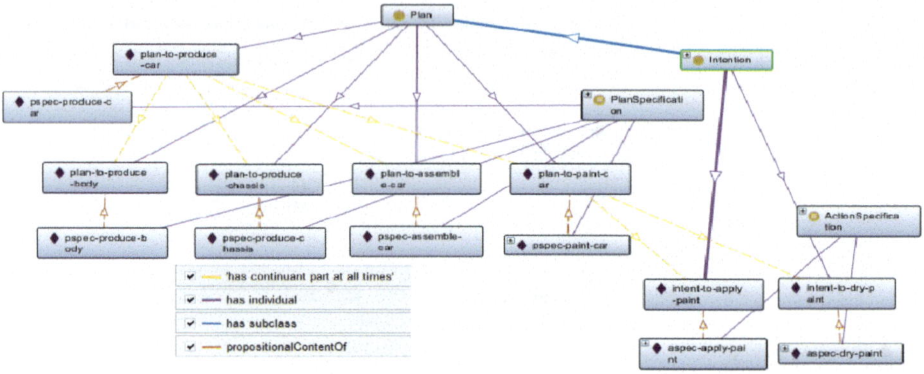

Fig. 4. Modeling hierarchical plan

example, two alternative plans are developed for going from A to B, one through C and another through D (Fig. 5).

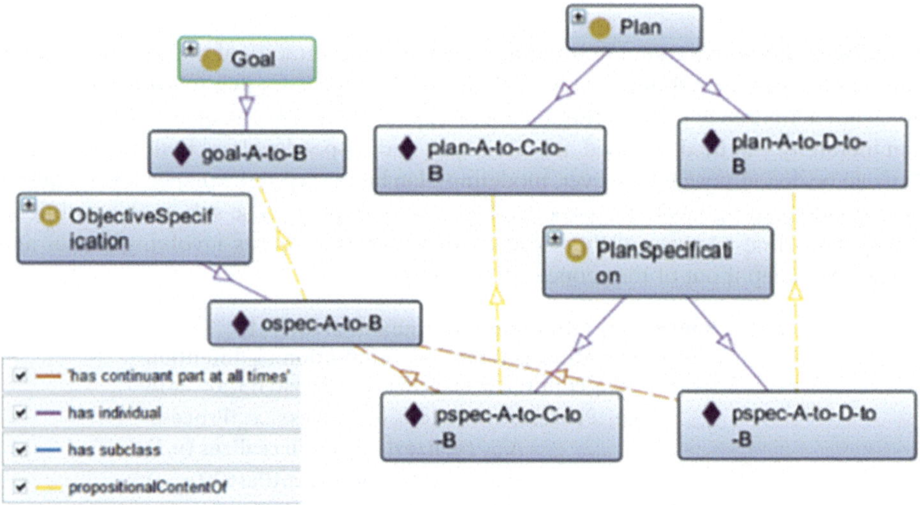

Fig. 5. Modeling alternative plans

2.3 Conformance of Executed Processes to Plan

As stated in Sect. 2.1, a plan is realized based on the conformance between the p-spec and a p-process occurrence. In turn, such conformance needs to recursively test conformance between the o-spec and a-spec that are part of the p-spec. The definition of a-spec given in Sect. 2 may be extended by qualifying their parts with other concepts to add other information as process details. We add *InputSpecification*, *OutputSpecification*, and *ResourceSpecification* to add required input and output conditions and required resources, e.g., machine, operator, as part of the a-spec, denoting such requirements for performing a process according to that a-spec. Both *InputSpecification, OutputSpecification* can be prescribed by a m-state and p-char. A *ResourceSpecification* can be prescribed by a *MaterialResource* or an *Operator*. These concepts are already present in IOF. If the a-spec, p-spec, or o-spec are *prescribedBy* on successful conformance as defined by the sufficient conditions given in Sect. 2.1, we can apply rules to assert that the corresponding Intention, Plan, and Goal are realized. However, the inclusion of details mentioned above in the rule is application-specifc. We define the following rules with the only process being *prescribedBy* the a-spec.

A13: ActionSpecification(a)∧Process(p)∧Intention(i)∧
prescribedBy(p,a)∧propositionalContentOf(a, i)→realizes(p, i)
A14: (ProcessCharacteristic(x)∨MaterialState(x))∧
ObjectiveSpefication(o)∧Goal(g)∧prescribedBy(o,x)∧
propositionalContentOf(o, g)→realizes(x, g)

Realizing a plan requires the corresponding p-spec to be *prescribedBy* a p-process. It is natural to assume that the p-process occurrence will have the processes, that prescribe all the containing a-spec, as occurrent part. Therefore, it is necessary for the intentions that the a-specs are propositional content to be realized for the p-process to conform to the p-spec. Additionally, the Goal that the containing o-spec is propositional content of to be also realized. Together it denotes that the occurrence of the p-process achieved the Goal as specified in the corresponding o-spec and the p-process is performed by executing all the underlying steps correctly. Not including the complexity of checking the conformance.

A15: PlanSpecification(s)∧ObejctiveSpecification(o)∧
continuantPartOf(o,s)∧propositionalContentOf(o,g)∧
realises(x,g)∧PlannedProcess(p)∧prescribedBy(p,s)
∧Plan(i)∧propositionalContentOf(s,i)→realized(p,i)

3 Application of the Ontology

In this section, we demonstrate the application of the proposed ontology by representing the information of a complete process plan. Aiming to produce Spur-gear, the plan consists of 5 processing steps, which need to be processed in the following order: turning and milling to cut the metal, then quality check for the dimension, then laser marking to imprint the part number, then a round-grinding to polish the gear and final inspection to check quality. We show the representation of the plan, considering turning and milling along with a quality check as part of the 'fabrication process', and grinding and final inspection as part of the 'finishing process'. The grouping is included to show the hierarchical process plan (Fig. 6).

Next, we show an example of the conformance of executed processes to the plan. In Fig. 7, we only show the corresponding a-spec and p-spec that the processes are prescribedBy (follows as planned) and the Intention and Plan instances they prescribe.

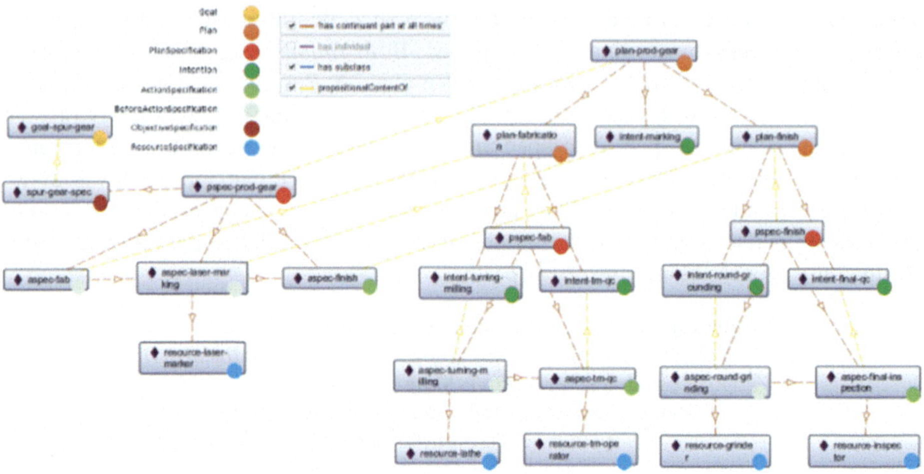

Fig. 6. Representation of spur-gear production plan

We include the goal is realized by a material state with a spur-gear. In this example, it is assumed that the executed processes and existing states are related to their respective specifications by prescribedBy after making sure they follow the details of the specifications. It could be observed that "plan-prod-gear", the top-level plan, is realized because every sub-plan is realized in the hierarchy, as dictated by axiom A12 and rule A15. Although we did not show the case where some executed production does not go as planned, it can be observed from the example in Fig. 7, that if a certain process, e.g., "round-grinding-process" fails to occur as prescribed by "aspec-round-grinding", then the corresponding Intention "intent-round-grinding" will also not be realized (see A13). If "intent-round-grinding" is not realized then "plan-finish" will also not be realized (A15), as "plan-finish" has "intent-round-grinding" as part. Ultimately, "plan-prod-gear" will not be realized for "plan-finish" not being realized. Also note, this is one of many ways "plan-prod-gear" may not be realized.

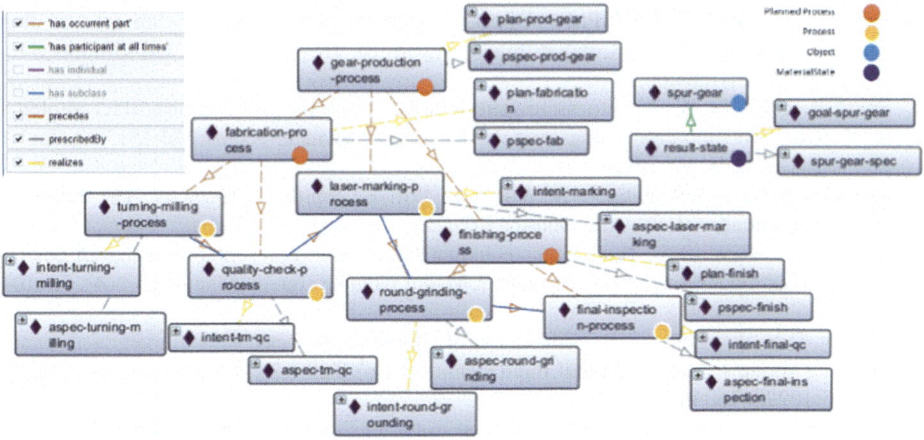

Fig. 7. Conformance of executed processes to plan

4 Conclusion

The main contribution of this article is to provide a model of plan based on psychological state of mind. At the same time, we show that the model, along with some extensions from existing concepts from IOF-Core, is able to capture detailed process plan typical to manufacturing, including temporal order among processes, hierarchical plan, and alternative plans. We also show that the model helps in building the conformance of executed processes, that when actual production happens following the plan, from bottom up starting from conformance of steps.

Many other details still need to be furnished for enhancing the capability of the model, e.g., representation of multiple temporal relationships that one process may have with other processes, the inclusion of details on the configuration of manufacturing machines and production systems as well as various domain-specific details particular to specific domains of manufacturing, e.g., metal-cutting, assembly, and batch production in the a-spec. Planning with uncertainty, resource constraints, and scheduling also may require some adjustment and improvement to the proposed model.

References

1. Huang, Z., Qiao, L., Anwer, N., Mo, Y.: Ontology model for assembly process planning knowledge. In : 21st International Conference on Industrial Engineering and Engineering Management, pp. 419–423 (2015)
2. Denkena, B., Dittrich, M.-A., Münch, G.V.: Ontology-based production planning under the consideration of system robustness. Procedia CIRP **104**, 547–552 (2021). https://doi.org/10.1016/j.procir.2021.11.092
3. Grüninger, M. : Ontology of the process specification language. Handb. Ontol. (2004). https://pdfs.semanticscholar.org/ea24/cd340a0662636b09339c0f2ee8494bb8f17e.pdf. Accessed 09 Oct 2018

4. Šormaz, D., Sarkar, A.: SIMPM – Upper-level ontology for manufacturing process plan network generation. Robot. Comput. Integr. Manuf. **55**, 183–198 (2019). https://doi.org/10.1016/J.RCIM.2018.04.002
5. Smith, B., Ceusters, W.: Aboutness: towards foundations for the information artifact ontology. In : CEUR Workshop Proceedings, vol. 1515 (2015). http://ceur-ws.org/Vol-1515/regular10.pdf. Accessed: Oct. 27, 2018
6. Drobnjakovic, M., Kulvatunyou, B., Ameri, F., Will, C., Smith, B., Jones, A.: The Industrial Ontologies Foundry (IOF) Core Ontology (2022)
7. Sormaz, D., Khoshnevis, B.: Process planning knowledge representation using an object-oriented data model. Int. J. Comput. Integr. Manuf.Comput. Integr. Manuf. **10**(1–4), 92–104 (1997)
8. Sormaz, D.N., Khoshnevis, B.: Generation of alternative process plans in integrated manufacturing systems. J. Intell. Manuf.Intell. Manuf. **14**(6), 509–526 (2003). https://doi.org/10.1023/A:1027302604072
9. Woodfield, A., Searle, J.: Intentionality--an essay in the philosophy of mind. Philos. Q. **36**(143), 300 (1986). https://doi.org/10.2307/2219777
10. Richetti, P.H.P., Baião, F.A.: A measurement ontology for beliefs, desires, intentions and feelings within knowledge-intensive processes. In: CEUR Workshop Proceedings, vol. 2228, no. 2005, pp. 44–55 (2018)

Production Line Measurements as Predictors: Towards Geometric Quality Assurance for Assembly

Déborah Sarria(✉), Anis Assad Neto, Fernando Deschamps, Alexandre Surkus-Castro, Swen Giese, and Juliano Rafael Dal Negro Costa

Industrial and Systems Engineering, Pontifical Catholic University of Paraná (PUC-PR), Curitiba, Paraná, Brazil
deborahsarria@gmail.com

Abstract. Many companies are already collecting data on the geometry of the components that they manufacture for use in the assembly of their products. This data can be used for better control of quality assurance by predicting the cumulative effect of geometric deviations that can have an impact on assembly. However, this data is not standardized, it is collected at different points in the process and stored in different systems. Therefore, a strategy is needed to facilitate using these geometric data by applying algorithms that allow using them to improve product assembly. To address this problem, in this article, we propose a technological framework to predict the quality defects that may occur in the production process, based on geometric deviations, using inspection data collected on the production line. This allows a virtual simulation of the assembly of the vehicles to prevent quality defects that may affect the assembly of parts on the vehicle.

Keywords: Virtual simulation · prediction · quality · assembly · point cloud

1 Introduction

In recent decades, studies on assembly accuracy analysis methods for evaluating assembly deviations due to manufacturing errors have been conducted [1]. For instance, artificial face models with geometry errors have been created to represent manufacturing errors and simulate the assembly state of mating faces, and inspection data may be incorporated into virtual product models to detect and correct non-conformities resulting from geometric deviations. These deviations, when found at the component level, tend to propagate throughout the assembly process as components are interconnected [2]. Such deviations lead to assembly problems and failure to meet functional requirements. Therefore, 3D assembly tolerance analysis methods are primarily used to predict the effects of part and component deviations on assembly accuracy and verify the assembly process [3]. Leveraging knowledge about the geometry of executed components, the assembly of other components can be simulated and optimized [2]. The digitalization era has introduced real-time control and optimization of the assembly process, inspiring concepts such as digital twins and cyber-physical factories [4].

M. Dragomir et al. (Eds.): ICPR 2023, LNPE, pp. 397–405, 2025.
https://doi.org/10.1007/978-3-031-92082-0_44

Advancements in metrology enable the representation of actual surface data as a point cloud, which is used to evaluate assembly accuracy before product assembly [1]. Technological tools in manufacturing facilitate a digital process that integrates modern information and communication technologies, enabling a continuous flow of information throughout the product life cycle. This wealth of information can be utilized to enhance the manufacturing process, anticipate product behavior, and improve product quality [5].

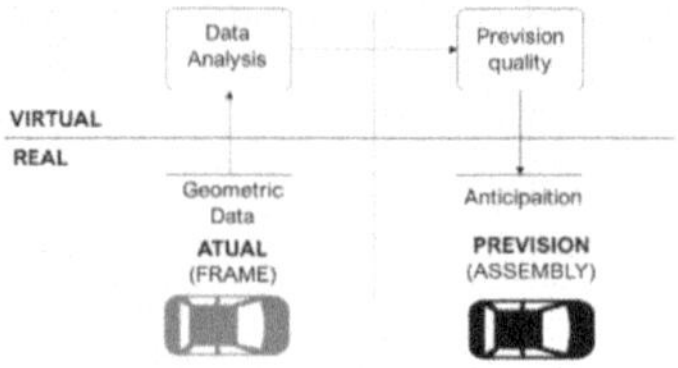

Fig. 1. Virtual simulation-based approach for managing geometric deviations during assembly.

This work presents a virtual simulation approach using data collected directly from the production line. The goal is to enhance quality control by addressing geometric deviations in product manufacturing (Fig. 1).

The manufacture of complex products involves manufacturing and assembly steps. At each step, some risk of introducing geometric deviations and variations may arise [6]. To improve quality control in the manufacturing of metallic assemblies, our proposal involves predicting defects that may occur in the assembly due to geometric deviations occurring during the vehicle's manufacturing. The rest of this paper is organized as follows; the 2nd section provides a literature review of geometric deviation and digital technologies. The 3rd Section explains the framework for the virtual simulation of the assembly process. The 4th Section explains the proposed model in detail and the testing model is described.

2 Literature Review

Assembled products often face geometric variations, requiring companies to manage these deviations throughout the product lifecycle [7]. Variation in assembled parts leads to geometric variation in the product [8]. The availability of vast data and automated production lines offers new opportunities in design and production [9]. Increasing digitalization in manufacturing ensures consistent geometries and codified variations [9].

Combining big data with fast simulation enables real-time optimization of products and production processes [10]. Cyber-physical systems, the Internet of Things (IoT), big data, and cloud computing drive these technological trends [7]. Data collection, transfer, and processing are vital for developing models used in real-time virtual simulation [7]. Measurement systems like camera-based or laser-based scanners are employed in the production line to collect data [11]. Scanned data maps the product surface to generate CAD geometry. Measurement data represents deviations from the stored geometry in a database [12]. Online measurement systems facilitate the digitization of part geometry using different sensors [13–15]. The resulting data from points clouds [1, 6, 7, 16, 17] offers the potential for assessing assembly accuracy prior to the process [1].

Real-time data from the physical world can be analyzed for real-time equipment and assembly process simulation [18]. The virtual assembly uses virtual reality, computer graphics, AI, and simulation technology for interactive analysis and assembly simulation [16]. The use of simulation models is facilitated by sensors and measurement equipment on the production line [10]. Point cloud scanning constructs the physical model, which can be compared to the CAD model for alignment [17].

2.1 Alignment

The points cloud of an object can be compared to its CAD model, facilitating alignment between the two elements. [17]. For the alignment, it is necessary to do a transformation of the element, which is necessary for the calculation of a rotation and a translation. During an alignment, a given entity is transformed into a given target entity. Depending on the alignment, this transformation follows different rules because the specified element and the target element are never perfectly aligned. A rigid body has six degrees of freedom in space, including three translations in x, y, and z and three rotations around these axes. To fixate the part for production processes, including the assembly process, these degrees of freedom should be locked by locators [19].

Best-fit
For finding the best fit of the points cloud to the CAD model, it is needed to estimate a rigid body transformation, consisting of a rotation and a translation vector, such that our cloud of data points fits the surface of the CAD model under the transformation [17]. The best-fit alignment gives a good matching between the representation of the measured shape and the CAD model.

3 Proposed Approach

3.1 Framework to the Assembly Simulation

Consider implementing a virtual simulation model. This is a digital representation of the product in relation to the physical world. The physical model uses data from the digitization of the product's surfaces. This allows the creation of a points cloud. This data will be used to virtually identify and anticipate failures before they occur in the assembly when used in the model of mathematical calculations, as shown in Fig. 2.

We propose a technological framework composed of three layers: a physical layer, which includes the measurement data from the inspection of the product surface; and digital and service layers that allow the prediction of quality failures through the control of deviations that can affect the parts in the assembly.

The physical layer for this proposal is focused on the technological tools that exist in the production process, e.g., the ERP system and sensors that are used to gather data from the production line. These tools allow the digitization process of the product's surface and generate a database with the measurement of the surface of the bodywork that allows the creation of a cloud of product points. The digital layer includes the

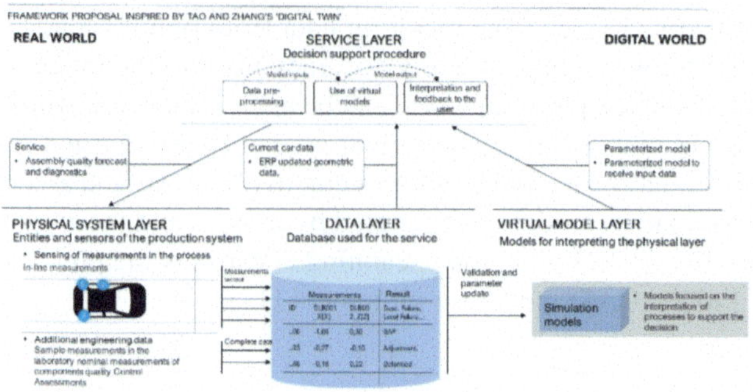

Fig. 2. Quality failure prediction technology framework based on geometric deviations.

database of digital models of the product, the CAD models of the theoretical values, and the simulation model. The service layer presents the result after the pre-processing of the measured data in the production line, using mathematical models for simulation and interpreting the results as feedback to the production area. In this way, it is possible to determine the assembly processes, and whether the bodywork should be segregated. We also propose a structure that allows data to be connected and transferred between each layer to obtain a fast and reliable result (Fig. 3).

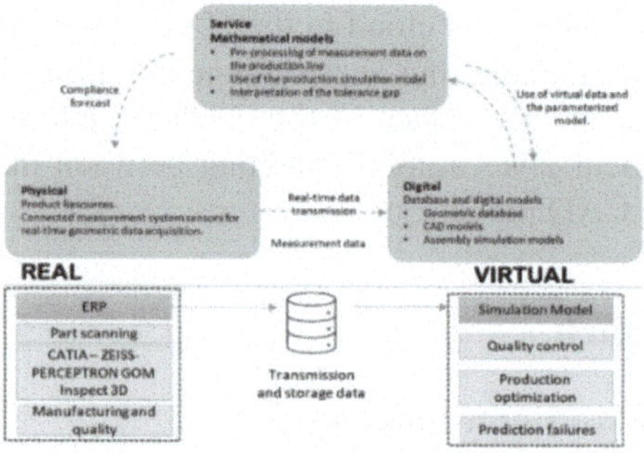

Fig. 3. Framework layers.

The points cloud and nominal values of the CAD model are the input data to be considered. The points captured by the scanner on the surface of the metal assembly make up the points cloud. For the modeling stage, it is proposed to superimpose the theoretical values of the product (CAD model) and the actual values that are collected in the production line, as well as the nominal values of the components, to perform an alignment of the best position of the components at the time of assembly on the bodywork. To perform an alignment between the values collected in real-time, in the production line, to the CAD model, it is necessary to apply the best-fit alignment with tolerance since the objective is to predict the assembly between the vehicle components and the bodywork.

3.2 Assembly Simulation

This article proposes a virtual assembly simulation based on data analysis of geometric deviations. It aims to improve product quality assurance control in the assembly process. To align the components, real points cloud data collected in the production line is transferred. Before using the data, a pre-analysis and transformation into a new database are necessary to ensure accurate and high-quality data. The 3D inspection analysis system imports the data from the organized database, forming a cloud of product points. It then performs a check between the CAD model and the real-time measurements collected on the production line. Alignment is achieved through an overlay of the real points cloud and the CAD model. Mathematical calculations measure 3D distances between points and vector calculations enable the definition of geometric deviations.

3.3 Best-Fit

We suggest using the best-fit alignment: first, the best fit by RPS (Reference Point System), then the Local Best-fit for a region of the bodywork, and finally, the Local best fit with tolerances to check part fit. This last alignment verifies gap and flush specification values between adjacent parts when connected to the bodywork. Figure 4 shows the alignment process.

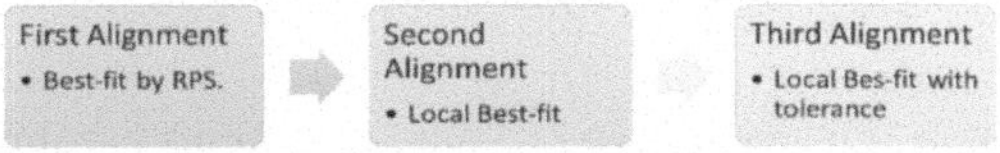

Fig. 4. Alignment process.

For each bodywork measured in the production line, the three stages of alignment are applied. The CAD models are always the same for both the bodywork and the components. For this reason, a spreadsheet has been created that allows agility and speed in updating information and generating a virtual simulation of the data collected in real time.

4 Case Study

In the car assembly environment, we tested this case on the windshield gap area to predict the optimal position for the glass. Virtual simulation ensures part fit and compliance with tolerance values. It also provides alerts for the sheet metal joining processes.

4.1 Assembly Simulation

The bodywork model is created through points cloud scanning. Real-time data from sensors during a part inspection on the production line is collected and stored in a database. This data is transformed to generate a new database, which is then imported into the 3D inspection analysis system (GOM software). Alignments are performed to achieve a local best fit with tolerances. Geometric deviations are visualized to assess component fitting and detect quality defects. The virtual model results from aligning the points cloud with the CAD model. Three alignments are used, starting with the Best-fit by RPS, followed by automated application of the other two alignments.

Best-fit by RPS
The RPS best-fit alignment is used for automated measurements. Real point clouds result from transforming the measurement database of each body's surfaces, aligning with the CAD model. This initial alignment is ideal as it is based on the RPS of the car. In the windshield opening area, a defined number of points speeds up the model, making calculations faster.

Local Best-fit
This alignment is limited to one region. In this case, it is applied to the region of the windshield gap. The software checks the vector direction of the surfaces to be located. If point vector setup is required, the point cloud normal vectors point away from the point cloud. This is useful because it allows comparison between the data of the CAD model to guarantee the geometry. As shown in Fig. 5.

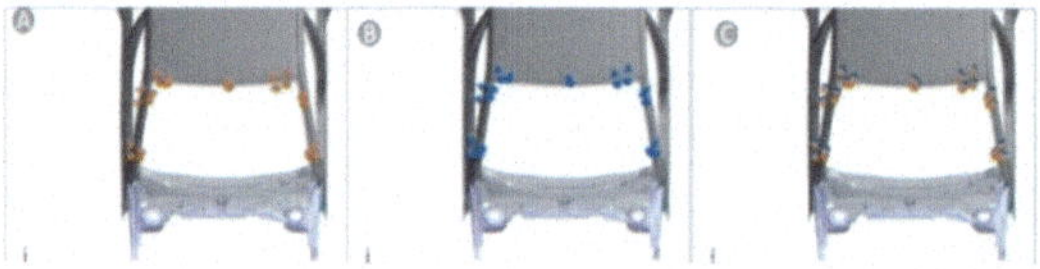

Fig. 5. Vector position of the bodywork for application of the Local Best-fit alignment.

To verify the fit between the components, it is necessary to add the CAD model of the windshield, which is located in the area of the windshield gap, during the car assembly process. Therefore, after carrying out the alignment between the actual measurements and the CAD model of the bodywork, it is necessary to add the CAD model of the component (CAD model of the windshield gap). Consequently, it is necessary to invert the direction of the normal vectors of the selected element (windshield glass) when aligning two parts to check their fit. As shown in the figure, the normal vectors will now point in the direction of the CAD model, as shown in Fig. 6.

Local Best-fit with tolerance
In this alignment, data selection isn't adjusted to the target element but to a tolerance zone, like a 3mm area around it (bilateral tolerance). The zone can be inside or outside the element (one-sided tolerance). Tests considered car design specifications, including gap and flush values on sides and roof. This provides alerts for frame area feedback and analyzes joining process parts to avoid quality failures due to non-compliance with specifications.

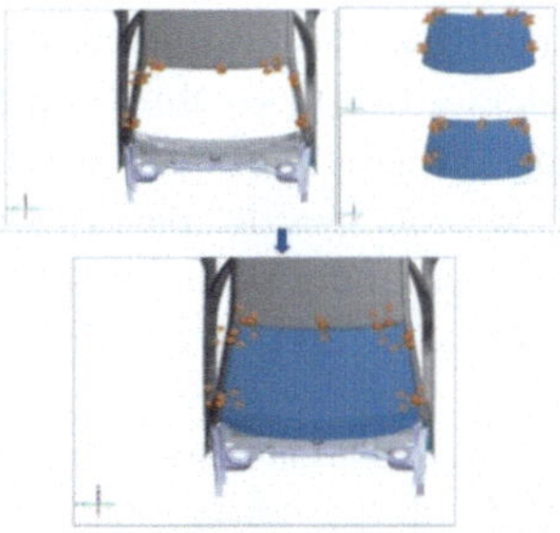

Fig. 6. Vector position of the bodywork and the windshield glass for application of the Local Best-fit with tolerance alignment.

5 Results

For the tests, the critical points of the windshield gap region were defined to provide the most important points of the points cloud to allow agility and speed in the model, during the performance of alignments. Real-time measured data has become a great opportunity to be processed and provides a good quality database.

Data is transferred to the measurement analysis system and alignments are calculated. The same CAD model of the bodywork is used for all measured objects, facilitating that the points cloud can be compared with the CAD model, and the alignment is done.

The 3D inspection software uses algorithms to align the measurement points cloud with the product model. It generates individual reports for each bodywork, showing the windshield's positioning and adherence to tolerances. A final report summarizes the trend of recent measurements and the behavior of each point in relation to process indicators (Fig. 7).

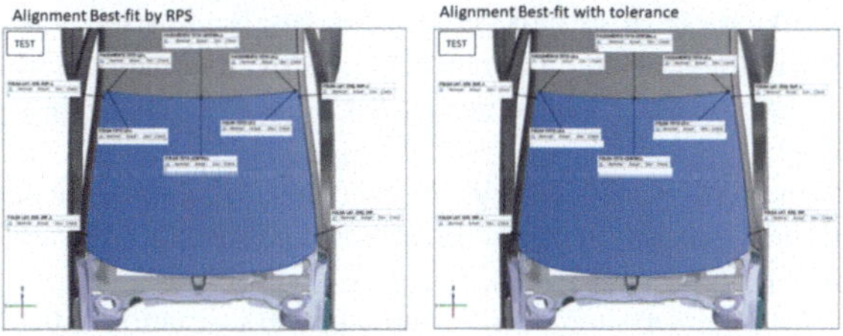

Fig. 7. Alignment between the windshield bodywork and the windshield glass, for the alignment by RPS and by Local Best-fit with tolerance.

6 Conclusion

A simulation of real-time data can confirm the possibility of aligning components accurately. This enables early identification of deviations and predicts assembly failures, allowing for timely reactions in the metallic assembly process. However, there are significant upcoming challenges that must be addressed. It is crucial to raise awareness about providing high-quality data that can be reused and updated in real-time during body manufacturing. The bottleneck lies in establishing a coherent connection system for data transfer and automating the generation of 3D inspection analysis from scanning systems. This study focused on a specific part of the bodywork, but future work can include virtual simulations for other car parts. Additionally, optimizing the framing process and simulating the joining of metal parts could be alternative approaches.

Acknowledgments. The Digital Transformation Team of the Pontifical Catholic University of Paraná would like to thank CNPq for the resources provided and the pilot factory team car manufacturer for supporting data and scanning systems and 3D monitoring analysis.

References

1. Wang, K., Liu, D., Liu, Z., Wang, Q., Tan, J.: An assembly precision analysis method based on a general part digital twin model. Robot. Comput. Integr. Manuf. **68**, 102089 (2021). https://doi.org/10.1016/j.rcim.2020.102089
2. Ma, Y., Zhou, H., He, H., Jiao, G., Wei, S.: A digital twin-based approach for quality control and optimization of complex product assembly. In: Proceedings - 2019 International Conference on Artificial Intelligence and Advanced Manufacturing, AIAM 2019, pp. 762–767 (2019). https://doi.org/10.1109/AIAM48774.2019.00157
3. Yi, Y., Yan, Y., Liu, X., Ni, Z., Feng, J., Liu, J.: Digital twin-based smart assembly process design and application framework for complex products and its case study. J. Manuf. Syst. **58**, 94–107 (2021). https://doi.org/10.1016/j.jmsy.2020.04.013
4. Tabar, R.S., Wärmefjord, K., Söderberg, R., Lindkvist, L.: Efficient spot welding sequence optimization in a geometry assurance digital twin. J. Mech. Des. Trans. ASME **142**(10), 102001 (2020). https://doi.org/10.1115/1.4046436
5. Schleich, B., Anwer, N., Mathieu, L., Wartzack, S.: Skin model shapes: a new paradigm shift for geometric variations modelling in mechanical engineering. Comput. Aided Des. **50**, 1–15 (2014). https://doi.org/10.1016/J.CAD.2014.01.001
6. Wärmefjord, K., Söderberg, R., Lindkvist, L., Lindau, B., Carlson, J.S.: Inspection data to support a digital twin for geometry assurance. In: ASME International Mechanical Engineering Congress and Exposition, Proceedings (IMECE), vol. 2 (2018). https://doi.org/10.1115/IMECE2017-70398
7. Schleich, B., Wärmefjord, K., Söderberg, R., Wartzack, S.: Geometrical variations management 4.0: towards next generation geometry assurance. Procedia CIRP **75**(January), 3 (2018). https://doi.org/10.1016/j.procir.2018.04.078
8. Jareteg, C., et al.: Geometry assurance integrating process variation with simulation of spring-in for composite parts and assemblies. J. Comput. Inf. Sci. Eng. **16**(3), 031003 (2016). https://doi.org/10.1115/1.4033726
9. Rezaei Aderiani, A., Wärmefjord, K., Söderberg, R., Lindkvist, L.: Developing a selective assembly technique for sheet metal assemblies. Int. J. Prod. Res. **57**(22), 7174–7188 (2019). https://doi.org/10.1080/00207543.2019.1581387
10. Söderberg, R., Wärmefjord, K., Carlson, J.S., Lindkvist, L.: Toward a digital twin for real-time geometry assurance in individualized production. CIRP Ann. **66**(1), 137–140 (2017). https://doi.org/10.1016/J.CIRP.2017.04.038
11. Esposito, C., Cosenza, C., Gerbino, S., Martorelli, M., Franciosa, P.: Virtual shimming simulation for smart assembly of aircraft skin panels based on a physics-driven digital twin. Int. J. Interact. Des. Manuf. (IJIDeM) **16**(2), 753–763 (2022). https://doi.org/10.1007/s12008-022-00875-4

12. Qie, Y., Anwer, N.: Data-driven deviation generation for non-ideal surfaces of skin model shapes. Procedia CIRP **109**, 1–6 (2022). https://doi.org/10.1016/j.procir.2022.05.205
13. Tabar, R.S., Lindkvist, L., Wärmefjord, K., Söderberg, R.: Efficient joining sequence variation analysis of stochastic batch assemblies. J. Comput. Inf. Sci. Eng. **22**(4), 040905 (2022). https://doi.org/10.1115/1.4054000
14. Sadeghi Tabar, R., Wärmefjord, K., Söderberg, R.: Rapid sequence optimization of spot welds for improved geometrical quality using a novel stepwise algorithm. Eng. Optim. **53**(5), 867–884 (2021). https://doi.org/10.1080/0305215X.2020.1757090
15. Tabar, R.S., Wärmefjord, K., Söderberg, R.: A method for identification and sequence optimisation of geometry spot welds in a digital twin context. Proc. Inst. Mech. Eng. C J. Mech. Eng. Sci. **233**(16), 5610–5621 (2019). https://doi.org/10.1177/0954406219854466
16. Sun, X., Bao, J., Li, J., Zhang, Y., Liu, S., Zhou, B.: A digital twin-driven approach for the assembly-commissioning of high precision products. Robot. Comput. Integr. Manuf. **61**, 101839 (2020). https://doi.org/10.1016/j.rcim.2019.101839
17. Bergström, P., et al.: Automatic in-line inspection of shape based on photogrammetry. In: 7th Swedish Production Symposium, pp. 1–9 (2016)
18. IUzhno-Ural'skiĭ gosudarstvennyĭ universitet and Institute of Electrical and Electronics Engineers, Proceedings. In: 2020 International Russian Automation Conference (RusAutoCon) : Sochi, Russia (2020)
19. Rezaei Aderiani, A., Wärmefjord, K., Söderberg, R.: Combining selective assembly and individualized locator adjustments techniques in a smart assembly line. In: Procedia CIRP, 2020, vol. 97, pp. 429–434 (2022). https://doi.org/10.1016/j.procir.2020.05.263. https://techguide.gom.com/en/gom-software2022/article/introduction_to_alignments.html#mark-0. ZEISS. Tec guide. Introduction to Alignments.

Approach for Integration of Model-Based Systems Engineering and Product Life Cycle Management

Benjamin Schneider[1](✉), Josip Zilic[1], Matei Rhaul Gideon[2], Stefan Giosan[1], Mehmet Kürümlüoglu[1], and Carmen Constantinescu[1]

[1] Fraunhofer Institute for Industrial Engineering IAO, Nobelstr. 12, 70569 Stuttgart, Germany
benjamin.schneider@iao.fraunhofer.de

[2] Institute of Human Factors and Technology Management IAT, University of Stuttgart, Allmandring 35, 70569 Stuttgart, Germany

Abstract. Advanced Systems Engineering (ASE) is a new paradigm for handling the complexity of todays and tomorrow's products and services. Advanced systems, in the focus of the ASE paradigm, are supposed to be autonomous, dynamically networked, optimize socio-technical interaction and offer combinations of products and services by the means of product-service-systems. To handle the resulting complexity, the ASE paradigm describes model-based systems engineering (MBSE) as a suitable approach. To ensure a digital continuity, MBSE models and structures, created in the early stages of product development, have to serve as a basis for further development steps and therefore need to be stored and managed in a central IT-system. A product life cycle management (PLM) system is such a system. This paper describes a concept for a prototype integration and bidirectional interface between a MBSE modelling environment and a commercial PLM system. The prototype development and evaluation of the interface is based on a use case from a research project with a background in automotive industry.

Keywords: Advanced Systems Engineering · Model-based Systems Engineering · Product Life Cycle Management

1 Introduction

Main challenges of manufacturing companies today are ever increasing complexity and rising demands towards flexibility. Main drivers are a multitude of crisis, regulatory requirements, customer demands for individualization, new technologies such as artificial intelligence and the networking of systems, as well as product-service systems. Digitalization is meant to solve most of these problems, but the transformation towards a fully digitalized company poses a big challenge for a lot of the established companies. (Dumitrescu et al. 2021).

A German research initiative (BMBF - Bundesministerium für Bildung und Forschung 2021), "Advanced Systems Engineering" addresses these challenges. The

M. Dragomir et al. (Eds.): ICPR 2023, LNPE, pp. 406–414, 2025.
https://doi.org/10.1007/978-3-031-92082-0_45

initiative describes Advanced Systems (AS), as a response to the demands of tomorrow's value creation. Advanced Systems are similar to cyber physical systems, known from the I4.0 context. Their main properties are autonomy, socio-technical interaction, dynamic networking as well as product-service systems. AS thereby are relevant to meet future market demands but at the same time increase complexity for product creation. (Dumitrescu et al. 2021).

Systems Engineering (SE) and Advanced Engineering (AE) provide the methods and tools to handle this increased complexity and cover all phases of the product life cycle, from the first idea, engineering, production and use phase, to the end of the product or system life cycle. (Dumitrescu et al. 2021).

SE describes the conventional systems engineering approach to structure projects, teams and companies but is carried out in a model-based manner (INCOSE 2014). Model-based Systems Engineering (MBSE), as defined by INCOSE (INCOSE 2004), is the "formalized application of modeling to support system requirements, design, analysis, verification and validation activities beginning in the conceptual design phase and continuing throughout development and later lifecycle phases.".

AE describes current and future best practices in design of organizations, agile and creative environments as well as digital technologies and artificial intelligence (AI). A focus is put on digital continuity or end-to-end integration or the digital thread. Data, information and according models are supposed to be created only once, stored in a central environment and be re-used and extended with additional data and models over the life cycle of a product or system. (Dumitrescu et al. 2021).

This kind of continuity is relevant in all phases of product creation. Seamless interconnection of data and models can be seen as the basis for efficient digitalization of a company. (Riedel et al. 2024).

Product life cycle management (PLM) is a management approach as well as a software solution for storing, managing and connecting all relevant data and models concerning the life cycle of a product (Eigner 2021).

A starting point in product engineering is usually the gathering of requirements and the systematic documentation of these inside a requirements model, which can then be exchanged via ReqIF data format (OMG ReqIF - formal/2016–07-01, 2016). ReqIF is a semantics-based description of requirements and meta data, meant to standardize the exchange of requirements independent of the software used. In MBSE requirements can be used to define the functional and then the logical and physical architecture of a product or system, following the RFLP (Requirements, Functions, Logic, Physical) approach (Friedenthal et al. 2014). Further on, the structures defined in the MBSE system model should be transferred to a PLM system and be used as a starting point for all further development activities (Dumitrescu et al. 2021; Eigner 2021). Subsequently the system model can serve as base for simulation, validation, verification, further detailing of the product under development (Eigner 2021). The system model can also be used as base structure for the digital twin of a product by providing the structure to allocate sensor data from the field to corresponding simulation models. The model of a product can subsequently be used to design or configure an according production system, as well as to initialize discrete event simulations for layout and planning of the production system (Haße et al. 2017).

This paper presents a prototype method and software script for bi-directional connection and interchangeability of early-phase MBSE Models of a production system, modeled in Capella (Eclipse) and simulation models, modeled in Plant Simulation, managed in Siemens Teamcenter. Section 2 describes the current state of the art in bi-directional model exchange in the field of MBSE and PLM. Section 3 highlights the research methodology and the use case, which is set in a living lab with a focus on automotive production. Section 4 presents the research results, followed by discussion (Sect. 5) and conclusion (Sect. 6).

2 State of the Art

The "Gesellschaft für Systems Engineering" (GfSE), a German chapter of INCOSE is working on a project called PLM4MBSE since 2015 (Adam et al. 2015). They describe ten theses stressing the need for an integration between MBSE and PLM. Most important factors described are (1) need for using MBSE models and the included information as a base for further activities over the whole life cycle of a product and (2) the need of future PLM systems to provide a holistic view on complex, multidisciplinary systems and (3) the need to rely on technology independent standards in terms of interoperability of systems. (Adam et al. 2015).

Haße et al. (Haße et al. 2017) describe an integration of MBSE models, created in Cameo Systems Modeler (commercial/ non open source) in the PLM system Siemens Teamcenter based on the "Teamcenter Visio PlugIn". This is done due to the fact that there was no direct integration of MBSE tools to PLM systems at the time. The interface is used to integrate the MBSE model with the models of product, process and simulation in the PLM environment. The coupling of the systems is understood as a proof of concept. Haße et al. (Haße et al. 2017) conclude, that the coupling of MBSE and PLM models is beneficial especially for simulation-based verification of architectural decisions. The simulations used are descrete event simulations concerning the expected value stream of the production system.

Exchange standards such as FMI (Junghanns et al. 2021), OSLC (OASIS Open Projects 2021) and SpecIF (GfSE 2021; Alt 2022) are supposed to standardize data, information and model exchange between different modeling and simulation environments.

FMI focusses on data exchange for dynamic simulation models. It is an open-source project. A lot of software tools support the standard. (Junghanns et al. 2021).

OSLC is meant to connect data and to realize the digital thread across domains, applications, and organizations. It follows a linked data approach based on the Resource Description Framework (RDF). Specific adaptors are required, which generally follow a mediator approach, meaning that data is translated in the specific format for the according tool, with the tool being responsible for the data storage. (OASIS Open Projects 2021).

SpecIF is a further development of the ReqIF standard. It uses semantics to describe and model data and information. The goal is to provide life cycle integration of models from the beginning of development. The current implementation is capable of automatically importing domain specific models from different data sources such as *.xls or *.xml. Imported data can then be visualized in a tool independent manner. (Alt 2022).

Obeo presents a direct integration between Capella and Siemens Teamcenter, called "Systems Modeling Workbench" (Obeo 2023).

To summarize there are several promising interfaces and exchange standards for connecting modeling systems and managing environments such as MBSE and PLM system. A previous analysis (Schneider et al. 2022) has shown that MBSE implementation is strongly dependent on the methods and tools used for modeling. Therefore, the following analysis and development steps need to analyze whether the described solutions can be used with the tools available to the research group inside the project context.

3 Use Case and Methodology

As shown in Chapter 2 there are several approaches for connecting early phase MBSE Models and PLM software environments. As detailed in the following, the project setting was based on Siemens Software. Which was clearly setting the boundary conditions for the prototype implementation of the bi-directional Interface. In the following first the use case and then the methodology for the development of the interface are described.

3.1 Use Case

The use case for the development of the interface is derived from the project setting of the I4Produktion project (Constantinescu et al. 2020). The use case and corresponding demonstrator are located in ARENA 2036, an automotive research environment at University of Stuttgart. The project focus of the "I4Produktion" project is on creating resilience in highly flexible production systems based on data continuity and cognitive assistance in the form of artificial intelligence. The physical demonstrator is based on a measurement cell equipped with several scanners and scanning technologies as well as autonomous mobile robots (AMR) and robot arms. Additional scanners mounted near the roof of the demonstration environment are capable of scanning the environment and based on AI, detecting changes in the production system, which might for example impact routing of the AMRs. For the production system to be resilient it must automatically adopt to changes resulting from internal or external factors. Such resilience should be based on (real-time) simulation models. The interface between MBSE and simulation models can be used (1) to initialize discrete event simulations based on the planning of the production system in MBSE models, (2) to update simulation and consequently also MBSE models to changes in the production system and (3) to reuse existing production models e. g. for redesign activities to save resources.

3.2 Methodology

The methodology for development and testing of the interfaces comprises the following steps.

- MBSE based modelling in Capella of a sample structure
- Analysis of the Capella model with a focus on the model information in the XML file
- Modelling of a structure in the PLM system

- Export and analysis of the code structure in the XML file generated by the PLM system
- Defining linking elements between both structures
- Implementing the interface
- Testing the interface
- Implementing and testing the actual use case

The first step of the methodology to develop an interface between MBSE and PLS structures is the creation of MBSE models. These are created according to the Arcadia Method in the Eclipse based, open-source software tool Capella (Capella 2023). A focus is put on the structure of the production system as well as the processes relevant for the simulation of the system. As shown in Fig. 1 Operational Capabilities are used as a synonym to processes and Operational Entities are used as synonyms to resources in the PLM environment.

Next the open-source nature of Capella and Eclipse was used to analyse the code structure of the MBSE model. As the model information is stored in XML files and objects are identifiable in plain text in the code, a script-based extraction of the model elements seemed to be a promising option.

Analysis of interfaces provided by the PLM system led to the conclusion, that only *.xml files can be used as input and output. As shown in Sect. 4 these files have also been analysed.

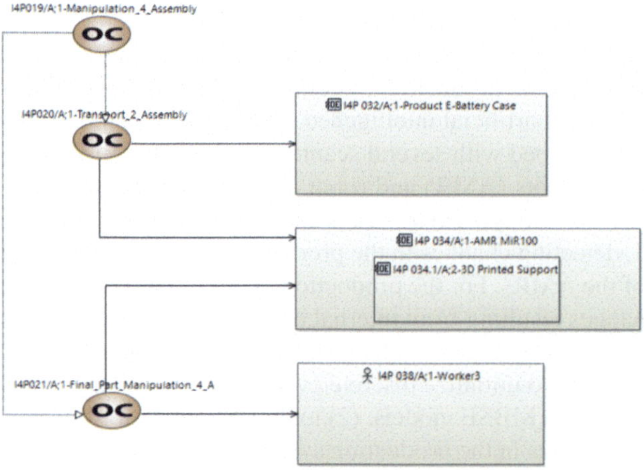

Fig. 1. MBSE process and resource model in Capella

4 Results

The main objective of the presented research is to demonstrate the benefits of data transfer between PLM data or models and an MBSE-based Capella models. The goal is to implement an exemplary solution between two standardized systems employing state of the art methods.

A Capella system model and corresponding diagrams can be the foundation of any production system. The diagrams are developed by a team of interdisciplinary experts collaborating to address all the challenges related to the production of the chosen product. The results of the presented research are focused as well on overcoming the problems of a seamless data transfer interface. The challenge is to create the PLM structure (Fig. 2) in relation to the Capella model (Fig. 1) automatically.

This chapter presents the solution to overcome this challenge, using open-source methods for an exemplary use-case validation.

Considering the flexibility of Python based solutions, a third-party script was developed to be employed as interface between the MBSE model and PLM system The common data format between this two systems is ".PLMXML", a standard XML (DENUMIRE) file.

The script enables three functionalities: XML file read, XML file write, logic for connecting the MBSE elements to the PLM items. The result is an interface capable to read a model file from MBSE/ Capella and write an import file for the PLM system and to update the MBSE/ Capella model based on updates in the PLM structure.

The Python software solution is making use of the "xml.etree.ElementTree" library. The main purpose of this library is to facilitate the read and write of XML files functionality of the script.

Therefore, the script can be divided in four sections:

- Initialization – Import of the library and declaration of variables:
- XML data read and process – This section reads and process the XML data from both PLM and MBSE/ Capella files to increase the efficiency and speed of the script
- XML data write – Creation of the import PLM file or Update of the MBSE/ Capella model
- Script interface – Display of the script functionalities and wait for user input.

The initial testing was performed with a small and easy to understand MBSE/ Capella structure data. The structure describes a process with two operations and six resource relations (Fig. 1).

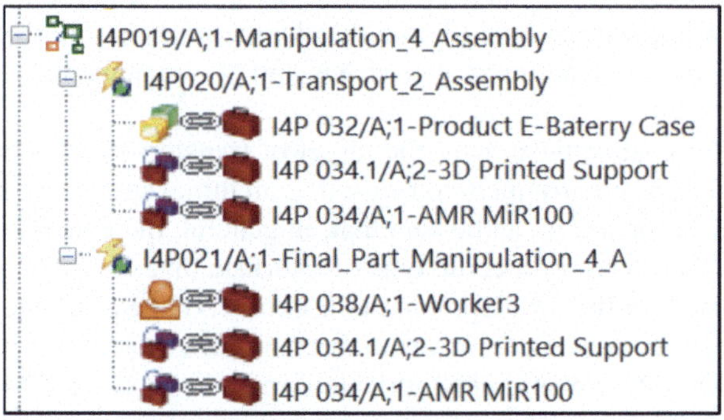

Fig. 2. Exemplary PLM structure.

The evaluation of this test is based on the capability of the script to read and write the XML data and to corelate the "ownedDiagramElements" (UML) to the "Process", "Operation" and "Product" PLM elements.

The use case for the validation of the presented methodology, is defined in relation to a production process for the quality analysis of manufactured products.

The initial situation, related to the product life cycle time frame, is defined during the setup segment. As result of history data for the quality analysis process, the decision is to change the route of the AMR to decrease energy consumption and increase productivity. This change is made on the UML diagram level and then propagated further with the PLM system – Path1 to Path2 (Fig. 3).

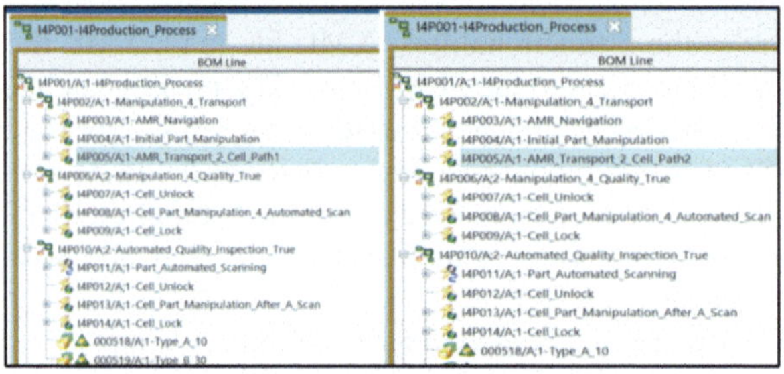

Fig. 3. Change propagation in a PLM structure.

Exemplary is demonstrated the functionality of the methodology and the overall potential for scalability to address changes for the entire PLM project structure.

5 Discussion

The developed interface, as described in the previous section, can be used for bi-directional data exchange between a MBSE and a PLM environment. The results will be discussed concerning (1) the achievement of the goals set in the beginning, (2) the method and according limitations, (3) the development of the interface in context of exiting standards and programming approaches and (4) implementation and testing of the described use case.

Concerning the goal achievement, the target of transferring structural data from a MBSE model to a PLM environment to be used as an initial structure for simulation has been achieved. In summary, it can be said that, in general, the goal was set correctly at the beginning, as there was no specific type of interface that should be used.

The method used for the development in the context of the available and selected IT systems has some limitations. Early on in the creation and export of the models, it became clear that the PLM system would only allow import and export via XML files. A direct integration between the two software tools was therefore not possible. Also, the exchange standards such as FMI, OSLC or SpecIF could not be used for the implementation of the interface. Although XML and UML files can be generated automatically based on the structures from the MBSE model and the PLM export, a direct interface was not possible.

Other PLM systems that support REST calls, for example, could offer possibilities for easier and more direct exchange of structures. This should be investigated in the future. A major limitation is that the models still need to be stored in both systems and updated via the interface's Python script. Linking the data would be a better solution in terms of a single source of truth and should be investigated in further research. Another option that should be mentioned here is the use of ontologies for the semantic mapping of the objects in the MBSE and PLM systems.

The programming itself and the selection of the programming language was mainly based on the experience of the authors as well as the available libraries for working with the XML structures. The development was straight forward, no limitations were encountered.

The tests and the implementation of the actual use case went as expected, based on the development and testing with the basic structure of the example model. As demonstrated, changes made in the Capella model, for example with respect to a different routing through the shop floor, can be propagated to the PLM system and used to update the discrete event simulation structures.

6 Conclusion

This paper presents a software script that provides a bi-directional interface between MBSE models of an early-stage production system and production system structures managed in a PLM system. In a larger context, the use of the structural information contained in an MBSE model throughout the entire life cycle of a product or system is crucial in terms of efficiency and continuity, since the information in an MBSE model can be used as the basic structure for all subsequent phases of the product life cycle and can be expanded by all subsequent models, simulations and real-world data.

For the current state of the interface described in this publication, there are several limitations that result from the software used in the project. Future work will focus on further developing the interface towards creating a single source of truth for the information used in the MBSE environment and the PLM system. Ontologies will be at the center of this approach.

Acknowledgments. This research and development project was funded by the German Federal Ministry of Education and Research (BMBF) (funding number 03INT710BF) and managed by the Project Management Agency Jülich (PTJ). The authors are responsible for the content of this publication.

References

Adam, A., et al.: 10 theses about MBSE and PLM - Challenges and Benefits of Model Based Engineering (MBE): PLM4MBSE Working Group Position Paper (2015). https://www.gfse.org/Dokumente_Mitglieder/ag_ergebnisse/PLM4MBSE/PLM4MBSE_Position_paper_V_1_1.pdf

Alt O (2022) SpecIF – Was ist das? https://github.com/GfSE/SpecIF-Presentations/raw/main/SpecIF_Tutorial_TdSE_2022/TdSE_2022_SpecIF_Tutorial.pdf

BMBF - Bundesministerium für Bildung und Forschung (2021) Zukunft der Wertschöpfung: Forschung zu Produktion, Dienstleistung und Arbeit. https://www.zukunft-der-wertschoepfung.de/img/ZdWs.pdf. Accessed 12 May 2022

Capella (2023) Eclipse Capella™: Open Source Solution for Model-Based Systems Engineering. https://www.eclipse.org/capella/

Constantinescu, C., Giosan, S., Matei, R., Wohlfeld, D.: A holistic methodology for development of Real-Time Digital Twins. Procedia CIRP **88**, 163–166 (2020). https://doi.org/10.1016/j.procir.2020.05.029

Dumitrescu, R., Albers, A., Riedel, O., Stark, R., Gausemeier, J.: Engineering in Deutschland - Status quo in Wirtschaft und Wissenschaft: Ein Beitrag zum Advanced Systems Engineering, Paderborn (2021)

Eigner, M.: System Lifecycle Management: Engineering Digitalization (Engineering 4.0), 1st edn. Springer eBook Collection. Springer Fachmedien Wiesbaden; Imprint Springer Vieweg, Wiesbaden (2021). https://doi.org/10.1007/978-3-658-33874-9

Friedenthal, S., Moore, A., Steiner, R.: A Practical Guide to SysML: The Systems Modeling Language. Morgan Kaufmann Publishers Inc, San Francisco (2014)

GfSE (2021) Specification Integration Facility (SpecIF): Public Released Specification. Version 1.1. https://github.com/GfSE/SpecIF/releases/download/release%2F1.1/SpecIF_Specification_1.1.pdf

Haße, A., Walecki, R., Fischer, J., Sinnwell, C.: Demonstrator 2 – Kollaborative Entwicklung cybertronischer Produkte und Produktionssysteme. In: Eigner M, Koch W, Muggeo C (eds) Modellbasierter Entwicklungsprozess cybertronischer Systeme. Springer Berlin Heidelberg, Berlin, Heidelberg, pp 209–226 (2017). https://doi.org/10.1007/978-3-662-55124-0_24

INCOSE (2004) Systems Engineering Vision 2020: INCOSE-TP-2004-004-02

INCOSE (2014) SE Vision (2025). https://www.incose.org/about-systems-engineering/se-vision-2025

Junghanns, A., et al.: The functional mock-up Interface 3.0 - new features enabling new applications. In: Proceedings of 14th Modelica Conference 2021, Linköping, Sweden, September 20–24, 2021, pp 17–26. Linköping University Electronic Press (2021). https://doi.org/10.3384/ecp2118117

OASIS Open Projects (2021) OSLC Core Version 3.0. Part 1: Overview. https://docs.oasis-open-projects.org/oslc-op/core/v3.0/os/oslc-core.html

Obeo (2023) System Modeling Workbench: An integrated MBSE approach to improving multi-domain product development. https://www.obeosoft.com/en/products/smw

OMG ReqIF - formal/2016-07-01 (2016) Requirements Interchange Format (ReqIF). Object Management Group

Riedel, O., Kürümlioglu, M., Schneider, B.: Advanced systems engineering. In: Riedel O, Hölzle K, Schlund S, Spath D (eds) Handbuch Unternehmensorganisation, pp 1–24. Springer, Heidelberg (2024). https://doi.org/10.1007/978-3-642-45370-0_132-1

Schneider, B., Riedel, O., Bauer, W.: Review: model-based systems engineering and artificial intelligence for engineering of sustainable systems – what contribution can systems engineering and artificial intelligence provide for the engineering of sustainable systems as of today? In: Plapper P (ed) Digitization of the Work Environment for Sustainable Production. GITO Verlag, pp 37–59 (2022). https://doi.org/10.30844/WGAB_2022_3

Application of Alternate Routes to Reduce Risk of Job Tardiness in Manufacturing Schedules

Saruda Seeharit, Dušan Šormaz(✉), Mohammad Milad Omar, and Mandvi Malik Fuloria

Department of Industrial and Systems Engineering, Ohio University, Athens, OH 45701, USA
sormaz@ohio.edu

Abstract. This paper reports on an application of alternate routings on manufacturing system's performance, defined as risk of completing production plans. In previous work an integrative methodology was developed in which process planning system IMPlanner was integrated with an FMS simulation module. IMPlanner's rule-based process selection system performed knowledge intensive task of generating alternative processing options for each feature for parts in production plan. In this paper an updated simulation model using Simio was developed and Simio's scheduling capability was utilized in order to determine the reduction of risk of job tardiness in the production plans when alternate routings are considered. Methodology is explained and new simulation model described. The experiments were performed on several case studies from machining domain, with a number of alternate routings as input parameters, and different due dates as performance criteria. Results from the updated simulation model were compared with the previous study in which only resource utilization were analyzed. The level of risk reduction with alternate routings is quantified.

Keywords: Simulation · Flexible manufacturing · Alternate routing · Routing policy · Job tardiness · Processing time

1 Introduction

Over the past decades, manufacturing industries have been pushed to modify their production structures as a result of the rise of mass customization and global competition. The growing competition in the manufacturing market has pushed industries from mass manufacturing through flexible manufacturing to lean manufacturing [1]. The optimum strategy to keep clients satisfied and stay one step ahead of competing companies is to satisfy each customer's demand on time and according to their unique requirements. Flexible manufacturing systems (FMS) have grown to be a well-liked remedy for these difficulties. Manufacturers can more effectively manage resources by using FMS to determine whether equipment or procedures are being over or underused. This contributes to waste reduction and production optimization, both of which serve to increase profitability.

M. Dragomir et al. (Eds.): ICPR 2023, LNPE, pp. 415–424, 2025.
https://doi.org/10.1007/978-3-031-92082-0_46

As production processes become more complex and customers' demand continues to change nowadays, quicker and informed decision making in the manufacturing cells plays a vital role in efficiency and profitability. Dynamic decision making is the ability to make real-time decisions in response to changes in the manufacturing processes. It involves detecting problems or opportunities via real-time data analysis and changing operations as necessary. As an example, if a machine in the manufacturing cell breaks down or stops functioning, dynamic decision making will allocate other available resources to ensure that production continues without any delay, also if demand for a certain product rise, the dynamic decision-making process will alter the production schedule to give priority to the product with the highest demand. Dynamic decision making in general enhances system resilience by allowing entities to dynamically navigate around obstacles or disruptions. When faced with unexpected events or changes in the environment, dynamic routing policies can quickly reroute parts or products to avoid delays or bottlenecks, ensuring smooth operation of the system.

This work integrates IMPlanner's process planning system which generates alternate routes for different products with unique specifications through different machines available in the system [2] with an FMS simulation model using the Simio software. The results are then analyzed to calculate the risk of not completing orders on time and minimize the risk by using alternative dynamic routings. The organization of this paper is outlined as follows. In Sect. 2, we present a brief summary of the prior literature and research in the field. In Sect. 3, we explained the methodology employed in this study. Section 4 provides a detailed account of the experiments conducted. Section 5 provides the analysis of the obtained results. Finally, in Sect. 6, we present the conclusions derived from this research endeavor and the following section.

2 Previous Work

The growing competition to produce a product on time has led manufacturing companies to use computer aided process planning (CAPP). The CAPP is a bridge between computer aided design (CAD) and computer aided manufacturing (CAM) where a design of an order is fed as input into the system and output is the production process plan to manufacture the order from the system. The CAPP system used in this research is IMPlanner and various alternate routes and plans are created [2]. The advantage of dynamic decision is being used for manufacturing the parts with various machines available in the system and the most important aim is to find the optimum routing or plan to create the part.

Flexible manufacturing is one of the new advances in manufacturing [1], a system is flexible when it can easily adapt to changes, the change can be in terms of production of parts or change in product itself. Flexibility is described in eight situations as: machine flexibility, process flexibility, routing flexibility, product flexibility, production, volume, expansion and operation flexibility [3], authors in paper [4] discuss the complexity of manufacturing process planning and scheduling and how they are related with each other. In paper [5] an experiment conducted on simulated FMS model with 14 dispatching rules and three performance measures showed that dynamic scheduling decisions improved system performance overall when compared to the performance produced by running the system with just one dispatching rule.

A simulation model that uses feature-based routing strategy to determine the optimum routing for manufacturing components is presented in [6] and [10]. Later, a different author developed an agent-based simulation model that reflects dynamic routing in manufacturing parts [7]. However, the literature has not examined the methods of employing flexible routing policies to reduce job tardiness in manufacturing cell which this study attempts to explore and quantify the risks associated.

3 Methodology

The entire methodology employed in this research is visually depicted in Fig. 1, providing a clear overview of the approach undertaken in this study. This study utilizes IMPlanner and the FMS model to generate a comprehensive process plan. The generated routes are then stored in an Excel file for further analysis and reference. The FMS model is built using simulation in Simio and then the data table from Excel is imported to Simio model to create the sequence table for analysis.

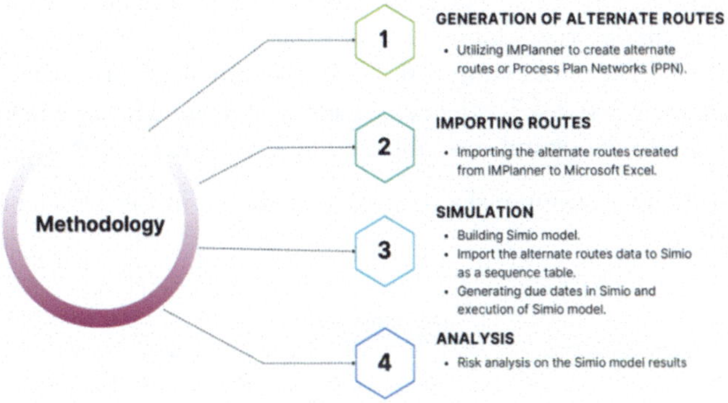

Fig. 1. Methodology

Each feature of a product has its own specific process plan, and they are created using IMPlanner. Feature is a geometric shape that is made of a single tool with the machining operation performed. The products are designed in CAD software and then they are imported to IMPlanner in order to generate the process plans using the rule-based system and several types of process plans are generated to manufacture a product, this ability to manufacture a product using various process design is what gives flexible manufacturing systems their name.

A process plan typically lists all the equipment (machines and tools) that is used to create a component, the amount of time it takes to make the product, and the various cutting tools needed to conduct machining operations on the parts. The FMS model is built in the Simio software and the objects used within the software are explained as follows [8]:

i. Entities: entities are the parts a manufacturing plant makes, they come as order in the simulation model with an arrival date and have a due date that the orders need to be shipped. Each individual part needs specific tools and machines to manufacture.

ii. Servers: They are the machines in the manufacturing plant and every machine has its own processing time to manufacture each feature of a part.
iii. Resources: The resources are tools that are used by machines as a secondary resource to manufacture features of the part.

The alternate routes that are previously generated by IMPlanner are imported into the Simio platform. Four different types of data tables are used in the Simio model, and they are described below:

i. Job types: This table stores the information regarding the different part types (primary keys) that come in to the system for manufacturing and the product mix which is the probability of receiving order of that specific part.
ii. Operations: This table consists of a single column that corresponds to the process name and serves as the table's primary key, storing data on the type of operation that must be performed on each feature of the part.
iii. Job Routing: The job routing contains details of the procedures that must be followed in order to complete the production of the part. It has a foreign key for the process name that is connected to the operations table and a foreign key for the part type that is connected to the *job types* table.
iv. Processing Task: The processing task table contains data on processing time, tool names, task names, and task sequences. In addition, it has a foreign key of the process name connected to the operations table and task sequence number.

A portion of the processing tasks data table is shown in Fig. 2. The details will be explained in Sect. 4.

Part Table | Part Routings | Processes | Processing Tasks | Part Arrival

	Task Name	Process	T...	Tool	RandomPro (Minutes)
1	AES94_1_HOLE_11	AES94_1_ssM2DrillFast	10	T507	Random.Triangular(2.295,2.55,2.805)
2	AES94_1_HOLE_10	AES94_1_ssM2DrillFast	20	T507	Random.Triangular(2.295,2.55,2.805)
3	AES94_2_SLOT_9	AES94_2_ssM8HMillFast	10	T606	Random.Triangular(1.089,1.21,1.331)
4	AES94_3_HOLE_8	AES94_3_ssM2DrillFast	10	T508	Random.Triangular(1.314,1.46,1.606)
5	AES94_4_SLOT_7	AES94_4_ssM8HMillFast	10	T609	Random.Triangular(1.539,1.71,1.881)
6	AES94_5_HOLE_6	AES94_5_ssM2DrillFast	10	T507	Random.Triangular(0.729,0.81,0.891)
7	AES94_5_HOLE_4	AES94_5_ssM2DrillFast	20	T507	Random.Triangular(0.729,0.81,0.891)
8	AES94_6_POCKET_3	AES94_6_ssM8HMillFast	10	T704	Random.Triangular(1.296,1.44,1.584)
9	AES94_6_SLOT_2	AES94_6_ssM8HMillFast	20	T608	Random.Triangular(1.962,2.18,2.398)
10	AES94_6_SLOT_1	AES94_6_ssM8HMillFast	30	T608	Random.Triangular(1.962,2.18,2.398)
11	AES94_Finish	AES94_Complete	10	None	Random.Triangular(0,0,0)

Fig. 2. The Portion of a Processing Tasks Data Table

Simio allows users to model and analyze risk by integrating variability into Simio's risk based planning and scheduling (RPS) this accomplished by running several replications and identify tardy orders [9]. The initial risk of completing each product stands at 50% when a simulation run begins. However, this risk is subject to change after each replication, due to random variations in processing times and other factors. A concise overview of this concept is illustrated in Fig. 3. At the start of the simulation, the probability of successfully fulfilling an order is set at 50%, After each replication, if order is complete on time risk is decreased, or, if order is late, the risk increases. The amount of increase or decrease in smaller after each new replication. Depending on processing time variability, due dates, and other factors for the same order this risk may also oscillate about some value, which after sufficient number of replications will lead to an overall risk measure.

4 Experiment

The manufacturing cell of this experiment is built of eight different machines and they are characterized based on their spindle power as Fast, Medium and Slow machines. The eight types are the following:

i. Drilling machines: Fast and Slow.
ii. Horizontal milling machines: Fast, Medium and Slow.
iii. Vertical milling machines: Fast, Medium and Slow.

Each machine has its own processing time and is connected with other machines in the manufacturing cell of Simio model. The machining tools as mentioned before are the secondary resources available for machines in order to perform the machining operations on different features. The simulation model layout is shown in Fig. 4.

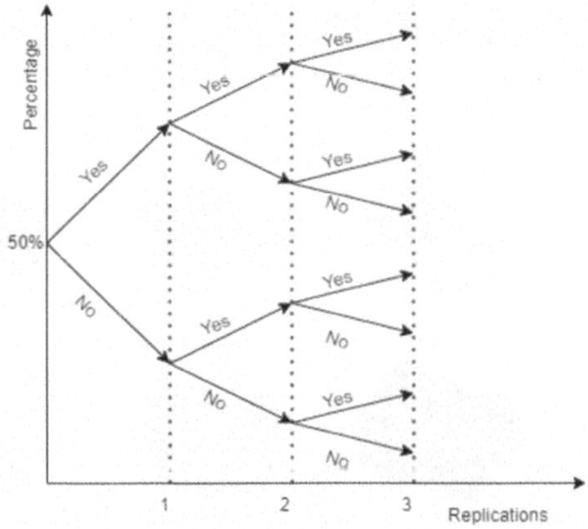

Fig. 3. Risk calculation explanation

There are six different types of parts that are used in this model, and they have their unique machining processes with different processing times. Figure 5 shows the part types and the number of features they have. In this model parts come into the system as entities, for example Slider (Fig. 5) is an entity, and they follow the sequence to go to the machines necessary to manufacture the part. Each part is given to a specific machine based on the machine capability to create each feature of the part for example one of the job types is AES94 and it has a feature called *SimpleHole* and this operation can be done by twist drilling which can be performed on two different machines *CNCDrillSlow* and *CNCDrillFast*, the tool they use is *T507* and processing time for each machine takes to do operation is 3.06 and 2.55 respectively, as shown in Fig. 2.

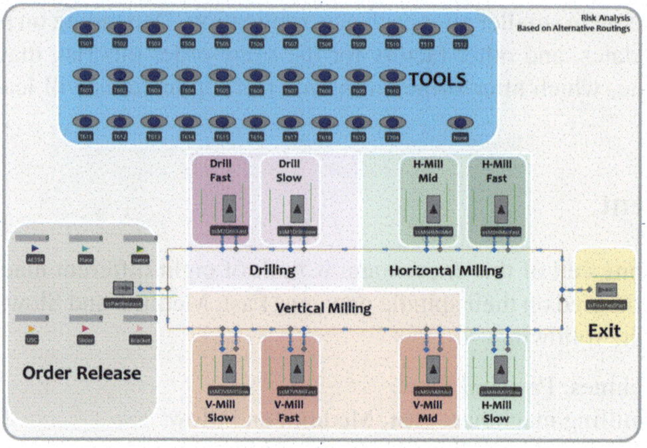

Fig. 4. Manufacturing cell layout

The purpose of the simulation experiment is to perform risk analysis and mitigate the risk of job tardiness within a manufacturing cell.

Each order entering the simulation model was characterized by its arrival time, due date, and order name. The objective of the experiment was to assess and evaluate risks using the planning feature provided by Simio software.

Name	AES94	Plate	Netex
Model			
Number of features	10	11	7
Name	**USC**	**Slider**	**Bracket**
Model			
Number of features	6	18	11

Fig. 5. Manufacturing parts

5 Results

The results obtained from the simulation model are exported to perform further analysis and evaluate the risks associated with each order. After the completion of simulation model, the system outputs from each simulation model and their respective replications are recorded and consolidated into a single file for further analysis. The model is run for fifteen replications to get reasonable data for statistical analysis and the results are visualized graphically.

The visualization of the order completion percentage can be effectively presented using a Gantt chart, as demonstrated in Fig. 6.

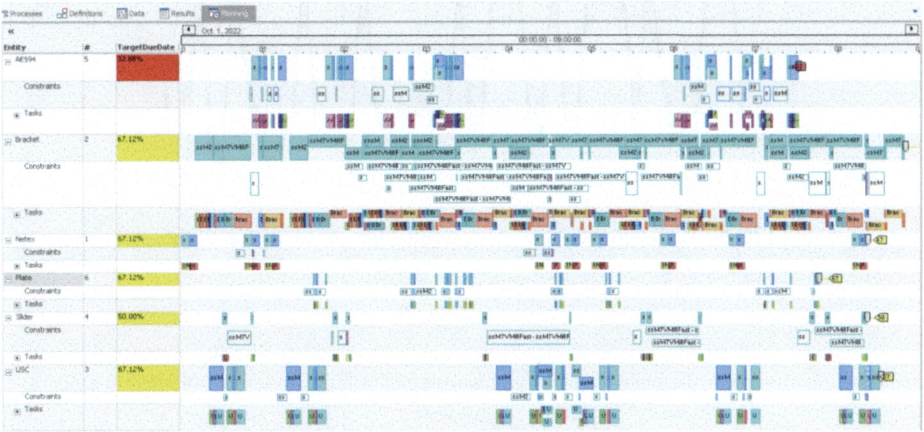

Fig. 6. Gantt chart for one replication

This chart provides a comprehensive overview of the progress made in fulfilling customer orders. In Fig. 6, the first column of the Gantt chart corresponds to the various part types involved in the manufacturing process, as illustrated in Fig. 5. The second column displays the number of orders received for each specific product, allowing for a clear understanding of the demand and popularity of different items. The third column in the Gantt chart provides essential information about the completion status of each part, presented in terms of a percentage.

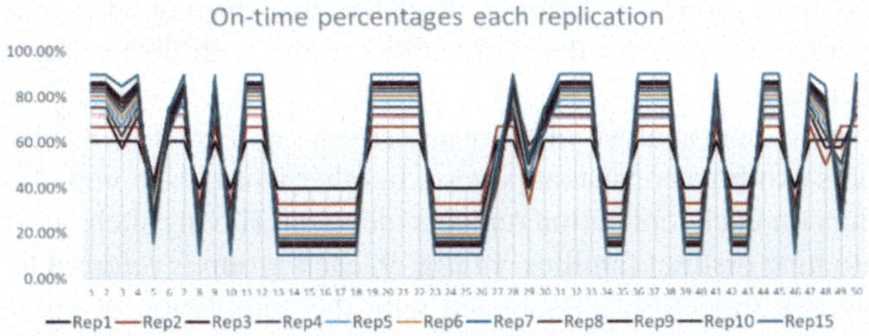

Fig. 7. On-time percentage of each replication for best routing

Figure 7 provides a visual representation of the on-time performance of each order for one set of part routes for each replication. Based on the analysis of these results, it can be concluded that the orders were successfully completed ~54% of the time.

The data presented in Fig. 8 illustrates the outcomes obtained from multiple replications conducted on a different routing configuration. The results indicate that, on average, only 40% of orders are delivered on time. This approach involves the random selection

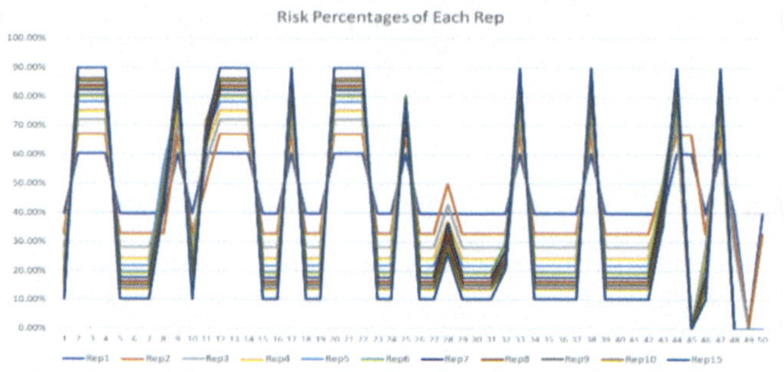

Fig. 8. Risk percentage of each replication for median routing

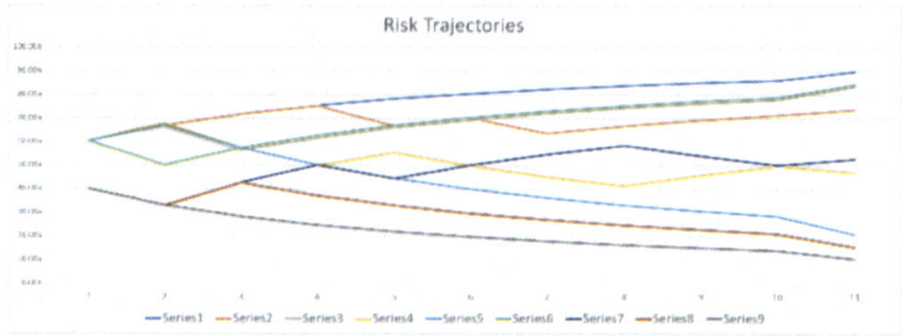

Fig. 9. Risk trajectories

of routes for product manufacturing, followed by simulation to gather relevant data. By employing this methodology, a comprehensive understanding of the performance and efficiency of different routing options can be obtained.

Figure 10 depicts the table that corresponds to the on-time completion percentage of each family in every replication simulated within the model. The data reveals that there are four distinct groups or families of orders that display an increasing trend in their on-time percentages across the replications, three with decreasing and two with alternating in between.

During the subsequent analysis of the results, from Fig. 7 and Fig. 8, it became evident that certain orders exhibit a consistent pattern. In other word, the completion percentage of different orders for each replication remains identical. Having such, we divided a total of 50 orders into nine distinct families. In Fig. 9, each group is referred to as a "series" representing the risk trajectories associated with the completion of orders within that particular group. Interestingly, we observed that the majority of orders (84%) exhibited a consistent trend in terms of their on-time percentage, either consistently increasing or decreasing. This finding highlights the prevalence of unidirectional patterns in the timely completion of orders within our dataset.

Rep1	Rep2	Rep3	Rep4	Rep5	Rep6	Rep7	Rep8	Rep9	Rep10	Rep15	Trend	Behavior	Intensity
60.33%	67.12%	71.92%	75.50%	78.28%	80.48%	82.28%	83.78%	85.04%	86.12%	89.81%	1	AlwaysU	46%
60.33%	67.12%	71.92%	75.50%	66.97%	70.32%	63.84%	66.89%	69.47%	71.67%	73.88%	2	UDUDU	4%
60.33%	67.12%	57.31%	62.75%	66.97%	70.32%	73.06%	75.33%	77.26%	78.90%	84.50%	3	UDU	2%
60.33%	67.12%	57.31%	50.00%	55.66%	50.00%	45.39%	41.56%	46.11%	50.00%	47.35%	4	UDUDUD	2%
60.33%	67.12%	57.31%	50.00%	44.34%	39.84%	36.16%	33.11%	30.53%	28.33%	20.81%	5	UD	2%
60.33%	50.00%	57.31%	62.75%	66.97%	70.32%	73.06%	75.33%	77.26%	78.90%	84.50%	6	DU	2%
39.67%	32.88%	42.69%	50.00%	44.34%	50.00%	54.61%	58.44%	53.89%	50.00%	52.65%	7	DUDUDU	2%
39.67%	32.88%	42.69%	37.25%	33.03%	29.68%	26.94%	24.67%	22.74%	21.10%	15.50%	8	DUD	2%
39.67%	32.88%	28.08%	24.50%	21.72%	19.52%	17.72%	16.22%	14.96%	13.88%	10.19%	9	AlwaysD	38%

Fig. 10. Grouping of orders

6 Conclusion

This research showcases the successful development of a dynamic routing policy and its integration with Simio, along with the incorporation of IMPlanner. The routes selected via IMPlanner are integrated with Simulation model. This integration enabled us to leverage the simulation model and perform risk analysis, allowing us to quantify the percentage of job tardiness in a manufacturing cell. By employing this comprehensive approach, we were able to gain valuable insights into the manufacturing processes and identify areas where improvements can be made to minimize tardiness and enhance overall efficiency.

While results are promising, further work may employ a procedure for optimization of dynamic decisions in order to maximize the likelihood of completing orders on time, or, which is equivalent, to minimize overall risk of completing tardy orders.

References

1. Chituc, C.M., Restivo, F.J.: Challenges and trends in distributed manufacturing systems: are wise engineering systems the ultimate answer. In: Second International Symposium Engineering System. MIT (2009)
2. Šormaz, D.N., Arumugam, J., Harihara, R.S., Patel, C., Neerukonda, N.: Integration of product design, process planning, scheduling, and FMS control using XML data representation. Robot. Comput.-Integr. Manuf. **26**(6), 583–595 (2010). https://doi.org/10.1016/j.rcim.2010.07.014
3. Browne, J.J., Dubois, D., Rathmill, K., Sethi, S., Stecke, K.: Classification of flexible manufacturing systems. FMS Mag. **2**(2), 114–117 (1984)
4. Shen, W., Norrie, D.H.: Agent-based systems for intelligent manufacturing: a state-of-the-art survey. Knowl. Inf. Syst. **1**(2), 129–156 (1999). https://doi.org/10.1007/BF03325096
5. Chan, F.T.S., Chan, H.K., Lau, H.C.W., Ip, R.W.L.: Analysis of dynamic dispatching rules for a flexible manufacturing system. J. Mater. Process. Technol. **138**(1), 325–331 (2003). https://doi.org/10.1016/S0924-0136(03)00093-1
6. Patel, C.R.: Development of the simulation based integrative decision support framework for flexible manufacturing system with real time process plan selection. Ohio University, Ohio (2010). http://rave.ohiolink.edu/etdc/view?accnum=ohiou1282413649. Accessed 16 Oct 2016
7. Ghosal, S.: Object Oriented Simulation of Agent Based Flexible Manufacturing System with Dynamic Routing and Scheduling. Ohio University (2015). https://etd.ohiolink.edu/pg_10?0::NO:10:P10_ACCESSION_NUM:ohiou1438812428. Accessed 31 Aug 2017
8. Kelton, W.D., Smith, J.S., Sturrock, D.T.: Simio & Simulation - Modeling, Analysis, Applications, 3rd edn. McGraw Hill (2014)

9. Simulation Model Risk-Based Scheduling Tool for High-Mix and Lowvolume Manufacturing Facility with Sequence-Dependent Setup Times | Simio. https://www.simio.com/resources/papers/WinterSim2019/Simulation-Model-Risk-Based-Scheduling-Tool-for-High-Mix-and-Low-Volume-Manufacturing-Facility-with-Sequence-Dependant-Setup.php. Accessed 29 May 2023
10. Sormaz, D., Patel, C.: Development and evaluation of feature-focused dynamic routing policy. Int. J. Adv. Manuf. Technol. **99**, 15–28 (2018)

Integration of Industry 4.0 Concepts into Lean Six Sigma Projects: A Case Study in the Ready-Mixed Concrete Industry

J. M. Sepúlveda(✉) and T. A. Droguett

Industrial Engineering Department, University of Santiago of Chile, 3769 Ecuador Avenue, Santiago, RM, Chile
juan.sepulveda@usach.cl

Abstract. This paper presents a case study for the achievement of operational excellence in a company branch dedicated to the production and distribution of ready-mix concrete. Lean Six Sigma (LSS) methodology is selected to carry out the objective by following the DMAIC stages. The main quality attribute for customer satisfaction is product docility upon delivery at customer's site, which is related to the water/cement ratio of the mixture. For this, the focus is placed on quality deviations within the production and transportation cycle. The deviations in the tolerance of the unloaded product and the performance indicators were measured to construct a regression model to find the main process variables affecting variations. In the Improvement stage, it is implemented a new process control system with GPS, water flow sensors installed at the mixer trucks and an APP for the driver to help control the process by regulating the mixer pressure and the amount of water to keep the docility of the product while in movement. Over a period of six months, an improvement of the main performance indicators was achieved.

Keywords: Lean Six Sigma · Industry 4.0 · Ready-mix concrete

1 Introduction

Concrete is a material with characteristics similar to a rock, which is obtained by means of a carefully proportioned mixture of cement, water, stone aggregates, additives and/or additions, which after a setting process, the mixture hardens, adapting to the dimensions and shape of its formwork [1]. The concrete plants are responsible for the production and distribution of the concrete to the construction site in compliance with the technical specifications of the calculus engineer. Given the perishable characteristics of the product, it is vital to control the quality of the process in the concrete mixing plant, during the travel and before unloading. This is particularly important in concrete batching plants where the material is mixed in transit in mixer trucks. The most common properties of concrete are: a) docility, b) resistance, and c) durability and permeability [1], and are governed by international standards (ASTM C94; BS EN 12350-2 2019) and local adoptions in each country. Workability is the main quality metric addressed in this article and is the

M. Dragomir et al. (Eds.): ICPR 2023, LNPE, pp. 425–432, 2025.
https://doi.org/10.1007/978-3-031-92082-0_47

ease of fresh concrete to be transported, placed and compacted without segregation; it is measured through the Abrams cone settlement method, which consists of placing and compacting a sample of fresh concrete in a slump cone and emptying it to measure the height it achieves [1]. If this variable is not under control the product is rejected by the customer.

1.1 Type of Processes

The production of ready-mix concrete is carried out according to two main types of plants: a) Mixing Plants - a facility in which the concrete components are completely measured, weighed and kneaded and loaded into the mixer truck transport equipment; b) Batching Plants -the facility in which the concrete components are measured, weighed and loaded into a rolling mixer. In this loading process, the mixer operator is responsible for homogenizing the mix and adjusting the docility of the concrete according to job requirements by adding water. Mixing plants are simpler and cheaper and production volume is only limited by the number and capacity of mixer trucks, as well as the weight per axle. This case study combining elements of information technology (ICT and Industry 4.0) with Lean Six Sigma (LSS) deals with this type of production plant.

1.2 Lean Six Sigma (LSS)

LSS is the fusion of Lean manufacturing and Six Sigma methodologies, which together allow waste reduction and statistical process control in production. This merger is necessary because: a) Lean by itself cannot bring the process under statistical control, and b) Six Sigma by itself cannot dramatically improve process speed or reduce capital investment [2]. LSS is based on process principles and methodologies; of these, DMAIC is the best known [3, 4]. DMAIC corresponds to the acronym for its 5 stages: Define, Measure, Analyze, Improve, and Control. In the study by [5], in which 2215 companies that combined the implementation of lean tools with Six Sigma projects, were surveyed, it was identified that certain lean practices had a greater influence on Six Sigma projects than others depending on how they are implemented. Thus, Lean and Six Sigma should be seen as complementary strategies.

1.3 Industry 4.0 (I4.0)

I4.0 relies on the synergy of mechatronic components linking the physical world (through sensors, actuators, robotics, and embedded systems) with the virtual world of data and information processing [6]. It integrates technologies, such as internet of things (IoT), industrial IoT (IIot), cloud computing, machine intelligence, big data, AI, and efficient and clean energy sources for stand-alone operation [7]. According to databases Scopus and Web of Science, the relationship between I4.0 and specific supply chain processes [8, 9] in several industries has been extensively studied during the last five years [10, 11].However, in the concrete industry, this topic shows few or no results. On the other hand, very few works relating LSS and the concrete industry have been found [12–14].From these findings, the present work comes to narrow the existing gap on the number of articles joining the concepts of LSS, I4.0 and the concrete industry, as compared to other industries.

2 Case Study

The company has 10 plants, of which, eight are batching plants and two are mixing plants, so the concrete production process in the batching plant depends to a large extent on the mixer operator in compliance with the operational procedures to adjust the workability of the concrete delivered to the client. The performance of the concrete plants is measured with respect to two main indicators: a) Strength Compliance Index (SCI), and b) Ratio Technical Dose/Average Cement Dose (CTD). The SCI corresponds to the resistance obtained at 28 days with respect to the required average or design resistance (f'cr). SCI $= f$ 28/f'cr. The optimal SCI of a branch must be between the values 1.00 and 1.03; under this range, the most recurring decision within the industry is to increase the dose of cement in the concrete design so that the resistance obtained is as close to the average resistance required; if the SCI is above 1.03, it is possible to consider a decrease in the cement dose in order to optimize the required design. This indicator is essential to optimize the dosage of cement within the design. In the CTD ratio (Kg Cement/m^3 Concrete), the technical dose is the average cement dose of certain products selected for analysis; while the average dose is the amount of cement consumed per branch with respect to the cubic meters dispatched in a month. Therefore, the optimization of the concrete design is strongly linked to the average technical dose in compliance with the strength of the concrete. Next is presented a brief description of the stages according to the DMAIC method of the LSS.

2.1 Application of LSS

In the Define stage (D), analysis of internal and external stakeholders (sales and quality managers, post sales service, branch managers, customers of different account sizes) and SIPOC mapping (Suppliers, Inputs, Processes, Outputs, Customers) were performed. The main topics were: a) to increase customer satisfaction: improve the shopping experience (ease of programming and modifications, products or services related to constructive solutions, service of the mixer operator at delivery, and service of the call center), punctuality of the dispatch, compliance with resistance and other technical specifications, b) to reduce costs and increase profitability, that is, optimizing mix design by reducing the cement dose, which can be achieved by controlling the variability of the delivered product, the reduction of the water/cement ratio, control of operational processes in the plant and on site, among others.

In the Measure (M) stage, the main focus was on recording the SC and CTD indices; for example 1.072 and 252.9 respectively in February 2021 in a particular branch. The SC over 1.03 indicates the possibility of reducing cement doses and therefore lowering costs.

In the Analyze (A) stage, tools such as: Cause and Effect Diagram (or Ishikawa Diagram), Pareto Diagram, 5 Why were used. For the design of the Cause and Effect Diagram, the work involved 20 key agents of the organization (operations assistant manager, branch administrators, quality managers, designers, relevant clients, mixer operators, among others), who through the brainstorming tool determined the most recurring problems and the causes associated with the Deviation in Mixture Quality as an effect. Main causes found, focused on Materials (low quality, storage conditions, decay, multiple suppliers); Man (training, experience, efficiency, salary and incentive policy); Machine (mixer malfunctioning, low maintenance); Methods (low supervision, mixing, curing, water/cement ratio, variable operating conditions). A Pareto diagram showed that the main problems associated with the quality deviation, were the workability and the quality of the concrete mix. The accumulated percentage line shows that 80% of the causes that affect quality are focused on six main reasons and 20% of them correspond to the workability of the mix, this closely related to its water/cement ratio.

By using 5 Why tool, the analysis of the deviation in product quality led to six actions: include storage areas in branches, increase maintenance standards, increase supervision of procedural compliance (in plant, in travel, at unloading), increase control of water addition by the mixer operator, deliver alternative dispatch routes, and provide recommendations for concrete installation under adverse conditions. The part of the project presented in this paper, focuses on two actions: a) increasing supervision of compliance with the Procedure, and b) increasing control of water addition by the mixer operator.

In the Improve (I) stage, and based on the previous analysis, it was decided to implement a digital and online measurement system that allows to control the addition of water inside the mixer's bitonera and the knead pressure to reach the docility of the product. In order to study the possibility of implementing the new control system, a pilot plan on six mixers was implemented. A study was carried out on the correlation of the variables: mixer pressure, location of the mixer and time while in travel (located by new GPS), with the docility of the mix upon unloading (the explained variable). The optimal results were coded in a proprietary algorithm interacting with the mixer´s operator with a mobile application.

3 Results

The results of the pilot plan allowed to conclude: a) the mixing pressure of the bitonera and the docility of the mixture are adequately correlated (r^2 ~80%); b) the correlation models adequately estimate the Abrams cone from the mixing pressure (80% of the estimated cones are in the ± 1 cm. deviation range and 98% in the ± 2 cm. range); c) the amounts of water incorporated into the mixtures to adjust the cone are correlated and coincide with the theoretical amounts of water known for this purpose. The new control system makes it possible to verify that: a) the pressure generated by the bitonera during mixing is correlated with the volume of mixing and the docility (cone settling) of the mix; b) the pressure of the concrete can be transmitted and captured remotely via GPS through the pressure sensors; c) the amount of water added to the concrete can be transmitted and captured remotely via GPS through water flow sensors; and d) it is possible to remotely control via GPS the quality of the concrete delivered to customers by measuring the workability and adding water to the concrete (Fig. 1).

After the pilot experience, it was decided to implement the new control system in the branch. With the implementation, the certainty in each of the stages is increased, eliminating the uncertainty conditions that the current model presented. The new control system reports the docility of the product loaded in the mixer truck, reducing the responsibility of the worker and thus avoiding the over-addition of water, diverting the water/cement ratio of the mix and therefore, jeopardizing the resistance of the final product.

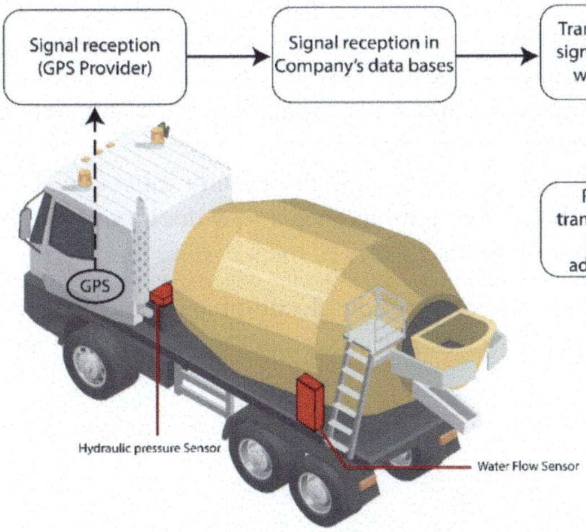

Fig. 1. System diagram

Figure 2 shows the information registered by the system that allows the mixer operator to control the pressure and water addition with the App while in travel and upon arrival to the work site.

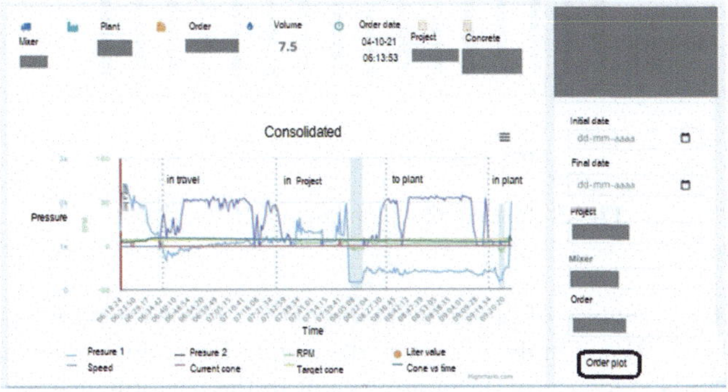

Fig. 2. Information system

The implementation of the new system requires a rigorous procedure by the mixer operator in plant and at delivery:

- Knead in the plant at 14 rpm for 6 min to ensure product homogeneity, this will also provide a signal to the new control system so that it can register the docility of the product in the mobile application of the mixer operator.

- A concrete docility adjustment before leaving the plant according to the route and environmental conditions to arrive at the site with the required cone, following what is indicated in the cone module in the mobile application.
- On site, the mixer operator must knead for at least one minute at 14 rpm or more, to send a signal to the new control system where the docility of the product will be recorded before unloading, ensuring that a product is delivered within the established tolerances without adding water that could jeopardize the water/cement ratio of the mix and with it the resistance of the final product.
- In case of adjusting the cone on site, the mixer operator must knead the product for at least two minutes so that the system updates a new docility record, thus avoiding the over-addition of water to the mix.

Figure 3 shows the evolution of procedural compliance and the tolerance of the cone measured in the actual deliveries per mixer over a period of 10 months between 2021–2022.

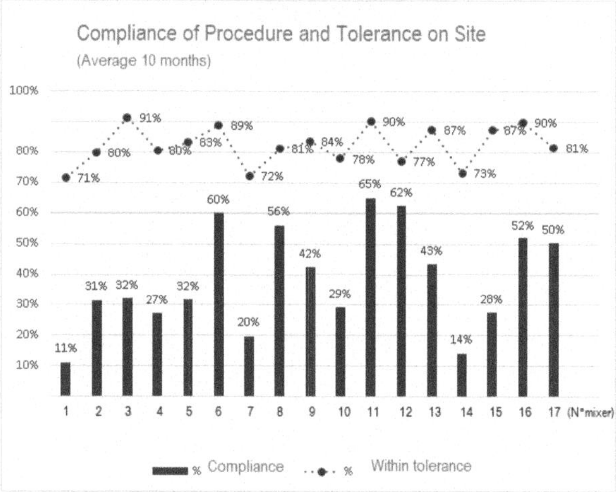

Fig. 3. Procedure Compliance and Tolerance

As shown, the evolution of procedural compliance is not yet stable as it remains in the range of 84% to 91%. On the other yet, the Sigma Level does not exceed the value of 2.87, but there is evidence of an improvement in the performance indicators of the branch. In fact, there is an increase in the SDI and a constant decrease in the average technical dose (CTD) as shown in Fig. 4.

The system is still under development. A weakness detected is that the new operational procedures do not manage to exceed 65% compliance (Fig. 2). Some operators do not follow the new defined procedure, so for some months it was not possible to study all the dispatches and the deviations in the tolerance of the unloaded product. Given the high relationship of compliance with experimental procedures with the rigor and commitment of the mixer operator, an incentive plan (bonus) is considered to increase staff motivation.

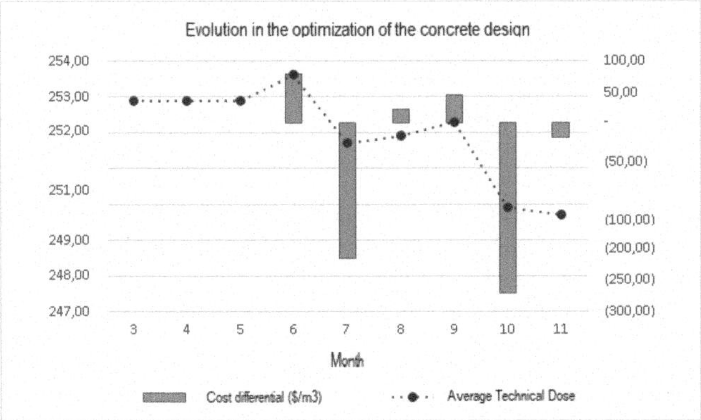

Fig. 4. Evolution of quality performance indicators

4 Conclusions

In this article, a case study of the application of the Lean Six Sigma methodology has been presented, incorporating an Industry 4.0 tool in the Improvement stage that allows reducing the factors of quality deviation in the delivery of concrete with mixer trucks.

It is possible to ensure that the improvement in the performance indicators of the branch, SCI and CTD, is due to the greater control provided by the implementation of the new system by the new operational control at the plant, during travel and at unloading the product. However, the optimization in the dose of cement for each cubic meter of concrete may be related to design variables, that is, to the performance of raw materials of the branch and other operational controls.

Although improvements has been observed in the main indicators, the adoption of the procedures has been gradual and more study is needed to get greater involvement of the mixer operators in order to arrive at a robust solution.

Given the potential optimization in the design of concrete, the reduction in risk due to eventual returns of the product, the reduction of costs and the increase in profitability associated with it, it is essential to work on operational gaps linked to empowering human resources linked to performance of the branch, increasing the commitment to the quality of the dispatch product and the increase in customer satisfaction associated with this.

For this, it is necessary that the operational supervisors carry out a daily follow-up of compliance with the procedures on site, as well as compliance with the tolerance of the unloaded product, through the use and publication of the indicators available for the control of the supervisors; this must be accompanied by incentive plans that seek to motivate the mixer operator in compliance with the operational procedures related to his responsibilities within the process.

Acknowledgements. The authors gratefully acknowledge the support of DICYT University of Santiago - Chile, Grant N° 061119SS, the Industrial Engineering Department and Mr. Gabriel Larraín for the support in the graphics.

References

1. Neville, A.: Properties of Concrete, 5th edn. Essex Pearson Education Limited (2011)
2. Vivekananthamoorthy, N., Shanmuganathan, S.: Lean Six Sigma. In: Coskun, A., Inal, T., Serteser, M. (eds.) Six Sigma: Projects and Personal Experiences, Rijeka, Intechopen, pp. 1–22 (2011)
3. Drohomeretski, E., Gouvea da Costa, S., Pinheiro de Lima, E., Andrea da Rosa, P.: Lean, Six Sigma and Lean Six Sigma: an analysis based on operations strategy. Int. J. Prod. Res. **52**(3), 804–824 (2014)
4. Chakravorty, S.: Six Sigma programs: an implementation model. Int. J. Prod. Econ. **119**, 1–16 (2009)
5. Salah, S., Rahim, A., Carretero, J.A.: The integration of Six Sigma and lean management. Int. J. Lean Six Sigma **1**(3), 249–274 (2010)
6. Sadiku, M.N.O., Wang, Y., Cui, S., Musa, S.M.: Cyber-physical systems: a literature review. Eur. Sci. J. **13**(36) (2017)
7. Hamilton Ortiz, J.: Industry 4.0 - Current Status and Future Trends. IntechOpen (2020)
8. Tortorella, G.L., Fettermann, D.: Implementation of Industry 4.0 and lean production in Brazilian manufacturing companies. Int. J. Prod. Res. **56**(8), 2975–2987 (2018)
9. Núñez-Merino, M., Maqueira-Marín, J.M., Moyano-Fuentes, J., Martínez-Jurado, P.J.: Information and digital technologies of Industry 4.0 and Lean supply chain management: a systematic literature review. Int. J. Prod. Res. **58**(16), 5034–5061 (2020)
10. Ciano, M.P., Dallasega, P., Orzes, G., Rossi, T.: One-to-one relationships between Industry 4.0 technologies and Lean Production techniques: a multiple case study. Int. J. Prod. Res. **59**(5), 1386–1410 (2021)
11. Busto Parra, B., Pando Cerra, P., Álvarez Peñín, P.I.: Combining ERP, lean philosophy and ICT: an Industry 4.0 approach in an SME in the manufacturing sector in Spain. Eng. Manag. J. 3**4**(4), 655–670 (2022)
12. Oguz, C., Kim, Y.-W., Hutchison, J., Han, S.: Implementing lean Six Sigma: a case study in concrete panel production. In: Proceedings of IGLC20: 20th Annual Conference of the International Group on Lean Construction: Are We Near a Tipping Point?, San Diego, CA, USA, 18–20 July (2012)
13. Ray, B., Ripley, P., Neal, D.: Lean manufacturing - a systematic approach for improving productivity in the precast concrete industry. PCI J. **51**(1) (2006)
14. Wang, C., Liu, M., Hsiang, S.M., Leming, M.L.: Causes and penalties of variation: case study of a precast concrete slab production facility. J. Constr. Eng. Manag. **138**(6) (2012)

Mapping Robotic Assembly Planning Documents: Towards Production Planning Automation

Ran Shneor(✉) and Sigal Berman

Department of Industrial Engineering and Management, Ben-Gurion University of the Negev, Beer-Sheva, Israel
shneorr@post.bgu.ac.il

Abstract. Automation of production planning for dexterous robotic processes requires the integration of information from multiple production documents. One of the widespread production documents used in industry is the computer-aided design (CAD) model. CAD models detail geometric topologies relevant to the assembly of rigid objects. However, deformable objects which are common in various industrial assemblies are not represented in CAD and their production definitions are detailed in multiple production documents. A prerequisite to automatic planning of robotic assembly suitable for industrial setups is mapping relevant production documents applied to real-world products in different industry sectors. This study aims at identifying production documents with meaningful information for robotic assembly planning of real-world products, A theoretical analysis was conducted to define robotic assembly planning features, followed by a field study in factories to map relevant production documents. A case study with a real-world electrical product is demonstrated. The datasheet, bill of materials, electric scheme, and job guide were pinpointed as meaningful production documents for robotic assembly planning. Combining production documents can identify production part characteristics and facilitates addressing their assembly requirements for automatic robotic assembly planning procedures suitable for industrial environments.

Keywords: robotic assembly planning · production documents · deformable objects

1 Introduction

Production process planning in the industry is comprised of an array of production documents. Production documents describe the materials, equipment, and activities involved in the production process. Robotic assembly is a core production activity in the industry. Robotic assembly planning, especially for products containing deformable objects requires the integration of information from several production documents for determining the assembly plan and for robotic programming. Automation of the production planning stages for robotic assembly is critical for shortening setup times. Shorter setup times are crucial for small-batch manufacturing facilities. A first step in facilitating automatic planning in industrial facilities is mapping the relevant production documents.

M. Dragomir et al. (Eds.): ICPR 2023, LNPE, pp. 433–439, 2025.
https://doi.org/10.1007/978-3-031-92082-0_48

Among the widespread production documentation used for production planning are computer-aided design (CAD) models [1, 2]. There are several formats applied in manufacturing industries, among which, most common are IGES (Initial Graphics Exchange Specification) and STEP (STandard for the Exchange of Product) [3]. CAD models are used for representing product and sub-assembly geometry and are used for assembly planning [4, 5]. However, CAD models typically represent only rigid objects [6, 7]. While in recent years, several libraries have been developed for representing deformable objects, e.g., Voxelyze [8]. Such a three-dimensional (3D) representation of deformable objects is less common in the industry.

Production documents document relevant information for various organizational purposes. One of the most relevant information is the materials involved in the production process. The materials or objects involved are important for different activities such as procurement, inventory control, and assembly planning. Robotic assembly of many industrial products requires assembly planning of both deformable and rigid objects. For example, robotic assembly with a rigid and deformable linear object requires accuracy for placing an object in a desired position. Objects such as gaskets, tubes, or wires require a higher level of accuracy due to the possible changes in the object's shape, size (i.e., deformation), and position during and even after the robotic manipulation [9, 10].

The deformable objects' distinct characteristics are documented in several separate production documents. The scattered documentation and deformable object characteristics led to robotic assembly with deformable objects to be performed manually or with partial robotic involvement, i.e., human-robot interaction [11]. Few studies performed robotic assembly planning of deformable linear objects based on geometric topologies (e.g., vertex, edge, surface) from CAD together with human demonstrations [12, 13]. Developing automatic process planning methods can assist in reducing current barriers to the widespread industrial implementation of robotic assembly. A prerequisite to automatic planning of robotic assembly suitable for industrial setups is mapping relevant production documents applied to real-world products in different industry sectors for extracting process information from such documents. This study aims at identifying production documents with meaningful information for robotic assembly planning toward production planning automation.

2 Method

The research methodology consisted of three phases: features definition followed by production documents mapping and demonstration on a real-world case study (Fig. 1).

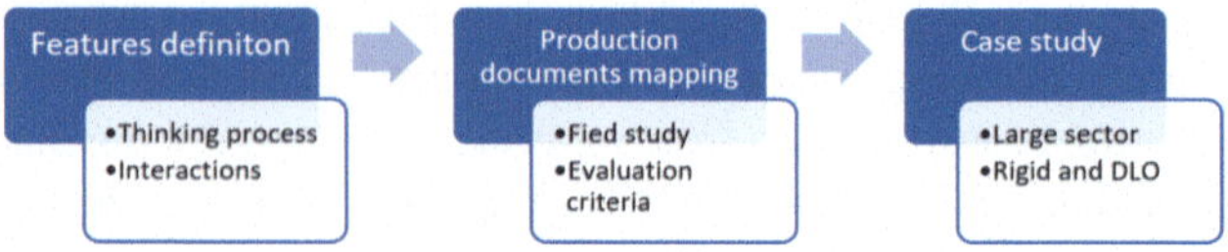

Fig. 1. Overview of the methodology (DLO-Deformable Linear Object).

In the features definition phase, a production management-oriented philosophy was applied to define relevant features for robotic assembly planning for products with deformable objects. The theory of constraints (TOC) thinking process was applied, and a dilemma in a smart factory regarding robotic assembly sequence planning of products containing deformable objects was defined [14]. An interaction entity was introduced as a potential solution to the robotic assembly planning dilemma. Interactions are defined based on physical, geometric, and manufacturing attributes.

In the production documents mapping phase, relevant production documents were mapped by a field study on industries from different sectors where various products containing deformable objects are assembled with rigid objects. As part of the field study, production documents were evaluated according to the following criteria:

- Standardization – whether acceptable widespread formats for the production document or content exist?
- Information level – what is the information level available in the document? single-part level or an assembly level? (i.e., different parts connected).
- Data accessibility – how easily is the document reached? (e.g., manufacturers' websites or commercially proprietary).
- Relevancy for robotic planning – what is the applicability of the document's information to robotic planning? (e.g., geometric topologies, points of interest).
- Representation – how do deformable objects are described in the document?

The combination of theoretical study on interactions together with field study on-site encounters of diverse industrial sectors facilitated a concretization of the features and production documents' definitions, thus addressing real-world scenarios.

The case study approach is widespread in advanced robotic environment studies [15]. It was applied in the current study to validate the theoretical features definition of interactions and to demonstrate the applicability of the identified production documents to extract interaction values for robotic assembly planning of a product with deformable objects in an industrial environment. In the case study phase, a real-world product from the electrical industry was chosen. Such a product represents a large industry sector with rigid and deformable objects assembled.

3 Results

3.1 Features Definition of Interactions

Interactions describe the steady state relationship of the deformable object with other objects in an assembly. Interactions are used for semantic high-level robotic assembly planning. The interactions between objects regarding deformable objects' characteristics (with attention to deformable linear objects) consist of the following categories [14].

- Dimension (objects' geometry) – the values are uni-parametric (1D), e.g., a deformable linear object such as a wire. Bi-parametric (2D), and tri-parametric (3D).
- Deformability (object's behaviour) – the values are rigid (r), elastic (e), plastic (p), and elasto-plastic (ep).

- Direction of interaction – the values are male (m), female (f), and other (o), where no object is superior to the other object.
- Connection type – how objects are in touch after assembly. The main values are peg-in-hole (PIH), self (SLF), thread-in-channel (TIC), peg-in-wrap (PIW).
- Affinity act – the way of establishing the connection type. The main values are adhesive (adhs), placing (plac), shrink (shrnk), solder (sldr), screw (scrw), tensile inward (tnsi), tensile outward (tnso).

The interaction categories are applicable in industrial scenarios, and some are not claimed to be exclusive to deformable objects.

3.2 Production Documents Mapping

The mapping of production documents is based on a field study conducted in 22 different industries (50% electrical, 41% medical, and 9% plastic). The production documents found, and their evaluation criteria values are presented in Table 1. The following documents were identified: CAD, bill of materials (BOM) contained in a product, electric scheme (ES) of wiring which describes the connections between ports (relevant for the electrical industry sector), material safety data sheet (MSDS) which lists information relating to occupational safety and health of humans having contact with the assembled objects. Additional documents include a patent description, a job guide (JG) describing the assembly procedure, and a datasheet (DS) describing the functional information of the object. According to the abovementioned criteria, the relevant documents for robotic planning are the CAD, BOM, electric scheme, job guide, and datasheet. These documents include relevant information for robotic assembly planning, such as coordinates, dimensions, and points of interest (e.g., holes for deformable linear object insertion). Electric schemes and job guides are less accessible online as they contain commercial proprietary information. Contrarily, CAD, and datasheets are available online on many original equipment manufacturers' (OEM) websites.

Table 1. Production document mapping (highlighted documents relevant to robotic planning).

Criterion\Document	CAD	BOM	Electric scheme (ES)	MSDS	Patent	Job guide (JG)	Datasheet (DS)
Standardization	Content, Format	Content	Content	Content	Content, Format	Semi-content	Semi-content
Information level (Single part/ Assembly)	Single, Assembly	Assembly	Assembly	Single	Single, Assembly	Assembly	Single
Data accessibility	High (OEM websites)	Medium	Low	High (OEM websites)	Medium	Low	High (OEM websites)
Relevancy for robotic planning	**High**	**High**	**High**	Low	Medium	**High**	**High**
Representation	Low	High	High	Medium-high	Medium-high	High	High

3.3 Case Study

A case study approach was selected to demonstrate the applicability and validity of production document mapping. The product chosen is an electrical panel designed by industry experts and applied in the lighting industry. It contains a wire with plastic deformation and a terminal and a controller on a metal board (Fig. 2a). The wire is inserted into the terminal and controller ports (Fig. 2b). The interaction dimensions are 1D and 3D based on datasheets and CAD, respectively (Table 2). The deformability is plastic and rigid derived from BOM and datasheets. The direction of assembly is male, meaning the peg (i.e., wire) is inserted into the hole (i.e., port) and it is based on the electrical scheme. The connection type is peg-in-hole, and the affinity act is snap, derived from the datasheet and job guide, respectively. The datasheet is a production document with valuable information regarding object characteristics. The datasheet aids in identifying the deformable object's dimensions, its level of deformation (together with BOM), and its connection type of it with other objects. CAD is a basic production document in industrial applications, and it aids in identifying rigid objects' dimensions and locations of points of interest, such as insertion points. The electrical scheme aids in identifying the connection direction between objects, and the job guide contributes to affinity act information.

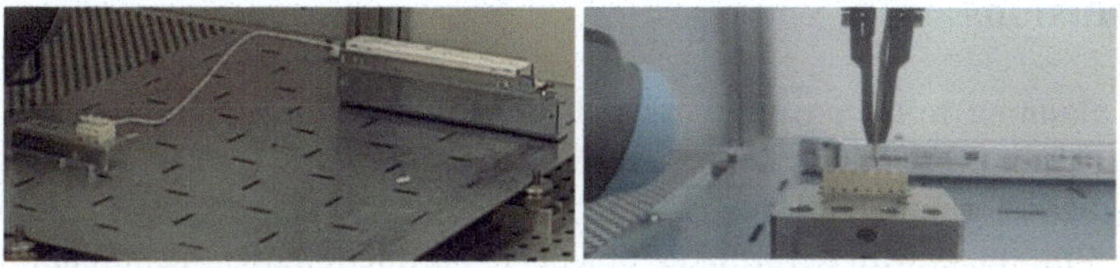

(a) (b)

Fig. 2. A case study of an electric panel (a) overall product, (b) interaction close-up.

Table 2. Case study interaction categories and production documents.

Interaction category	Dimensions	Deformability	Direction	Connection type	Affinity act
Interaction value	1D-3D	Plastic (p) – Rigid (r)	Male (m)	Peg-in-hole (PIH)	Snap
Production document	DS, CAD	BOM, DS	ES	DS	JG

4 Discussion

Applying a production-oriented management philosophy aided in describing high-level aspects of robotic assembly planning of products with deformable objects and led to the definition of interaction and the categories constituting it. An interaction combines manufacturing, geometric and physical characteristics appearing in production documents. The combination of various production documents aided in addressing the lack of deformable objects standardized representation.

CAD files are fundamental production documents for manufacturing, mainly for rigid components and rigid assemblies. As deformable objects usually are not represented in 3D drawings, the interaction categories appear in other production documents. The BOM is a production document widespread in many industries. It lists the components involved in the product, and in some cases, the description (or catalog number) can pinpoint the level of object deformation. The deformability of objects can often be derived from the DS, where detailed information on the deformable object appears. There are some industry sectors (e.g., medical) where the level of deformation is specifically described and measured (e.g., Shore Hardness values). Also, in the datasheet, the connection type can be derived (e.g., the type of connecting the deformable object to the rigid object). The electrical scheme directs which ports are connected by the wires, and the datasheet indicates the port number. In the absence of an electrical scheme (i.e., other industry sectors), such information can be derived from the job guide or CAD in case the deformable objects are represented. The job guide is also a source for the affinity act between the objects.

5 Conclusions

Industrial planning relies on production documentation. Documentation standardization is critical for production planning automation, especially for robotic assemblies of products with deformable objects. A prerequisite for robotic assembly planning for the assembly of products with deformable objects is production document mapping. Interaction contains manufacturing, geometric and physical features appearing in production documents. Combining different production documents can identify deformable object interaction characteristics and address their uniqueness within automatic robotic assembly planning procedures suitable for industrial environments. Automating the feature extraction is left for future work.

Acknowledgments. This research was partially supported by the Ben-Gurion University of the Negev through the Agricultural, Biological, and Cognitive Robotics Initiative, the Marcus Endowment Fund, and the W. Gunther Plaut Chair in Manufacturing Engineering. The authors thank Dr. Israel Tirkel for his valuable contribution to the field study.

References

1. Ou, L., Xu, X.: Relationship matrix based automatic assembly sequence generation from a CAD model. Comput. Aided Des. **45**(7), 1053–1067 (2013). https://doi.org/10.1016/j.cad.2013.04.002
2. Belhadj, I., Trigui, M., Benamara, A.: Subassembly generation algorithm from a CAD model. Int. J. Adv. Manuf. Technol. **87**(9–12), 2829–2840 (2016). https://doi.org/10.1007/s00170-016-8637-x

3. Mohammed, S.K., Arbo, M.H., Tingelstad, L.: Leveraging model based definition and STEP AP242 in task specification for robotic assembly. Procedia CIRP **97**, 92–97 (2021). https://doi.org/10.1016/j.procir.2020.05.209
4. Neb, A.: Review on approaches to generate assembly sequences by extraction of assembly features from 3D models. Procedia CIRP **81**, 856–861 (2019). https://doi.org/10.1016/j.procir.2019.03.213
5. Neb, A., Hitzer, J.: Automatic generation of assembly graphs based on 3D models and assembly features. Procedia CIRP **88**, 70–75 (2020). https://doi.org/10.1016/j.procir.2020.05.013
6. Fakhurldeen, H., Dailami, F., Pipe, A.G.: Cara system architecture - a click and assemble robotic assembly system. In: 2019 International Conference on Robotics and Automation (ICRA) (2019). https://doi.org/10.1109/icra.2019.8794114
7. Tariki, K., Kiyokawa, T., Nagatani, T., Takamatsu, J., Ogasawara, T.: Generating complex assembly sequences from 3D CAD models considering insertion relations. Adv. Robot. **35**(6), 337–348 (2020). https://doi.org/10.1080/01691864.2020.1863258
8. Hiller, J., Lipson, H.: Dynamic simulation of soft multimaterial 3D-printed objects. Soft Rob. **1**(1), 88–101 (2014). https://doi.org/10.1089/soro.2013.0010
9. Jimenez, P.: Survey on model-based manipulation planning of deformable objects. Robot. Comput.-Integr. Manuf. **28**(2), 154–163 (2012). https://doi.org/10.1016/j.rcim.2011.08.002
10. Sanchez, J., et al.: Robotic manipulation and sensing of deformable objects in domestic and industrial applications: a survey. Int. J. Robotics Res. **37**(7), 688–716 (2018). https://doi.org/10.1177/0278364918779698
11. Billard, A., Kragic, D.: Trends and challenges in robot manipulation. Science **364**(6446) (2019). https://doi.org/10.1126/science.aat8414
12. Ghalyan, I.F.J.: Force-Controlled Robotic Assembly Processes of Rigid and Flexible Objects (2016)
13. Acker, J., Kahl, B., Henrich, D.: Environment guided handling of deformable linear objects: from task demonstration to task execution. Vdi Berichte **1956**, 231 (2006)
14. Shneor, R., Berman, S.: Assembly sequence planning with deformable linear objects in the smart factory: dilemmas and injections. IFAC-PapersOnLine **55**(10), 2457–2462 (2022). https://doi.org/10.1016/j.ifacol.2022.10.077
15. Dawarka, V., Bekaroo, G.: Building and evaluating cloud robotic systems: a systematic review. Robot. Comput.-Integr. Manuf. **73**, 102240 (2022). https://doi.org/10.1016/j.rcim.2021.102240

Recent Robotic Research and Development Trends

Ran Shneor(✉) and Yael Edan

Department of Industrial Engineering and Management, Agricultural, Biological, Cognitive Robotics Initiative, Ben-Gurion University of the Negev, Beer-Sheva, Israel
shneorr@post.bgu.ac.il

Abstract. Robotics research is advancing at an increased pace. The global Covid-19 pandemic further increased the need for robotization of processes worldwide. This work aimed to characterise recent trends in robotics research and development (R&D) to direct efforts both in academia and industry. Emerging and declining hot topics were identified by a quantitative and qualitative analysis of scientific sources and reports from the International Federation of Robotics (IFR) and a bibliometric analysis of keywords co-occurrence in leading robotic conferences. Two-time frames (2010–2011, 2020–2021) were compared using the VOSviewer tool. Major increases, decreases, and new entries of robotic topics were identified, leading to the establishment of a keywords chart comparing trends between the two periods. Word clouds analysis was conducted on reports by IFR. The main findings from the keywords chart indicate that artificial intelligence and human-robot interaction (HRI) are major trends. A decrease in tangible keywords (e.g., "locomotion") was identified along with an increase in "equationless" and less tangible keywords (e.g., "social robotics"). The IFR sources pinpointed additional trends (e.g., "semantic intelligence" and "swarm robotics"). The combination of scientific and representative market data sources highlights robotic R&D trends from different perspectives. In the future, more sources (e.g., patents) and text-mining techniques should be added to identify additional trends in robotics.

Keywords: keywords chart · robotics trends · VOSviewer.

1 Introduction

Robotics has ignited individuals' imagination to automate human activities. There is an impressive endeavour in industry and academia to realize such ambitious dreams. The global Covid-19 pandemic further increased the automation and digitization of processes worldwide with increased robotics applications penetrating the market. There has been a continuous increase in robotics research and applications in recent years with a 636% increase from 2000 to 2020 (Fig. 1). To direct and guide efforts in academia and industry and identify emerging and declining hot topics it is important to analyse recent trends and themes in robotics research and development (R&D).

Quantitative and qualitative analysis is commonly applied for hot topics and research trend identification [1, 2]. One of the widespread tools for hot topics and research trend

M. Dragomir et al. (Eds.): ICPR 2023, LNPE, pp. 440–446, 2025.
https://doi.org/10.1007/978-3-031-92082-0_49

identification is the VOSviewer software [3]. This software package is applied for bibliometric analysis of scientific sources from various knowledge domains, e.g., reconfigurable manufacturing systems [4], and reinforcement learning-based robotics [5]. The keywords co-occurrence is an acceptable metric for trend identification [6–8]. Such a metric has been applied in various robotics reviews, e.g., robotics in construction [6], service robots [7], and collaborative robots in logistics [8]. However, it is important to be able to identify trends as they emerge without restriction to a specific robotic domain. Previous trends identification has been made based on robotic scientific journals solely [9] or on designated domains such as hot topics identifications in the medical robotics domain (e.g., robotic laparoscopy) [10].

Contrary to scientific sources highly analysed in robotics R&D trends identification, market data are less frequently applied as an information source. The International Federation of Robotics (IFR) is a non-profit organization aiming to promote research, development, use, and international cooperation in robotics [11]. IFR provides data and designated papers on different robotics issues, and its database has been applied for analysing trends in robot prices and classifying robotic application domains [12]. Combining scientific and such market data sources can shed light from various perspectives on robotics trends and contribute to rich and comprehensive results.

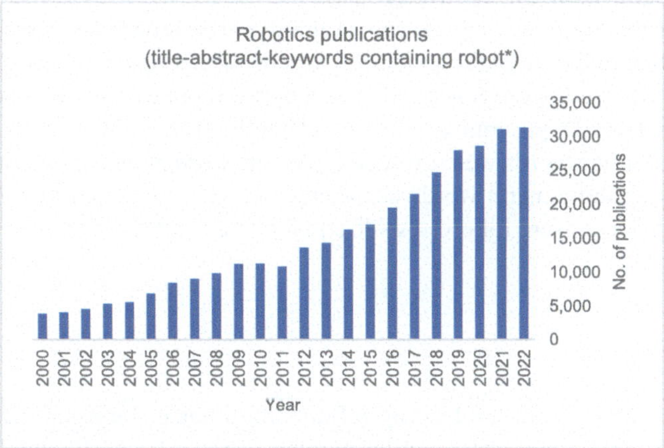

Fig. 1. Robotics publications 2000–2022* (*Based on title, abstract, and keywords containing *robot** retrieved from the Web of Science database, May 7th, 2023).

2 Method

The research methodology included quantitative and qualitative analyses of scientific and the IFR as market data sources. The scientific sources include a bibliometric analysis of keywords co-occurrence in leading robotic conferences publications retrieved from the Web of Science database. Six leading robotic conferences were chosen as adequate quick-responding indicators for emerging and declining trends.

Four robotic conferences were chosen based on the 2022 Google Scholar top publications ranking in the robotic category. Two additional conferences were included based on CORE ranking [13] and robotic experts' recommendations. The robotic conferences selected were IEEE International Conference on Robotics and Automation (ICRA) [14] which is ranked as the leading robotic avenue (conferences and journals) in the Google Scholar robotics category, IEEE/RSJ International Conference on Intelligent Robots and Systems (IROS) [15], which is ranked in the fourth place in Google Scholar ranking. Other conferences are Robotics: Science and Systems (RSS) [16], ACM/IEEE International Conference on Human-Robot Interaction (HRI) [17] (ranked ninth and twelfth, respectively in the abovementioned Google Scholar ranking). The International Symposium on Robotics Research (ISRR) [18] and IEEE International Symposium on Robot and Human Interactive Communication (ROMAN) [19] are the additional conferences included.

Two-time frames a decade apart were compared: 2010–2011 and 2020–2021. The first time frame was before the Covid-19 pandemic and the second shortly after. Keyword co-occurrences were exported from the VOSviewer software and compared. The keywords comparison was conducted using the total link strength of a keyword's links with other keywords, which was calculated and compared for the two-time frames. Major increases, decreases, and new entries of robotic topics were identified based on the total link strength establishing the keywords chart.

The market data sources comprised word clouds analysis on IFR reports. Word clouds have been applied in the literature to identify trends in robotics domains such as service robots [20]. Word clouds were generated on recent IFR reports on world robotics research and development (R&D) programs and artificial intelligence (published in December and February 2022). Some words were excluded from the word clouds for clarity (e.g., verbs such as *will*, *can*, and general words according to report title such as *robot*, *research*, etc.). The word clouds were followed with a qualitative analysis of the above-mentioned documents.

3 Results

The scientific sources from the leading robotic conferences between the years 2010–2011 and 2020–2021 contained a similar number of records (3,639 and 3,983, respectively). The keywords chart based on the two-time frames and bibliometric analysis are described in Table 1. Keywords that appeared in 2010–2011 conferences and 2020–2021, with higher values of total link strength in 2020–2021, were defined as "major increases". For example, major increases include *human-robot interaction (HRI)* and *social robotics*. Similarly, keywords that appeared in 2010–2011 conferences and 2020–2021, with lower values of total link strength in 2020–2021, were defined as "major decreases". For example, major decreases include *locomotion* and *kinematics*. The "new entries" category contains keywords that appeared in 2020–2021 and did not appear in 2010–2011 records with total link strength higher than the median value. For example, new entries include *trust* and *anthropomorphism*. Keywords with total link strength lower than the median value were termed "potential new entries?" as these terms can continue and become dominant in future years or otherwise. For example, the keywords chart pinpoints *assistive robotics* and *microrobots* as potential new entries. Artificial intelligence and especially, *deep learning* keywords appeared as new entries in the keywords chart.

Some of the robotics conferences are being conducted for more than three decades with considerable scientific publications on robotics annually and some with fewer publications are conducted bi-annually, e.g., ISRR (Table 2). Throughout the years designated conferences on specific robotic domains have evolved, e.g., HRI.

Table 1. Robotics R&D keywords chart (TLS-Total Link Strength).

Major increases (TLS 2020–21, TLS 2010–11)	Major decreases (TLS 2020–21, TLS 2010–11)	New entries (TLS2020–21) Appeared in 2020–21 and NOT in 2010–11; TLS > median (10)	Potential new entries? (TLS 2020–21) Appeared in 2020–21 and NOT in 2010–11; TLS < median (10)
Human-robot interaction (91,32)	Vision (15,44)	Trust (34,-)	Assistive robotics (8,-)
Social robotics (38,6)	Locomotion (28,56)	Anthropomorphism (26,-)	Game (8,-)
Localization (47,21)	Coordination (5,32)	Soft robotics (18,-)	Microrobots (6,-)
Recognition (42,21)	Kinematics (10,33)	Child-robot interaction (15,-)	Augmented reality (5,-)
Fusion (23,5)	Stability (18,38)	Deep learning (14,-)	
Emotion (18,8)		Engagement (11,-)	

The IFR market data sources word cloud analyses of the world R&D programs report (Fig. 2a) emphasizes *manufacturing*, *service*, and *autonomous* themes. Additional topics highly mentioned in different countries' R&D programs are *unmanned*, *medical*, and *rehabilitation*. A qualitative review of the IFR world R&D programs surfaced in different directions planned by different countries. Some countries emphasized defence and space domains, thus investing in unmanned robotics. Others, apparently driven by the aging population and Covid-19 insights, emphasized medical and rehabilitation robotics. The artificial intelligence document (Fig. 2b) emphasizes production-oriented themes such as *inspection*, *safety*, and *assembly*.

The IFR paper on artificial intelligence elaborates on market directions on three timelines: the next five years, five to ten years, and a longer time period. The next five years' commercial themes are concentrated mainly on common manufacturing activities such as *pick and place*, *quality inspection*, and *recycling*. The five to ten years are concentrated on intricate manipulations such as *clothing manufacturing*, *service assistant robots*, and *construction robots*. The long-term is less detailed and refers, in general, to *personal* and *household assistive robots*.

The research directions refer to the following themes in robotics and artificial intelligence: *Cost-effective aspects*, i.e., training artificial intelligence algorithms faster and

Table 2. Robotics conferences data.

Conference name	Average publications per year (2010–2021)*	Establishment year
IEEE International Conference on Robotics and Automation (ICRA) [14]	959	1984
IEEE/RSJ International Conference on Intelligent Robots and Systems (IROS) [15]	938	1988
Robotics: Science and Systems (RSS) [16]	65	2005
ACM/IEEE International Conference on Human-Robot Interaction (HRI) [17]	177	2006
The International Symposium on Robotics Research (ISRR) [18]	44	1986
IEEE International Symposium on Robot and Human Interactive Communication (ROMAN) [19]	207	1992

* Based on retrieval of conference name from the Web of Science database, May 30th, 2023.

a. b.

Fig. 2. Word clouds of IFR world robotic R&D programs (a) and artificial intelligence report (b).

enabling them to generalize better. *Learning from demonstration*, thus enabling robots to generate their program by watching a demonstration. *Semantic intelligence* is another research direction aiming at enabling the robot to understand the properties and characteristics of the object or person it is interacting with and make appropriate decisions (e.g., the robot responding differently to an elderly person than to a child). Yet such full-scale scenarios are assumed to take decades to become commercially viable. An additional research direction is *swarm* in which robots coordinate their movements among themselves without input from a higher control system (e.g., excavation of disaster sites or agriculture).

4 Discussion and Conclusions

Concentrating on leading robotic conferences seems like a quick response and potential "sensor" for trend identification. The establishment of a keywords chart is based on a simple and structured process as well as on an empirical database. By doing so, the keywords chart can be applied in literature reviews, roadmaps, and strategic planning in various domains and not limited to robotics R&D trends.

The main findings from the scientific sources pinpoint "the rise of HRI" as a major trend developing and leading to *child-robot interaction* to be identified as a new entry for the upcoming years. Soft and less tangible aspects regarding the interaction of humans and robots seem to become dominant in future years. Issues such as *emotion*, *trust*, and *engagement* with robots were identified as prominent in the robotics keywords chart. On the contrary, there is a decrease in issues based on high levels of mathematical rules (e.g., *dynamics*, *vision*). Maybe it is an evolutional development of the robotics domain. After building the foundations of basic robotics actions (sensing, design, acting) mainly in well-controlled and predefined scenarios such as industrial layouts, the next stage is advancing toward interaction with humans (adults and children). These domains involve social aspects and should be further explored in the upcoming years.

The market sources presented a very modest pace estimation regarding the advancement of robotics. This may be because they are commercially oriented, and additional considerations (e.g., regulations) besides development challenges influence the implementation of robotics in everyday surroundings. The world R&D programs highlighted different directions according to countries' visions and characteristics. Some countries concentrate on leading in unmanned military and space domains, whereas others concentrate on aging populations and the potential benefits of robots in assisting them. Further analysis and comparisons between different world regions and states are left for future research.

The combination of scientific and market data sources highlights robotic R&D trends from different perspectives. Artificial intelligence is a major trend identified by both.

Social aspects are emphasized in the scientific reports supported also by the entrance of several new focused conferences (e.g., HRI [17]). However, additional themes were identified based on the market data sources not found in the scientific sources (e.g., *swarm, semantic intelligence*).

In the future, additional sources should be considered (e.g., patents), and advanced artificial intelligence techniques such as text-mining should be incorporated as part of the keywords chart.

Acknowledgments. This research was partially supported by the Ben-Gurion University of the Negev through the Agricultural, Biological, and Cognitive Robotics Initiative, the Marcus Endowment Fund, and the W. Gunther Plaut Chair in Manufacturing Engineering.

References

1. Adamides, G., Edan, Y.: Human–robot collaboration systems in agricultural tasks: a review and roadmap. Comput. Electron. Agric. **204**, 107541 (2023). https://doi.org/10.1016/j.compag.2022.107541
2. Carter-Templeton, H., Frazier, R.M., Wu, L., Wyatt, T.H.: Robotics in nursing: a bibliometric analysis. J. Nurs. Scholarsh. **50**(6), 582–589 (2018). https://doi.org/10.1111/jnu.12399
3. Visualizing Scientific Landscapes. VOSviewer. https://www.vosviewer.com/. Accessed 07 May 2023
4. Pansare, R., Yadav, G., Nagare, M.R.: Reconfigurable manufacturing system: a systematic bibliometric analysis and future research agenda. J. Manuf. Technol. Manag. **33**(3), 543–574 (2021). https://doi.org/10.1108/JMTM-04-2021-0137
5. Khan, M.-M., et al.: A systematic review on reinforcement learning-based robotics within the last decade. IEEE Access **8**, 176598–176623 (2020). https://doi.org/10.1109/access.2020.3027152
6. Xiao, B., Chen, C., Yin, X.: Recent advancements of robotics in construction. Autom. Constr. **144**, 104591 (2022). https://doi.org/10.1016/j.autcon.2022.104591
7. Yörük, T., Akar, N., Özmen, N.V.: Research trends on guest experience with service robots in the hospitality industry: a bibliometric analysis. Eur. J. Innov. Manag. (2023). https://doi.org/10.1108/EJIM-09-2022-0530
8. Atzeni, G., Vignali, G., Tebaldi, L., Bottani, E.: A bibliometric analysis on collaborative robots in Logistics 4.0 environments. Procedia Comput. Sci. **180**, 686–695 (2021). https://doi.org/10.1016/j.procs.2021.01.291
9. Yu, L., Yang, W., Duan, Y., Long, X.: A study on the application of coordinated topsis in evaluation of robotics academic journals. Math. Probl. Eng. **2018**, 1–7 (2018). https://doi.org/10.1155/2018/5456064
10. Dupont, P.E., et al.: A decade retrospective of medical robotics research from 2010 to 2020. Sci. Robot. **6**(60) (2021). https://doi.org/10.1126/scirobotics.abi8017
11. IFR. International Federation of Robotics. IFR International Federation of Robotics. https://ifr.org/. Accessed 07 May 2023
12. Jurkat, A., Klump, R., Schneider, F.: Tracking the rise of robots: the IFR database. Jahrbücher für Nationalökonomie und Statistik **242**(5–6), 669–689 (2022). https://doi.org/10.1515/jbnst-2021-0059
13. Conference portal - core. http://portal.core.edu.au/conf-ranks/. Accessed 10 Apr 2023
14. ICRA. IEEE Robotics and Automation Society. https://www.ieee-ras.org/conferences-workshops/fully-sponsored/icra. Accessed 31 May 2023
15. IROS. IEEE Robotics and Automation Society. https://www.ieee-ras.org/conferences-workshops/financially-co-sponsored/iros. Accessed 31 May 2023
16. Robotics: Science and Systems. Robotics. https://roboticsconference.org/. Accessed 31 May 2023
17. HRI conference – home. ACM Digital Library. https://dl.acm.org/conference/hri. Accessed 31 May 2023
18. ISRR International Symposium of Robotic Research. ISRR - International Foundation of Robotics Research. http://ifrr.org/isrr. Accessed 31 May 2023
19. International Symposium on Robot and Human Interactive Communication (RO-MAN). IEEE Robotics and Automation Society. https://www.ieee-ras.org/conferences-workshops/financially-co-sponsored/ro-man. Accessed 31 May 2023
20. Savin, I., Ott, I., Konop, C.: Tracing the evolution of service robotics: insights from a topic modeling approach. Technol. Forecast. Soc. Chang. **174**, 121280 (2022). https://doi.org/10.1016/j.techfore.2021.121280

The Role Played by Digital Technologies in Implementing Remanufacturing for the Circular Economy

L. Taquetti[1,2](✉), F. Deschamps[2], E. P. de Lima[3], and S. E. G. da Costa[3]

[1] Universidade Federal do Oeste da Bahia, Luís Eduardo Magalhães, BA, Brazil
larissa.taquetti@ufob.edu.br

[2] Industrial and Systems Engineering Graduate Program (PPGEPS), Pontifícia Universidade Católica do Paraná, Curitiba, PR, Brazil

[3] Universidade Tecnológica Federal do Paraná, Curitiba, PR, Brazil

Abstract. The circular economy is a promising approach to sustainability, and the remanufacturing process is one of the most appropriate practices, as it occupies the highest level of a product's value chain. In this context, digital technologies are increasingly recognized as a powerful tool to support the transition to the circular economy. However, there is still a lack of studies that address the integration between remanufacturing and digital transformation and how digital technologies can impact the circular transition. This article reviews the literature that addresses this research gap. The results suggest that, although digital technologies are modifying business processes and models bringing potential benefits, there are risks from economic, social, technological and political perspectives.

Keywords: Circular economy · sustainability · remanufacturing · digital transformation · digital technologies

1 Introduction

The dilemma for more sustainable operations has caused changes in supply chains. The traditional supply chain model is predominantly linear: resources are extracted, transformed by organizations, consumed by society and, finally, discarded [1]. As a result, there is scarcity of resources, environmental impacts and economic concerns [2]. In this context, the circular economy (CE) was born. The idea is that materials are recovered from old products and reused as inputs for as long as possible [4], merging operations such as reduce, reuse and recycle [3, 5].

Among the various CE operations, especially in the field of recycling, remanufacturing is considered essential for the preservation of resources and the mitigation of impacts on the environment [6, 7] due to the enormous potential for extending the useful life of the product [8, 9]. Only the remanufacturing approach allows the quality conditions of the product to be similar or even better than that of a newly manufactured product [10].

Remanufacturing systems face greater complexity when compared to traditional manufacturing systems, so it is crucial that new solutions are adopted to contain these

M. Dragomir et al. (Eds.): ICPR 2023, LNPE, pp. 447–454, 2025.
https://doi.org/10.1007/978-3-031-92082-0_50

challenges. In this sense, digital technologies can help facilitate the implementation of CE and remanufacturing [11, 12]. Although the literature has shown advances in the analysis and application of digital technologies and CE, there are still gaps to be investigated in the integration between the two fields of research [11, 13–15]. Specifically for remanufacturing, these search streams are even rarer [16, 17]. Therefore, this work contributes to fill this identified gap. Through a systematic literature review, this study investigates how digital technologies can impact the circular transition through remanufacturing, allowing companies to better organize their resources and identify emerging opportunities for the sustainability of operations.

2 Research Context

The CE represents an opportunity to take sustainability to a higher level. The Ellen MacArthur Foundation (EMF) treats CE as a systemic approach to economic development designed to benefit business, society and the environment [18]. CE incorporates a closed-loop model, so that products circulate at their highest utility for the longest possible period, minimizing waste generation [14].

The CE implementation is often characterized by so-called R-principles. The 10R-principles are characterized as refuse, rethink, reduce, reuse, repair, refurbish, remanufacture, repurpose, recycle and recovering [7]. In particular, remanufacturing is more efficient than other CE strategies because it retains the core and used product, reducing the need for new products, materials and production inputs [19].

Remanufacturing restores a non-functional product to like-new condition. Since each component of a product can be recycled differently, the process faces several challenges [17]. The used product goes through a progression of steps such as disassembly, sorting and cleaning, scanning, refurbishment or replacement, reassembly and quality tests before its introduction into the market [10]. There is a growing interest in using digital technologies to achieve greater efficiency and to be able to exploit the full potential of remanufacturing [20, 21]. Thus, the digital transformation in organizations has helped in driving sustainable operational decisions and in taking advantage of opportunities for the growth of the remanufacturing industry.

Continuous innovations provided by digital technologies challenge the way production systems work, leading them to a disruptive approach with implications for all stakeholders [22]. As information technologies become more accessible and increase the availability of real-time data, new opportunities are identified to improve remanufacturing activities and manage the uncertainties of this process [23].

3 Methodology

This study adopted the systematic literature review (SLR) to identify how digital technologies can impact the circular transition from the point of view of remanufacturing. SLR is a robust, detailed and rigorous scientific method used for mapping, evaluating and developing a relevant research base, to generate new knowledges and consolidate discoveries [24]. The systematic process followed the SLR development steps proposed by [25].

The search for publications was carried out in the Scopus database. This database was chosen because it contains a wide range of peer-reviewed articles and because it is used in several existing literature reviews on CE and digital transformation [12, 14]. The search engine used combined the terms "remanufactur*" AND "digital" OR "4.0" AND "circular economy" in the search by title, abstract or keywords. Only articles and conference papers in English were included. This research resulted in the identification of an initial sample of 64 documents. However, after excluding papers that were not available in full or that, after reading them in full, were not aligned with the objective, the final selection reduced the number of papers to 31. These articles were analyzed and recorded so that it was possible to discuss the information found.

4 Results

The annual number of publications resulting from the combination of keywords is shown in Fig. 1. The distribution of articles over the years shows that, despite not containing expressive numbers, the trend is for growth and justifies research opportunities. The 2023 quantity did not consider the entire year, since data collection took place in the first half of the year.

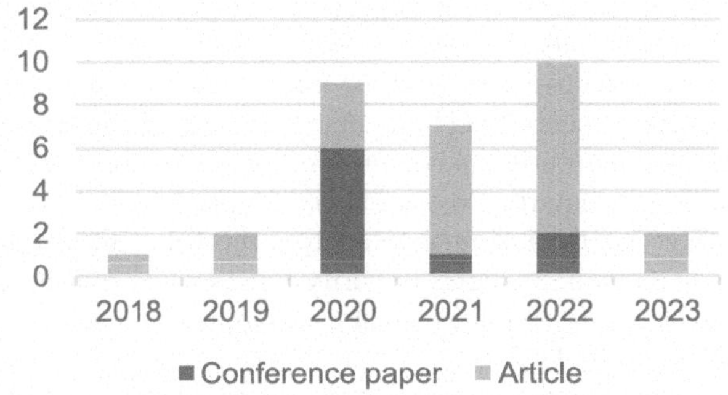

Fig. 1. Distribution of literature by year

Regarding the countries where the researchers of the sample articles are located, in a total of 23 countries, United Kingdom and Italy are the only countries with more than 5 publications, representing, together, a quarter of the total number of articles sample. India, France and South Africa also stand out for the number of publications. The researchers with the highest number of publications in the sample (3 articles each) work at universities in the United Kingdom (D. T. Pham and A. Tiwari) and India (S. Bag). The journals with more publications in the sample were Procedia CIRP and Sustainability, with 5 articles each. The most explored sector is manufacturing and electric motors, with contributions to the supply chain.

The results demonstrate that digital transformation acts as an accelerator of CE development and the improvement of collaborations between actors in an ecosystem. The cooperation between manufacturers, suppliers, logistics companies and consumers allows the creation of value for reverse logistics activities, recycling, dismantling and reprocessing of materials. This relationship requires a strong communication system to share the flow of information and materials that enable circular business [29]. In particular, the remanufacturing industry is undergoing a digital transformation because this process increases the ability to diagnose and predict the condition of components and products, making them more reliable and available to circulate in an intelligent supply chain [26]. In this way, digital technologies have acted as facilitators of remanufacturing. The combination of digital technologies adds value to operations management, indicating new opportunities and innovative solutions for remanufacturing.

As for the types of digital technologies used, the publications address the use of, in order of manifestation, internet of things (IoT), additive manufacturing (AM) and 3D printing, augmented reality, digital twin, big data analytics (BDA), cloud technology, artificial intelligence (AI), blockchain and robotics. How these technologies impact the circular transition from a remanufacturing perspective is discussed below.

- IoT: is the most adopted technology in the selected documents. It is generally seen as the starting tool for unlocking the full potential of circular manufacturing. IoT has been used to collect data as close to real time as possible, throughout the product lifecycle, so that companies can turn this data into insights and effective actions [26], monitor the state of resources and the evolution of the work, establish connections between physical and virtualized resources [27], determine the ideal treatments to be applied after product recovery [28] and digitize processes that guarantee the collection and distribution of materials and inputs [29].
- AM and 3D priting: effectively support the implementation of remanufacturing and reconditioning of components for cellular manufacturing. It can be used to bring worn parts or cores back to their original conditions [26] and to provide spare parts for the industry, in addition to requiring less resources for production [30].
- Augmented reality: has the potential to analyze the behavior of complex systems, predict failures, test repair strategies or verify functionality during remanufacturing [31]. It can make learning activities, such as training and machine assembly, more productive and less costly [26].
- Digital twin: it is fundamental to support decisions related to the processing strategy, system conditions and post-use returned products, and to evaluate the product's revenue potential [9]. Digital twin is able to determine the possible alternatives to the behavior of the material flow and indicate the right moment to carry out the remanufacturing [32].
- BDA: collect, store and process large amounts of data. They allow reviewing the data collected by the IoT in order to optimize the decision-making process and create value for the customer in the service offering [33]. By accessing data on the conditions of materials and equipment, it is possible to improve resource management.
- Cloud technology: deal with the same set of data as IoT and big data. The software allows data sharing to carry out the remanufacturing process, creating a distributed network managed by a central system. It introduces a new approach in which resources are made available on the internet for worldwide access [28]. This facilitates standardized data exchange in order to maintain knowledge throughout the lifecycle [30]. Cybersecurity is essential for implementing cloud technology

- AI: Enables the digitization of physical assets and improves analytics across the entire supply chain. AI techniques increase prediction accuracy, quality of produced parts and productivity due to greater optimization and analysis of real-time data [30]
- Blockchain: all supply chain activities can be monitored using this technology. The use of blockchain allows the traceability and location of parts from manufacturing to remanufacturing [34]. The use of algorithms can be used to solve lot-sizing problems for remanufactured products and for planning and scheduling the production of these products [32].
- Robotics: robot-assisted approaches can be used in disassembly processes to achieve more efficiency and reduce operating costs, especially in more hazardous environments [14, 20].

Studies indicate that the application of digital technologies is positively related to the ability to remanufacture products. Figure 2 shows the relationship between digital technologies and remanufacturing. In a general context, the adoption of technologies is critical for all businesses. However, some challenges need to be overcome for organizations to advance in this practice. Technical and human barriers must be considered for a successful transition to CE [20].

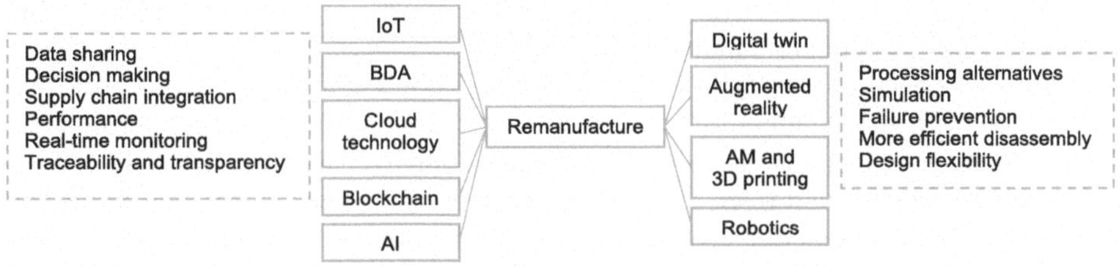

Fig. 2. Relation between remanufacture and digital technologies

Barriers to technology adoption can lead to lack of access to data and this can result in missed business opportunities. Decision making based on inaccurate product lifecycle data can generate operational inefficiency, hindering reverse logistics and end-of-life strategies [26]. Uncertainties regarding the quality and quantity of returned products, disassembly hierarchy and predominance of manual operations that generate high costs and reduced efficiency can be treated with the aforementioned technologies [27, 32]. Remanufacturing requires greater automation in processes to enable more effective use of digital transformation. This requires significant investments in infrastructure, equipment and workforce qualification. It is important to develop operator training to ensure the effectiveness of the workforce. In addition to the workforce, an environment dedicated to carrying out the activities involved in remanufacturing is also necessary, but the high cost of this investment can discourage performance improvement strategies in remanufacturing [27].

Another aspect identified in the literature is the use of digital technologies as a way of offering opportunities to optimize supply chain operations in the suppliers-purchasing relationship. IoT, AI and cloud computing can change the way purchasing decisions are made and improve their performance [35]. Faced with the uncertainties and risks of remanufacturing, obtaining greater availability of market intelligence data and global trends in input prices will provide advantages over cost control in the supply chain and more agile procurement [36]. Studies also point to barriers on the consumer side, as they may identify the perceived quality as inferior to new products, or not even be aware of the environmental benefits related to the reduction in the use of materials and energy [27]. In this sense, changes in attitudes can engage in sustainable purchasing behavior.

Finally, given the barriers to technology adoption, the articles warn of the need for cooperation between companies, society, researchers, investors, regulatory bodies and policy makers to integrate technological advances into environmental standards and legislation and ensure the appropriate circulation of resources. Among ecosystem actors.

5 Conclusions

This study analyzed the role of digital technologies in implementing remanufacturing for the CE. The results indicate that digital technologies act as enablers for the circular transition and are the key to the propagation of stronger and more sustainable remanufacturing [17, 30]. The main benefits involve the identification of opportunities and value propositions in the product lifecycle, automation of activities and access to data for better visualization and mitigation of risks in the process of extending the useful life of products and enabling remanufacturing.

Institutions need to have an orientation towards digital innovation, data sharing and cooperation with stakeholders to build an environment prone to remanufacturing [26]. Policy makers need to understand the role of digitization to ensure that CE is understood and exploited quickly and responsibly [29]. Transparent processes enabled by digital technologies allow interaction and cooperation, collection and exchange of data to support circular strategies. This study contributes to the understanding of the role of digital technologies in the practice of remanufacturing. Future research can practically and in-depth investigate the extent to which digital technologies contribute to the adoption of other circular practices.

References

1. Oliveira, C.T., Oliveira, G.G.A.: What Circular economy indicators really measure? An overview of circular economy principles and sustainable development goals. Resour. Conserv. Recycl. (2023). https://doi.org/10.1016/j.resconrec.2022.106850
2. Sohal, A., De Vass, T.: Australian SME's experience in transitioning to circular economy. J. Bus. Res. **142**, 594–604 (2022)
3. Sohal, A., Nand, A.A., Goyal, P., Bhattacharya, A.: Developing a circular economy: an examination of SME's role in India. J. Bus. Res. **142**, 435–447 (2022)
4. Safarzynska, K., Domenico, L., Raberto, M.: The circular economy mitigates the material rebound due to investments in renewable energy. J. Clean. Prod. (2023). https://doi.org/10.1016/j.jclepro.2023.136753
5. Rasi, R.Z., Ismail, H., Shahbaz, M.S., Sundram, V.P.K.: Interdisciplinary challenges in the circular supply chains: a systematic literature review. Heliyon (2023). https://doi.org/10.1016/j.heliyon.2023.e15225

6. Hartono, N., Ramírez, F.J., Pham, D.T.: Optimisation of robotic disassembly plans using the Bees Algorithm. Robot. Comput.-Integr. Manuf. (2022). https://doi.org/10.1016/j.rcim.2022.102411
7. Bag, S., Gupta, S., Kumar, S.: Industry 4.0 adoption and 10R advance manufacturing capabilities for sustainable development. Int. J. Prod. Econ. (2021). https://doi.org/10.1016/j.ijpe.2020.107844
8. Stavropoulos, P., Papacharalampopoulos, A., Athanasopoulou, L., Kampouris, K., Lagios, P.: Designing a digitalized cell for remanufacturing of automotive frames. Procedia CIRP **109**, 513–519 (2022)
9. Karin, M., Hartono, N., Pham, D.T.: Optimising remanufacturing decision-making using the bees algorithm in product digital twins. Sci. Rep. **13**(701) (2023)
10. Singhal, D., Tripathy, S., Jena, S.K.: Remanufacturing for the circular economy: study and evaluation of critical factors. Resour. Conserv. Recycl. (2020). https://doi.org/10.1016/j.resconrec.2020.104681
11. Teixeira, E.L.S., Tjahjono, B., Beltran, M., Julião, J.: Demystifying the digital transition of remanufacturing: a systematic review of literature. Comput. Ind. (2022). https://doi.org/10.1016/j.compind.2021.103567
12. Trevisan, A.H., Lobo, A., Guzzo, D., Gomes, L.A.V., Mascarenhas, J.: Barriers to employing digital technologies for a circular economy: a multi-level perspective. J. Environ. Manag. **332** (2023)
13. Toth-Peter, A., Oliveira, R.T., Mathews, S., Barner, L., Figueira, S.: Industry 4.0 as an enabler in transitioning to circular business models: a systematic literature review. J. Clean. Prod. **393** (2023)
14. Cagno, E., Neri, A., Negri, M., Bassani, C.A., Lampertino, T.: The role of digital technologies in operationalizing the circular economy transition: a systematic literature review. Appl. Sci. **11**(8) (2021)
15. Kerin, M., Pham, D.T.: Smart remanufacturing: a review and research framework. J. Manuf. Technol. Manag. **31**(6), 1205–1235 (2020)
16. Chau, M.Q., et al.: Prospects of application of IoT-based advanced technologies in remanufacturing process towards sustainable development and energy-efficient use. Energy Sources, Part A: Recovery, Utilization, and Environmental Effects (2021). https://doi.org/10.1080/15567036.2021.1994057
17. Ellen MacArthur Foundation. The Circular Economy in Detail (2023). https://ellenmacarthurfoundation.org/the-circular-economy-in-detail-deep-dive. Accessed 14 Mar 2023
18. Burmaoglu, S., Gungor, D.O., Kirbac, A., Saritas, O.: Future research avenues at the nexus of circular economy and digitalization. Int. J. Prod. Perform. Manag. (2022). https://doi.org/10.1108/IJPPM-01-2021-0026
19. Roda-Sanchez, L., et al.: Comparison of RGB-D and IMU-based gesture recognition for human-robot interaction in remanufacturing. Int. J. Adv. Manuf. Technol. **124** (2023)
20. Kerin, M., Pham, D.T.: A review of emerging industry 4.0 technologies in remanufacturing. J. Clean. Prod. (2019). https://doi.org/10.1016/j.jclepro.2019.117805
21. Marcon, E., Le Dain, M., Frank, A.G.: Designing business models for Industry 4.0 technologies provision: changes in business dimensions through digital transformation. Technol. Forecast. Soc. Change (2022). https://doi.org/10.1016/j.techfore.2022.122078
22. Kerin, M., Hartono, N., Pham, D.T.: Optimising remanufacturing decision-making using the bees algorithm in product digital twins. Sci. Rep. **13** (2023)
23. Tranfield, D., Denyer, D., Smart, P.: Towards a methodology for developing evidence-informed management knowledge by means of systematic review. Br. J. Manag. **14**, 207–222 (2003)
24. Khan, K.S., et al.: Five steps to conducting a systematic review. J. R. Soc. Med. **96**(3), 118–121 (2003)

25. Nham, N.T.H., Ha, L.T.: Making the circular economy digital or the digital economy circular? Empirical evidence from the European region. Technol. Soc. **70** (2022)
26. Subramoniam, R., Sundin, E., Subramoniam, S., Huisingh, D.: Riding the digital product life cycle waves towards a circular economy. Sustainability **13**(16) (2021)
27. Caterino, M., Fera, M., Macchiaroli, R., Pham, D.T.: Cloud remanufacturing: remanufacturing enhanced through cloud technologies. J. Manuf. Syst. **64**, 133–148 (2022)
28. Delpla, V., Kenné, J., Hof, L.A.: Circular manufacturing 4.0: towards internet of things embedded closed-loop supply chains. Int. J. Adv. Manuf. Technol. **118** (2022)
29. Krstic, M., Agnusdei, G.P., Miglietta, P.P., Tadić, S.: Logistics 4.0 toward circular economy in the agri-food sector. Sustain. Futures (2022). https://doi.org/10.1016/j.sftr.2022.100097
30. Pratapa, P., Subramoniam, R., Gaur, J.: Role of standards as an enabler in a digital remanufacturing industry. Sustainability **14**(3) (2022)
31. Blomeke, S., et al.: Recycling 4.0–mapping smart manufacturing solutions to remanufacturing and recycling operations. Procedia CIRP **90**, 600–605 (2020)
32. Tiwari, D., Miscandlon, J., Tiwari, A., Jewell, G.: A review of circular economy research for electric motors and the role of Industry 4.0 technologies. Sustainability **13** (2021)
33. Bressanelli, G., Saccani, N., Perona, M., Baccanelli, I.: Towards circular economy in the household appliance industry: an overview of cases. Resources **9**(128) (2020)
34. Tang, Y.M., Chau, K.Y., Fatima, A., Waqas, M.: Industry 4.0 technology and circular economy practices: business management strategies for environmental sustainability. Environ. Sci. Pollution Res. **29** (2022)
35. Franco, C.W., et al.: A contingency-configurational view of purchasing operations: the mediating role between supplier relationship and firm performance. J. Purchasing Supply Manag. **29** (2023)
36. Surajit, B., Pavitra, D., Shivam, G., Uthayasankar, S.: Examining the role of procurement 4.0 towards remanufacturing operations and circular economy. Prod. Plan. Control **32**(16), 1368–1383 (2021)

Trends Within Creativity Support Tools: Remote Collaboration Within New Product Development

Reto Wechner[1] (✉), Verena Kaschub[2], Michael Gräf[3], Benjamin Wingert[1], Matthias Bues[1], and Oliver Riedel[1]

[1] Fraunhofer Institute for Industrial Engineering (IAO), Nobelstr. 12, 70569 Stuttgart, Germany
reto.wechner@iao.fraunhofer.de
[2] GSaME University of Stuttgart, Nobelstr. 12, 70569 Stuttgart, Germany
[3] Institute of Human Factors and Technology Management (IAT), University of Stuttgart, Nobelstr. 12, 70569 Stuttgart, Germany

Abstract. The importance for the generation of new, innovative, technical products has increased significantly in recent years. This derives from an increase in competition. However, the development of innovative, technical products is time-consuming and depends to a certain level on the creativity of engineers. CSTs can help product developers to be more creative within the development and design task and can support them all along the product development process. Furthermore, such tools can enable remote collaboration to fit within the future challenges of new product development.

In this paper, a comprehensive examination of over 900 creativity-related publications from the ACM Digital Library and ScienceDirect Library, using a keyword search, is conducted. The results show an overview of identified tools based on criteria proposed in the paper. The analysis is used to identify trends of CSTs for future technical product development, regarding especially aspects of local and remote collaboration. These trends show current development activities as well as existing fields of action for CSTs within technical product development, to further increase creativity.

Keywords: Creativity Support Tools · CST · Innovation process · Remote collaboration · New product development

1 Introduction

1.1 Motivation

The rapid change in the creation process of new products has a direct impact on the entire value chain from idea generation to planning, development and production. To maintain competitiveness, the focus is primarily on finding new, creative and innovative solutions and ideas [1]. In the next ten to 15 years, the introduction of automation and AI technologies, as well as technologies that allow remote collaboration will therefore change the engineering workplace, as people will increasingly interact with more intelligent machines. As a result, product developers gain more freedom. This will change

M. Dragomir et al. (Eds.): ICPR 2023, LNPE, pp. 455–464, 2025.
https://doi.org/10.1007/978-3-031-92082-0_51

the requirements of the product developer skills [2]. The engineers will increasingly take on the role of a creative problem solvers [3]. Finding new, creative, and innovative solutions and ideas thus forms an essential basis for the development of future products.

1.2 Creativity Support Tools

Creativity Support Tools (CST) can be used to support the need for engineers to deal with complex environments and to fulfil the role of innovative and creative problem solvers. The use of new, digital tools in the context of creativity is reflected in human-computer interaction (HCI) research. The science of HCI has been predominantly concerned with analogue, specialized creativity tools (e.g. creativity toolkits and books). In recent years, however, HCI research has also sought to harness the potential of intelligent, digital tools to support user creativity in a meaningful way. This can be seen in the introduction of various digital tools in the form of Creativity Support Tools (CST).Creativity support tools are defined differently in the literature [4–7]. The different definitions are due to the complexity of the description. Depending on perception, any technical aids can be defined as CSTs. Frich et al. (2019) conducted a meta-analysis on CSTs. As part of his work, CSTs were analyzed in terms of their application form and devices used.

For a reproducible analysis, this paper uses a modified definition from Frich et al. (2019). The definition of Frich et al. (2019) only considers digital tools. The focus of the investigation in this analysis is on digital and analog systems. In this work, a CST is therefore defined as follows:

'A Creativity Support Tool runs on one or more analog or digital systems, includes one or more creativity-oriented functions, and is used to positively influence the user with varying levels of expertise in one or more different phases of the creative process.' - adapted from Frich et al. (2019), p.10.

1.3 CSTs for Remote Collaboration

Remote tools are getting more important in creativity sessions for engineers due to several reasons. The globalization of engineering teams and the need for collaboration across distances make remote tools essential for effective communication and idea sharing and creation. They provide flexibility, enabling engineers to work from anywhere. Therefore, in the context of new work, they can also maintain a better work-life balance. Remote tools also promote inclusivity by allowing individuals with disabilities or geographical limitations to actively participate. The lockdowns during the corona pandemic even accelerated this trend towards remote collaboration [8].

Nevertheless, Gabriel et al. (2016) stresses that CSTs are lacking the focus on remote settings. First approaches such as Balters et al. (2021) show that remote settings could have a negative impact on aspects such as communication and collaboration. Furthermore, a remote setting offers an "endless" space (e.g., with whiteboard tools), which in turn increases complexity and can overload cognitive capacities [11].

According to Balters et al. (2021), there is a need for further studies that take a closer look at the influence of remote settings on creative processes and innovation potential.

1.4 Research Question

The motivation behind this study is to determine the state-of-the-art of remote collaboration with CSTs identify trends of CSTs for future technical product development, focusing on aspects of local and remote collaboration. The derived research questions therefore are:

- What is the current state of the art of remote collaboration using creativity support tools within product development?
- What trends within creativity support tools for remote collaboration can be derived from this analysis of the state of the art?

2 Method

To be able to answer the research questions a literature analysis on CSTs was carried out. Based on the two existing publications by Frich et al. (2019) and Gabriel et al. (2016), a total of 192 publications were included. In addition, a literature analysis was executed, and further publications were added. For this analysis, the methodological approach of Frich et al. (2019) provides a suitable starting point. Frich et al. (2019) builds on the work of Frich et al. (2018). For a coherent, temporal continuation of the literature analysis of the search results from Frich et al. (2019), the keywords used from Frich et al. (2019) were applied for an extended time period.

To complement the literature analysis from the ACM Digital Library, the keyword search in the literature analysis was extended to include the Science Direct Library.In the context of the literature analysis, it makes sense to extend the keyword search beyond the ACM Digital Library to pay particular attention to publications within the context of technical product development. The extended consideration can ensure that a large part of the scientific publications on the topic of CST in technical product development are covered. To extend the analysis according to the underlying research of Frich et al. (2019), a search based on the keywords *creativity support tool* and *creativity tool* was linked with the Boolean operator *AND* in the Science Direct Library and applied within the subject area: *engineering*. The period from 01.01.1900 to 31.10.2022 is also considered. 207 publications were recorded for these keywords.

Using the five defined literature sources, 947 publications were extracted. Of these 947 publications, 91 publications were recorded as not scientifically relevant from the ACM Digital Library keyword search (see Fig. 1). In the following, a methodical overview of the literature review is given:

For the analysis of the state of the art of CSTs in technical product development, further clustering and evaluation was necessary. First, a distinction was made between publications that describe a specific CST and publications that do not describe CSTs or do not focus specifically on a CST. For this purpose, the previously elaborated definition of a CST was used as a basis. According to the definition established, CSTs are special tools that are developed to support the creative process of users. They can be executed using analog or digital systems and include one or more functions designed to enhance the user's creativity. CSTs are used at different stages of the creative process and can help users with different levels of expertise improve their creative skills. In this work, the identification of CSTs based on the definition is done by analyzing the abstract and summary. In the first step of the process, the 947 publications included in the paper were divided into three groups: Publications that contain CSTs, publications that do not contain CSTs, and publications that can be considered not scientifically relevant.

A review of the publications showed that CSTs are characterized by several features. These characteristics include the type of equipment used, the form of application, and the opportunities for collaboration. Within the literature review, these characteristics were ranked and evaluated (present vs. absent) in relation to the publications to gain a deeper understanding of CSTs and the associated status quo. Due to the focus of the paper, the focus here is only on engineering CSTs and publications. Figure 1 provides an overview of the evaluation and consideration criteria as well as the analysis process.

A clustering by three different aspects is intended to help identify and understand certain trends in CSTs.

For the analysis, the absolute numbers and distributions are first examined to obtain an overview of the development situation. Then a temporal analysis is performed by examining the development of the numbers over time. In this way, statements can be made about changes in the development over time.

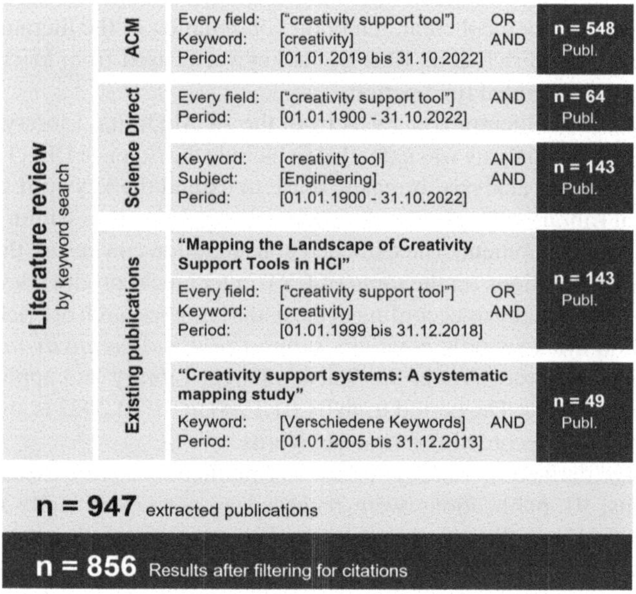

Fig. 1. Literature review by keyword search.

3 Results

The analysis of the 947 publications shows that 454 publications are recorded with a CST and 402 publications without a CST. 31 publications turned out to be duplicate descriptions of a CST already recorded. Thus, 916 publications are examined in the paper, of which 423 publications are recorded with a CST. The classification of the publications is based on the established definition. This results in an exclusion of 91 publications in the analysis.

Overall the absolute numbers of publications containing a CST increased significantly since 2018. 174 of the recorded 423 CSTs can be assigned to a product development context. Within these 174 publications 145 CSTs can be assigned to multiple subjects and 29 CSTs are assigned exclusively to the engineering field.

When the 174 CSTs in the product development context are differentiated according to the equipment used, the following distribution emerges. 123 CSTs are assigned to web and computer systems. Of these, 24 CSTs contain components of AI. 36 CSTs are assigned to the analog category, of which ten CSTs are assigned to the *Cards-CST* category, seven CSTs are assigned to whiteboards, and another seven represent CSTs with construction kits. Fifteen analog CSTs also contain digital tool properties and therefore cannot be clearly distinguished from the digital CSTs. 67 CSTs are assigned exclusively to digital CSTs, although some of these can also be assigned to web and computer systems (Fig. 2).

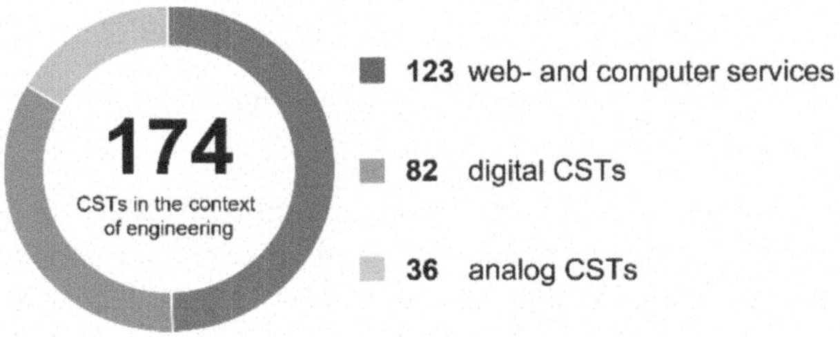

Fig. 2. Devices used in engineering CSTs.

When taking a further look at the digital CSTs, 7 CSTs are assigned to the *Phone* category, 17 CSTs to *Tabletop,* 16 CSTs to *Wall-displays*, 18 CSTs to *Tablets*, 51 CSTs to *Laptops*, 29 CSTs to *Tangibles*, and seven CSTs to the *VR/AR* category.

The temporal analysis shows that there is an increase in digital CSTs compared to analog ones. The temporal analysis of CSTs based on web and computer systems shows an increase in the period from 2004 to 2020. Since 2020, a large proportion of CSTs have been based on web and computer systems (Fig. 3).

Examining the **form of use** of the 174 CSTs shows that 135 CSTs are assigned to the **individual** form of use. 67 CSTs have characteristics for **collective** use. 28 CSTs can be used both individually and collectively.

The development of collective CSTs increased slowly but steadily from 2005 to 2012, with a peak in 2010. From 2013 to 2018, the number of CSTs with a collective application has remained about the same. Since 2018, the number of CSTs for individual use has increased and peaked in 2020 (Fig. 4).

By examining the 174 CSTs, 76 CSTs **collaboration** tools can be identified. This corresponds to a share of 44% of the examined CSTs. Of these 76 collaboration tools, 42 CSTs exhibit remote capabilities, representing a 55% share. Of these 42 remote-capable CSTs, eleven CSTs (26%) exhibit a crowdsourcing approach.

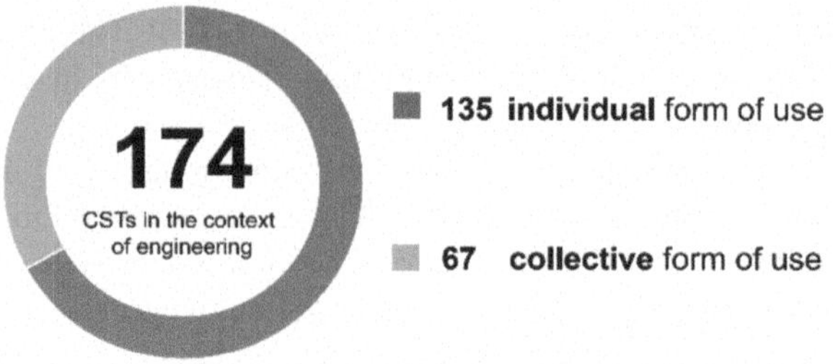

Fig. 3. Forms of use of engineering CSTs.

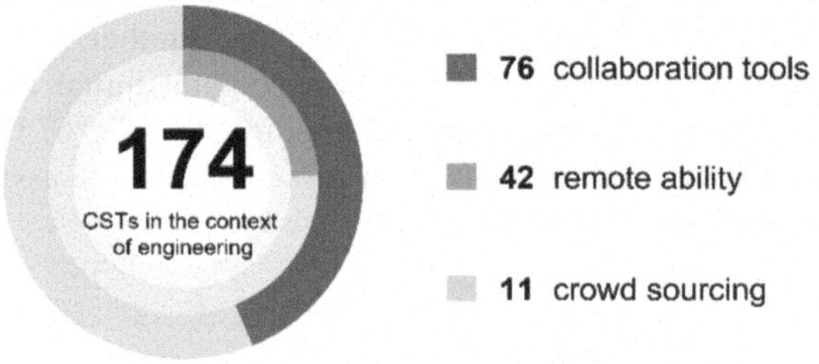

Fig. 4. Collaboration in engineering CSTs.

Collaboration tools are primarily designed to facilitate collaboration and improve communication within a team. Collaboration tools differ from collective application forms in the concrete feature of collaboration. CSTs of the collective application form do not imply an inevitable focus on collaboration. Nevertheless, a collaboration characteristic is obvious. Tools with an individual application form can also represent a collaboration tool.

The results show that the majority of the considered CSTs do not support collaborative aspects. More than half of the collaboration tools also have remote functionality. The fact that more than a quarter of remote-enabled CSTs have a crowdsourcing approach demonstrates the growing importance of an open innovation approach. Crowdsourcing allows users or companies to leverage the creativity and innovation of a large number of people. This leads to users getting new perspectives and ideas from multiple people.

4 Discussion

Looking at the trend analysis of remote CSTs within product development, it can be assumed that both the quantitative number of publications with the selected search terms and the resulting number of CSTs to be recorded will increase in the future. Looking at the data of used **devices**, it can be assumed that the recorded increase of web and computer services CSTs will continue in the coming years. The trend for physical, digital CSTs is downward. The trend for analog CSTs shows no clear direction of development.

Web and computer systems based CSTs have gained importance over time, today these systems make up the majority of CSTs. This finding could be due to different reasons. Web- and computer-based CSTs might be easier to use and share. They can be accessed via the Internet and accessed from anywhere. In addition, the use of Web- and computer-based CSTs often involves lower costs. Physically tangible, analog and digital CSTs, on the other hand, are locally bound and can only be applied on-site. The maintenance of these CSTs is associated with more effort. Depending on the CST, it may not be possible to update the hardware, making it unusable in the event of further developments in the technology. In such cases, CSTs based on web and computer systems could offer an advantage. They can be further developed to adapt to changing user needs and requirements.

An examination of the devices used reveals a small number of **VR and AR** based CSTs. VR and AR based CSTs can be particularly important in technical product development as they allow prototypes and designs to be experienced and tested in a realistic, immersive environment. This can help design decisions and changes be made more quickly and effectively. VR and AR based CSTs provide the ability to create and visualize three-dimensional models or simulations of engineering products. This fosters a better understanding and idea of the final product. Using VR and AR in prototype status can have positive effects in creative activities [12]. VR and AR could also foster creativity among developers. Within a development team by allowing them to work and discuss together in a virtual environment [13]. VR and AR-based CSTs can improve collaboration and communication by leveraging informal factors such as body language, something that classic collaboration tools such as MS Teams have so far failed to do [14, 15].

The reasons why there are few VR and AR based CSTs in engineering product development may vary. The development of such CSTs can be complex and costly. In addition, integration into development activities can be difficult, leading to little interest in such CSTs. In addition, the acquisition of hardware for VR and AR-based systems involves high costs [16]. Another reason could be that VR and AR technologies are not yet mature for technical product development and thus difficulties can arise during integration. Furthermore, it is also possible that there is not yet enough experience and knowledge in the field of developing CSTs with VR and AR to develop effective and useful tools. The reasons for the low number of such CSTs can thus be manifold. Further research into the causes could provide insights into this.

The decline in **analog** tools may indicate a fundamental, increasing digitization of CSTs in engineering product development, and may also indicate that digital CSTs are more suitable to support certain aspects of new product development creative processes than analog CSTs.

In the years from 1998 to 2013, the proportion of CSTs that can be attributed to **individual and collective use** is similarly divided. Since 2013, the proportion of CSTs used individually has increased, while the proportion used collectively has decreased. While CSTs for individual use have been gaining in importance in recent years, tools for collective use have been losing relevance. Only since 2020 has a trend reversal become apparent. CSTs for individual application have reached a high point in 2020. Despite the trend reversal, it can be assumed that the majority of CSTs will continue to be used individually in the future.

One possible explanation for the high number of CSTs for individual use can be found by looking at the devices used. The analysis of the devices used revealed that CSTs based on web and computer services in particular account for a large proportion of the CSTs examined. These systems are usually used independently of each other. This becomes clear when analyzing the web and computer systems regarding the type of application. Regarding the 123 CSTs that can be assigned to Web and Computer Systems, 103 CSTs have characteristics for an individual application. Of which 15 CSTs can be used both individually and collectively. Only 20 CSTs can be used exclusively in the collective.

Collaboration tools have become increasingly important in recent years, and it can be expected that their significance in the development of new tools will continue to grow in the future. The ability to work on new technical product ideas in distributed collaboration scenarios has become more important in recent years. Here, an increase in CSTs can be expected in the coming years. Crowdsourcing has become more important over the last few years. Here, too, an increase in CSTs with this approach can be expected.

The analysis of collaboration shows an increase in collaboration tools and an increase in collaboration tools with remote properties. This is due to an increasing spread of home offices and flexible working models. The development makes it possible to work collaboratively with talent from different regions without having to bring the participants together at a fixed location.

5 Conclusion

The analysis shows that there are less collective CSTs and the trend is falling. On the other hand, Web and Computer Services are gaining importance while most of them can only be used individually. This does not exclude that they can be used collaboratively. The number of collaboration CSTs and CSTs that enable remote collaboration is growing as well.

This shows a tendency to use CSTs with Web- and Computer Services to also enable remote collaboration. Even if they are not remote enabled, they could be used remotely by using screensharing techniques. While such systems (Web- and Computer Services) are easier to prototype, implement and conduct studies with, they bring a few disadvantages.

Firstly, it is important to note that there are still advantages of physically tangible analog and digital CSTs. For example, users may have certain user preferences that can still play a role. The generation of creative ideas may be more pronounced through physical interaction, as it is easier to capture participants' gestures and facial expressions [17].

The current trend is that individually used CSTs are given more importance than CSTs for collective application. CSTs applied collectively can help fewer creative individuals to generate more creative ideas through interaction [18]. The interaction of people in teams or groups can lead to new perspectives and ideas that would not have emerged individually. Most of the individually used CSTs do not account for this effect.

While VR- and AR-CSTs could overcome some of these downsides, currently the number of such CSTs are still quite low. Such systems would still be considered individual CSTs, but they could enable a more productive communication than remote tools that only support communication via video and audio.

Some Web based CSTs offer virtual collaboration spaces. These virtual spaces are often *endless* and therefore situational awareness is quite limited.

VR- and AR-Solutions could solve some of these downsides and improve implementations of remote collaboration in CSTs. They could offer a confined virtual collaboration space and by using avatars improving situational awareness and communication. Using the appropriate hardware, they can be used individually and collectively and even in hybrid scenarios.

Although these kind of remote CST tools offer a variety of advantages further research is needed to ensure that user acceptance, situational awareness are comparable to collocated CSTs.

References

1. Zoche, P., Kimpeler, S., Joepgen, M.: Mobilität BT - Virtuelle Mobilität: Ein Phänomen mit physischen Konsequenzen? Zoche, P., Kimpeler, S., Joepgen, M. (eds.) Zur Wirkung der Nutzung von Chat, Online-Banking und Online- Reiseangeboten auf das physische Mobilitätsverhalten, pp. 7–24. Springer, Heidelberg (2002). https://doi.org/10.1007/978-3-642-56234-1_2
2. Bughin, J., et al.: Skill shift: automation and the future of the workforce. McKinsey Global Institute. McKinsey & Company (2018)
3. Lisa Diez-Holz I VDI. Wie werden Ingenieure 2030 arbeiten? (2020). https://www.ingenieur.de/technik/wirtschaft/arbeitsmarkt/wie-ingenieure-2030-arbeiten/
4. Frich, J., MacDonald Vermeulen, L., Remy, C., Biskjaer, M.M., Dalsgaard, P.: Mapping the landscape of creativity support tools in HCI. In: Proceedings of the 2019 CHI Conference on Human Factors in Computing Systems, pp. 1–18 (2019)
5. Schilit, B., Adams, N., Want, R.: Context-aware computing applications. In: Mobile Computing Systems and Applications - Workshop Proceedings, pp. 85–90 (1995). https://doi.org/10.1109/WMCSA.1994.16
6. Sielis, G.A., Tzanavari, A., Papadopoulos, G.A.: Enhancing the creativity process by adding context awareness in creativity support tools. Lecture Notes in Computer Science (including subseries Lecture Notes in Artificial Intelligence and Lecture Notes in Bioinformatics), vol. 5616, pp. 424–433 (2009)
7. Vocke, C., Bauer, W.: Status quo and quo vadis: creativity techniques and innovation methods for generating extended innovation processes. Hum. Factors Syst. Interact. **52** (2022)
8. Mark, G., Kun, A.L., Rintel, S., Sellen, A.: Introduction to this special issue: the future of remote work: responses to the pandemic. **37**, 397–403 (2022). https://doi.org/10.1080/07370024.2022.2038170
9. Gabriel, A., Monticolo, D., Camargo, M., Bourgault, M.: Creativity support systems: a systematic mapping study. Think Skills Creat **21**, 109–122 (2016)
10. Balters, S., Baker, J.M., Hawthorne, G., Reiss, A.L.: Inter-brain synchrony and innovation in a zoom world using analog and digital manipulatives. In: Meinel, C., Leifer, L. (eds.) BT - Design Thinking Research: Translation, Prototyping, and Measurement, pp. 9–32. Springer, Cham (2021). https://doi.org/10.1007/978-3-030-76324-4_2
11. Wolferts, D., Stein, E., Bernards, A.-K., Reiners, R.: Differences between remote and analog design thinking through the lens of distributed cognition. Front Artif Intell **5**, 255 (2022)
12. Vogel, J., Schuir, J., Thomas, O., Teuteberg, F.: Gestaltung und Erprobung einer Virtual-Reality- Anwendung zur Unterstützung des Prototypings in Design-Thinking-Prozessen. HMD Praxis der Wirtschaftsinformatik **57**(3), 432–450 (2020)

13. Schenk, J., Kurik, J., Gelberg, J., Lischka, A.: Wissensmanagement in virtuellen Welten: Wissensgenerierung in Virtual Reality- Umgebungen. HMD Praxis der Wirtschaftsinformatik **59**(1), 159–176 (2021)
14. Mueller, J., Hutter, K., Fueller, J., Matzler, K.: Virtual worlds as knowledge management platform – a practice-perspective. Inf. Syst. J. **21**, 479–501 (2011)
15. Piumsomboon, T., Lee, Y., Lee, G., Billinghurst, M.: CoVAR: a collaborative virtual and augmented reality system for remote collaboration. In: SIGGRAPH Asia 2017 Emerging Technologies, SA 2017 (2017). https://doi.org/10.1145/3132818.3132822
16. Rief, S., Hofmann, J., Bauer, W.: Connected Work Innovation Hub (2022). https://doi.org/10.24406/PUBLICA-32
17. Vyas, D., Heylen, D., Nijholt, A., Van Der Veer, G.: Collaborative practices that support creativity in design. In: ECSCW 2009 - Proceedings of the 11th European Conference on Computer Supported Cooperative Work, pp. 151–170 (2009). https://doi.org/10.1007/978-1-84882-854-4_9
18. Xue, H., Lu, K., Hao, N.: Cooperation makes two less-creative individuals turn into a highly-creative pair. Neuroimage **172**, 527–537 (2018)

Co-located Collaborative Virtual Reality to Accelerate the Engineering Process

B. Wingert[1(✉)], M. Bues[1], J. Hofmann[2], and O. Riedel[1]

[1] Visual Interactive Technologies, Fraunhofer IAO, Nobelstraße 12, Stuttgart, Germany
benjamin.wingert@iao.fraunhofer.de

[2] University of Stuttgart IAT, Allmandring 35, Stuttgart, Germany

Abstract. Collaboration is vital in product development, involving stakeholders from various disciplines, teams, locations, and management levels. Virtual reality (VR) systems can optimally support collaboration in various stages of the product development process, particularly for tasks like CAD review. VR goggles and multi-viewer powerwalls both enable collaborative CAD reviews, with multi-viewer powerwalls offering multiple 3D perspectives. However, both technologies have their advantages and disadvantages. A key distinction is that multi-viewer powerwalls allow people to see each other in reality, while VR goggles use virtual avatars lacking facial expressions and gestures, strongly limiting communication in co-located scenarios.

This paper explores the suitability of VR goggles and multi-viewer powerwalls for collaborative CAD review, comparing their advantages and disadvantages through a qualitative user study. Results show that multi-viewer powerwalls, such as the CoLEDWall, significantly enhance communication and teamwork, fostering more intensive collaboration in the product development process, surpassing the currently predominant use of VR goggles.

Keywords: Co-located Collaboration · Virtual Reality · Multi-Viewer Virtual Reality · Visual Engineering · Product Development · Human-Computer Interaction

1 Introduction

The engineering process and the associated amount of structural and geometrical information increase significantly as products become more complex. Consequently, effective management of engineering and manufacturing data has become a crucial task for contemporary organizations. Addressing this challenge and enabling engineering and production teams to handle large amounts of data and effectively solve problems can be supported by virtual reality (VR), particularly collaborative VR systems, across all domains. In the context of spatial problems, a visual 3D interface can be particularly valuable for supporting both teams and individuals. Additionally, tackling larger problems and making decisions often requires the involvement of multiple parties across various interdisciplinary teams, locations, and domains.

M. Dragomir et al. (Eds.): ICPR 2023, LNPE, pp. 465–473, 2025.
https://doi.org/10.1007/978-3-031-92082-0_52

In recent years, significant advancements were made in the field of VR, with head-mounted displays (HMDs) leading the way in giving people access to virtual environments and spatial data. These HMDs offer a cost-effective means of experiencing VR; however, they have limitations when it comes to adapting VR to remote or co-located collaborative scenarios.

New virtual reality technologies such as multi-viewer powerwalls have their own advantages, particularly facilitating co-located collaborations.

This paper aims to provide an overview and comparison of different technologies, specifically comparing the widely used head-mounted displays (the predominant VR technology today) with the CoLEDWall – a multi-viewer virtual reality powerwall. The focus of the comparison will be on co-located collaborative spatial problem-solving scenarios.

2 Related Work

Virtual reality (VR) and local collaboration have been explored using various technologies, including large screen displays, head-mounted displays (HMDs), and powerwall systems. Recently, new technologies, such as multi-viewer systems, have emerged in this field.

2.1 Large Screen Displays

In meetings or design reviews, large screen spaces are commonly used to share content through displays like flat panel LCD display walls or projections. Large resolution displays are used for data-driven decision-making [3], data analysis and sensemaking [4] and shared interaction and data manipulation [12]. While these solutions enable content sharing, they are limited to displaying flat images and lack the ability to convey depth and spatial information. Additionally, interaction is often restricted to a single user connected to the display, limiting collaborative and parallel input from other users.

2.2 Head Mounted Displays

Head-mounted displays have become widely available and offer an immersive experience in virtual environments. Widespread use cases are reviewing 3D CAD models [8], learning environments for engineering assembly tasks [9], and enhancing the remote design review process [14, 15]. However, they are limited to a single user's perspective, making it impossible for other users to see the virtual environment without wearing their own HMDs. In a co-located scenario, where users are in the same room, the use of HMDs can create a sense of distance and isolation, thereby strongly hampering communication and collaboration between users.

2.3 Powerwall- and CAVE-Systems

Powerwall and CAVE systems allow a user to view a virtual environment through large stereoscopic projections [5]. Before the advent of the current generation of HMDs, these were the most widely used VR setups in industry. [17, 18] However, these systems can only generate one correct perspective, resulting in skewed and distorted images for the other users. Consequently, albeit being frequently used in multi-user scenarios, these types of CAVE and Powerwall are essentially single-user systems.

2.4 Multi-viewer Powerwalls

To overcome the limitation of Powerwalls as described above, multi-viewer display solutions have emerged providing individual, perspective correct stereoscopic views for more than one user within the same Powerwall display space. Display technologies used are projection systems [1, 10] as well as LED-based matrix displays [7] to overcome the limitations of existing powerwalls and CAVE systems. These innovative approaches employ image separation through time multiplexing [1, 7, 10] and polarization [10] techniques to generate multiple stereo image pairs, enabling multiple users to experience the correct perspective simultaneously. With these systems, users can maintain awareness of each other standing in front of the screen while also having the ability to see the other persons around them without the need of virtual avatars.

2.5 Virtual Reality Collaboration

Doing collaborative work in VR requires suitable software. Remote collaboration scenarios showed that HMDs improve the sense of remote collaboration, spatial presence and co-presence compared to non-VR remote collaboration. [2] Also first proofs of concept for remote collaborative architectures for modifying parametric CAD data where shown, such as in [13]. However, this research and development rarely addresses co-located collaboration specifically, which however would be required to fully understand and harvest the benefits offered by collocated multi-viewer VR systems.

3 Co-located Collaborative Virtual Reality

Co-located collaboration in the sense of this paper includes ad-hoc meetings to address immediate issues, group meetings for decision-making, and reviews, all of which are common during the engineering and production phases when dealing with complex problems, often with spatial components. Compared to the well-known remote collaboration, co-located collaboration offers unique advantages.

3.1 Head-Mounted Displays

Head-mounted displays (HMDs) are immersive devices that enable users to disconnect from their physical surroundings and enter a virtual environment. This provides a high degree of presence in the virtual world. Typically, these devices are mobile and connected to a workstation PC or notebook in engineering and production settings to visualize complex product or production data. However, one inherent drawback of HMDs is that the display is worn directly in front of the user's face, thereby completely blocking out the surroundings and co-workers. Consequently, HMDs are primarily single-user devices, despite efforts to represent co-workers as avatars for those wearing HMDs. However, avatars mostly lack natural expression and visual accuracy, making them inadequate substitutes for real face-to-face collaboration.

3.2 Collaborative Powerwalls

Collaborative Multi-Viewer Powerwalls, such as the CoLEDWall which consists of a large LED-based display that provides multiple users with an individual 3D perspective to the same virtual environment while simultaneously seeing each other. This is achieved through display technology that sequentially display multiple images at a high refresh rate. For example, to accommodate two users, the display cycles through the four individual images (user 1 left/right, user 2 left/rIght) at a refresh rate of 240 Hz, enabling four distinct views to be flicker-freely separated by fast LC-shutter glasses. A spatial tracking system locates the users in relation to the display, and the 3D rendering software continuously adjusts the perspective for each individual image to align with the viewer's eye position. As a result, users experience a stereoscopic, immersive display, with the added benefit of accessing the same virtual environment while maintaining visual contact with each other, without the need for avatars.

4 Evaluation

Current HMDs are widely used for remote collaboration. In order to identify the most suitable technology for co-located collaboration, we conducted a user study in which head-mounted displays are compared with the emerging technology of multi-viewer powerwalls.

4.1 Hardware Systems – HMD and CoLEDWall

The VR hardware system used in the study was the *HP Reverb G2*, which includes a VR Head-Mounted Display (HMD) and two spatial motion controllers. The HMD features two 2160×2160 pixel LC-displays with a refresh rate of 90 Hz, providing a wide field of view of 98° horizontally and 90° vertically. This translates to a pixel density of 23.58 pixels per degree horizontally and 23.77 pixels per degree vertically, resulting in a visually immersive experience. Inside-out tracking is employed, utilizing four integrated cameras and an Inertial Measurement Unit (IMU) to track the position and orientation of the HMD and controllers. Two workstations and HP Reverb G2s were used for the evaluation involving two users in the same room (Fig. 1).

The *CoLEDWall*, on the other hand, is a display wall based on LED technology. Its dimensions are 2.80×2.25 m, with a pixel pitch of 1.25 mm, resulting in a resolution of 2240×1800 pixels. The CoLEDWall is designed to support high refresh rates of 240 Hz. For tracking, an OptiTrack system with five cameras was utilized to locate the shutter glasses equipped with retroreflective markers in front of the screen. Due to the fixed display size and users' movement in front of the wall, the field of view varies. The closer a user stands to the CoLEDWall, the larger the field of view becomes, albeit at the expense of reduced angular resolution (Figs. 2 and 3).

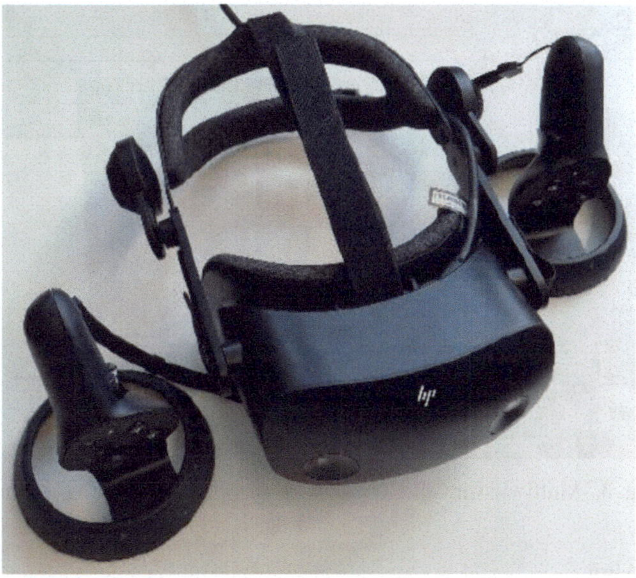

Fig. 1. HP Reverb G2 VR headset with two controllers

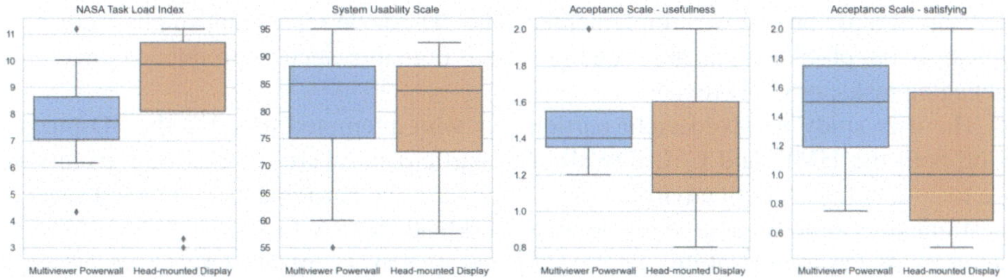

Fig. 2. Results of the NASA TLX, System Usability Scale and Acceptance Scale for the preliminary user study

4.2 Software Systems – CoVR

The software system employed in this study is called CoVR, a virtual reality platform specifically designed for local and remote collaboration. CoVR utilizes the Unity engine for clients and NodeJS for the server, enabling secure collaboration across multiple devices and locations. The architecture of CoVR allows collaborative work on 3D computer-aided design (CAD) data without the need for exchanging CAD and structural information between CoVR connections. This software enables the use of the same virtual environment for both the HMD and CoLEDWall, facilitating a fair and accurate comparison between the two technologies.

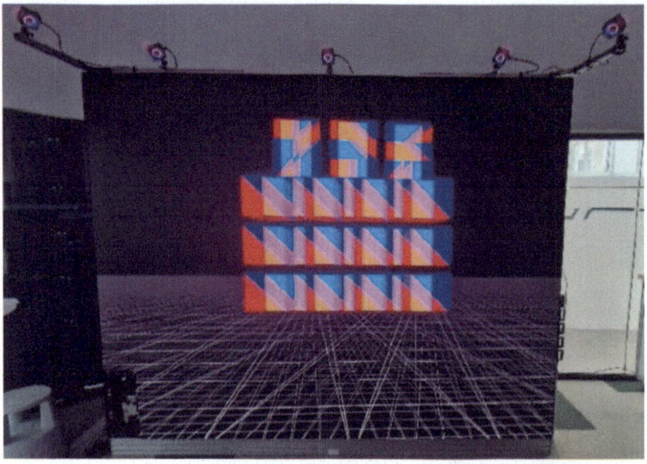

Fig. 3. Multi viewer CoLEDWall with an Optitrack tracking system

4.3 Study Design

To compare the two proposed display systems and determine which system is better suited for spatial tasks in a co-located collaborative scenario, a multi-user spatial task in the form of a 3D puzzle was implemented. The 3D puzzle represents an abstraction of typical spatial problems that are found in various domains, such as product engineering, packaging tasks, or CAD review.

The user study involved eight participants who attempted to solve the 3D puzzle using both the HMD and CoLEDWall. The participants were divided into groups of two to collaborate in both environments. Afterward, the participants completed several questionnaires aimed at assessing cognitive workload (NASA Task Load Index) [6], system usability (System Usability Scale) [11], and system acceptance (Acceptance Scale) [16]. Additionally, a questionnaire was conducted to gather feedback on which system was better suited for communication, perception of space, and ease of task handling. Finally, participants were asked to choose which technology they preferred for future CAD reviews.

4.4 Results

The results of the NASA Task Load Index (TLX) questionnaire shows a non-normal distribution for the data set. The results suggest a lower cognitive load for the CoLEDWall (median = 7.75, interquartile range [IQR] = 1.59) compared to the HMD (median = 9.84, IQR = 2.58).

Regarding the System Usability Scale, the collected data was also non-normally distributed. The CoLEDWall received a median score of 85 (IQR = 13.13), while the HMD setup scored a median of 83.75 (IQR = 15.625), indicating slightly higher usability for the CoLEDWall.

The Acceptance Scale yielded two dimensions: usefulness and satisfaction. The collected data for both dimensions were not normally distributed. The CoLEDWall received a median usefulness score of 1.4 (IQR = 0.2) and a median satisfaction score of 1.5 (IQR = 0.56). In comparison, the HMD setup scored a median usefulness of 1.2 (IQR = 0.5) and a median satisfaction of 1.0 (IQR = 0.88). These results further indicate a slight preference for the CoLEDWall in terms of both usefulness and satisfaction.

The subjective questionnaire aimed at determining which system was better suited revealed that 100% of the participants favored the CoLEDWall for communication, while 62.5% chose the HMD over the CoLEDWall for better perception of space. Regarding ease of task handling, 75% of participants preferred the CoLEDWall, while the remaining 25% perceived no significant difference between the systems. When asked about their preference for future CAD reviews, 75% of participants chose the CoLEDWall, while the remaining 25% opted for the HMD setup. These questionnaire results indicate a strong inclination towards the CoLEDWall in all cases, except for perception of space, where the HMD received relatively higher preference.

5 Discussion

The obtained data demonstrate that the CoLEDWall, a multi-viewer powerwall system, yields comparable, and in some instances, superior results compared to the HMD system. This suggests that the CoLEDWall offers enhanced system usability and user acceptance in co-located tasks. Although the results are not definitive, they do exhibit a discernible inclination towards the CoLEDWall.

Specifically in communication tasks, the CoLEDWall outperforms the HMD, providing a clear indication of the potential of multi-viewer powerwall systems. It is worth noting that the HMD setup, with its greater and constant field of view, enables unrestricted changes in view orientation, which contributes to its superiority in spatial perception. In contrast, the CoLEDWall, consisting of a single wall, lacks this capability. However, it is plausible that this difference between the two systems would significantly diminish if multiple walls and floor displays were utilized.

Overall, the findings strongly suggest a preference for the CoLEDWall in co-located product engineering, packaging tasks, or CAD review cases.

6 Future Work

This study aimed to assess the potential of a multi-viewer powerwall system in comparison to co-located collaborative head-mounted display (HMD) setups. The obtained results indicate that the multi-viewer powerwall system offers advantages for local collaboration. Consequently, for future investigations, we propose evaluating a wider range of spatial tasks involving a larger participant pool. This expanded study will provide a more comprehensive understanding of the technological benefits associated with a multi-viewer powerwall system, such as the CoLEDWall. The anticipated findings will facilitate informed decision-making regarding technology selection.

7 Conclusion

The utilization of virtual reality (VR) can facilitate natural access to virtual spatial information and its integration into the engineering process. However, conventional VR systems often lack effective collaboration capabilities in co-located scenarios. These systems either do not provide dedicated views for all users or represent users as avatars. The CoLEDWall, a novel multi-viewer powerwall technology, addresses this limitation by enabling the display of multiple stereoscopic images for multiple users. This allows users to access virtual spatial information, such as CAD data, while maintaining visibility of all other users in the virtual environment.

In our experiment, we compared state-of-the-art head-mounted display (HMD) based co-located solutions with the CoLEDWall. The results demonstrate the promising effects of the CoLEDWall, particularly in the areas of collaborative communication and task handling. Additionally, it slightly outperforms the existing solutions in terms of cognitive load, system usability, and user acceptance. Consequently, this innovative technology has the potential to expedite the local co-located engineering process.

References

1. Agrawala, M., Beers, A.C., McDowall, I., Fröhlich, B., Bolas, M., Hanrahan, P.: The two-user responsive workbench. In: Proceedings of the 24th Annual Conference on Computer Graphics and Interactive Techniques, SIGGRAPH 1997, pp. 327–332 (1997)
2. Bayro, A., Ghasemi, Y., Jeong, H.: Subjective and objective analyses of collaboration and co-presence in a virtual reality remote environment. In: 2022 IEEE Conference on Virtual Reality and 3D User Interfaces Abstracts and Workshops (VRW), pp. 485–487 (2022)
3. Belkacem, I., Tominski, C., Médoc, N., Knudsen, S., Dachselt, R., Ghoniem, M.: Interactive Visualization on Large High-Resolution Displays: A Survey (2022)
4. Bradel, L., Endert, A., Koch, K., Andrews, C., North, C.: Large high resolution displays for co-located collaborative sensemaking: display usage and territoriality. Int. J. Hum. Comput. Stud. **71**(11), 1078–1088 (2013)
5. Cruz-Neira, C., Sandin, D.J., DeFanti, T.A.: Surround-screen projection-based virtual reality. In: Proceedings of the 20th Annual Conference on Computer Graphics and Interactive Techniques, pp. 135–142 (1993)
6. Hart, S.G., Staveland, L.E.: Development of NASA-TLX (Task Load Index): Results of Empirical and Theoretical Research, pp. 139–183 (1988)
7. Hergert, S., Bues, M., Riedel, O.: 50-2: LED-based next generation immersive virtual reality. In: SID Symposium Digest of Technical Papers, vol. 52, no. 1, pp. 687–689 (2021)
8. Horvat, N., Škec, S., Martinec, T., Lukačević, F., Perišić, M.M.: Comparing virtual reality and desktop interface for reviewing 3D CAD models. In: Proceedings of the Design Society: International Conference on Engineering Design, vol. 1, no. 1, pp. 1923–1932 (2019)
9. Huang, W.: Evaluating the effectiveness of head-mounted display virtual reality (HMD VR) environment on students' learning for a virtual collaborative engineering assembly task. In: 2018 IEEE Conference on Virtual Reality and 3D User Interfaces (VR), pp. 827–829 (2018)
10. Kulik, A., et al.: C1x6: a stereoscopic six-user display for co-located collaboration in shared virtual environments. ACM Trans. Graph. **30**(6), 1–12 (2011)
11. Lewis, J.R.: The system usability scale: past, present, and future. Int. J. Hum.-Comput. Interact. **34**(7), 577–590 (2018)

12. Liu, C., Chapuis, O., Beaudouin-Lafon, M., Lecolinet, E.: Shared interaction on a wall-sized display in a data manipulation task. In: Proceedings of the 2016 CHI Conference on Human Factors in Computing Systems, pp. 2075–2086 (2016)
13. Okuya, Y., Ladeveze, N., Gladin, O., Fleury, C., Bourdot, P.: Distributed architecture for remote collaborative modification of parametric CAD data. In: 2018 IEEE Fourth VR International Workshop on Collaborative Virtual Environments (3DCVE), pp. 1–4 (2018)
14. Tea, S., Panuwatwanich, K., Ruthankoon, R., Kaewmoracharoen, M.: Multiuser immersive virtual reality application for real-time remote collaboration to enhance design review process in the social distancing era. J. Eng. Des. Technol. **20**(1), 281–298 (2022)
15. de Casenave, L., Lugo, J.E.: Design review using virtual reality enabled CAD. In: 37th Computers and Information in Engineering Conference, vol. 1 (2017)
16. van der Laan, J.D., Heino, A., de Waard, D.: A simple procedure for the assessment of acceptance of advanced transport telematics. Transp. Res. Part C: Emerg. Technol. **5**(1), 1–10 (1997)
17. Tesch, A., Dörner, R.: Expert performance in the examination of interior surfaces in an automobile: virtual reality vs. reality. In: Proceedings of the 28th ACM International Conference on Multimedia, pp. 2673–2681 (2020)
18. Kovar, J., Mouralova, K., Ksica, F., Kroupa, J., Andrs, O., Hadas, Z.: Virtual reality in context of Industry 4.0 proposed projects at Brno University of Technology. In: 2016 17th International Conference on Mechatronics – Mechatronika, pp. 1–7 (2016)

Author Index

M. Dragomir et al. (Eds.): ICPR 2023, LNPE, pp. 475–477, 2025.
https://doi.org/10.1007/978-3-031-92082-0

The manufacturer's authorised representative in the EU is Springer Nature Customer Service Centre GmbH, Europaplatz 3, 69115 Heidelberg, Germany. If you have any concerns regarding our products, please contact ProductSafety@springernature.com

Printed and bound by CPI Group (UK) Ltd, Croydon, CR0 4YY
01/10/2026
02225913-0019